OCCUPATIONAL BIOMECHANICS

OCCUPATIONAL BIOMECHANICS

DON B. CHAFFIN
The University of Michigan
Center for Ergonomics
Ann Arbor, Michigan

GUNNAR B. J. ANDERSSON
Department of Orthopedic Surgery I
Sahlgren Hospital
Göteborg, Sweden

A Wiley-Interscience Publication

JOHN WILEY & SONS

New York/Chichester/Brisbane/Toronto/Singapore

Library of Congress Cataloging in Publication Data:

Chaffin, Don B.
  Occupational biomechanics.

  "A Wiley-Interscience publication."
  Bibliography: p.
  Includes index.
  1. Work—Physiological aspects. 2. Human mechanics.
I. Andersson, Gunnar, 1942– . II. Title.

QP301.C525 1984 612′.7 84-3500
ISBN 0-471-87634-8

Printed in the United States of America

10 9 8 7 6 5 4 3 2 1

FOREWORD

Human well-being, considered by many even today as primarily a question for political and social action, cannot be separated from working life and its consequences. For countless millions the consequences of their work are borne, in later years, by themselves in the form of restrictions on their activities and on their opportunities.

These restrictions are very evident when they limit activity as a result of musculoskeletal problems. In these days of sophisticated electronics and mechanics it is commonly believed that physical effort is no longer a component of work. Industrial sickness figures belie this naive belief and at the very basic level of people as force producers, the design of work has, until very recently, been woefully inadequate.

Research and application to improve this situation have, in no small measure, been stimulated by the work of these two authors. Their research efforts to increase the understanding of biomechanics, and to develop methods for measurement and design in the field, have contributed to the creation of an applied discipline with widespread utility in the whole field of work design. Not least in importance are the changes that this development brings to the ways of thinking about work loads and body forces. Increased knowledge permits more effective design and greater understanding of the causes and effects of inadequate work situations.

The availability of this book now permits this important discipline for the design of work to be presented as a coherent structure both for teaching and practice. It can take its place alongside engineering anthropometry and work physiology to provide the foundations for the study and design of physical work activities. Indeed, these foundations may now be considered, for the first time, to be firmly in place and others may now build with confidence.

E. N. Corlett

University of Nottingham
May 1984

PREFACE

This textbook is dedicated to the belief that a sound understanding and application of biomechanical principles is important both in the prevention of musculoskeletal disorders in industry and to improve manual working conditions in general. The book condenses an extensive amount of recent biomechanical literature along with the authors' personal research findings. It reflects over 35 years of combined experience in applying biomechanics in various industries and work situations.

Occupational biomechanics can be defined as a science concerned with the mechanical behavior of musculoskeletal tissues when physical work is performed. As such, it seeks to provide an understanding of the physics of manual activities in industry. Similarly, psychologists over the last 50 years have provided the behavioral knowledge necessary to define a human factors discipline useful in the evaluation and design of information displays and controls. Likewise, exercise physiologists have provided the basic concepts necessary to define a work physiology discipline that predicts the metabolic, respiratory, and cardiovascular effects of prolonged, strenuous activities in industry. In a very real sense, occupational biomechanics complements the existing psychological and physiological knowledge which is the foundation of contemporary ergonomics.

Today the need to apply occupational biomechanics principles to improve different work situations appears to be even greater than in past years. There is an increasing awareness by both management and labor organizations that musculoskeletal harm and reduced human performance capabilities are often the result of a physical mismatch between workers and the manual tasks required of them. Unfortunately, the conditions that dictate the existence of such a mismatch are usually complex. The early identification and modification of such adverse conditions requires special expertise. Such expertise requires the combined knowledge of mechanics and musculoskeletal anatomy and physiology, along with knowledge of industrial work specification and practice.

In working with various industries we became convinced that a large amount of appropriate literature existed, but only in the form of various research papers and technical reports. Furthermore, the information was published in a diverse set of journals in several different disciplines, and some reports were accessible only in university research libraries. The growing need for expertise in this field and the lack of a comprehensive textbook that would join research methods and industrial applications prompted us to write this book. Our hope is that it will stimulate students in engineering, medicine, and occupational health and safety to study and contribute to this growing and important field in future years.

As a final note, we believe there are additional research findings and industrial problems that should have been included in this book. We invite the readers of this first edition to comment on their experience with the book and suggest improvements to it for future editions.

DON B. CHAFFIN
GUNNAR B. J. ANDERSSON

Ann Arbor, Michigan
Göteborg, Sweden
May 1984

ACKNOWLEDGMENTS

Over the two-year period of writing this textbook many different people and organizations were of immense assistance. Though space does not permit us to acknowledge all of these, a few deserve particular note. First we'd like to thank Barbara Chaffin, who not only tolerated a preoccupied husband, but who drew over 200 outstanding illustrations for this book, and Kerstin Andersson who inspired in times of despair. Certainly the people who typed and retyped drafts of the book deserve special recognition, specifically Pat Terrell, Pam Swick, Gail Miller, Kerstin Månsson, Birgitta Sjöberg, and Ingrid Melkersson. We'd also like to thank Professors Al Schultz, Gary Herrin, Thomas Armstrong, and Dwight Culver, who contributed to our efforts through challenging discussions of occupational biomechanics. Certainly, this text would not be possible without the institutional support of our home Universities, and in particular both the Department of Industrial and Operations Engineering and Center for Ergonomics at The University of Michigan and The Department of Orthopaedic Surgery I at the Sahlgren Hospital, Göteborg, Sweden. Finally, we'd like to express our gratitude to the Southern California Center for Occupational Health at the University of California, Irvine, who hosted Don Chaffin, and the Orthopedics Department of Rush-Presbyterian-St. Lukes Medical Center, Chicago, who hosted Gunnar Andersson, permitting both of them to write major parts of this book while on their sabbaticals.

D. B. C.
G. B. J. A.

CONTENTS

1. OCCUPATIONAL BIOMECHANICS AS A SPECIALTY **1**

1.1 Definition of Occupational Biomechanics, 1
1.2 Historical Development of Occupational Biomechanics, 2
1.2.1 Kinesiological Developments, 3
1.2.2 Biomechanical Model Developments, 5
1.2.3 Anthropometric Developments, 5
1.2.4 Mechanical Work-Capacity Evaluation Methods, 5
1.2.5 Bioinstrumentation Developments, 6
1.2.6 Motion Classification and Time-Prediction Systems Development, 6
1.3 The Need for an Occupational Biomechanics Specialty, 7
1.3.1 Epidemiological Support of Occupational Biomechanics, 7
1.3.2 Social/Legal Support of Occupational Biomechanics, 9
1.3.3 Ergonomic Support of Occupational Biomechanics, 9
1.4 Who Uses Occupational Biomechanics, 10
1.5 Organization of Book, 11
References, 11

2. THE STRUCTURE AND FUNCTION OF THE MUSCULOSKELETAL SYSTEM **14**

2.1 Introduction, 14
2.2 The Connective Tissues, 14
2.2.1 Ligaments, Tendons, and Fascia, 15

2.2.2 Cartilage, 18
2.2.3 Bone, 19
2.3 Skeletal Muscle, 26
2.3.1 The Structure of Muscles, 26
2.3.2 The Muscle Contraction, 30
2.3.3 Mechanical Aspects of the Muscle Contraction, 34
2.3.4 Muscle Fatigue, 39
2.3.5 The Action of Muscles, 43
2.4 Joints, 44
2.4.1 The Synovial Joint, 45
2.4.2 Joint Lubrication, 47
2.4.3 Osteoarthritis, 48
2.4.4 Intervertebral Discs, 48
References, 51

3. ANTHROPOMETRY IN OCCUPATIONAL BIOMECHANICS 53

3.1 Measurement of Body-Segment Physical Properties, 53
3.1.1 Body-Segment Link Length Measurement Methods, 53
3.1.2 Body-Segment Volume and Weight, 59
3.1.3 Body-Segment Mass Center Locations, 61
3.1.4 Body-Segment Inertial Property Measurement Methods, 63
3.2 Anthropometric Data for Biomechanical Studies in Industry, 68
3.2.1 Segment-Link Length Data, 68
3.2.2 Segment Weight Data, 72
3.2.3 Segment Mass-Center Location Data, 72
3.2.4 Segment Moment of Inertia and Radius of Gyration Data, 73
3.3 Summary of Anthropometry in Occupational Biomechanics, 75
References, 76

4. MECHANICAL WORK-CAPACITY EVALUATION 78

4.1 Introduction, 78
4.2 Joint Motion — Methods and Data, 78
4.2.1 Measurement Methods of Joint Motion, 81
4.2.2 Normal Ranges of Joint Motion, 85
4.2.3. Factors Affecting Range-of-Motion Data, 85
4.3 Muscle Strength Evaluation, 89
4.3.1 Muscular Strength Definition, 89
4.3.2 Static and Dynamic Strength-Testing Methods, 91
4.3.3 Normal Muscle Strength Values, 97
4.3.4 Personal Factors Effecting Strength, 103

4.4 Summary and Limitations of Mechanical Work Capacity Data, 107
References, 108

5. Bioinstrumentation for Occupational Biomechanics 111

5.1 Introduction and Measurement System Criteria, 111
5.2 Human Motion Analysis Systems, 112
5.2.1 Basis for Measurement of Human Motion, 112
5.3 Applied Electromyography, 122
5.3.1 Theory of the Use of Electromyography in Occupational Biomechanics, 122
5.3.2 EMG Measurement System, 125
5.4 Muscle Strength-Measuring Systems, 128
5.4.1 Localized Static Strength-Measurement Systems, 128
5.4.2 Whole-Body Static Strength-Measuring System, 130
5.4.3 Whole-Body Dynamic Strength-Measuring System, 132
5.5 Intradiscal Pressure Measurement, 132
5.5.1 Measurement Development Technique, 132
5.5.2 Intradiscal Pressure-Measurement System, 136
5.5.3 Applications and Limitations in Occupational Biomechanics, 136
5.6 Intra-Abdominal (Intragastric) Measurements, 137
5.6.1 Measurement Development, 137
5.6.2 Measurement System, 138
5.6.3 Applications and Limitations in Occupational Biomechanics, 138
5.7 Force Platform System, 141
References, 142

6. OCCUPATIONAL BIOMECHANICAL MODELS 147

6.1 Why Model?, 147
6.2 Planar Static Biomechanical Models, 148
6.2.1 Single-Body Segment Static Model, 148
6.2.2 Two-Body Segment Static Model, 155
6.2.3 Static Planar Model of Nonparallel Forces, 158
6.2.4 Planar Static Analysis of Internal Forces, 160
6.2.5 Multiple-Link Coplanar Static Modeling, 166
6.3 Static Three-Dimensional Modeling, 171
6.4 Dynamic Biomechanical Models, 176
6.4.1 Single-Segment Dynamic Biomechanical Model, 178

6.4.2 Multiple-Segment Biodynamic Model of Load Lifting, 182
6.4.3 Coplanar Biomechanical Models of Foot Slip Potential While Pushing a Cart, 187
6.5 Special Purpose Biomechanical Models of Occupational Tasks, 190
6.5.1 Low-Back Biomechanical Models, 191
6.5.2 Biomechanical Models of the Wrist and Hand, 213
6.5.3 Muscle Strength Prediction Modeling, 221
6.6 Future Developments in Occupational Biomechanical Models, 227
References, 229

7. METHODS OF CLASSIFYING AND EVALUATING MANUAL WORK **233**

7.1 Traditional Methods, 233
7.1.1 Historical Perspective, 233
7.2 Traditional Work Analysis System, 236
7.2.1 MTM-1: An Example of a Predetermined Motion–Time System, 237
7.2.2 Benefits and Limitations in Contemporary Work Analysis Systems, 243
7.3 Contemporary Biomechanical Job Analysis Systems, 246
7.3.1 Physical Stress Checklists and Surveys, 246
7.3.2 Manual Lifting Analysis, 246
7.3.3 Job Static Strength Analysis, 251
7.3.4 Job Postural Evaluation Method, 254
7.3.5 Upper Extremity Postural Analysis, 257
7.3.6 Trunk Flexion Analysis, 258
7.4 Future Impact of Occupational Biomechanics on Work Measurement Systems, 260
References, 261

8. MANUAL MATERIALS HANDLING LIMITS **263**

8.1 Introduction, 263
8.2 Lifting Limits in Manual Materials Handling, 264
8.2.1 Scope of NIOSH Work Practices Guide for Manual Lifting, 267
8.2.2 Definition of Lifting Hazard Levels, 268
8.2.3 NIOSH Recommendations to Control Lifting Hazards, 272
8.2.4 Comments on the Status of the NIOSH Lifting Guide, 274

8.3 Load Pushing and Pulling Capabilities, 275
8.3.1 Foot-Slip Prevention, 279
8.4 Asymmetric Load Handling, 279
8.5 Summary of Manual Materials Handling Limits, 286
References, 286

9. GUIDELINES FOR SEATED WORK **289**

9.1 General Considerations in Sitting Postures, 289
9.2 Anthropometric Aspects of Seated Work, 294
9.3 Comfort Aspects of Seated Work, 295
9.4 The Spine and Sitting, 298
9.4.1 Clinical Aspects of Sitting Posture, 298
9.4.2 Radiographic Data, 299
9.4.3 Disc Pressure Data During Sitting, 302
9.4.4 Muscle Activity, 305
9.4.5 Discussion of the Spine in Sitting Postures, 308
9.5 The Shoulder and Sitting, 309
9.6 The Legs and Sitting, 309
9.7 The Sitting Workplace, 313
9.7.1 The Chair, 313
9.7.2 The Table (Work Surface), 317
9.8 Summary, 319
References, 319

10. BIOMECHANICAL CONSIDERATIONS IN MACHINE CONTROL AND WORKPLACE DESIGN **324**

10.1 Introduction, 324
10.1.1 Localized Musculoskeletal Injury in Industry, 324
10.2 Guidelines for Workplace and Machine Control Layout, 329
10.2.1 Shoulder-Dependent Overhead Reach Limitations, 331
10.2.2 Shoulder- and Arm-Dependent Forward Reach Limitations, 338
10.2.3 Neck/Head Posture Work Limitations, 345
10.2.4 Torso Postural Considerations in Workbench Height Limitations, 347
10.2.5 Biomechanical Considerations in Design of VDT Workstation, 350
10.3 Summary, 352
References, 353

11. HAND TOOL DESIGN GUIDELINES **355**

11.1 The Need for Biomechanical Concepts in Hand Tool Design, 355

11.2 Hand Tool Shape and Size Considerations, 357
11.2.1 Tool Shape to Avoid Wrist Deviation, 357
11.2.2 Tool Shape to Avoid Shoulder Abduction, 358
11.2.3 Tool Shape to Assist Grip, 360
11.2.4 Size of Tool Handle to Facilitate Grip, 363
11.2.5 Finger Clearance Considerations, 365
11.3 Hand Tool Weight and Use Considerations, 365
11.4 Summary, 368
References, 368

12. GUIDELINES FOR WHOLE-BODY AND SEGMENTAL VIBRATION 369

12.1 Definition and Measurement, 369
12.2 The Effects of Vibration on the Human Body, 376
12.3 Whole-Body Vibration, 379
12.3.1 General Effects, 379
12.3.2 Effects on the Spine, 383
12.4 Segmental Vibration, 384
12.4.1 Vibration Syndrome, 385
12.4.2 Transmission of Vibration in the Upper Extremity, 386
12.5 Vibration Exposure Criteria, 387
12.5.1 Whole-Body Vibration, 387
12.5.2 Segmental Vibration, 392
References, 396

13. WORKER SELECTION AND TRAINING CRITERIA 399

13.1 Worker Selection, 399
13.1.1 Introduction, 399
13.1.2 History and Physical Examination, 400
13.1.3 Radiographic Pre-Employment Examination, 402
13.1.4 Quantitative Physical Pre-Employment Screening, 403
13.2 Pre-Employment Training, 405
13.3 Summary, 407
References, 407

14. SUMMARY 411

APPENDIX A

Part 1: Anatomical and Anthropometric Landmarks, 413
Part 2: Glossary of Anatomical and Anthropometric Terms, 416

APPENDIX B

Population Weight and Mass-Center Data, 423

Review Questions 427

Glossary 434

Index 445

OCCUPATIONAL BIOMECHANICS

1

OCCUPATIONAL BIOMECHANICS AS A SPECIALTY

1.1 DEFINITION OF OCCUPATIONAL BIOMECHANICS

In a recent book (Frankel and Nordin, 1980), the general field of biomechanics was defined as follows:

> *Biomechanics uses laws of physics and engineering concepts to describe motion undergone by the various body segments and the forces acting on these body parts during normal daily activities.*

By this definition, biomechanics is a multidisciplinary activity. It requires combining knowledge from the physical and engineering sciences with knowledge from the biological and, to a lesser extent, behavioral sciences. As we shall describe, this effort is not new, nor is it only academic. A large variety of human disorders and performance limitations have been shown to be amenable to biomechanical interpretation and resolution. A discussion of some of them took place at a recent NATO Advanced Study Institute (Akkas et al., 1980).

This book does not attempt to describe all the varied developments and areas of application in the general field of biomechanics. Rather, it deals with the study of one set of human disorders and performance limitations—those produced or aggravated by the mismatching of human physical capacities and human performance requirements. Such mismatching exists at present in a variety of manual tasks in industry. *Occupational Biomechanics* is the emerging discipline concerned with this mismatching in industry. *Occupational Biomechanics* can be defined as the study of the physical interaction of workers with their tools, machines, and materials so as to enhance the worker's performance while minimizing the risk of future musculoskeletal disorders.

In this context, occupational biomechanics is an applied discipline in the general field of biomechanics. Because of its application orientation it is necessary for one who specializes in occupational biomechanics to have both an appreciation of the parent field of biomechanics and an understanding of how various biomechanical principles can be applied within different engineering and management functions in industry.

1.2 HISTORICAL DEVELOPMENT OF OCCUPATIONAL BIOMECHANICS

The general field of biomechanics has a long distinguished history. Professor Y. C. Fung has listed many contributors from both the engineering and life sciences (Fung, 1981). In the late 1500's, the physicist Galileo Galilei used the concept of a constant pendulum oscillation period to measure heart rates. William Harvey followed Galileo's concepts of physical fluid measurement and, against great odds at the time, was able to demonstrate in 1615 the necessity of capillaries connecting veins and arteries, even though they were not seen by the crude microscopes of the time until 1661 by Marcello Malpigi.

Shortly after Malpigi, Stephen Hales (1677–1761) was able to measure arterial pressure and correlate it with both hemorrhage and ventricular forces in the heart. He also demonstrated how the elastic properties of the aorta converted the pulsatile flow from the heart to a smooth flow. Professor Fung continues by stating (Fung, 1981):

> *Otto Frank worked out a hydrodynamic theory of circulation. Starling proposed the law of mass transfer across biological membranes and clarified the concept of water balance in the body. Krogh won his Nobel prize on the mechanics of microcirculation. Hill won his Nobel prize on the mechanics of the muscle.*

Regarding biomechanical concepts related to the musculoskeletal system, one must also be impressed with the concern that Leonardo da Vinci displayed over the function of muscles and bone. In fact, in reviewing da Vinci's work, Professor A. Seireg concluded that da Vinci deserved the title of "Biomechanician" and states . . . "Leonardo's penetrating biomechanical studies were far in advance of the state of the art at his time (1452–1519) and provided inspiration for future generations" (Seirig, 1969).

The accumulated knowledge of biomechanics was not readily applied to the improvement of the work situation, however, despite the following observation of the great physician Ramazzini in the 1700's (Tichauer, 1978):

> *Manifold is the harvest of diseases reaped by craftsman . . . As the . . . cause I assign certain violent and irregular motions and unnatural postures . . . by*

which . . . the natural structure of the living machine is so impaired that serious diseases gradually develop.

Tichauer rationalizes that the lack of concern for minimizing human mechanically induced trauma in the workplace before 1900 was based on economics—a manual laborer was cheap and easily replaced if injured (Tichauer, 1978). It must also be conceded that the biomechanical knowledge needed to improve working conditions was limited and restricted to the aristocracy and intellectuals of the time. The commercial class that could have used the information to improve its businesses was rarely informed.

In addition, much of the biomechanical information available before the turn of the century was empirical and descriptive, which restricted its generality and application in the workplace. In other words, for occupational biomechanics to develop as a discipline, not only must there be a broad set of general biomechanical principles, but also the management of an organization has to be sophisticated (educated) and motivated enough to assimilate the principles into a management structure. Thus, two developments were necessary—one concerning the scientific basis for the field and one concerning the necessity for its application in industry.

Regarding the first point, the scientific knowledge or methodological basis for occupational biomechanics has rapidly expanded. What follows is a brief attempt to depict some of the more important methodological developments. To accomplish this, six different methodological areas have been arbitrarily defined and are illustrated at the top of Figure 1.1. Further descriptions of these methodological developments are given in subsequent chapters.

1.2.1 Kinesiological Developments

In this book, kinesiology is considered to be a subdiscipline of biomechanics. Kinesiology embraces the whole area of human movement. It can be divided into *kinematics,* which describes motion of the whole body or of major body segments independent of the forces that cause the movement, and *kinetics,* which describes the forces causing the movement. Kinematic variables include angular and linear displacements, velocities, and accelerations; kinetic variables include both internal and external forces.

Kinesiology originated when anatomists attempted to describe musculoskeletal actions and the resulting body motions. In fact, the terms kinesiology and functional anatomy are used synonymously today. From research, a large volume of data has emerged describing human motions in everyday activities as well as in sports and physical rehabilitation.

Through classifications of body segment motions and identification of muscle actions responsible for these motions, kinesiology provides descriptive models upon which quantitative biomechanical models can be formulated. In a very real sense, kinesiological knowledge is a prerequisite to the development of biomechanical models and their application. For this reason,

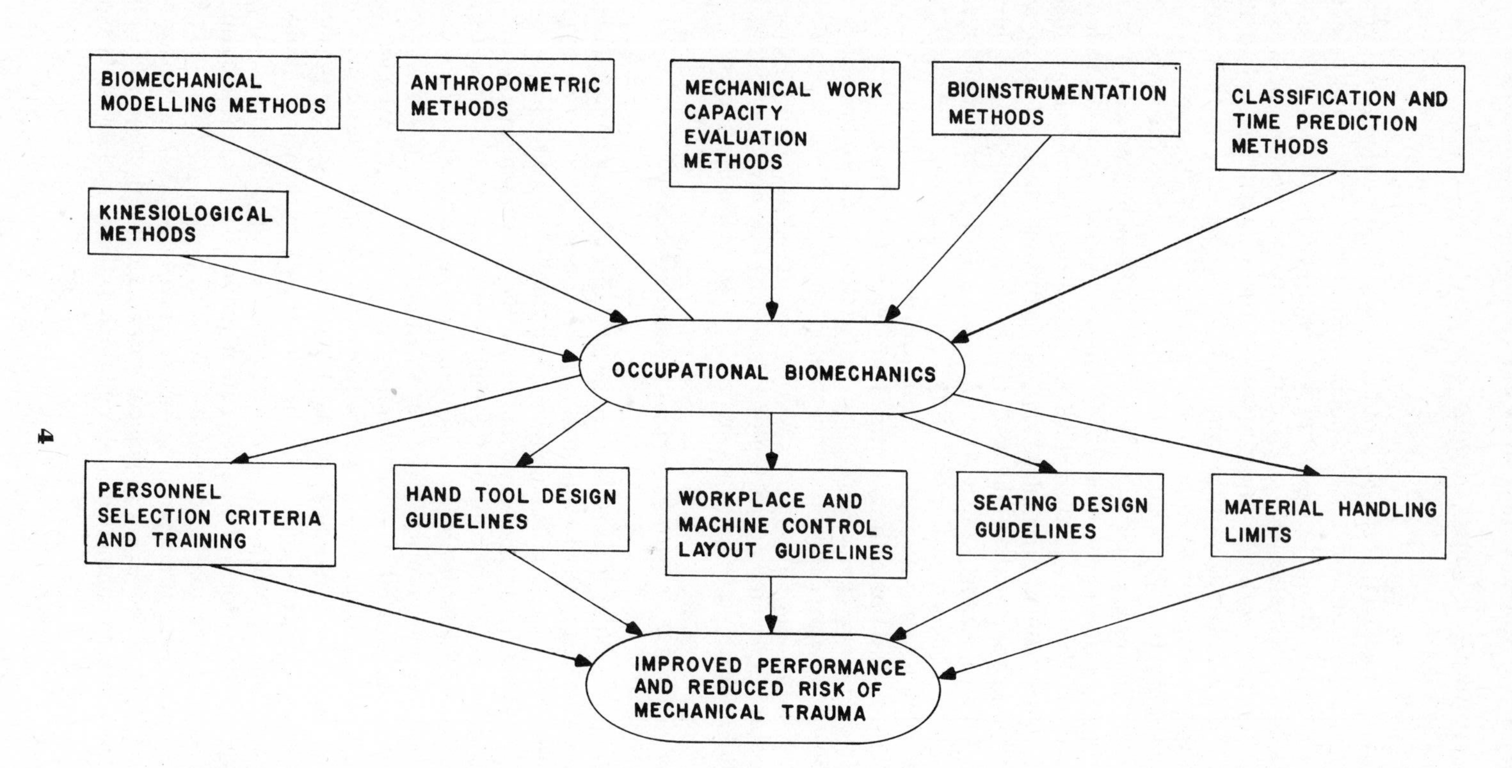

FIGURE 1.1 The six methodological areas of occupational biomechanics (top), with some major application areas (bottom.

kinesiological concepts are presented throughout the text, not in a separate chapter.

1.2.2 Biomechanical Model Developments

A direct result of kinesiological investigations of body kinematics is the development of quantitative biomechanical models of the forces operating on the human body while a person is performing common manual tasks. At first, military activities were studied. Braune and Fischer (1889), for example, studied German military populations while carrying loads, and Dempster (1955) described the mass and inertia properties of U.S. military population in various postures and motions. With the advent of the high-speed digital computer in the 1960's comprehensive, multiple-body segment, kinetic models were possible. These whole-body models of humans performing various lifting, pushing, and pulling tasks have been published since the early 1960s and serve as a major methodological development in occupational biomechanics.

1.2.3 Anthropometric Developments

Anthropometry is the empirical science that attempts to define reliable physical measures of a person's size and form for anthropological comparison. *Engineering anthropometry* stresses the application of these measurements in developing and evaluating engineering drawings and produces mock-ups to assure that the drawings are suitable for the intended user (Roebuck, Kroemer, and Thomson, 1975). Work-space designs are largely based on such data. At present, over 300 different human size and form variables have been statistically tabulated for U.S., European, and Asian populations (Webb Associates, 1978). The accuracy of an analysis clearly depends on the reliability of the anthropometric measures.

1.2.4 Mechanical Work-Capacity Evaluation Methods

In the discussion of mechanical trauma due to physical mismatching of a worker and the job demands, it must be readily conceded that the physical capacities of a normal population can vary greatly, depending on genetic factors, fitness, skill, and many other factors. It has been a traditional role of occupational medicine to devise methods for determining a person's capacity to safely perform certain types of work. For instance, the extensive developments in the field of exercise and work physiology have provided the scientific basis necessary to determine a person's capacity to perform sustained dynamic work via cardiopulmonary stress tests (Åstrand and Rodahl, 1977). Our text does not concern itself with work physiology in this context. In addition, specialists in physical medicine and sports medicine have developed the methods to reliably evaluate a person's physical strength and the flexibility of his or her joints (Miller and Nelson, 1976). Through this latter development, the capability to measure and predict the strength and joint mobility of any selected population now exists in the field of occupational biomechanics.

1.2.5 Bioinstrumentation Developments

A very important development in the field of occupational biomechanics in recent years is in the area of bioinstrumentation—data acquisition as well as analysis. The development of (1) kinematic measurement and computer analysis techniques, (2) multidirectional force transducers and force plates, and (3) electromyographic multichannel recording techniques and processing techniques to indirectly estimate muscle forces permits experimental determination of human performances in a variety of laboratory and field conditions. Such performance data now allow occupational biomechanics to be highly quantitative.

1.2.6 Motion Classification and Time-Prediction Systems Development

At the turn of this century, F. W. Taylor proposed that labor activities be standardized and coordinated to increase productivity. Managers since then have been developing the necessary scientific basis (Barnes, 1949). This effort has resulted in various work classification systems that describe human activities as a set of standard tasks or elements—for example, reach, grasp, move, position. For each element identified in a job, a ''normal time'' to perform it has been assigned, based on observations of skilled operators in different work situations. The sum of these standard element performance times then becomes the predicted standard time necessary for a skilled individual to perform a job. These work classification and time prediction systems have become quite sophisticated, largely due to the pioneering efforts of Frank and Lillian Gilbreth in the 1920's and 30's and many others since then. These systems provide a means by which human activities in industry can be described and modified (improved) as appropriate biomechanical principles are developed. In this sense, these formal work classification systems are a mechanism by which new principles in occupational biomechanics can be used by managers and engineers to improve working conditions.

In summary, biomechanics as a general field of study is as old as engineering and medical science. The development of biomechanical principles in work design, however, is a relatively new concept. It requires one to combine knowledge from several disciplines, that is, functional anatomy, mechanics, human physiology, bioinstrumentation, and work methods analysis. This book presents these methods as they relate to occupational biomechanics.

1.3 THE NEED FOR AN OCCUPATIONAL BIOMECHANICS SPECIALTY

As mentioned in Section 1.2, it is not sufficient to have just a scientific basis for a discipline to develop. One must also have an interest in such an endeavor. Several major social, legal, and economic trends appear to support the need for a special interest in occupational biomechanics today. These are briefly described in the following subsections.

1.3.1 Epidemiological Support of Occupational Biomechanics

It must first be conceded that people rarely die as a direct result of diseases or injuries involving the musculoskeletal system. Thus, mortality statistics are not relevant to the question of the importance of reducing mechanical trauma at the workplace. It is important, however, to realize that the health and quality of life is greatly reduced for a large proportion of the population because of acute and chronic musculoskeletal disorders. J. K. Kelsey and colleagues (Kelsey et al., 1978) recently published a report summarizing the results of several health surveys in the United States. They conclude that (Kelsey et al., 1978):

About 20 million people in the United States have musculoskeletal impairments.

Musculoskeletal conditions rank second only to diseases of the circulatory system on total economic cost, and are first among all disease groups in cost attributed to lost earnings and non-fatal illnesses . . . about $20 billion per year in 1972. (This is about 10 percent of the total annual cost of diseases in the U.S.)

In the United States, at least 85,000 workers receive (permanent) disability allowances for musculoskeletal conditions each year, and in California alone the total cost to insurance carriers is over $200 million per year (in 1972 U.S. dollars).

The majority of the population is at some time affected by back pain, which is the most frequent cause of activity limitation in persons under 45 years of age, and accounts for about one-sixth of all occupational injuries.

Musculoskeletal conditions rank third in frequency of occurrence among acute conditions, second in number of visits to physicians, fifth in number of visits to hospitals, and third in number of operations in hospitals.

The incidence of low back pain in different musculoskeletal disorders and injuries is high. According to the U.S. National Safety Council, low back pain accounts for about 400,000 workers compensation claims each year, with an average cost per claim (in 1974 dollars) of about $4,000.00 (National Safety Council, 1978). In some industries, particularly health care, tire and rubber manufacturing, iron and steel production, construction, and lumber processing, low back claims account for between 40 and 50% of all workers compensation claims (Bureau of Labor Statistics, 1977).

British statistics indicate that low back pain was responsible for 3.6% of all sickness absence days in 1969–1970 (Benn and Wood, 1975) and that during that period the number of sickness absence episodes was 22.6/1,000 men and 11/1,000 women. The annual loss of working days was 747/1,000 employed males. Taylor (1976) found that 1 in 25 men changed their occupation or work activities because of back problems.

Hult (1954) estimated the lifetime incidence of low back pain in Sweden to be about 60%, based on a survey of more than 1000 male workers in several different occupations. About 4% were work incapacitated by their pain for more than six months. In a Swedish population of 40–47-year-old men, 61% were found to have had low back pain at some time in life, while 31% actually had pain at the time of the survey—one month prevalence (Svensson and Andersson, 1982). The same investigators later studied a group of 35–60-year-old women, and again similar data emerged (Svensson and Andersson, 1982). Chronic and more severe disabilities were found in about 4% in both surveys. These data are close to those presented from Denmark by Bierring-Sørensen (1982), but somewhat higher than data from the Netherlands where Valkenburg and Haanen (1982) report that 51% of men and 57% of women in a large population survey had experienced low back pain.

The relationship of low back pain to mechanical stress and work has been reviewed by Andersson (1981), Snook (1982), and others. Although the precise relationships are unknown, there is considerable evidence to show that work load is important. Workers involved in heavy lifting, for example, have been found to have about eight times the number of low back injuries than those with more sedentary work (Chaffin and Park, 1973).

The low back is not the only anatomical region to be stressed in industry. Shoulder and neck symptoms appear to increase in society. Allander (1974) found clinical signs and symptoms from the shoulder region in 7–26% of a randomly selected group of Swedish people, the prevalence increasing with age. Waris (1979) summarized the effect of occupation on these conditions.

Other musculoskeletal problems of significant magnitude and clearly related to work are lateral elbow pain (tennis elbow), which occurs in about 1 to 3% of the population (Allander, 1974; Kurppa et al, 1979) and median nerve-compression syndrome causing numbness to the hands (Armstrong and Chaffin, 1979). There is also evidence that the incidence of tendinitis (tendon inflammatory disease) and other musculoskeletal disease is associated with jobs requiring repetitive and forceful manual exertions (Hymovich and Lindholm, 1966; Hadler, 1977). Further, working in an unnatural posture because of an inadequate seat and work-place design has been found to result in increased pain in the neck, shoulder, upper back, lower back, and leg, as documented by Grandjean (1980), Tichauer (1978) and others.

In short, there is ample evidence to indicate that musculoskeletal problems are quite prevalent and costly. Further, these incidents can be caused or aggravated by mechanical trauma in the workplace. In fact, it is now recognized in legal statutes in the United States that medical disability compensation should be paid to workers for cumulative trauma due to repetitive physical exertions over a period of time (Hershenson, 1979).

1.3.2 Social/Legal Support of Occupational Biomechanics

One major social (and often legislated) trend in industrialized countries is the attempt to accommodate individual workers with diverse physical capabilities, and sometimes disabilties, in the work force. This trend has meant that simple rules denying people of certain age, gender, race, or apparent disability a fair opportunity to perform a job are no longer valid (Miner and Miner, 1978). For instance, an allowable weight-lifting standard recommended in the middle 1960's by the International Standards Organization is no longer valid in many countries because it discriminates against women and people of advanced age by stating limits for these groups that are far below those for a younger male. It has been realized in recent years that a person's physical attributes, such as strength, flexibility, and endurance vary greatly within gender and age groups. Further, a person who is disabled in one task may excel in other tasks.

For these reasons it has become necessary to know with as much precision as possible the functional capacity of an individual relative to a job's performance requirements to assure that the person is not unjustly denied a job. This is also good medical practice and, indeed, is required by some occupational health and safety statutes. In this latter sense, the use of functional capacity tests can assist in assuring that a job applicant will not be over-stressed when placed on a new job. For instance, recent studies have shown that people who were not able to demonstrate an isometric strength sufficient to adequately perform the strength requiring tasks in various jobs assigned to them had a three-fold increase in both the incidence and severity of musculoskeletal injuries compared with their stronger peers (Chaffin, 1974; Chaffin, Herrin and Keyserling, 1978; and Keyserling, et al., 1980).

The development and validation of appropriate functional capacity tests of both job applicants and people who are seeking to return to work after suffering a musculoskeletal injury is a major occupational biomechanics effort. Because of this it will be discussed at length later in this volume.

1.3.3 Ergonomic Support of Occupational Biomechanics

Ergonomics, which can be defined as fitting the work to the person, has a primary goal of improving worker performance and safety through the study and development of general principles that govern the interaction of humans and their working environment. As such, ergonomics tends to be a much broader discipline than occupational biomechanics. The two disciplines are highly complementary, however, as indicated in the following quote from the keynote address by Professor J. Wartenweiler at the founding meeting of the International Society on Biomechanics (Nelson and Morehouse, 1974):

> *Earlier (human) motor research work of physiologists was usually limited to industrial movements. Although these (manual) processes are increasingly being changed due to automation, ergonomics is still very much interested in biomechanics.*

Indeed, some of the most frequently cited early work in ergonomics dealt with problems of a biomechanical nature. Examples are the development of design criteria for aircraft pilot and secretarial seats; the establishment of permissible control forces and locations in aircraft and industrial crane cabs and turning lathes and other machines; recommending allowable loads to be lifted and carried in various postures; and design guides for hand tools (Shackel, 1976). In fact, a recent book by Professor E. R. Tichauer is entitled *The Biomechanical Basis of Ergonomics* (Tichauer, 1978). In this context, occupational biomechanics can be viewed as a supporting subspecialty within the broader field of ergonomics.

1.4 WHO USES OCCUPATIONAL BIOMECHANICS

It should be clear from the preceding that the scientific basis for occupational biomechanics resides in several different disciplines. Likewise, in practice, occupational biomechanics is used by a variety of specialists. The following examples are proposed to illustrate how occupational biomechanics can be useful in different situations and who will use the information obtained.

If a company is in an expansion phase or is redesigning its facilities or manufacturing processes, occupational biomechanics expertise can provide engineering guidelines regarding machines, tools, and workplace layouts. Thus, occupational biomechanics can be regarded as an engineering specialty, with applications executed through industrial, production, and manufacturing engineering functions.

If a company is not rapidly changing its processes, knowledge of occupational biomechanics can be used to (1) evaluate the extent to which existing jobs place physical demands on the workers, (2) simulate alternative work methods and determine potential reduction in physical demands if new work practices were instigated, and (3) provide a basis for employee selection and placement procedures. These activities are most often executed through the personnel functions of a firm, i.e., within the labor relations, safety, hygiene, or medical departments.

If an organization does not have a large, multifunction staff, an ergonomist can be hired to serve in both the engineering and personnel roles. Obviously the ergonomist must be broadly educated, with occupational biomechanics being only one of several disciplines in the person's training.

1.5 ORGANIZATION OF BOOK

The book is divided into two parts. The first part provides a description of the structure and function of the musculoskeletal system, followed by an introduction to each of the five methodological areas discussed earlier that define the science of occupational biomechanics. These methodologies are, in order of presentation:

1. Anthropometry (Chapter 3)
2. Work-Capacity Evaluation (Chapter 4)
3. Bioinstrumentation (Chapter 5)
4. Biomechanical Models (Chapter 6)
5. Work Classification and Time Prediction (Chapter 7)

The second part presents six contemporary applications of occupational biomechanics necessary to establish:

1. Materials Handling Limits (Chapter 8)
2. Guidelines for Seated Work (Chapter 9)
3. Machine Control and Workplace Layout Guidelines (Chapter 10)
4. Hand-Tool Design Guidelines (Chapter 11)
5. Guidelines for Whole-Body and Segmental Vibration (Chapter 12)
6. Worker Selection and Training Criteria (Chapter 13)

REFERENCES

Akkas, N., W. Goldsmith, and R. M. Kenedi, "Biomechanics: Applications, Costs and Benefits, Priorities," *J. Biomechanics,* **13,** 737–743 (1980).

Allander, E., "Prevalence, Incidence, and Remission Rates of Some Common Rheumatic Diseases and Syndrome," *Scand. J. Rheumatol.* **3,** 145–163 (1974).

Andersson, G. B. J., "Epidemiological Aspects of Low Back Pain in Industry," *Spine* **6,** 53–60 (1981).

Armstrong, T. J. and D. B. Chaffin, "Carpal Tunnel Syndrome and Selected Personal Attributes," *J. Occup. Med.,* **21**(7), 481–486 (1979).

Åstrand, P. O. and K. Rodahl, *Textbook of Work Physiology,* McGraw-Hill, New York, 1977.

Barnes, R. M., *Motion and Time Study,* Wiley, New York, 1949.

Benn, R. T. and P. H. N. Wood, "Pain in the Back: An Attempt to Estimate the Size of the Problem," *Rheum. and Rehab.* **14,** 121–128 (1975).

Biering-Sørensen, F., "Low Back Trouble in a General Population of 30, 40, 50 and 60 Year Old Men and Women," *Dan. Med. Bull.* **29,** 289–299, 1982.

Braune, W. and O. Fischer, *The Center of Gravity of the Human Body as Related to the German Infantryman,* Leipzig, Germany (AT1138, 452, available from U.S. National Technical Information Office), 1889.

Bureau of Labor Statistics, *Supplemental Data Systems Annual Report,* U.S. Department of Labor, Washington, D.C., 1977.

Chaffin, D. B., "Human Strength Capability and Low-Back Pain," *J. Occup. Med.,* **16**(4), 248–254 (1974).

Chaffin, D. B., G. D. Herrin, and W. M. Keyserling, "Preemployment Strength Testing," *J. Occup. Med.,* **20**(6), 403–408 (1978).

Chaffin, D. B. and K. S. Park, "A Longitudinal Study of Low-Back Pain as Associated with Occupational Weight Lifting Factors," *Amer. Ind. Hyg. J.,* 34, 513–525 (1973).

Dempster, W. T., *Space Requirements of the Seated Operator,* WADC-TR-55-159, Aerospace Medical Research Laboratories, Wright-Patterson AFB, Ohio, 1955.

Frankel, V. H. and M. Nordin, *Basic Biomechanics of the Skeletal System,* Lea and Febiger, Philadelphia, 1980, pp. ix.

Fung, Y. C., *Biomechanics,* Springer-Verlag, New York, 1981, pp 1–10.

Grandjean, E., *Fitting the Task to the Man,* Taylor and Francis, London, 1980, pp 41–62.

Hadler, N., "Industrial Rheumatology: Clinical Investigations Into the Influence of the Pattern of Usages on the Pattern of Regional Musculoskeletal Disease," *Arthritis. Rheum.,* **20,** 1019–1025 (1977).

Hershenson, A., "Cumulative Trauma: A National Problem," *J. Occup. Med.,* **21**(10), 674–676 (1979).

Hult, L., "Cervical, Dorsal and Lumbar Spinal Syndromes," *Acta. Orthop. Scand.,* Suppl. 17 (1954).

Hymovich, L. and M. Lindholm, "Hand, Wrist and Forearm Injuries—The Result of Repetitive Motions, *J. Occup. Med.,* **8,** 573–577 (1966).

Kelsey, J. L., H. Pastides, and G. E. Bisbee, *Musculo-Skeletal Disorders,* Prodist, New York, 1978.

Keyserling, W. M., D. G. Herrin and D. B. Chaffin, "Isometric Strength Testing as a Means of Controlling Medical Incidents on Strenous Jobs," *J. Occup. Med.,* **22**(5), 332–336 (1980).

Kurppa, K., P. Waris and P. Rokkanen, "Peritendinitis and Tenosynovitis. A Review," *Scand. J. Work, Environ. and Health* Suppl, **13,** 19–24 (1979).

Miller, D. I. and R. C. Nelson, *Biomechanics of Sport,* Lea and Febiger, Philadelphia, 1976.

Miner, M. G. and J. G. Miner, *Employee Selection Within the Law,* Bureau of National Affairs, Washington, D.C., 1978.

National Safety Council, *Accident Facts,* 1978 edition, Chicago, 1978.

Nelson, R. C. and C. A. Morehouse, *Biomechanics IV,* University Park Press, Baltimore, 1974.

Roebuck, J. A., K. H. E. Kroemer, and W. G. Thomson, *Engineering Anthropometry Methods,* Wiley-Interscience, 1975, pp 4–7.

Seirig, A., "Leonardo da Vinci—The Biomechanician," in D. Bootzin and H. C. Muffley, Eds., *Biomechanics,* Plenum, New York, 1969, pp. 65–74.

Shackel, B., *Applied Ergonomics Handbook,* Vol. 3, IPC Press, Guildford, England, 1976.

Snook, S. H., "Low Back Pain in Industry," *Symposium on Idiopathic Low Back Pain,* Mosby, St. Louis, 23–38, 1982.

Svensson, H.-O. and G. B. J. Andersson, "Low Back Pain in 40–47 Year Old Men: Frequency of Occurrence and Impact on Medical Services," *Scand. J. Rehab. Med.,* **14,** 47–53 (1982).

Svensson, H.-O. and G. B. J. Andersson, "Low Back Pain in 40–47 Year Old Men: Work History and Work Environment Factors," *Spine*, 8, 272–276 (1983).

Svensson, H.-O. and G. B. J. Andersson, "Low Back Pain in Swedish Women 38–65 Years Old," Paper read at the *International Society for the Study of the Lumbar Spine,* Toronto, 1982.

Taylor, D. G., "The Costs of Arthritis and the Benefits of Joint Replacement Surgery," *Proc. Roy. Soc. Lond. 192,* 144–155 (1976).

Tichauer, E. R., *The Biomechanical Basis of Ergonomics,* Wiley-Interscience, New York, 1978.

Valkenburg, H. A. and H. C. M. Haanen, "The Epidemiology of Low Back Pain," *Symposium on Idiopathic Low Back Pain,* St. Louis, pp 9–22, 1982.

Waris, P., "Occupational Cervicobrachial Syndromes," *Scand. J. Work, Environ. and Health,* Suppl. 3, 3–14 (1979).

Webb Associates, *Anthropometric Source Book,* Vol. II, NASA Reference Publication 1024, Washington, D.C., 1978.

2

THE STRUCTURE AND FUNCTION OF THE MUSCULO-SKELETAL SYSTEM

2.1 INTRODUCTION

Application of the laws of physics and engineering concepts to the human body requires knowledge of the basic structure and function of the musculoskeletal system. It is the purpose of this chapter to introduce the different component structures of the system and their mechanical functions. The main function of these components is to support and protect the body and its different organs and to provide motion.

For the whole body to function normally, each substructure must function normally. The six major substructures are: tendons, ligaments, fascia, cartilage, bone, and muscle, each to be discussed separately.

Joints are necessary to allow motion between the body segments. Their role in the entire musculoskeletal system is so important that joints are often referred to as the musculoskeletal system's *functional units* (Rosse and Clawson, 1980).

2.2 THE CONNECTIVE TISSUES

All connective tissue is made up of both *cells* and an *extracellular matrix* composed of *fibers* and *ground substance*. The cells produce the matrix, which determines the physical properties of the different connective tissues.

Fibroblasts are the special cells of loose connective tissue (adipose tissue, etc.), skin, tendons, and ligaments, while the cells of cartilage are called chondrocytes. Osteocytes are cells that produce the matrix of the bone.

There are two different types of fibers in connective tissue: *collagen fibers* and *elastic fibers*. The proportion of these in the different tissues has a major influence on their mechanical properties. Collagen is the most prevalent of the two fibers, existing in at least three different types. It is a material of high tensile strength and quite resistent to deformation. Type I collagen is found in tendons, ligaments, bone, skin, and loose connective tissue. Type II is mainly found in cartilage. Type III is present in the walls of large blood vessels. The elastic fibers are, as the name indicates, elastic, and their tensile strength is low.

The most important components of the ground substance are the proteoglycans—large molecules of glycosaminoglycans linked to a protein core. The proportion of proteoglycans to other constituents vary considerably. As an example, proteoglycans constitute only a few percent of the ground substance in bone, compared with more than 25% of the ground substance in cartilage. Other constituents of ground substance are water, glycoproteins, lipids, and in bone calcium.

2.2.1 Ligaments, Tendons, and Fascia

Ligaments and *tendons* are dense connective tissues similar in morphology and function. Ligaments connect bone to bone, providing stability at joints, while tendons attach muscles to bone, transmitting forces from the muscles. *Fascia* is also a dense connective tissue that covers organs or parts of organs and separates them from each other. Examples of fascia tissue are the *intramuscular septa* that separate the limb muscles.

The collagen fibers are organized in parallel bundles in tendons, with some non-parallel fibers in the ligaments (Figure 2.1). Cells, elastic fibers, and connective tissue are sparse. In fascia, a more open organization of collagen fibers is found, and the proportion of elastic fibers is greater.

The fiber arrangement as well as the proportion of the two types of fibers are important to the mechanical properties of the different tissues. Collagen fiber bundles, when tested in tension, at first elongate slightly and then become increasingly stiffer until they yield (Figure 2.2). This is because the fibers at rest have a somewhat wavy configuration that is straightened out at low load. Once straightened the deformation is small, ranging from 6 to 8% (Abrahams, 1967). The ultimate strength of collagen fibers is considerable, about 50% of that of cortical bone tested in tension (Frankel and Nordin, 1980).

The elastic fibers, on the other hand, are weak and brittle. When tension is applied, they deform or strain greatly at low loads, increasing their length by more than 100%. As they approach their maximum strain they suddenly become stiff and fail without much further deformation (Figure 2.3).

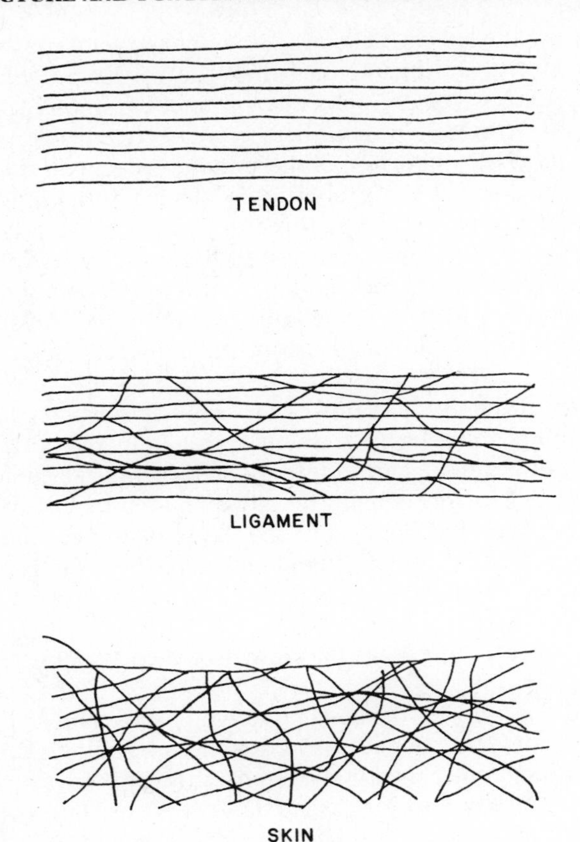

FIGURE 2.1 Schematic diagram of the structural orientation of the fibers of tendon, ligament and skin. (Adapted from Frankel and Nordin, 1980.)

It should be clear that the mechanical properties of the composite tissue will vary greatly depending upon the proportions of collagen to elastic fibers. The less-structured orientation of the collagen in the ligaments compared with tendons provides additional elastic properties to the ligaments, even more so in fascia. In a ligament, where about 90% of the fibers are collagen, the load deformation curve under tensile load will be almost identical with that of collagen bundles. This is illustrated in Figure 2.4 for an anterior cruciate ligament.

Tendons are often surrounded by sheaths of fibrous tissue in areas where friction would otherwise be a problem. The sheath has an inner lining, called a *synovium,* similar to the inner lining of joints (see Section 2.4). The synovial membrane produces *synovial fluid,* which facilitates gliding of the tendon. To further aid mechanical motion and force transmission, ligaments exist in parts of the sheath, providing the function of pulleys and tendon guides.

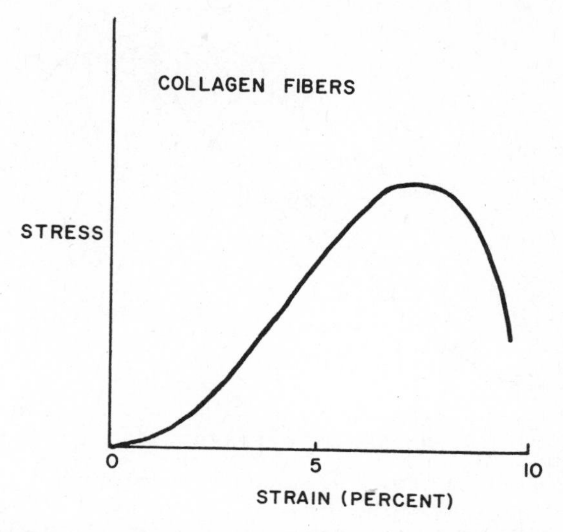

FIGURE 2.2 Stress–strain curve for collagen fiber bundle tested in tension to failure. (Adapted from Abrahams, 1967.)

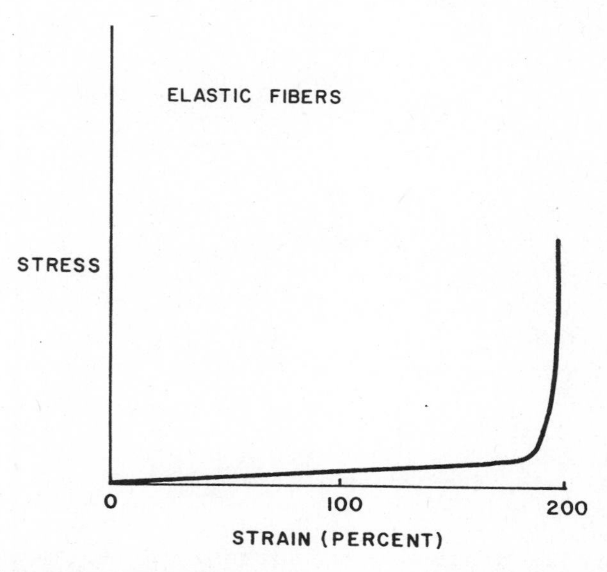

FIGURE 2.3 Stress–strain curve for elastic fiber bundles tested in tension to failure. (Adapted from Carton et al., 1962.)

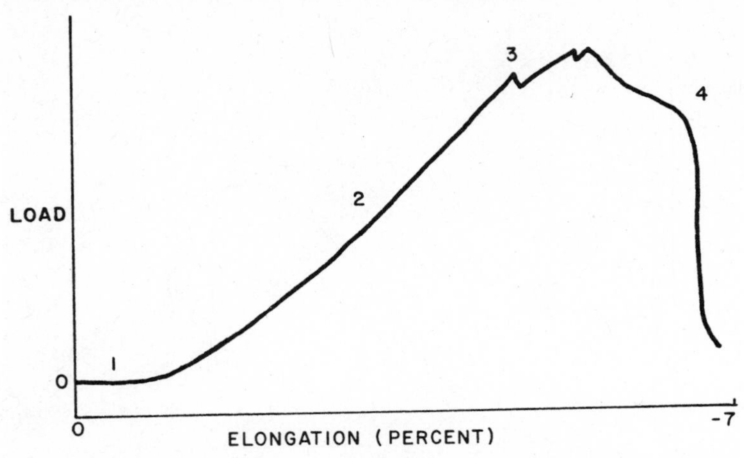

FIGURE 2.4 Stress–strain curve for a human anterior cruciate ligament tested in tension to failure. This ligament has about 90% collagen fibers. (Adapted from Noyes, 1977). 1. The wavy collagen fibers straighten out. 2. Deformation occurs. 3. Progressive failure of collagen bundles (elongation 6 to 8%). 4. Complete failure.

Thus the tendons can function around corners, as in the finger joints illustrated in Figure 2.5.

2.2.2 Cartilage

Cartilage covers articular surfaces and is also present in a few other organs—ear, nose, respiratory tract, and intervertebral discs. Three main types of

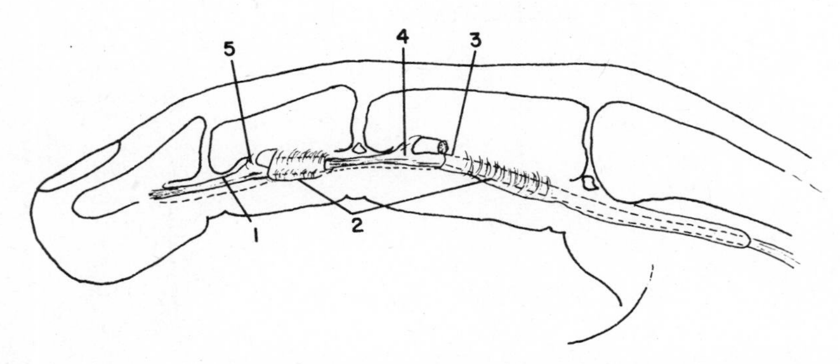

FIGURE 2.5 Schematic drawing of the flexor digitorum profundus tendon (1) of the human hand showing the location of the most important pulleys (2), which are heavy fibrous condensations of the tendon sheath. The sheath (3) has been omitted in places (dotted lines to show the tendon), the vinculum longus (4), which originates in the vinculum brevis of the superficialis tendon (not shown), and the vinculum breve (5). The primary function of the vincula is to carry a blood supply to the tendon within the tendon sheath. (Adapted from Frankel and Nordin, 1980.)

cartilage are distinguished: *hyaline cartilage,* which is present in growth plates and on articular surfaces as well as in the respiratory tract; *fibrocartilage,* which is present in intervertebral discs; and *elastic cartilage,* found in the ear and epiglottis.

Hyaline cartilage has a highly organized collagen fiber network and a homogenous matrix (see Section 2.4). Fibrocartilage is made up of islands of hyaline cartilage in a dense collagen mass of fibers, and elastic cartilage contains elastic fibers. Cartilage is a unique tissue in that it is devoid of nerves and blood vessels. It receives its nutrients by diffusion, which limits its thickness and influences healing following injury.

The function of joint cartilage will be discussed later in this chapter in Section 2.4.

2.2.3 Bone

Bone can be considered both in the context of whole bone and as a tissue. Whole bone actually consists of several tissues providing skeletal support and maintaining mineral homeostasis. As much as 99% of the body calcium is present in bone, and up to 80% by weight of cortical bone tissue is mineral. To aid in understanding bone and its mechanical properties, some terms are now introduced. Generally, bones are divided into two groups; the long bones of the extremities, and the axial (appendicular) skeleton, the latter containing the skull, vertebra, sternum, ribs, and pelvis. The bones of the axial skeleton are flat, while the long bones are round made up by a shaft, the *diaphysis,* and two expanded ends, the *epiphyses* (Figure 2.6). Both types of bone consist of *cortical* and *cancellous* bone. The cortical bone or *compact* bone has a much higher proportion of bone tissue to other tissue than cancellous bone. It provides the cortex (lining) of the bone. Cancellous bone (or spongy bone) is found mainly in the epiphyses of long bones and between the cortices of bones of the axial skeleton. It is a three-dimensional lattice of bone *trabeculae*.

2.2.3.1 The Structure of Bone

To understand the structure of bone, first consider its development. Bone matrix is formed by *osteoblasts,* which gradually transform into *osteocytes*. These cells then become isolated inside the mineralized matrix in so-called lacunae. Non-mineralized bone is called *osteoid*. In addition to the two cells mentioned, a third cell exists in bone—the *osteoclast,* a cell with considerable importance in bone remodelling. Bone formation and resorption occurs simultaneously and constantly throughout life. In this sense, bone is different from all other connective tissue.

The process of bone formation is called *ossification*. In a long bone, bone forms within and upon a cartilage model similar in shape to the final bone. Briefly, the sequence of events, as illustrated in Figure 2.7, is that at first

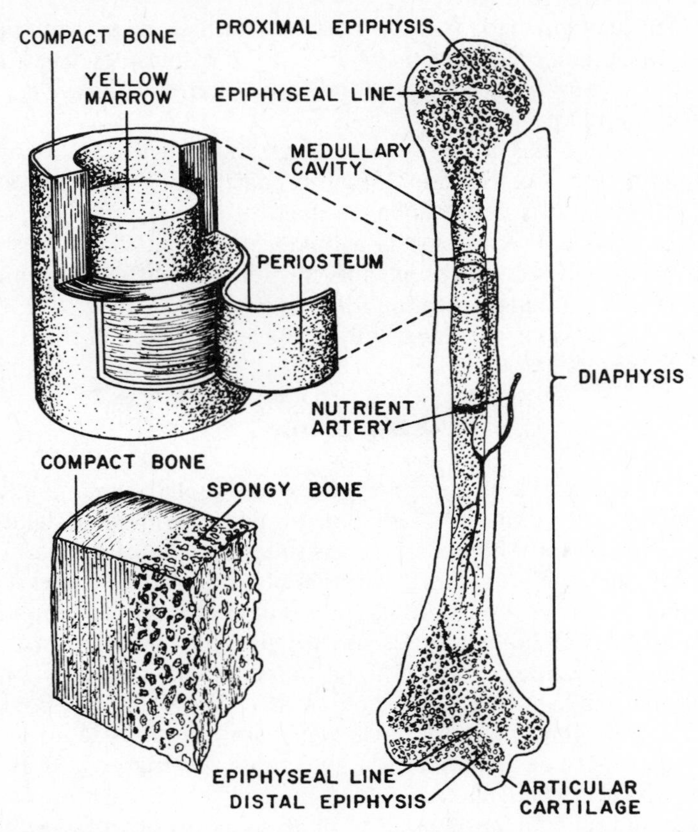

FIGURE 2.6 Diagram of a long bone shown in longitudinal section, with representative magnifications.

the osteoid matrix is deposited on the surface of the shaft. The osteoid is subsequently mineralized and penetrated by blood vessels. Osteogenic (bone producing) cells follow the vessels to form so-called *primary ossification centers* in the diaphyses. These centers expand toward the bone ends, leaving a hollow area, the *medullary cavity*. Subsequently, *secondary ossification centers* appear within the cartilage at the end of the long bone, the area called the *epiphysis*. Between the primary and secondary ossification centers remain areas of cartilage called the *epiphyseal plate* or the growth-plate. These areas are responsible for the longitudinal growth of the bone, and they remain open until growth is ended. At that point, the growth-plates ossify. Circumferential growth of whole bone, on the other hand, depends on *subperiosteal bone formation,* the *periosteum* being the fibrous outer lining of the bone.

The first bone, referred to as *woven bone,* is deposited at the periosteum in the ossification centers. Following injury it is deposited at the healing

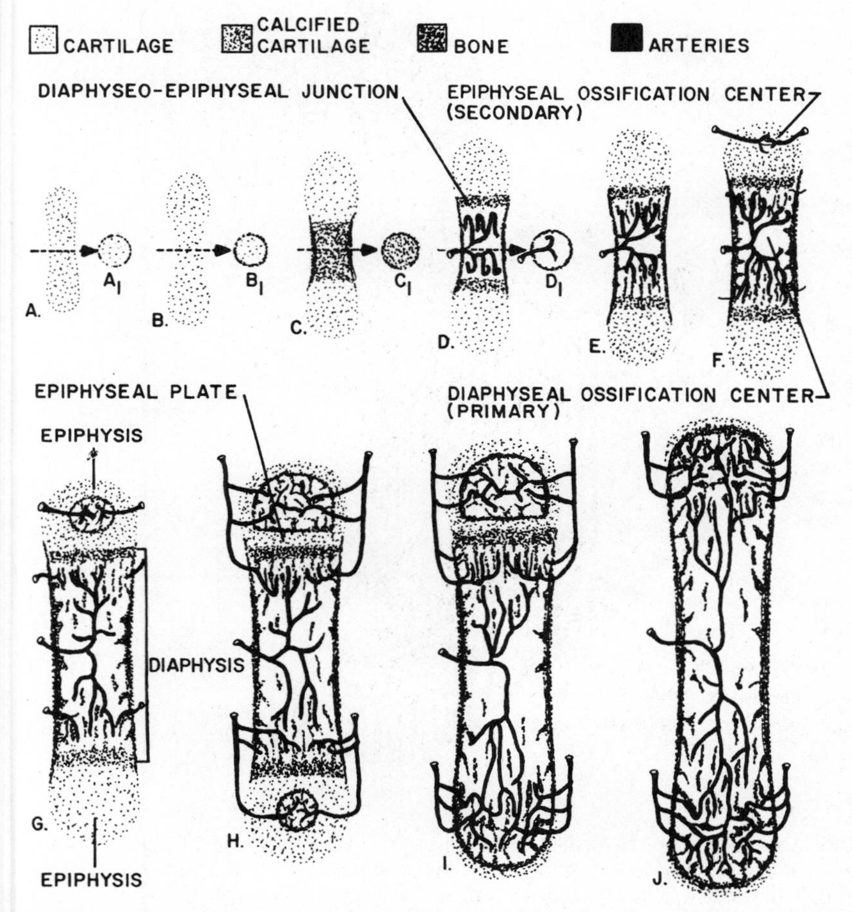

FIGURE 2.7 Schematic representation of endochondral ossification and the development of a long bone. A to J are longitudinal sections and A_1 to D_1 are cross sections at the levels indicated. A. Cartilage model of the bone. B. Subperiosteal collar of bone appears. C. Cartilage begins to calcify. D. Vascular osteogenic mesenchyme enters the calcified cartilage and the primary ossification center is established. E. Calcification of cartilage advances toward the ends of the bone primordium, followed by ossification. The medullary cavity is established. F. Blood vessels and osteogenic mesenchyme enter the upper epiphysis. G. The epiphyseal or secondary ossification center is established, with epiphyseal plate interposed between it and the diaphyseal (primary) ossification center. H. A secondary ossification center develops in the lower epiphyseal cartilage. I. The lower epiphyseal plate is ossified but growth proceeds at the upper epiphyseal plate. J. The upper epiphyseal plate is ossified. (Adapted from Moore, K. L.: The developing human, Philadelphia, W. B. Saunders, 1977.)

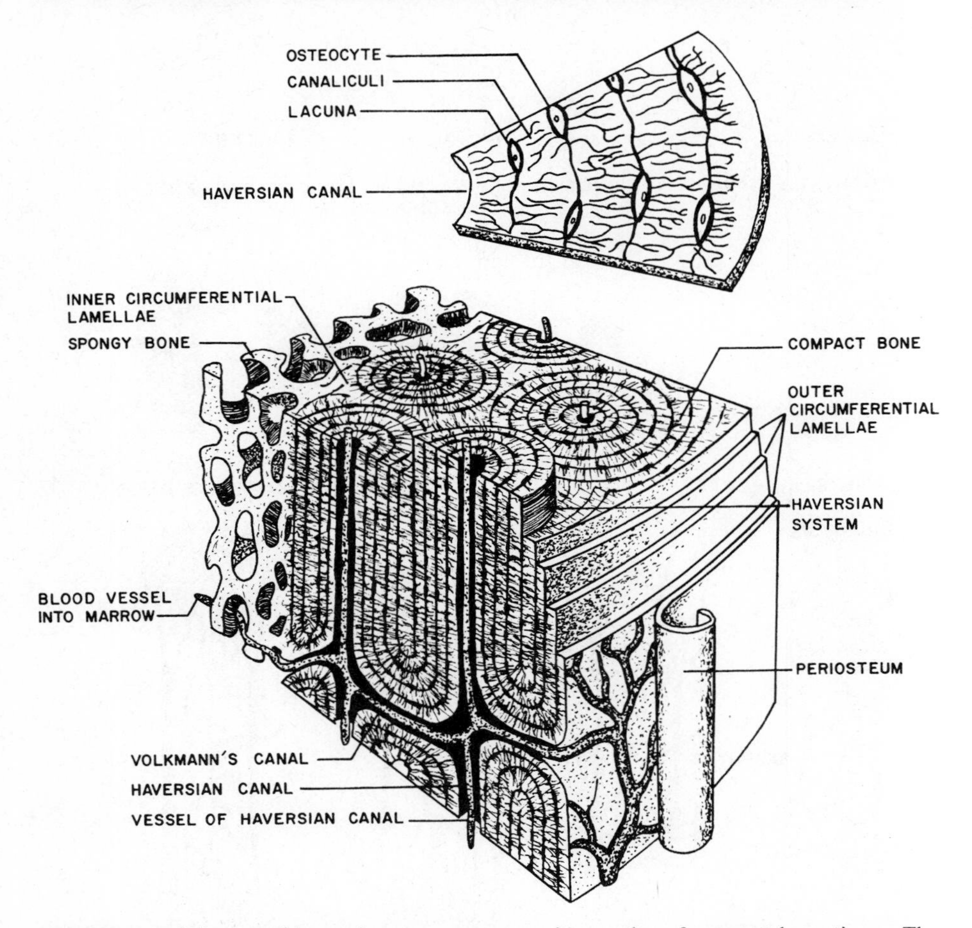

FIGURE 2.8 Diagram of haversian systems as seen in a wedge of compact bone tissue. The periosteum has been peeled back to show a blood vessel entering a Volkmann's canal. The upper right illustration shows osteocytes lying within lacunae. Canaliculi permit interstitial fluid to reach each lacuna.

area of a fracture. This type of bone, which is sometimes referred to as immature bone, is later replaced by so called *lamellar bone*. Lamellar bone is made up of sheets of bone tissue in which collagen fibers and hydroxyapatite crystals are arranged in regular parallel lamellae (Figure 2.8). The replacement procedure requires that the woven bone be resorbed, which occurs by osteolysis and phagocytosis by the previously mentioned osteoclasts, which are large cells with several nuclei.

In cancellous bone, the trabulae are made up of a few lamellae. The orientation of these trabulae is dependent on mechanical stress (Figure 2.8). At the bone perimeter, the outer layers are also made up of parallel lamellae, while the main part of the cortical bone shell is made up of *osteons,* also called *haversian systems* (Figure 2.8). Each such system consists of a central blood vessel—the haversian canal—and surrounding concentrical lamellae, up to about 20. The osteons are usually oriented in the long axis of the bone; between osteons, interstitial lamellae fill out the remainder of the bone, which, when constructed of osteons, is called *compact bone*. On the inside of the shell, circumferential lamellae appear again, connecting onto the trabeculae of cancellous bone.

2.2.3.2 The Mechanical Properties of Bone

The macrostructure of bone discussed previously is important to the mechanical properties of whole bones. The cortical shell of a long bone provides it with considerable strength and stiffness, while the cancellous trabecular network at the metaphysis is efficient in transferring the high loads applied to the articular cartilage covering the metaphysis to this cortical shell. To determine the material properties of bone tissue, small pieces can be taken from the whole bone. A summary of the test procedures and results obtained is given by Hayes and Carter (1979). Since cancellous and cortical bone have quite different porosites—5 to 30% for cortical bone, compared with 30% to over 90% in cancellous bone—large differences exist between these two types of bone. Both are *anisotropic*—that is, their mechanical properties are different when loads are applied in different directions (Figure 2.9).

Examination of the whole bone as a composite is more meaningful for the purpose of occupational biomechanics. The most important mechanical properties are the strength and stiffness of the bone. When a bone is loaded to failure, it fractures. Fractures in cancellous bone occur mainly from tensile and compressive forces. Such tensile forces can be produced by vigorous muscle contractions dislodging the tendon and adjacent bone, while compression fractures often occur because of external loads. An example of the latter is a skull fracture from the impact of a fall. Fractures due to shear forces occur also mainly in areas of cancellous bone. Cortical bone is generally stronger in compression than in tension and stronger in tension than in shear. Most fractures of bone occur, however, as a result of bending forces, from torsional forces, or from combined forces.

Bone has a significant energy-storage capacity, dependent on velocity. Figure 2.10 shows how energy storage is almost doubled by increasing the speed of loading. The curve also shows that the fracture load is doubled, and that the bone stiffness increases with higher loading speed.

Bone can also fracture from repetitive loading. Although fatigue fractures are uncommon, they can occur from repetitive loading in vocational activities. Three factors are important in this respect—the amount of load, the number of repetitions, and the frequency of loading. Carter and Hayes (1977) have shown that bone tested *in vitro* fatigues rapidly when loaded close to its yield strength. It should be remembered, however, that bone is a living

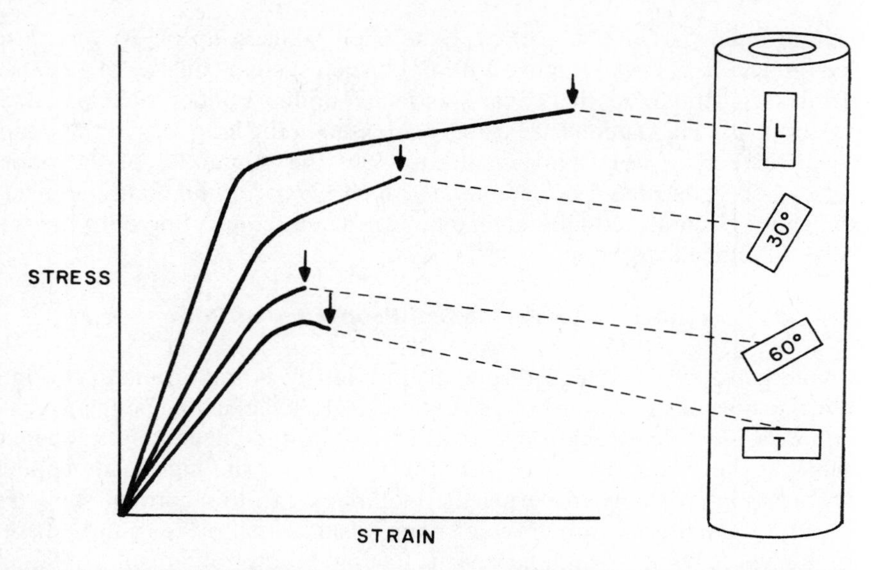

FIGURE 2.9 Anisotropic behavior of cortical bone specimens from a human femoral shaft tested in tension in four directions: longitudinal (L), tilted 30° with respect to the neutral axis of the bone, tilted 60°, and transverse (T). ↓ denotes point of failure. (Adapted from Frankel and Nordin, 1980.)

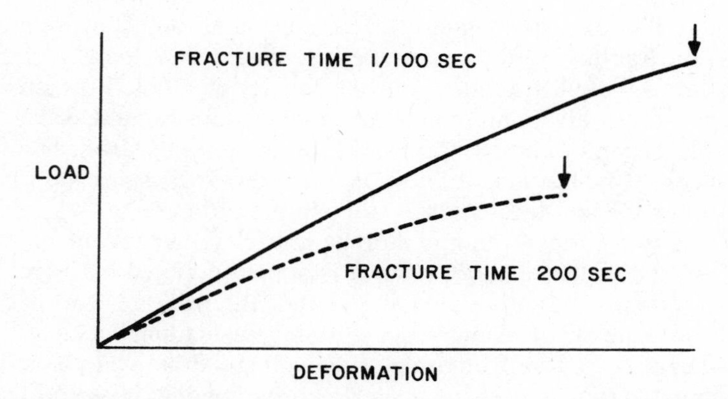

FIGURE 2.10 Energy storage in paired dog tibiae tested at high and low loading speeds. In the one instance the fracture time was $\frac{1}{100}$ sec; in the other, 200 sec. Fracture time could be given as a strain rate—dividing the strain (deformation) by the fracture time. Energy is the area under the curve. The load to failure and the energy stored to failure almost doubled at the higher speed. ↓ denotes point of failure. (Adapted from Sammarco et al., 1971.)

material capable of repairing small cracks given time, so that the repetition rate or recovery period after exertions can become significant factors.

It should also be noted that because muscles act to distribute the major stress within bone and absorb inertial energy which otherwise could be transferred directly to the bone, continuous strenuous activity can cause abnormally high stresses on bones as the muscles fatigue and lose their contractile capacity.

2.2.3.3 *Bone Remodeling*

As mentioned, bone is a dynamic tissue. Though specific mechanisms of remodeling are not discussed here, an excellent summary is given by Albright and Skinner (in Albright and Brand, 1979). Remodeling is important to occupational biomechanics because bone remodelling is stress dependent and age-related changes occur in bone.

Bone changes its shape, size, and structure as a result of the mechanical demands placed on it over time. The relationship between bone remodeling and mechanical load was expressed as early as 1892 by Wolff and is commonly known as *Wolff's law*. This states that bone will be deposited where needed and resorbed where not needed. Although Wolff suggested that the changes occurring from mechanical stress (or strain) could be described mathematically, the precise relationships are still unknown. Figure 2.11 illustrates the effect of immobilization on bone strength and stiffness. Converseley, bone hypertrophy occurs readily in areas where stress and strain are increased (Figure 2.12).

From the beginning of the ossification process until skeletal maturity is reached, bone growth occurs. The maximum skeletal mass is attained at about age 30, after which bone loss occurs continuously and with accelerated rate as we get older. At first there is little difference in the rate of bone loss between men and women, but after menopause a sharp increase in the rate at which bone mineral loss occurs in women has been noted. The gradual aging process changes "normal" bone into "osteoporotic" bone. The age related changes are characterized by:

1. A progressive decrease in bone mineral content.
2. A reduction in cortical bone thickness.
3. An increase in the outer diameter of long bones, resulting in an increase in the moment of inertia around the long bone axis.
4. A decrease in the number of trabeculae in cancellous bone.

The resulting bone is weaker, and as a consequence fractures are much more common after trivial trauma in older people, particularly in older women.

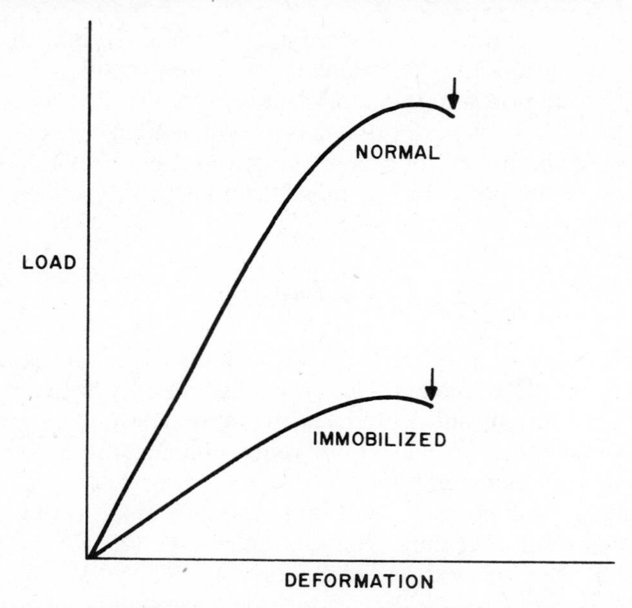

FIGURE 2.11 Load-deformation curves for vertebral segments (L5-L7) from normal and immobilized Rhesus monkeys. ↓ denotes point of failure. (Adapted from Kazarian and Von Gierke, 1969.)

2.3 SKELETAL MUSCLE

Skeletal muscle constitutes almost 50% of the body weight. There are several hundred muscles within the body, each with a special function. In general, they all serve to generate moments about the joints. Each muscle is its own separate organ, with a structure to be described later. Muscles are attached to bone by tendons, and cross one or several joints. They are under the direct control of the *voluntary* nervous system, sometimes referred to as the *somatic nervous system*.

The unique property of muscle is that it can contract. When contraction is initiated through the nervous system, the muscle either shortens, producing a movement at the joint, or contracts without shortening, providing stability.

2.3.1 The Structure of Muscles

The muscle is covered by a fascia, the *epimysium*, from which connective tissue septa extend inward, subdividing the muscle into muscle fiber bundles, *fasciculi*. These fasciculi are further subdivided into individual fibers that are again surrounded by connective tissue membranes (Figure 2.13). The connective tissue is important to muscle as it provides a pathway for nerves

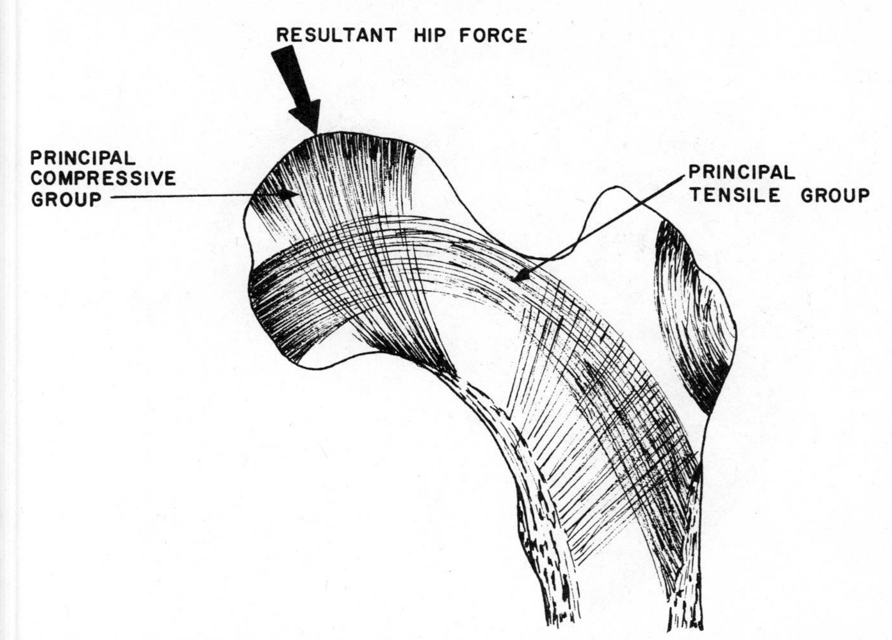

FIGURE 2.12 Organization of bone trabeculae in the upper end of the femur in response to the forces acting on the femoral head.

and blood vessels and contributes to the mechanical properties of the muscle. The nerves and vessels enter into the muscle at one or two areas and then distribute between fasciculi and fibers.

The structural and functional unit of nervous tissue is the nerve cell, also called the *neuron*. Neurons consists of cell bodies and several extending processes. Processes conducting impulses away from the cell bodies are termed *axons*. Each neuron has only one axon, while there are usually several *dendrites,* which are processes carrying impulses to the cell body. *Nerve fiber* refers to a long axon and its surrounding sheath of tissue. The speed with which an impulse is transmitted along a nerve fiber is approximately proportional to the diameter of the fiber. In peripheral nerves, several nerve fibers are united by a surrounding connective tissue. These nerves are mixed—that is, they contain both *sensory* or afferent nerve fibers, leading impulses to the central nervous system, and *motor* or efferent nerve fibers, leading impulses from the central nervous system. In the muscle, the individual muscle fibers are each innervated by a terminal branch of an efferent nerve fiber or axon. The group of muscle fibers innervated by branches of the same efferent neuron axon are called a *motor unit* (Figure 2.14).

The *motor unit* is the functional unit of the muscle. It varies in size (number of fibers) depending upon the muscle. Motor units are small in muscles where precise control is important and larger in coarse-acting muscles. Thus,

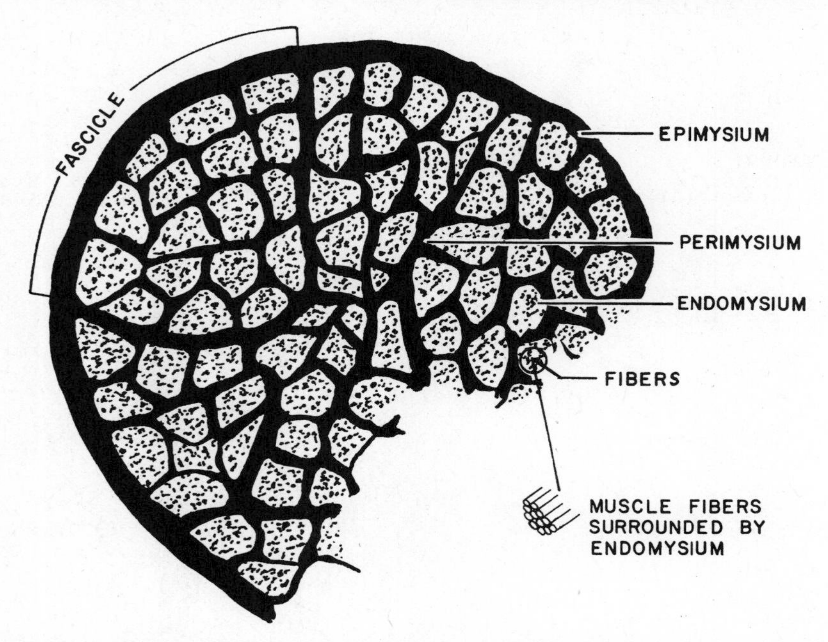

FIGURE 2.13 Portion of a muscle in transverse section. The muscle is enclosed in a fibrous tissue sheath, the epimysium. Fascicles are surrounded by the perimysium, and individual fibers by the endomysium.

in the eye muscles, there are only a few fibers in each motor unit, while in the limb muscles, one single nerve fiber can supply more than a thousand muscle fibers. Motor units act according to the law of "all or nothing." When an impulse reaches a motor unit, all the muscle fibers in the unit contract almost simultaneously.

The muscles are attached to bone by tendons. These are formed by condensation of the connective tissue in the muscle at the *myotendinal junction,* and then the tendon fibers insert into the bone matrix as so-called *Sharpey's fibers*.

The muscle fibers are long cylindrical cells, each containing several nuclei. Most of the fiber volume is taken up by the longitudinally arranged *myofibrils,* which are the contractile elements. The myofibrils show a banding pattern of cross striations with alternating white and dark segments. Due the arrangement of the fibrils, the pattern is transferred to the whole of the fiber. This transverse striation is the reason why skeletal muscle is sometimes referred to as striated muscle. The different bands of a muscle fiber are illustrated in Figure 2.15. The dark bands are called A-bands; the light, I-bands. Through the center of the A-band there is a narrow lighter band, the H-band; transecting this band is the so called m-line. The I-band also has a central dense transverse band, the Z-line. The relative length of each

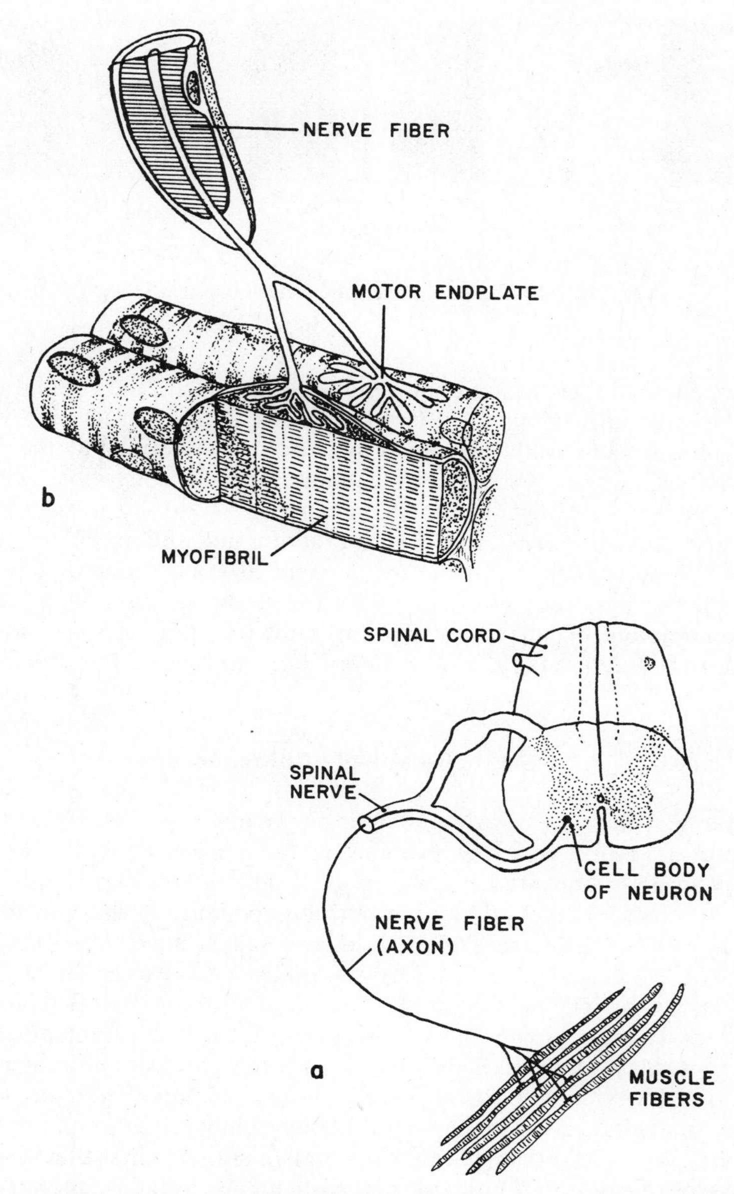

FIGURE 2.14 Scheme of (A) motor unit and (B) neuromuscular junction (motor endplate) in greater detail. (Adapted from Basmajian, 1980.)

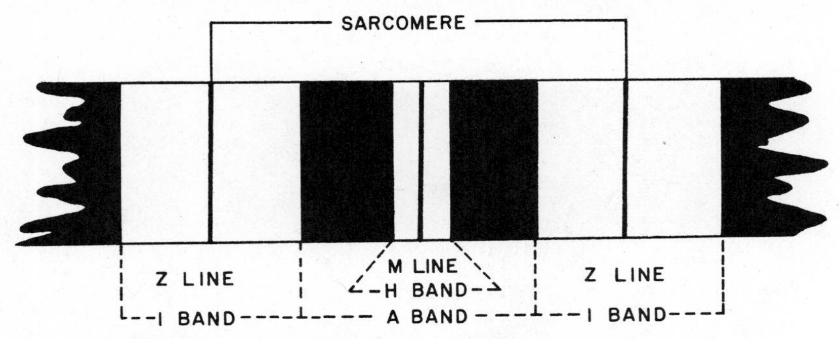

FIGURE 2.15 Banding pattern of skeletal muscle. See text.

band depends on the state of contraction of the muscle; the A-band remains constant in length, while the I-band shortens with contraction. The contractile unit in the myofibril is called the *sarcomere,* defined by the distance between two Z-lines. There are two subunits within the myofibrils, the thick myofilaments consisting mostly of the protein *myosin,* and the thin ones containing *actin* (Figure 2.16). The interaction and sliding of the filaments relative to each other is a basic mechanism of muscle contraction. Thus, muscle shortening does not occur because of shortening of the individual myofilaments but because they slide, as shown in Figure 2.17. This model of muscle contraction is called *the sliding filament model of muscle* (Huxley, 1974).

2.3.2 The Muscle Contraction

The sliding of the two myosin and actin filaments occurs because of a complex interaction between the proteins of the thick and thin filaments. In addition to actin and myosin, two more proteins, *tropomyosin B* and *troponin,* are also involved. They are both components of the thin filaments. Myosin is by far the largest of the four proteins. It makes up about 50% of the muscle mass and consists of two proteins coiled together in a double helical structure. The myosin molecule resembles a rod with two small globular heads (Figure 2-16). The heads protrude from the thick filaments at regular intervals along its length. They are the parts of the molecule that serve as cross-bridges to the actin. The actin, which is the major protein of the thin filaments, consists of two polymers: globular actin and fibrous actine. They are arranged so that the appearance of the actin is that of a twisted double row of beads. Tropomyosin B exists mainly in the I-zone, in complex with fibrous actine. Troponin is a minor component of the thin filament (10%), but the key to the contraction process. It has high affinity to calcium. Muscle contraction starts with the release of calcium, which immediately complexes with troponin. This shifts the protein conformation of the thin

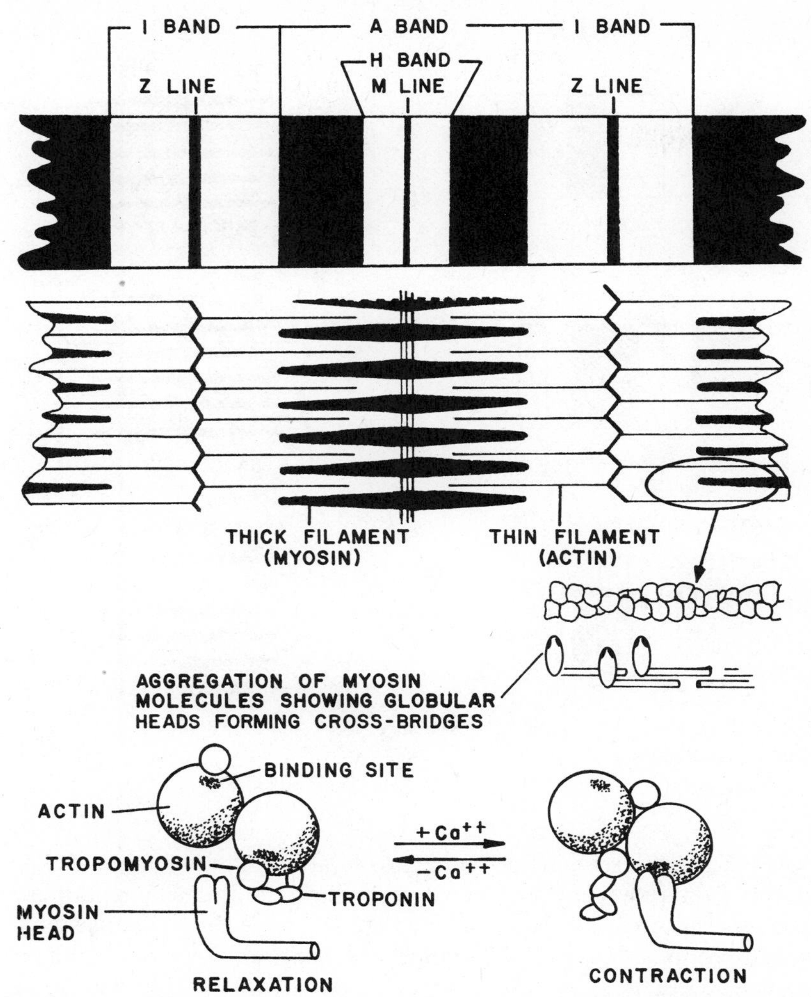

FIGURE 2.16 (Top) Relationship of cross banding to the thin and thick filaments. Shown enlarged is the interdigitation of thick and thin filaments and the relationship of muscle proteins to each other. (Bottom) Schematic representation of how Ca^{++} controls actin and myosin interaction in some muscles. On the left, in absence of Ca^{++}, tropomyosin hides the binding site from myosin and the muscle is relaxed. On the right, in the presence of Ca^{++}, calcium binds to troponin and the tropomyosin moves into the groove exposing the binding site, allowing interaction, and the muscle contracts. (Adapted from Rosse and Clawson, 1980.)

filaments so that reactive sites on the actin molecules come in contact with the globular heads of the myosin, forming cross-bridges between the two filaments. The actin molecule moves up along the myosin filament as cross-bridges are broken and re-established along the actin molecule. The driving force to this sliding is due to the attachment of the cross-bridges from the

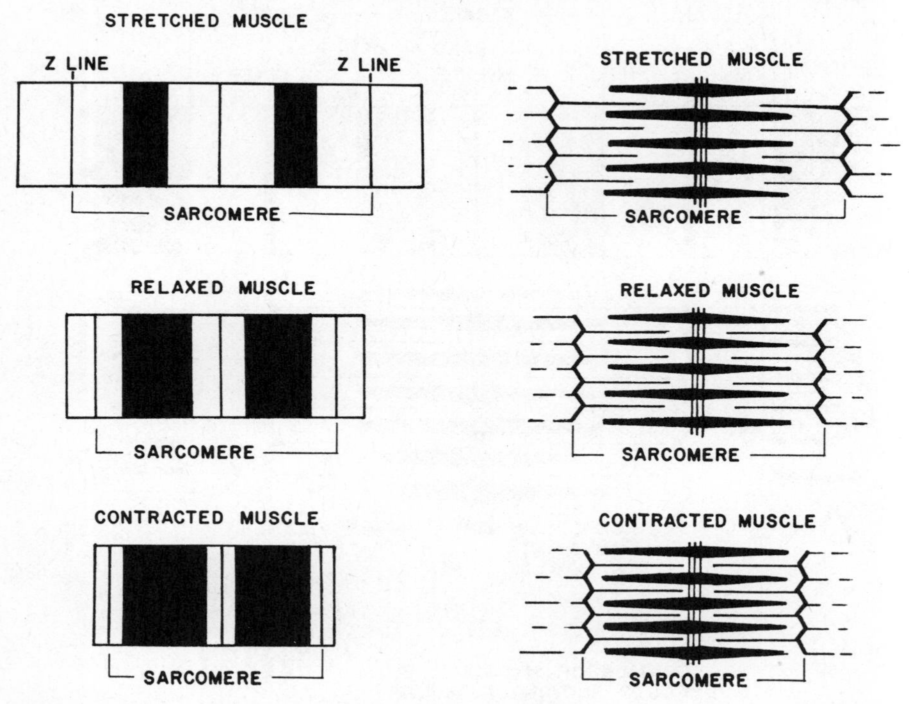

FIGURE 2.17 Banding pattern and arrangement of thick and thin filaments in stretched, relaxed, and contracted muscle.

molecules of the thick filament, which are established at an angle to the thin filament and then swivel to a new angle, pulling the thin filaments past the thick ones. The cross-bridges on opposite sides of the H-band swivel in opposite directions. Thus, the distances between the Z lines decrease and the muscle contracts.

The contractile process requires energy. The immediate source of energy is *adenosine triphosphate* (ATP), which is split to a lower energy form of ADP (*adenosine diphosphate*), releasing the energy used in contraction. Within less than a millisecond after a single contractile cycle ends, ADP is converted back to ATP by a reaction with *phosphocreatine,* another high-energy phosphate available in the muscle. This allows the contraction to continue, but the process described does not produce ATP at a rate needed for sustained contractions. The phosphocreatine is depleted after 15 to 30 sec in strenuous exertions. Resynthesis of ATP then occurs in one of two ways. One way is by anaerobic metabolism (not requiring oxygen) of glycogen (sugar). This process produces lactate as its end product. The second way is by resynthesis through an oxidative aerobic metabolic process, referred to as the citric acid cycle or *Krebs cycle,* producing CO_2 and H_2O as

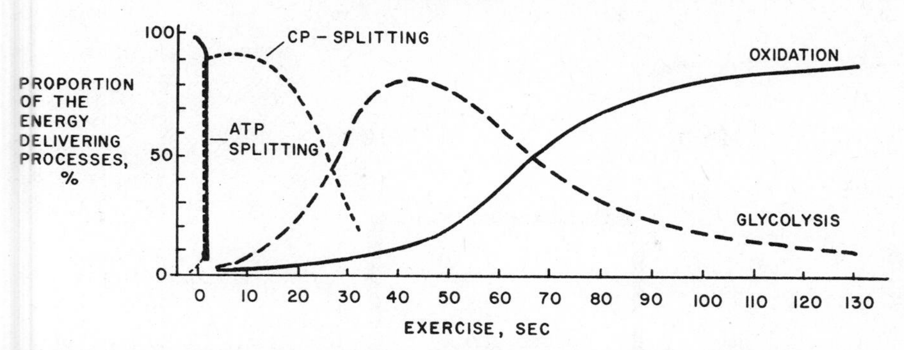

FIGURE 2.18 Schematic representation of substrates supplying energy for muscle contraction at maximum voluntary contraction level. Stored ATP sources supply only immediate needs for initial few muscle contractions. The next source is creatine phosphate, allowing 30 to 40 sec of muscular activity. As the creatine stores become depleted, anaerobic glycolysis becomes the next major source of high-energy nucleotides. With progression of activity, oxidative metabolism of carbohydrates takes over as the major energy source. (Adapted from Keul, Doll, and Kepler, 1972.) Two figures, 2.18 and 2.19, were obtained from Albright & Brand, but the copyright belonged to Keul, Doll, and Kepler: Energy Metabolism of Human Muscle, S. Karger, 1972.

end products. During moderate activity the oxygen supply is often sufficient to allow the aerobic supply of high-energy phosphates to occur. When an individual is working at such a level, the situation is referred to as *steady state*.

At higher levels of activity, the energy requirement to produce ATP must be met by the anaerobic process. This leads to an accumulation of lactic acid, and a person is considered to have an *oxygen debt*. If this debt, for any reason, is not repaid, *muscle fatigue* develops. The energy sources relative to the time of sustained muscle contraction can be summarized as in Figure 2.18, while Figure 2.19 illustrates the relative major sources of energy available. The chemical reactions produce large amount of heat. In fact, during normal dynamic activities, only about 20 to 25% of the liberated energy is converted to work. The rest is expended as heat.

The electrical events occurring in muscle that can be measured for biomechanical studies are discussed in Chapter 5. Because of differences in ions on the inside and outside of a nerve or muscle cell membrane, an electrical *potential* always exists. The nerve impulse generated in the efferent or motoneurons of the central nervous system manifests itself as a change in the selective permeability of the membrane, a so-called *depolarization*. The resulting *nerve action potential* is transmitted along the axon to the motor endplates. This causes a *muscle action potential* to develop along the muscle fiber membranes. The neuromuscular transmission at the endplates is chemically mediated. *Acetylcholine* (ACh) is released from the nerve-axon terminal branches, a release requiring the presence of calcium ions. The ACh then causes the muscle fiber membranes to depolarize, resulting in the

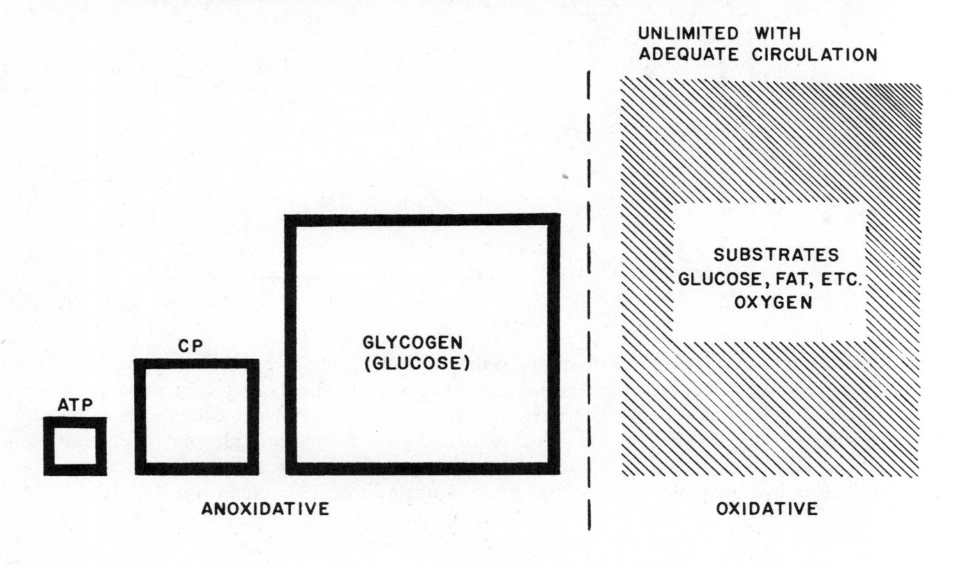

FIGURE 2.19 Major energy sources utilized during muscle contraction. (Adapted from Keul, Doll, and Kepler, 1972.) Two figures, 2.18 and 2.19, were obtained from Albright & Brand, but the copyright belonged to Keul, Doll, and Kepler: Energy Metabolism of Human Muscle, S. Karger, 1972.

muscle action potential. As mentioned in Section 2.3.1, the neural stimulus causes an ''all or nothing'' response that is, all muscle fibers of the motor unit contract. An ''all or nothing'' response also occurs in each muscle fiber in the sense that the whole of each fiber contracts.

Before discussing the mechanical properties of muscle contractions, a few terms should be defined. When a muscle is stimulated by a single nerve action potential, a contraction called a *twitch* occurs. The short period from stimulation to the actual contraction is the *latent period,* the time of shortening is the *contraction period,* and the time of lengthening is the *relaxation period.*The duration of these periods vary considerably from muscle to muscle (Figure 2.20).

2.3.3 Mechanical Aspects of the Muscle Contraction

Two types of contraction are often distinguished: *isometric,* meaning that the length of the muscle is constant during the contraction; and *isotonic,* indicating that the muscle length changes while its tension (force output) remains unchanged (Figure 2.21). In its pure sense, isometric contraction produces tension but no useful mechanical work.

The response of a whole muscle to a stimulus depends on several factors—for example, the size and frequency of the stimulus, the fiber composition

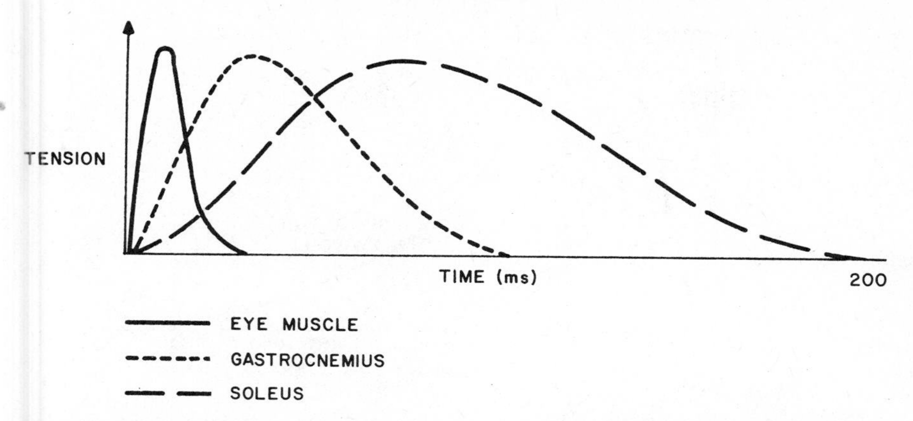

FIGURE 2.20 Relative durations of simple twitches of different types of skeletal muscles. The medial rectus of the eye is rapid in its responses. The gastrocnemius is intermediate. The soleus twitch is of long duration.

of the muscle, the length of the muscle, and the velocity of the muscle contraction.

When an individual muscle nerve is stimulated, a certain number of motor units will contract, depending on the size of the stimulus. As the neural stimulation increases, additional units are recruited, until a maximum contraction occurs. If a second neural stimulus is delivered before the muscle response from the first is ended, a greater contraction force will occur. This phenomenon is known as *summation*, and occurs even if the second stimulus is equal to the first, and even if all motor units have contracted as a result of the first stimulus (Figure 2.22). When a series of neural impulses are delivered regularly at a rapid enough frequency, a sustained maximal contraction occurs, known as *tetanus*. The rate at which stimuli must be applied to cause tetanus varies in muscles from about 30 per second in the soleus to more than 300 per second in the eye muscles. The resulting muscle tension will be maintained at higher than maximal single-twitch levels as long as the train of stimuli continues, or until fatigue develops.

Understanding why some muscles respond faster or are more resistant to fatigue requires a further study of muscle physiology. Though it is an over-simplification, two types of muscle fibers can be distinguished by the duration of the twitch—slow-twitch fibers, or *Type I*, and fast-twitch fibers, or *Type II*. Some researchers further subdivide the fibers, while other believe that the fiber can change from one type to another with specific use. Regardless, the Type I fibers are smaller, with a high capacity for aerobic metabolism, and thus are good at sustaining low levels of exertion. They are referred to as slow fibers (or tonic fibers) because they twitch with low peak tension and with a long rise time to peak tension. The Type II (or fast phasic fibers), on the other hand, rely greatly on anaerobic metabolism and have large peak tensions with a short rise time to peak tension. Histologi-

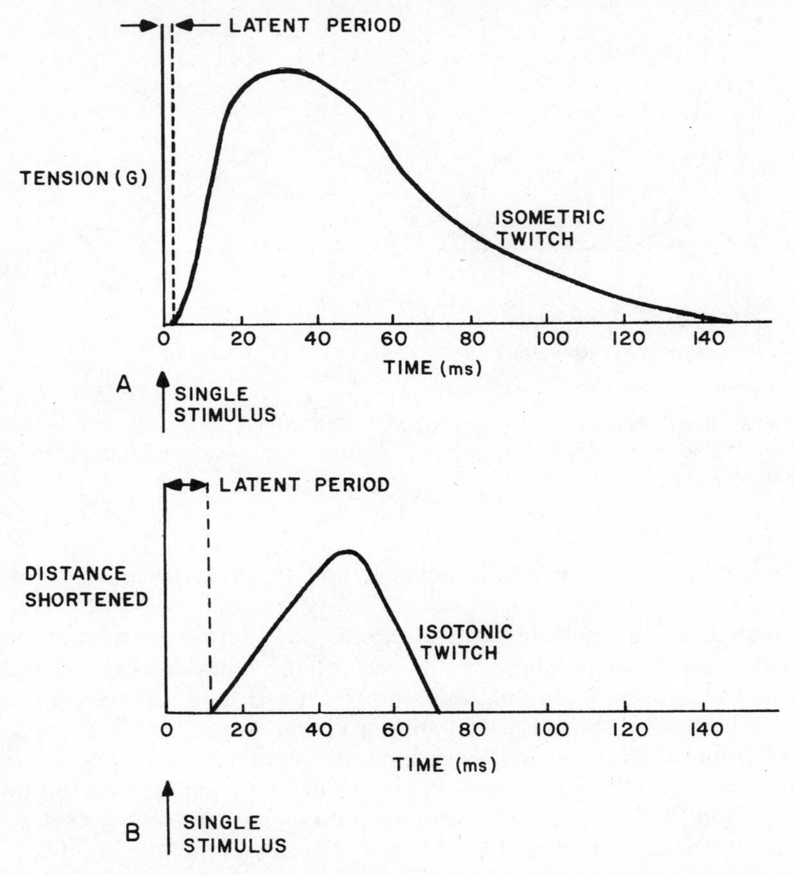

FIGURE 2.21 Isotonic and isometric muscle contraction. The mechanical response of the muscle to a single stimulus of approximately 1 msec duration is a twitch recorded as tension (A) or as shortening (B). (Adapted from Vander, A. J., Sherman, J. H., and Luciano, D. S., 1970.)

cally, the fibers are also distinguished on the basis of histochemical staining techniques, usually with ATP-ase stain giving a dark pattern for slow-twitch fibers and a light pattern for fast-twitch fibers. Human muscles have a mix of both fibers. The eye muscles are examples of fast-twitch muscles, while the soleus muscle of the leg is an example of a slow-twitch muscle (a postural muscle). The duration of the twitch in these two muscles is 10 and 100 msec, respectively.

2.3.3.1 *The Length-Tension Relationship in Muscle*

The length of the muscle is important to its ability to produce tension. This is partly because of the muscle connective tissue and partly because of the geometric relationships of contractile elements.

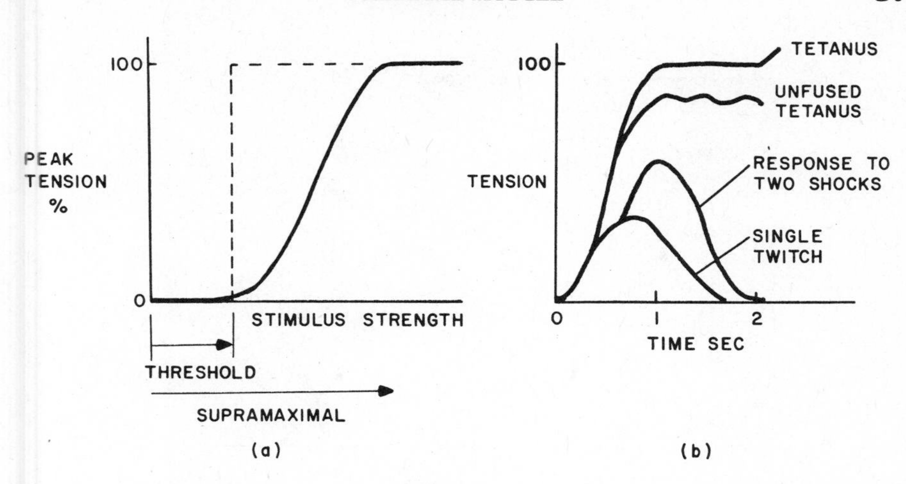

FIGURE 2.22 (A) The relation between strength of stimulus and size of response. The interrupted line shows an all-or-nothing response; the continuous line, a graded one. (B) The summation of responses following repeated stimulation. Frog sartorius, 0°C.

A resting muscle is elastic because of its connective tissue stroma. When an external force is applied, the muscle stretches and stiffens. The resting elasticity arises mainly from the connective tissue meshwork within the muscle. When the muscle is stretched, the fibers become progressively taught. The length–tension curve of a resting muscle is largely determined by the amount of connective tissue in the muscle and the properties of that tissue. Since the connective tissue fibers and muscle fibers are parallel to each other, tension from both tissues must be summed to give the *total tension* when a muscle is stimulated. The length–tension curve in Figure 2.23 demonstrates this summation of passive and active (contractile) tensions.

When a muscle changes in length from its so-called resting length, a further complication arises. The sliding-filament model discussed in Section 2.3.1 implies that the maximum number of cross-bridges between the thick and thin filaments exist when the muscle is about at its resting length—i.e., where passive tension begins to occur because of the elastic elements. A maximum contraction or active tension is possible only at that length. With further lengthening, there is a decreasing area of overlap between the filaments, reducing the number of cross-bridges; thus, active tension decreases. Similarly, when the muscle shortens, an overlap occurs between the filaments, causing interference with the formation of the cross-bridges, and active tension decreases. Thus the length–tension curve of an individual fiber will be as shown in Figure 2.24.

The overall resulting force of a contracting muscle to be transmitted by the tendon is illustrated as a function of muscle length in Figure 2.25. Since the amount of active tension is under voluntary control, the resulting re-

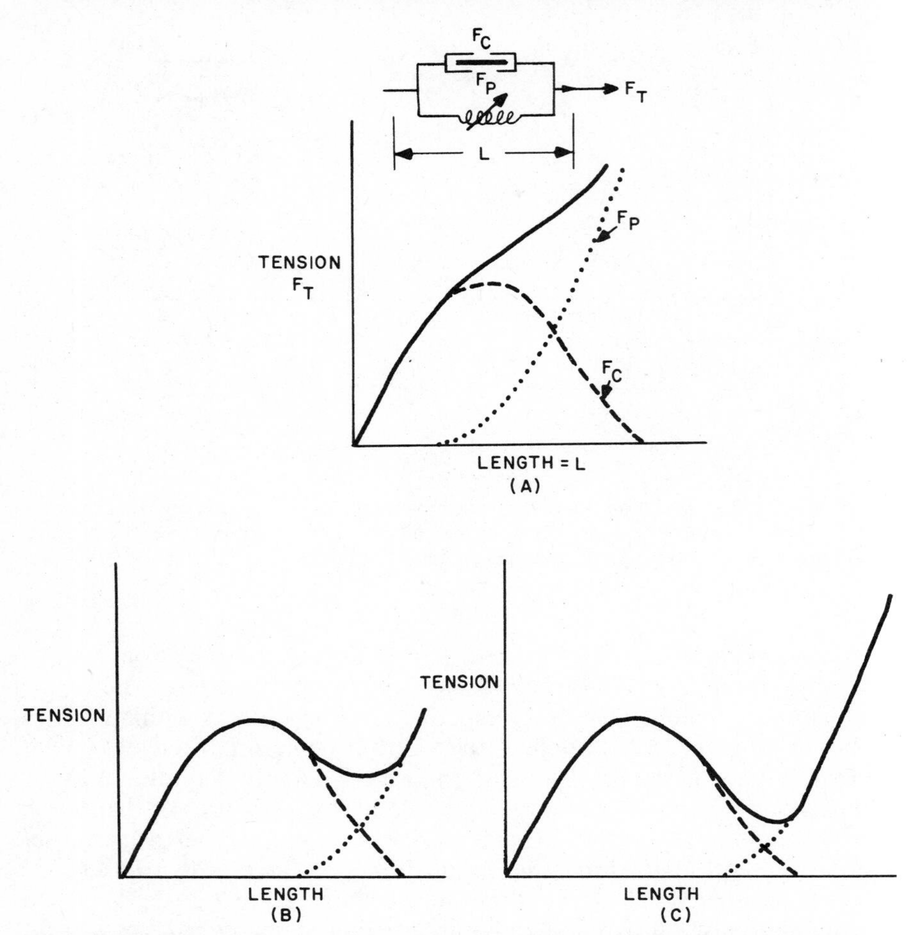

FIGURE 2.23 Illustration of the variation in the tension–length curve that results from having progressively less connective tissue, from A to C. The dotted line shows the length–tension curve of the resting muscle, which is largely determined by the amount of connective tissue. The total tension recorded on tetanizing the muscle is shown by the solid line, and the extra tension developed on stimulation (= solid minus dotted) is shown by the interrupted line. Note that even if the resting tension curve and the curve of tension developed are of constant shape as in the example, the curve of tension measured varies considerably, depending on the length at which resting tension begins to be developed. (Adapted from Carlsson and Wilkie, 1974.)

sponse will be largely dependent on the level of stimulus—i.e., the active contraction force dominates.

2.3.3.2 *Velocity–Tension Relationships*

The discussion of force–tension relationships was primarily concerned with isometric contractions. When movement is involved, which is often the case,

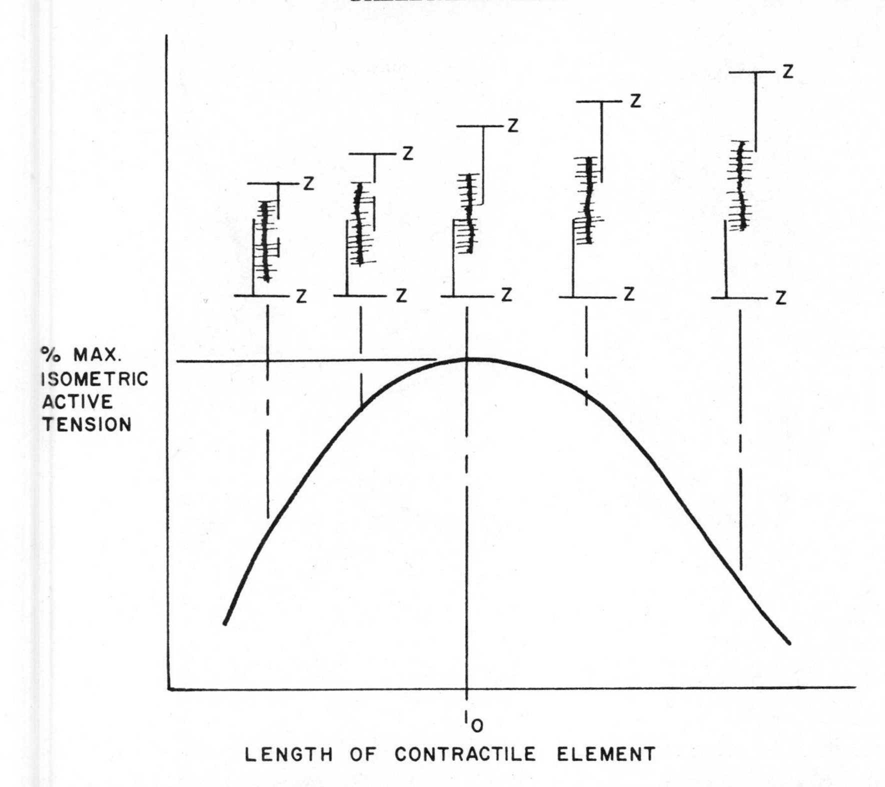

FIGURE 2.24 Active tension produced by a muscle as it changes its length. The top part of the illustration shows the influence of muscle length on the number of active cross-bridges. (Adapted from Winter, 1979.)

velocity effects must also be considered. Movements that cause shortening of muscles are called *concentric,* while those causing lengthening are called *eccentric*. When a muscle shortens, its tension-producing capability decreases because of two factors—a loss in tension due to an inefficient coupling at the cross-bridges as the filaments slide quickly past each other, and the fluid viscosity of the muscle causing viscous friction (Figure 2.26). The latter effect is believed to be a minor reason for the speed of the shortening effect on tension. Eccentric contractions cause somewhat different and much less well-established force–velocity curves. Usually the force increases slightly as the velocity of lengthening increases.

2.3.4 Muscle Fatigue

The different sources of energy of the muscle have been discussed previously. Table 2.1 summarizes these sources and the time of isometric tetanic contraction (maximum isometric contraction) that each energy source can

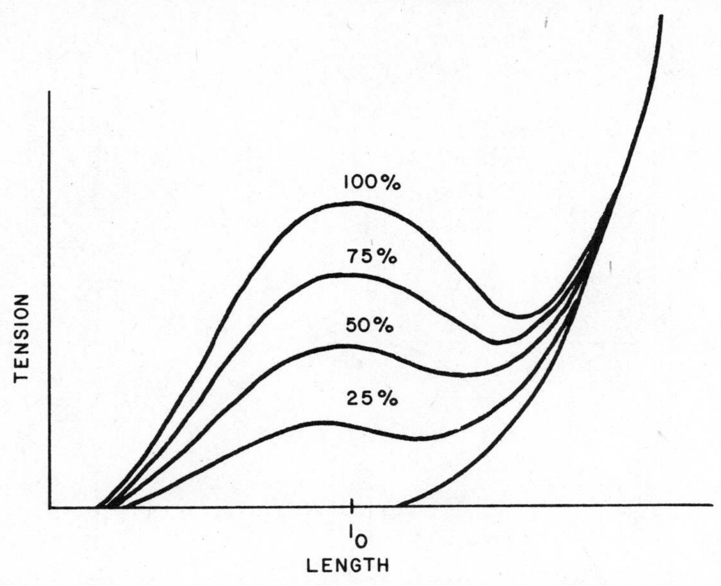

FIGURE 2.25 Tendon tension at various levels of muscle contraction. 0 denotes resting length. (Adapted from Winter, 1979.)

support. Since the anaerobic metabolism produces energy for only a small period of contraction, the oxygen supplied to the muscle becomes a critical factor in allowing continuous intensive muscle contraction. Muscle fatigue is not due only to depletion of energy stores, however (Edwards, 1981), because it has been shown that fatigue at high force levels occurs with less energy expenditure than fatigue at low long-lasting contraction levels. Fatigue is also not due simply to an accumulation of lactate (Karlsson et al, 1975). In general, however, muscle fatigue is influenced greatly both by energy depletion and lactate accumulation.

Endurance time, the time of a muscle to fatigue, is certainly a function of contraction force. Figure 2.27 shows the so-called *Rohmert curve* (von Rohmert, 1960) an important curve in exercise physiology and occupational biomechanics. The fiber-type content of a muscle is important to its fatigue reaction because of the metabolic differences between fibers. The Type I fibers (slow-twitch fibers) are richly surrounded by capillaries and have great potential to store and use oxygen. A high content of these fibers, therefore, will render the muscle more resistant to fatigue. Muscles used to maintain sustained postures are dominated by the Type I fibers for this reason. Type II fibers are common in more dynamic muscle actions requiring short bursts of high-exertion levels. Training influences the fiber composition of a muscle considerably. While there is dispute as to whether one fiber type can actually change over to another, there is clear evidence that one fiber type can be

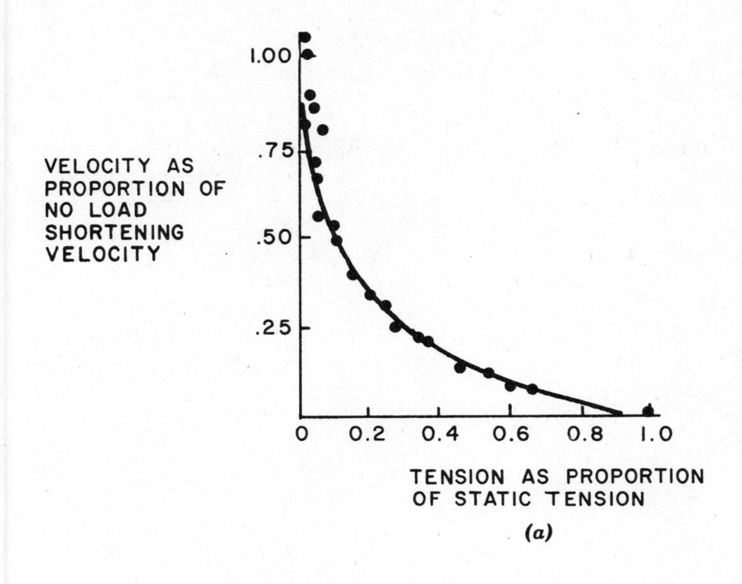

(a)

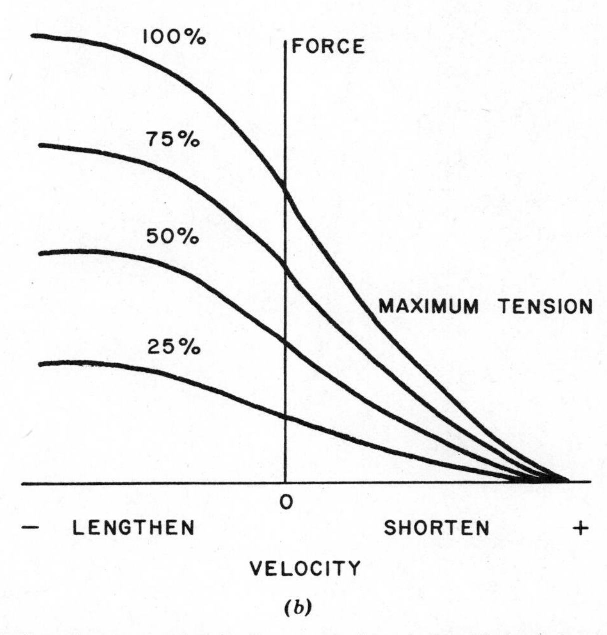

(b)

FIGURE 2.26 (a) Simplified graph representing the force–velocity relationship in skeletal muscle. The maximum velocity of contraction occurs at zero tension, while maximum force of contraction occurs at zero velocity. (Adapted from Best and Taylor, 1973.) (b) Force velocity characteristics of skeletal muscle showing decrease of tension as muscle shortens and increase as it lengthens for different stimulus levels. (Adapted from Winter, 1979.)

TABLE 2.1
Sources of Energy in a Muscle Fiber

Energy Stored as	Energy Producing Process	Approximate Duration of Contraction Energy (by Each Energy Source)
ATP	ATP → ADP	2 sec
Creatine phosphate	CP + ADP → ATP	15 sec
Glycogen	Glycolysis	2 min
Glycogen	Citric acid cycle (oxidative)	50 min or indefinite if circulation adequate at lower levels of contraction

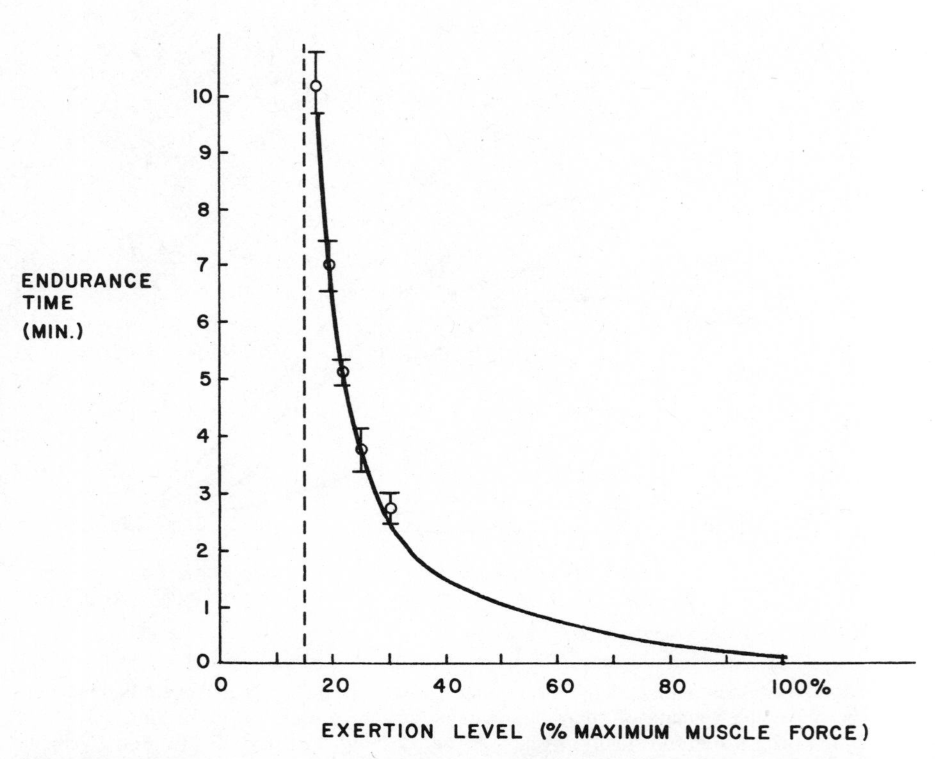

FIGURE 2.27 Static muscle endurance–exertion level relationship with ± 1 SD ranges depicted. (Adapted from Rohmert, 1968.)

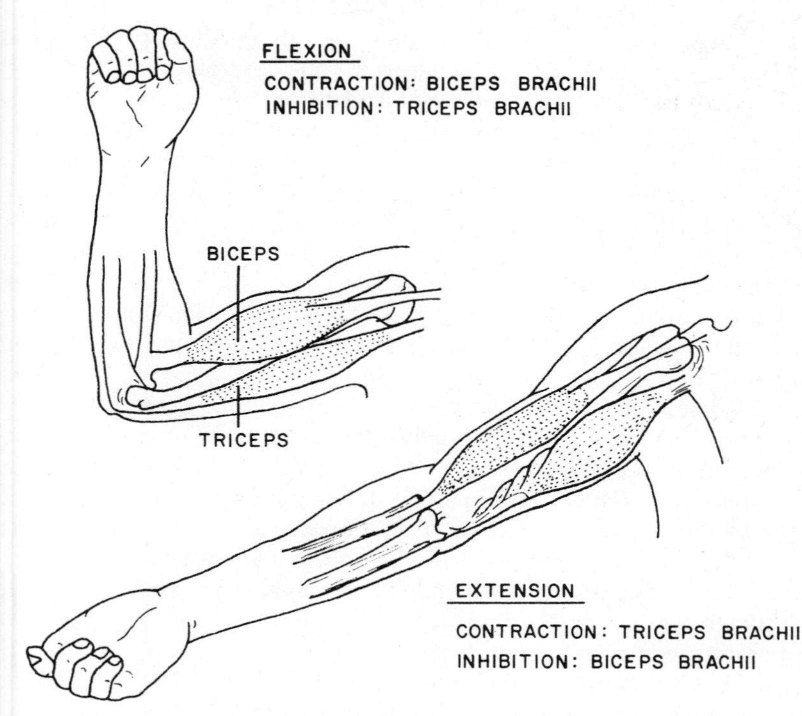

FIGURE 2.28 Diagram of coordinated movement. In reality, other muscles are also involved in these movements.

made to dominate. Thus, selective hypertrophy (growth in thickness) of muscle fibers occurs following strength training; changes in the muscles' ability to store and use oxygen and changes in capillary blood supply to the fibers occur with endurance-type of training. When muscles are not used, on the other hand, muscle atrophy results, first affecting the Type II fibers and causing weakness and loss of muscle volume.

2.3.5 The Action of Muscles

Most muscles are arranged in groups around joints, so that one or several (the *prime mover* or agonist [s]) are responsible for an activity, while others, the *antagonist* (s), have an opposite activity. Coordinated muscle function requires that the antagonist relaxes when the prime mover acts, as illustrated in Figure 2.28. Usually, however, the situation is more complex than that illustrated. In most activities, several muscles act together, as *synergists*; sometimes several joints are involved, so that some muscles control (stabilize) one joint while others move a joint distal or proximal to the stabilized joint. It is important to appreciate that the muscles provide not only movements at a joint, but also prevent the joint from dislocating, providing stability and control. The force generated by a muscle contraction across a joint can be divided into two vectors—one along the moving bone, causing compression at the joint, and another perpendicular to the moving bone, causing that bone to rotate about the other. The two vector components of force have been called *spurt* and *shunt* actions of the muscles. Thus, if a muscle has its anatomical and mechanical axes more parallel to the long axis of the moving bone, it is a shunt muscle at that particular line, while a spurt muscle has its action more aligned to the perpendicular vector. The implication of this is that the spurt muscles are those first called upon to move a joint, while the shunt muscles maintain stability and protect the joint against dislocation.

Muscle and tendon have sensory nerve fibers. Some afferent fibers are not only pain sensors but have other properties that are unique to muscle and tendon. These fibers are called *muscle spindles* or *stretch receptors* and *Golgi tendon organs*. The muscle spindles detect muscle length and the speed of change of muscle length; the Golgi tendon organs sense tension in the tendon. The two are different in their threshold levels to strain in that the muscle spindles are very sensitive, while the tendon organs are less so. When a muscle is stretched above threshold, a reflex is excited from the spindles that lie parallel to the fibers, causing the muscle to contract—the so-called *stretch reflex*. Impulses from the tendon organs, on the other hand, inhibit muscle contraction. When a load is applied to a muscle, it responds by a reflex contraction, which is maintained until the load is removed. This contraction increases as the applied load increases, or actually as the strain on the muscle increases. This is the system used to control posture. The muscles are strained by the force of gravity and respond by appropriate contraction "subconsciously" to maintain equilibrium. The stretch-reflex contraction together with the elastic properties of the muscle gives the muscle a resistance to passive movement, called *muscle tone*. A sudden stretch applied to a muscle—for example, by a reflex hammer hitting a tendon—causes a depolarization of a large number of muscle spindles; a brief strong twitch results, the so-called *tendon jerk*. This reflex is used clinically to determine the integrity of the stretch reflex.

2.4 JOINTS

A joint is a union of two or more bones. Depending on the structure, joints are classified as *synovial joints*, where there is no tissue between the *articular* (joint) *surfaces*; *fibrous joints*, where fibrous tissue bridges the joint; and *cartilaginous joints*, where cartilage bridges the joint. Most joints are synovial. Examples of fibrous joints are the connections between the bones of the skull, while cartilaginous joints exist as primary joints serving the purpose of growth, and secondary joints, such as intervertebral discs.

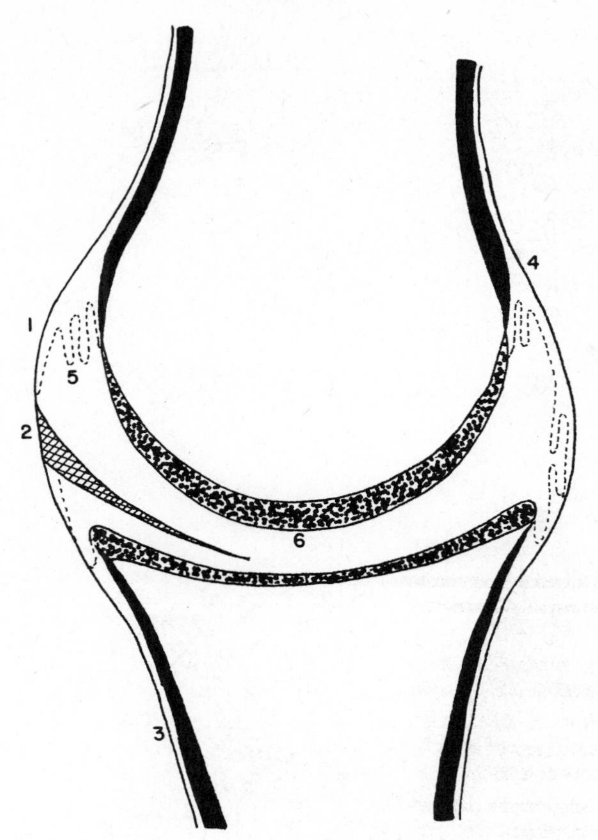

FIGURE 2.29 Synovial joint. 1. Subsynovial bursa, and 2. Intraarticular wedge or meniscus. 3. Periosteum, and 4. Fibrous capsule. 5. Synovial membrane. 6. Articular cartilage.

2.4.1 The Synovial Joint

A typical synovial joint is illustrated in Figure 2.29. It consists of the bone ends, covered by *articular cartilage* (hyaline cartilage), separated by a *joint cavity* (synovial cavity) bounded outward by a *synovial membrane*, and completely surrounded by the *joint capsule*. All structures inside the capsule are considered intra-articular. Many joints have discs of fibrocartilage to protect and support them. Examples are the *menisci* of the knee joints, which serve to distribute load and protect the articular cartilage.

The hyaline cartilage discussed previously is relatively acellular and has a matrix of collagen fibers imbedded in a gel-like substance of proteoglycans and water. The chondrocytes are arranged in layers, as shown in Figure 2.30. This allows the cartilage proper to be divided into three different zones—superficial, middle, and deep—separated from a calcified cartilage

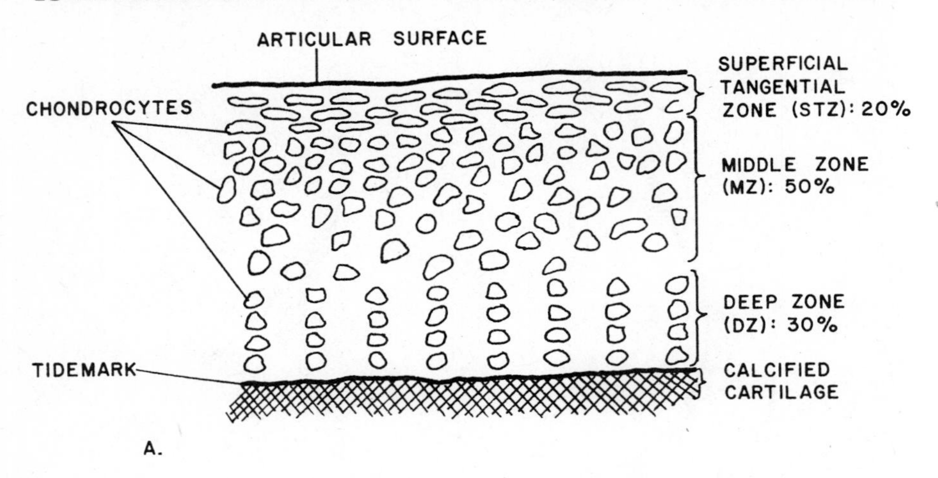

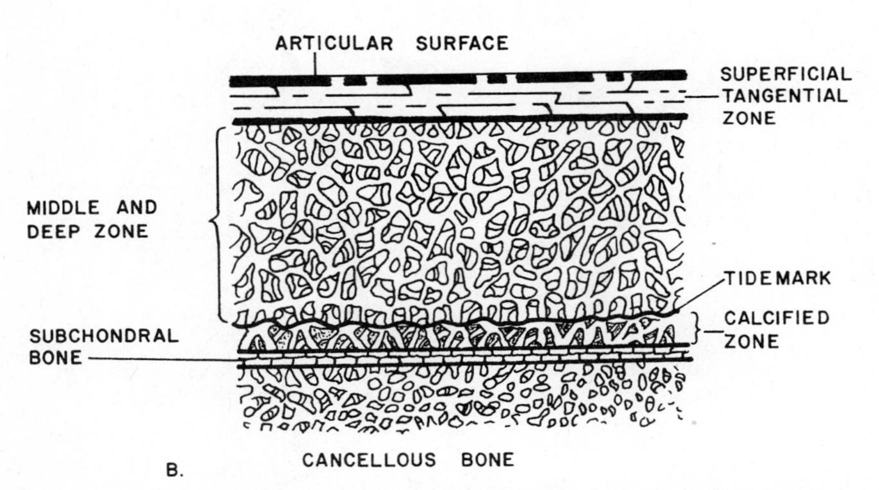

FIGURE 2.30 (A) Arrangement of chondrocytes throughout the dept of noncalcified cartilage. In the STZ, the chondrocytes are oblong with their long axes aligned parallel to the articular surface. In the MZ, the condrocytes are "round" and are randomly distributed throughout. In the DZ, the chondrocytes are virtually isolated from the rest of the physiological system. (B) The ultra-structural arrangement of collagen fibers throughout the depth of articular cartilage. Note the correspondence between the chondrocyte distribution and this collagen fiber architecture. (Adapted from Mow et al., 1974.)

zone by the so-called *tidemark*. The collagen network is also distributed differently in these zones, being more parallel to an articular surface in the superficial zone and more random beneath (Figure 2.30). The proteoglycans in the matrix are strongly hydrophilic. They are trapped within the collagen network and prevented by this network from fully expanding. This causes

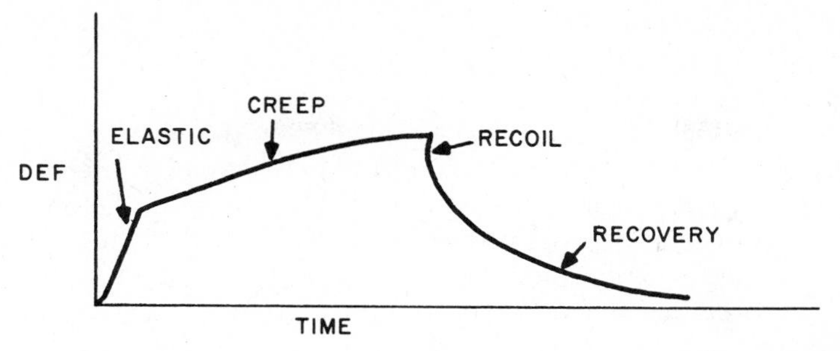

FIGURE 2.31 Articular cartilage, has a combination of elastic and viscoelastic properties. As load is applied, deformation increases with time, first in an elastic fashion, then with a slow creep. With the removal of load, there is an elastic recoil and then a slow recovery to the base line.

an osmotic swelling pressure, so that the collagen fibers are in tension when no stress is placed on the cartilage. When an external load is applied, an internal pressure develops that is greater than the osmotic pressure, causing water to be squeezed out of the cartilage. There is also a change in the collagen fiber pattern from a more criss-cross orientation to a flattened pattern.

Mechanically, joint cartilage can be considered as having both elastic and viscoelastic properties. On compression there is first an elastic deformation, and with time a slow creep. The recovery after compression shows a similar pattern (Figure 2.31). Cartilage being found where it is, its compressive properties have been the most frequently investigated. Cartilage also has tensile strength. Kempson (1973) found these properties to be related to the orientation of the collagen fibers, decreasing from the cartilage surface downward into the tissue. Cartilage provides the joints with sliding surfaces of very low frictional resistance. Cartilage on cartilage lubricated by synovial fluid has a coefficient of friction of 0.002, compared with 0.03 for ice on ice lubricated by water and 0.2 for steel on steel lubricated by oil.

2.4.2 Joint Lubrication

The articular surfaces are highly lubricated. Several theories have been proposed to describe the type of lubrication occurring. It is clear that it is complex, and present theory is that it includes both *boundary* lubrication and *fluid film* (or squeeze-film) lubrication. Boundary lubrication implies that a lubricant reacts physically or chemically with the articular cartilage so that a layer is formed, permitting the surfaces to move without adhesion and abrasion. There is evidence that synovial fluid acts as a lubricant in the joint, but boundary lubrication is probably important in situations where the joint is heavily loaded. Under less severe loading conditions, fluid-film or squeeze-film lubrication is more probable. In the latter situation, the bearing surfaces are constantly separated by a fluid film. The fluid film is generated by an external pressure, as in hydrostatic lubrication, where fluid is thought to be physically pumped into the gap between articulating surfaces.

In the case of cartilage, it is the cartilage itself that is both the pump and the reservoir. Loading the joint pressurizes the fluid in the pores of the cartilage and pushes it out at high pressure to soak the surfaces. Unloading the joints lets the cartilage expand again, resorbing the fluid the way a sponge surface resoaks. During joint motion of a tangential type (translation), a wedge is created, causing a lifting pressure (*hydrodynamic lubrication*) that further distributes the synovial fluid.

In compressive movement of a joint, on the other hand, fluid must be squeezed out between the surfaces (Figure 2.32). As the joint rotates, fluid will be squeezed out from the articular cartilage at the area of contact and in front of it, and as soon as peak stress has passed, the fluid starts to resorb (Figure 2.33). In normal walking, for example, a fluid film will develop in a hip joint during the swing phase and be squeezed out during the stance phase. During prolonged standing, on the other hand, the fluid film will generally disappear, leaving only the boundary lubricant to protect the surfaces.

2.4.3 Osteoarthritis

Cartilage has a poor ability to repair and regenerate because it is void of capillaries. As age progresses, a gradual degeneration of all joints occurs, called *osteoarthritis* (OA). Primary OA is a condition in which there is no obvious previous abnormality in the joint, while secondary OA has some previous detectable cause. This detectable cause is usually mechanical in nature. In other words, secondary osteoarthritis occurs when joints are deformed by disease or injury. Mechanical factors appear to be important in joint osteoarthritis not only in the secondary situation, however, but also in primary osteoarthritis. Recurrent trauma, joint instability, stress magnitudes, and their geometric concentration are important, but the precise relationships and mechanisms remain unclear.

2.4.4 Intervertebral Discs

The intervertebral disc is formed by two histologically chemically and mechanically distinct structures: the *nucleus*, an incompressible watery gel contained within an elastic wall, and the *annulus*, in which fibrous lamellae are arranged differently from layer to layer (Figure 2.33). The different properties of the two parts of the disc produce its load-bearing characteristics. When the disc is loaded, the nucleus deforms and transfers the force in all directions away from the nuclear center. The fiber arrangement of the annulus is particularly well adapted to accept the axial tensile stresses so produced, which decreases the risk of structural failure.

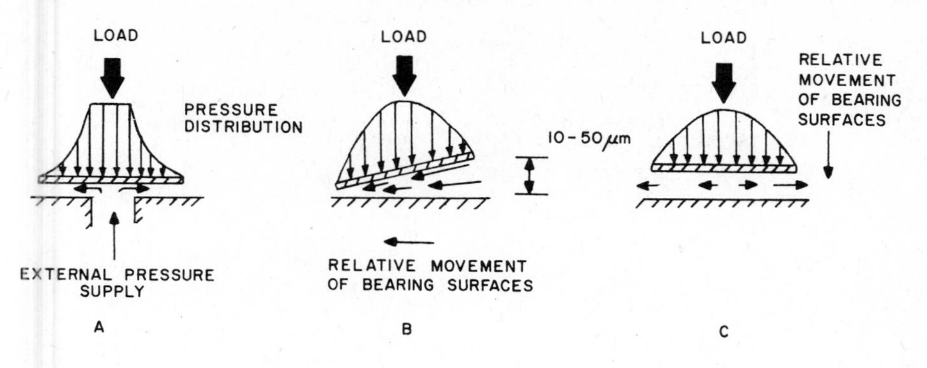

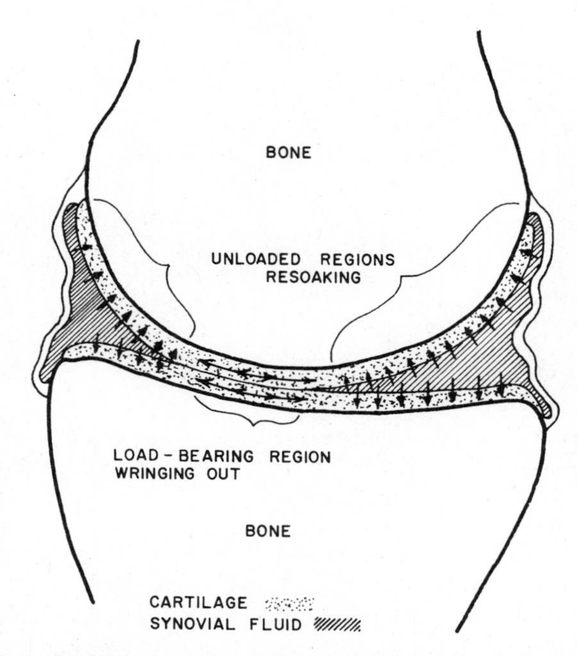

FIGURE 2.32 Diagrams illustrate the load-carrying capacity of a fluid via (A) hydrostatic lubrication mechanism, (B) hydrodynamic lubrication mechanism, and (C) squeeze film lubrication mechanism. (Adapted from Frankel and Nordin, 1980.) While loaded cartilage is being wrung out, cartilage elsewhere in the joint resoaks. (Adapted from McCutchen, in Simon, 1978.)

The disc is separated on both sides from the vertebral bodies by hyalin *cartilage endplates*. These influence the nutrition of the disc, which, because of its lack of vascular tissue, is achieved by diffusion only. The diffusion is also influenced by mechanical factors. When the load on the disc increases, there is an outflow of fluid; when it decreases, there is an influx. In addition to the nutritional aspects of this load-related fluid exchange, it has also been

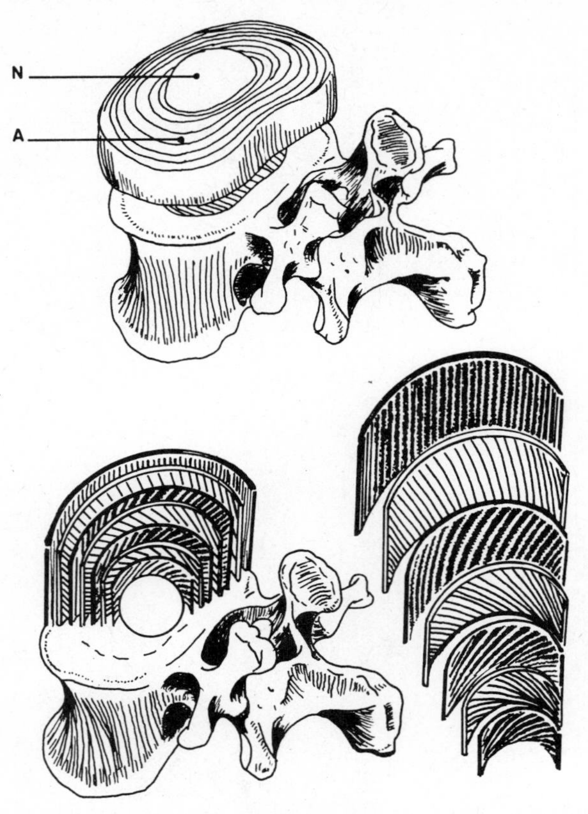

FIGURE 2.33 The disc consists of a central watery gel—the nucleus pulposus (N) surrounded by layers of hyaline fibers—the annulus fibrosus (A). (Adapted from Kapandji, 1974).

shown that the fluid content of the disc influences its mechanical properties. An increase in fluid content increases the stiffness of the disc.

The response of the disc to compression has been studied extensively. These studies show that the disc is quite flexible at low loads, but provides increasing resistance to load at high loads.

Extrusion of the disc nucleus through the annulus has not been produced experimentally, although after fracture of the endplate, disc material has been found to spread into the vertebral bodies. Some studies indicate that the disc is particularly at risk when subjected to bending and rotation, though the limits are not well established.

When a compressive load is applied to a *lumbar motion segment*—the name given to two adjacent vertebra and their intervening disc—failure occurs first in the endplate then in the vertebral bodies, and only then in the disc proper. The moment rotation and force deformation properties of the motion segments vary greatly between individuals and between age groups of individuals. They are also effected by degeneration (aging). Because of the importance of the disc as a biomechanical entity, it will be discussed further in different contexts in later chapters.

REFERENCES

Abrahams, M., "Mechanical Behavior of Tendon *In Vitro*. A Preliminary Report." *Med. Biol. Engng.*, **5,** 433 (1967).

Albright, J. A. and R. A. Brand, *The Scientific Basis of Orthopaedics,* Appleton-Century-Crofts, New York, 1979.

Basmajian, J. V., *Muscles Alive,* Williams and Wilkins, Baltimore, 1978.

Best C. H. and N. B. Taylor, *Physiological Basis of Medical Practice,* Williams and Wilkins, Baltimore, 1973.

Carter, D. R. and W. C. Hayes, "The Compressive Behavior of Bone as a Two-Phase Porous Structure," *J. Bone Joint Surgery,* **59A,** 954–962 (1977).

Carton, R. V., J. Dainauskas, and J. W. Clark, "Elastic Properties of Single Elastic Fibres." *J. Appl. Physiol.,* **17,** 547 (1962).

Carlson, F. D. and D. E. Wilkie, *Muscle Physiology,* Prentice-Hall, Englewood Cliffs, 1974, pp. 7–86.

Edwards, R. H. T., "Human Muscle Function and Fatigue." In *Human Muscle Fatigue: Physiological Mechanisms*. Pitman Medical, London, 1981, pp. 1–18.

Frankel, V. H. and M. Nordin, *Basic Biomechanics of the Skeletal System,* Lea and Febiger, Philadelphia, 1980.

Hayes, W. C. and D. R. Carter, "Biomechanics of Bone," in D. J. Simmons and A. S. Kunin, Eds., *Skeletal Research,* Academic Press, New York, 1979, pp. 263–300.

Huxley, A. F., "Muscular Contraction." *J. Physiology,* 243, 1 (1974).

Kapandji, I. A., *The Physiology of Joints,* Vol. 3, Churchill, Livingstone, Edinburgh, 1974.

Karlsson, J., B. Funderburk, B. Essen and A. R. Lind, "Constituents of Human Muscle in Isometric Fatigue." *J. Appl. Physiol.,* **38,** 208–211 (1975).

Kazarian, L. E. and H. E. von Gierke, "Bone Loss as a Result of Immobilization and Chelation. Preliminary Results in Macaca Mulatta," *Clin. Orthop.,* **65,** 67–75 (1969).

Kempson, G. E., "Mechanical Properties of Articular Cartilage," M. A. R. Freeman, Ed., in *Adult Articular Cartilage,* Pitman Medical London, 1973.

Mow, V. C., W. M. Lai, and I. Rodler, "Some Surface Characteristics of Articular Cartilage. 1. A Scanning Electron Microscopy Study and a Theoretical Model for the Dynamic Interaction of Synovial Fluid and Articular Cartilage, *J. Biomed.,* **7,** 449 (1974).

Noyes, F. R., "Functional Properties of Knee Ligaments and Alterations Induced by Immobilization," *Clin. Orthrop.,* **123,** 210–242 (1977).

Noyes, F. R. and E. S. Grood, "The Strength of the Anterior Cruciate Ligament in Humans and Rhesus Monkeys. Age-Related and Species-Related Changes," *J. Bone Joint Surgery,* **58A,** 1074 (1976).

von Rohmert, W., Ermittlung von Erholungspausen fur Statische Arbeit des Menschen, *Int. Z. Angew. Physiol.,* **18,** 123–124 (1960).

von Rohmert, W., *Arbeitsmedizin, Sozial medizin, Arbeitshygiene,* **22,** 118 (1968).

Rosse, C. and D. K. Clawson, *The Musculoskeletal System in Health and Disease*. Harper and Row, Hagerstown, 1980, pp. 3–88.

Rush, T. C. and H. D. Patton, *Physiology and Biophysics,* 9th ed., W. B. Saunders, Philadelphia, 1965.

Sammarco, G. J., A. H. Burstein, W. L. Davis, and V. H. Frankel, "The Biomechanics of Torsional Fractures: The Effect of Loading on Ultimate Properties." *J. Biomech.*, **4,** 113–117 (1971).

Simon, W. D. (ed.), *The Human Joint in Health and Disease,* Univ. Penn. Press, 1978, pp. 3–8. Anatomy and Development of Joints (M. Harty), 71–80, Mechanical Properties and Wear of Articular Cartilage (W. H. Simon), pp. 81–87, Cartilage Lubrication (C. W. McCutchen).

Vander, A. J., J. H. Sherman, and D. S. Luciano, *Human Physiology,* 2nd ed., McGraw-Hill, New York, 1970.

Winter, D. A., Biomechanics of Human Movement, Wiley, New York, 1979, pp. 1–202.

Wolff, J., *Das Gesetz der Transformation der Knochen,* Hirschwald, Berlin,1892.

3

ANTHROPOMETRY IN OCCUPATIONAL BIOMECHANICS

Anthropometry is the science that deals with the measure of size, weight, and proportions of the human body. It is empirical in nature and has developed quantitative methods to measure various physical dimensions and other properties of subject populations. The results are statistical data describing human size and form. Anthropometric data are fundamental to occupational biomechanics. Without it, biomechanical models to predict human reach and space requirements cannot be developed.

What follows is a brief description of some classical measurement methods used to acquire anthropometric data and a summary of resulting data relevant to occupational biomechanics.

3.1 MEASUREMENT OF BODY-SEGMENT PHYSICAL PROPERTIES

In performing a biomechanical analysis of a person performing a task, the human body is considered to be a system of mechanical links, with each link of known physical size and form. Anthropometry defines these size and form properties as presented in the following subsections. For additional details on the measurement methods, refer to *Engineering Anthropometry Methods* by Roebuck, Kroemer, and Thomson, 1975.

3.1.1 Body-Segment Link Length Measurement Methods

The measurement of the length of various body segments in a linkage system assumes that the segments are connected at easily identifiable joints. Clearly

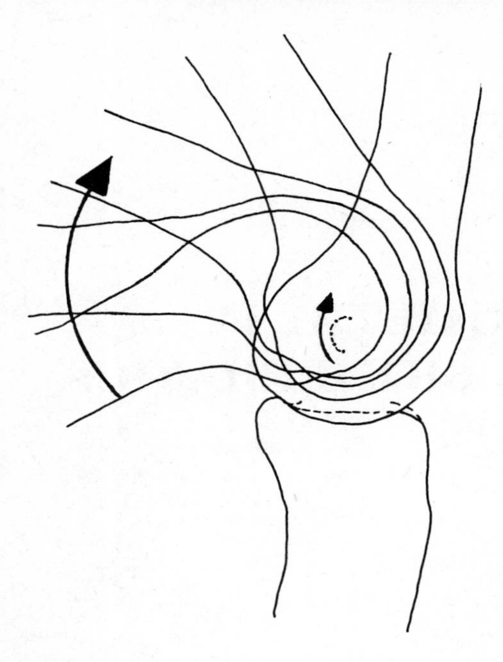

FIGURE 3.1 Change in the instant center location of normal tibiofemoral knee joint moving from flexion to extension (adapted from Frankel and Nordin, 1980, p. 121).

this assumption is better for the limbs than for the torso, neck, and head segments. Even so, the identification of segment joints for the limbs can be difficult, because the bony landmarks are often covered by muscle and fat tissue, especially at the shoulder and hip joints.

Anthropometrists have carefully dissected cadavers and estimated the location of joint centers-of-rotation over the last 100 years (Braune and Fischer, 1889; Dempster, 1955; and Snyder, Chaffin, and Schultz, 1972). For more simple hinge-type joints (e.g., distal finger, elbow, and knee joints) it is possible on living subjects to project approximate centers-of-rotation by moving the adjacent body segments through the range of motion of the joint. The intersection of the long axis center-line of the segments during such movements defines the joint's approximate center-of-rotation. This is most effectively accomplished by use of serial radiographs of the joint during a motion. Tracings from such radiographs are shown in Figure 3.1 for the knee. None of the joints are in fact simple single-axis hinge joints, but can be considered as such for most kinematic analyses. The resulting error in the link length estimate is less than $\pm 5\%$ from this approximation.

Once the joint center-of-rotations are known, link lengths can be defined as the distance between the projected centers. These link lengths have been correlated with distances measured between palpable bony landmarks located near the joint centers-of-rotation of interest. These landmarks have been carefully defined in the literature (Webb Associates, 1978). Major landmarks are presented in Figure 3.2, with additional reference points presented

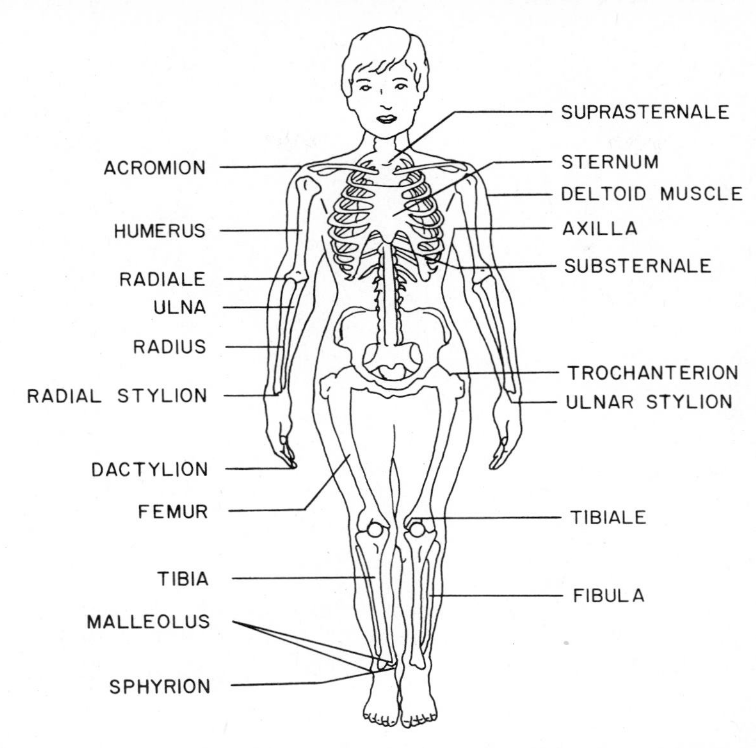

FIGURE 3.2 Common anatomical and anthropometric landmarks (Webb Associates, 1978).

in Appendix A. Thus, if one is attempting to define the length of a person's forearm link, using palpable bony landmarks at either end of the ulna or radius, an estimate acceptable for most occupational biomechanical studies can be obtained by measuring between these reference points and multiplying by a given proportionality factor. Table 3.1 provides one set of proportionality values that can be applied to the estimates of upper and lower extremity links (Dempster, 1955). In these data, the foot- and hand-link termination points are defined as the center-of-mass locations for each. For the hand link this represents a biomechanical end-point in the linkage system. When grasping an object to exert a strong force, a power-grip hand posture is most often used, and in so doing the hand is coupled to the object at a point that approximates the location of the hand center-of-mass. The foot link is considered to be a triangle, with its apex at the ankle and its base bounded by the lower posterior border of the calcaneum (base of the heel) and the lower distal aspect of the first metatarsal bone (ball of the foot). This triangular foot segment thus provides support along its base when in contact with the floor.

The linkage centers-of-rotation for the torso have also been determined on both cadavers and living subjects, using x-ray and photographic methods

TABLE 3.1
Extremity Link Length to Bone Length Ratios (Dempster, 1955)

Segment	Link-to-Length Ratio (%)
Humerus link Humerus length	89.0
Radius link Radius length	107.0
Hand link (wrist center to center of mass) Hand length	20.6
Femur link Femur length	91.4
Tibial link Tibial length	110.0
Foot link (talus center point to center of mass) Foot length	30.6

(Dempster, 1955; and Snyder, Chaffin, and Schutz, 1972). Once again, palpable bony reference points have been related to the two major centers-of-rotation while the person is in a specific posture. Table 3.2 presents some of these relationships.

A list of the major joint centers-of-rotation is given in Table 3.3, along with the definition of a linkage system defined for clothing and human operator reach and space requirements in the United States Air Force (Webb, 1978). Figure 3.3 illustrates this linkage system. It should be noted that in existing biomechanical models used to study industrial tasks, the torso is

TABLE 3.2
Select Torso Joint to Surface Point Relationships (from Snyder, Chaffin, and Schutz, 1978)

Joint to Body Surface Point	Distance Mean (cm)	Vector Mean Angle from 12:00 (Vertical) with + Clockwise[b]
L5/S1 disc to L5 spine surface point[a]	9.04	−84°
L2/L3 disc to L2 spine surface point	8.79	−92°
T12/L1 disc to T12 spine surface point	7.77	−102°
T8/T9 disc to T8 spine surface point	8.03	−104°
C7/T1 disc to C7 spine surface point	7.47	−65°
C5/C6 disc to C5 spine surface point	7.59	−77°
C2/C3 disc to C2 spine surface point	8.03	−90°

[a] Spinal surface points are the most superior dorsal palpable points on the spinous process.
[b] Angles were measured with subjects in erect seated postures.

TABLE 3.3
Joint Center Locations and Link Definitions (Webb Associates, 1978)

Joint Centers-of-Rotation	
Head/Neck	Midpoint of the interspace between the occipital condyles and the first cervical vertebra.
Neck/Thorax	Midpoint of the interspace between the 7th cervical and 1st thoracic vertebral bodies.
Thorax/Lumbar	Midpoint of the interspace between the 12th thoracic and 1st lumbar vertebral bodies.
Lumbar/Sacral	Midpoint of the interspace between the 5th lumbar and 1st sacral vertebral bodies.
Sternoclavicular	Midpoint position of the palpable junction between the proximal end of the clavicle and the sternum at the upper border (jugular notch) of the sternum.
Claviscapular	Midpoint of a line between the coracoid tuberosity of the clavicle (at the posterior border of the bone) and the acromioclavicular articulation (or the tubercle at the lateral end of the clavicle); the point, however, should be visualized as on the underside of the clavicle.
Glenohumeral	Midregion of the palpable bony mass of the head and tuberosities of the humerus; with the arm abducted about 45° relative to the vertebral margin of the scapula, a line dropped perpendicular to the long axis of the arm from the outermost margin of the acromion will approximately bisect the joint.
Elbow	Midpoint on a line between (1) the lowest palpable point on the medial epicondyle of the humerus, and (2) a point 8 mm above the radiale (radiohumeral junction).
Wrist	On the palmar side of the hand, the distal wrist crease at the palmaris longus tendon, or the midpoint of a line between the radial styloid and the center of the pisiform bone; on the dorsal side of the hand, the palpable groove between the lunate and capitate bones, on a line with metacarpal bone III.
Hip	(Lateral aspect of the hip). A point at the tip of the femoral trochanter 0.4 in. anterior to the most laterally projecting part of the femoral trochanter.
Knee	Midpoint of a line between the centers of the posterior convexities of the femoral condyles.
Ankle	Level of a line between the tip of the lateral malleolus of the fibula and a point 5 mm distal to the tibial malleolus.
Link System Definition	
Head Link	The straight line between the occipital condyle/C1 interspace center and the center of mass of the head.
Neck Link	The straight line between the occipital condyle/C1 and C7/T1 vertebral interspace joint centers.

TABLE 3.3 (*Continued*)

	Joint Centers-of-Rotation
Torso Link	The straight line distance from the occipital condyle/C1 interspace joint center to the midpoint of a line passing through the right and left hip joint center.
Thorax Links	*Thoraco-sternum*—A closed linkage system composed of three links. The right and left transthorax are straight line distances from the C7/T1 interspace to the right and left sternoclavicular joint centers of rotation. The transsternum link is a straight line distance between the right and left sternoclavicular joint centers of rotation. *Clavicular*—The straight line between the claviscapular and glenohumeral joint centers. *Thoracic*—The straight line between C7/T1 and T12/L1 vertebral body interspace joint centers.
Lumbar Link	The straight line between the T12/L1 and L5/S1 vertebrae interspace joint centers.
Pelvis Link	The pelvis is treated as a triangular-shaped linkage system composed of three links. The right and left iliopelvic links are straight lines between the L5/S1 interspace joint center and a hip joint center. The transpelvic link is a straight line between the right and left hip joint centers.
Upper Arm Link	The straight line between the glenohumeral and elbow joint centers of rotation.
Forearm Link	The straight line between the elbow and wrist joint centers of rotation.
Hand Link	The straight line between the wrist joint center of rotation and the center of mass of the hand.
Thigh Link	The straight line between the hip and knee joint center of rotation.
Shank Link	The straight line between the knee and ankle joint centers of rotation.
Foot Link	The straight line between the ankle joint center of rotation and the center of mass of the foot.

often described as a simple two-link system composed of the triangular pelvic link (connecting the two hip centers with the lumbosacral center at the L5/S1 disc) and a triangular-shaped lumbar-thoracic link (connecting the lumbosacral center-of-rotation directly with the two glenohumeral shoulder joints). Once again, such simplifications are justified on the basis that inclusion of additional torso links in the analysis would not aid in understanding the mechanical trauma resulting from the large variety of postural and force requirements imposed on workers of different anthropometry in

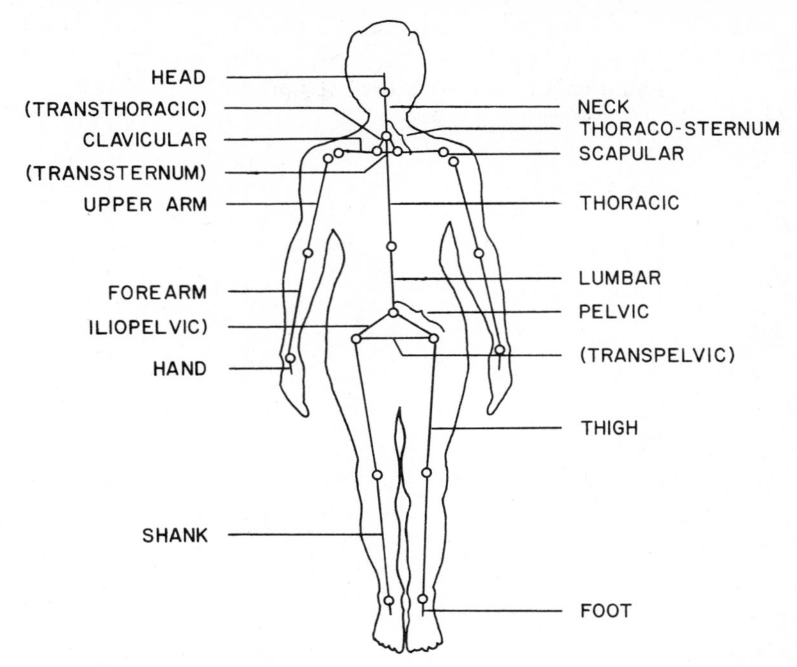

FIGURE 3.3 Common linkage system (Webb Associates, 1978).

industry. Clearly, as the means become available to better define job postural and force requirements in industry, and as clinical symptoms and signs of mechanical trauma are better classified, more elegant biomechanical models will be utilized to better understand the underlying cause of these musculoskeletal problems.

Fortunately, anthropometric studies have produced the means to estimate link lengths for various biomechanical representations. A summary of some data resulting from different population studies is given later in this chapter.

3.1.2 Body-Segment Volume and Weight

Not only are body tissues mechanically stressed by external loads carried, pushed or pulled on the job, but the effect of gravity on the mass of the body segments creates body-weight-related stresses. These weight-related stresses can be of considerable magnitude in certain postures. A simple example is holding an arm straight out. Quite rapidly the shoulder muscles become fatigued and the posture cannot be maintained. Knowledge of the distribution of body mass throughout the body, therefore, is of importance in determining the effect of gravity on various musculoskeletal areas.

Body-segment mass and volume are related by the density of the segment. This has been determined by immersing cadaver body segments (of known

TABLE 3.4
Body-Segment Densities from Cadaver Studies in g/cm³ (from Miller and Nelson, 1976)

Segment	Harless[a] (1860)	Dempster[b] (1955)
Head and Neck	1.11	1.11
Trunk	—	1.03
Upper Arm	1.08	1.07
Forearm	1.10	1.13
Hand	1.11	1.16
Thigh	1.07	1.05
Lower Leg	1.10	1.09
Foot	1.09	1.10

[a] Based on the dissection of five cadavers.
[b] Based on the dissection of eight cadavers.

weight) in water and measuring the volume of water displaced. The resulting equation is

$$D = \frac{M}{V} = \frac{W/g}{V} \tag{3.1}$$

where

D is the average density, mass per unit volume (g/cm^3)
M is the mass of the body segment (g or kg)
V is the volume of water displaced (cm^3)
W is the weight of the body segment (N)
g is gravitational acceleration constant (m/sec^2)

Studies by Harless and Dempster (as summarized by Miller and Nelson, 1976) for five and eight cadavers indicate that average body-segment densities vary as shown in Table 3.4.

Body-segment volumes can also be measured on living subjects by immersion of the segment with reference to anthropometrically defined palpable reference marks. A double-tank system for such measurement was developed by Drillis and Contini (1966) to provide a means of controlling and measuring the water volume (Figure 3.4, from Miller and Nelson, 1976). Thus, by measurement of body segment volumes, and by substitution of the

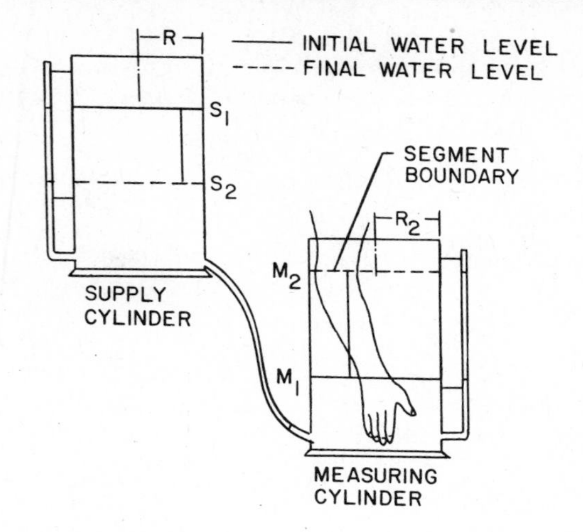

SEGMENT VOLUME = Δ MEASURING CYLINDER VOLUME - Δ SUPPLY CYLINDER VOLUME
SEGMENT VOLUME (CC) = $(M_2 - M_1)^* \pi R_2^2 - (S_2 - S_1)^* \pi R_1^2$
SEGMENT MASS (GM) = SEGMENT VOLUME (CC)* SEGMENT DENSITY (GM/CC)

FIGURE 3.4 Segment volume measuring system (Miller and Nelson, 1976).

density values in Table 3.4, the mass of body segments can be determined by

$$\text{Segment Mass (g)} = \text{Segment Volume (cm}^3\text{)} \times \text{Segment Density (g/cm}^3\text{)} \quad (3.2)$$

These techniques or similar ones have been used by investigators since the mid 1800's to measure body density or mass (Roebuck, Kroemer, and Thomson, 1975). The results indicate that body-segment weights can be expressed as a proportion of whole-body weight without great loss of accuracy. The resulting proportionality values, averaged from several studies by Webb Associates (1978), are shown in Table 3.5. Body-segment weight values will be presented for different populations, as derived from the proportionality values, later in this chapter (3.2.2).

3.1.3 Body-Segment Mass Center Locations

It is not sufficient to know only the weight of a body segment to perform a biomechanical analysis. One must also be able to locate within a segment where the gravitational effect of the distributed segment mass can be considered to act. In other words, if a body segment is suspended from only one attachment, where would that attachment point be located so that the gravitational effect is equal (or balanced) on either side of the attachment,

TABLE 3.5
Percentage Distribution of Total Body Weight According to Different Segmentation Plans (from Webb Associates, 1978)

Grouped Segments, % of Total Body Weight	Individual Segments, % of Grouped Segments Weight
Head and neck = 8.4%	Head = 73.8%
	Neck = 26.2%
Torso = 50.0%	Thorax = 43.8%
	Lumbar = 29.4%
	Pelvis = 26.8%
Total arm = 5.1%	Upper arm = 54.9%
	Forearm = 33.3%
	Hand = 11.8%
Total leg = 15.7%	Thigh = 63.7%
	Shank = 27.4%
	Foot = 8.9%

regardless of the orientation of the segment in space? The resulting point on the segment is known as its *mass center* (referred to in some earlier books as the *mass center-of-gravity*).

Body-segment mass center locations have been measured on cadavers by a number of investigators using a suspension technique (Roebuck, Kroemer, and Thomson, 1975). The frozen section was suspended from a pin that was systematically reinserted until a point of balance was determined. In living subjects, two methods can be used. The more traditional method requires the subject to assume various positions while supported on a force platform (see Chapter 5 for operating details of a force platform). The force platform provides a means to determine the location of the segment mass centers by applying the static equilibrium principle that requires the sum of moments around any point in a force system, while in equilibrium, to be zero. By having a person assume two different positions on the force platform and by knowing the segment weights, one can solve for the segment center of mass using a procedure described in detail by Williams and Lissner (LeVeau, 1977). Figure 3.5 presents the basic procedure for determining the mass center location for the combined shank and foot segments.

A second method for determining mass center locations is more involved. It is a modification of the immersion method of determining the volume of a segment. In this technique, described in detail by Miller and Nelson (1976), the segment is immersed in known discrete intervals r_i, while the volume V_i is measured. Figure 3.6 describes the procedure and calculations.

As Miller and Nelson note, a simple method of determining the approximate center of mass on the long axis of a segment is to immerse the limb to its proximal end using a second catch tank to measure the volume of water

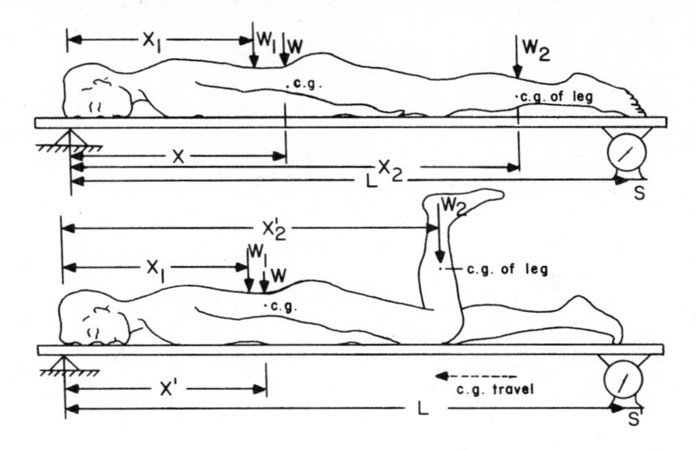

IF MASS CENTER LOCATIONS ARE KNOWN, THEN SEGMENT WEIGHT PREDICTED FOR FOOT AND SHANK BY:

$$W_2 = \frac{L(S - S')}{(X_2 - X_2')}$$

IF SEGMENT WEIGHT IS KNOWN, THEN MASS CENTER (C.G. LOCATION) CAN BE PREDICTED BY BALANCING SEGMENT OVER JOINTS, WHICH FOR FOOT AND SHANK YIELD:

$$X_2 = \left[\frac{L(S - S')}{W_2} + X_2' \right]$$

FIGURE 3.5 Estimation of body segment weights and mass center locations using moment subtraction method described by Williams and Lissner (LeVeau, 1977, p. 214).

displaced. Withdraw the limb slowly until half the volume first displaced is returned. The water level at this point on the limb, assuming uniform segment density, bisects the mass center.

Segment mass-center locations have been obtained for several different populations using these methods. In general, because the shape of segments are not greatly altered from one person to another, though size varies greatly, the current practice is to present the data as a proportion of the segment link length. Table 3.6 does this for three different studies, as compiled by LeVeau (1977). The differences can be attributed to different measurement techniques.

3.1.4 Body-Segment Inertial Property Measurement Methods

Knowledge of the mass center location in a body segment, along with its weight and link length, is sufficient to perform a static analysis of the forces

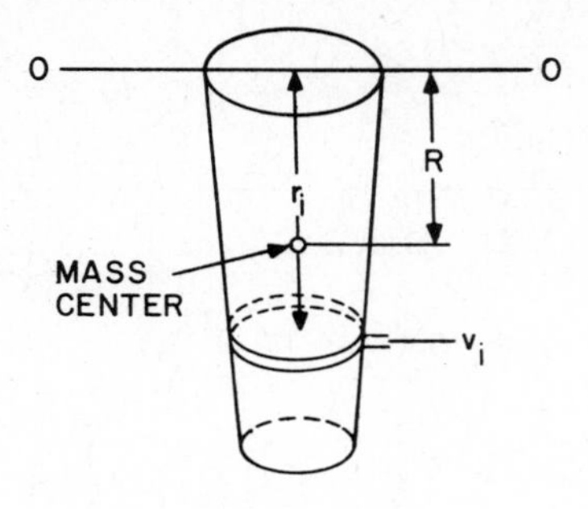

MASS = DENSITY * VOLUME

$\sum m_i = \rho v_i$

LOCATION OF THE MASS CENTER WITH RESPECT TO THE OO AXIS:

$$RM = \sum_{i=1}^{N} m_i r_i$$

$$R = \left[\frac{\sum_{i=1}^{N} m_i r_i}{M} \right]$$

IN WHICH $M = \sum_{i=1}^{N} m_i$ = TOTAL MASS OF THE SEGMENT.

MOMENT OF INERTIA WITH RESPECT TO THE OO AXIS:

$$I_{oo} = \sum_{i=1}^{N} m_i r_i^2$$

FIGURE 3.6 Segmental zone approach to the immersion technique (Miller and Nelson, 1976).

and moments at each joint for a given body posture. When a person moves a body segment, the inertial property of the segment must also be considered in the analysis. This property is referred to as the *moment-of-inertia*, and is defined as

$$I_0 = \sum_{i=1}^{N} m_i r_i^2 \quad \text{or} \quad \int r^2 \, dm \tag{3.3}$$

where

m_i are discrete quantities of mass

r_i is the perpendicular distance each quantity of mass is located from a given axis of rotation of the segment.

TABLE 3.6
Center of Mass/Segment Length Ratios to Proximal End from Several Cadaver Studies[a] (from LeVeau, 1977).

Source	Braune and Fischer (1889)	Dempster (1955)	Clauser et al (1969)
Total body	—%	—%	41/2%
Head	—	43.3	46.6
Trunk	—	—	38.0[b]
Arm	47.0	43.6	51.3
Forearm	42.1	43.0	39.0
Hand	—	49.4	48.0[b]
Total arm	—	—	41.3
Forearm and hand	47.2	67.7[b]	62.6[b]
Thigh	44.0	43.3	37.2[b]
Calf	42.0	43.3	37.1
Foot	44.4	42.9	44.9
Total leg	—	43.3	38.2[b]
Calf and foot	52.4	43.7	47.5

[a] Adapted from Clauser et al., 1969.
[b] These values are not directly comparable owing to variations in the definition of segment length used by the different investigators.

Figure 3.7 from Miller and Nelson (1976) illustrates the concept with three different axes of rotation assumed. In biomechanical studies the moments-of-inertia are considered to act around either the joint or the mass centers. The moment-of-inertia value, when multiplied by the angular acceleration of the segment, provides an estimate of the instantaneous moments and forces acting at a joint due to the acceleration or deceleration of the segment during the motion. These are then added to the moments and forces obtained by a static analysis of the work postures to produce a complete kinetic analysis of a task. Chapter 6 describes these biomechanical modeling procedures in detail.

Several methods have been used to determine body-segment moments-of-inertia. In cadaver studies, the frozen body segments were swung as a pendulum around anatomically identified axes. The resulting period of the oscillation was used to estimate the moment-of-inertia. The calculation used in such a procedure is

$$I_0 = \frac{WL}{4\pi^2 f^2} \tag{3.4}$$

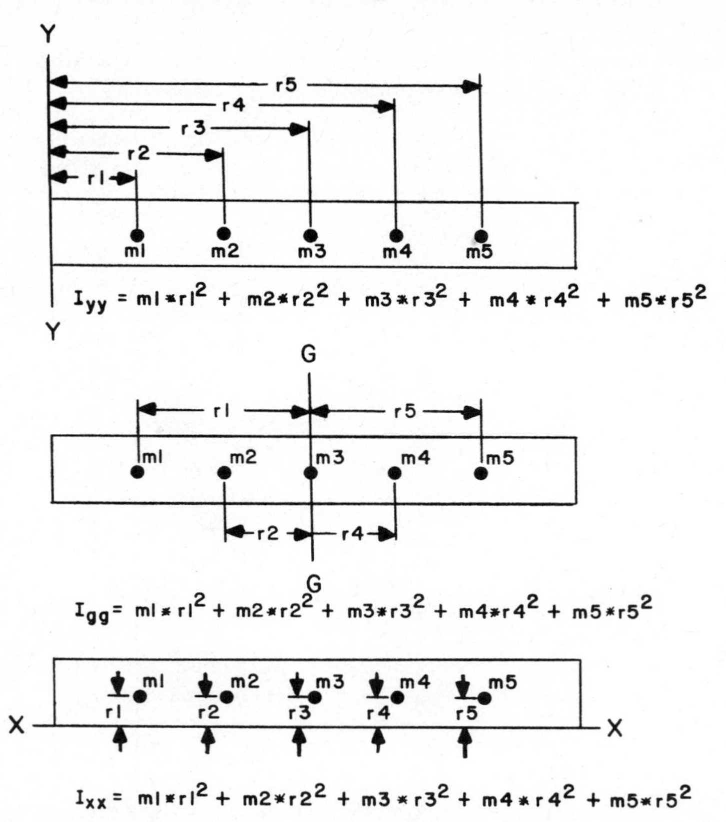

FIGURE 3.7 Moments of inertia of an idealized body in respect to three different axes (Miller and Nelson, 1976).

where

I_0 is the moment of inertia measured about an axis at one end of the segment

W is the segment weight

L is the distance of the center of mass to the axis of suspension

f is the frequency of oscillation

The moment of inertia at the center of mass I_{cm} is found by the parallel axis theorem equation:

$$I_{cm} = I_0 - ML^2 = I_0 - \left(\frac{W}{g}\right) L^2 \tag{3.5}$$

The detailed procedure is well described by Roebuck, Kroemer, and Thomson (1975).

A second method, which can be used on living subjects, requires the incremental submersion technique developed by Cleaveland in 1955 to be modified. Recall that the incremental submersion technique was described in the preceding subsection as a method for determining the location of the mass center also (see Figure 3.6). By carefully measuring the incremental volumes obtained as a body segment is submerged in discrete steps, the moment of inertia can be estimated. The calculations for this are included in Figure 3.6.

Drillis and Contini (1966) proposed a third method, referred to as a quick release technique by Miller and Nelson (1976). This technique requires the subject to exert a measured torque against a restraining strap. The strap is suddenly released and the instantaneous angular acceleration of the limb is measured (via accelerometers attached to the segment). The moment-of-inertia is computed as follows:

$$I_0 = \frac{T}{\alpha} = \frac{Fd}{\alpha} \tag{3.6}$$

where

- I_0 is the moment-of-inertia of segment about its proximal joint
- T is the torque developed by the subject just before release, which is equal to the force F on the restraining strap multiplied by the distance d that the strap is from the proximal joint
- α is the angular acceleration of segment at instant of strap release

Clearly the latter two methods (the incremental submersion and quick release methods) are most appropriate for extremity measurements. Unfortunately, it is difficult to obtain body-segment physical parameter data on large numbers of subjects, for the head, neck, and torso, and thus the cadaver-based data must be used.

In biomechanical computations of the moment of inertia effect, the concept of a radius of gyration K is often useful. It is a derived variable that expresses the radial distance from the axis of rotation at which the mass of the segment can be concentrated without altering the moment of inertia of the segment. Recall from equation 3.3, that $I = mr^2$; i.e., the mean radii of a quantity of mass particles comprising a segment. The radius of gyration is computed by substituting K for r and thus

$$K = \sqrt{\frac{I}{M}} \tag{3.7}$$

Using moments of inertia data obtained from earlier cadaver pendulum stud-

TABLE 3.7
Radius of Gyrations as a Percentage of a Segment Length[a] (Plagenhoef, 1966)

Segment	Radius of Gyration From Proximal End	From Distal End
Head, neck, and trunk	49.7%	67.5%
Arm	54.2	64.5
Forearm	52.6	64.5
Hand	58.7	57.7
Upper limb[a]	64.5	59.6
Forearm and hand[a]	82.7	56.5
Thigh	54.0	65.3
Leg	52.8	64.3
Foot	69.0	69.0
Lower limb[b]	56.0	65.0
Leg and foot[b]	73.5	57.2

[a] To the ulnar styloid.
[b] To the medial malleolus.

ies of Dempster (1955), Plagenhoef (1966) calculated the radius of gyration for the major body segments. These are given in Table 3.7 as a percentage of the length of the segment. A general method of estimating these values will be given later in this chapter.

3.2 ANTHROPOMETRIC DATA FOR BIOMECHANICAL STUDIES IN INDUSTRY

As presented before, several physical properties of various body segments must be known in order to perform a biomechanical analysis of a task. What follows is a synthesis of the data available from various populations that can be used in performing such studies.

3.2.1 Segment Link Length Data

Link length data have been derived on living subjects with reference to surface landmark measurements. The resulting values have also been statistically regressed onto a subject's stature. Thus values can be given as percentiles of the population and as proportions of stature. Figure 3.8 expresses the latter form, as derived by Drillis and Contini (1966) and presented in Roebuck, Kroemer, and Thomson (1975). Specific regression equations

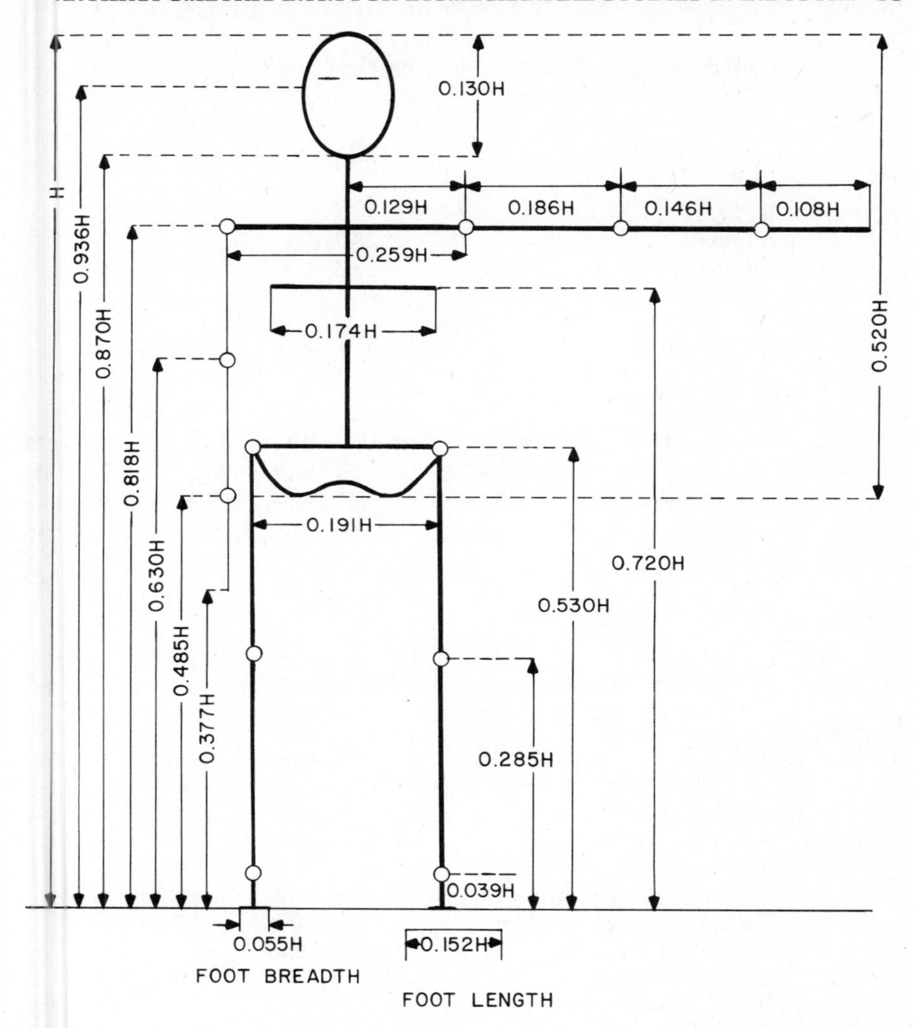

FIGURE 3.8 Body segment lengths expressed as proportion of body stature by Drillis and Contini (Roebuck, Kroemer and Thomson, 1975).

for the bone lengths of the extremities as a function of stature have been derived by Webb Associates (1978). These indicate the standard error estimate to be approximately 1.0 cm when using stature as the basis for the bone-length estimates. This error estimate would be increased slightly in converting the bone lengths to link lengths, due to the error in defining the precise center-of-rotation for the link, as discussed earlier.

A tabular estimate of extremity lengths is given in Table 3.8 from data synthesized by Webb Associates (1978). Some specific mean dimensions

TABLE 3.8
Link Length Values in Centimeters (from Webb Associates, 1978)

	Male White			Female White		
Limb	5th	50th	95th	5th	50th	95th
Upper arm link	28.6	30.4	32.3	26.1	27.8	29.5
Forearm link (ulna)	25.6	27.1	28.7	22.7	24.1	25.5
Forearm link (radius)	25.9	27.5	29.2	22.7	24.1	25.5
Thigh link	40.4	43.2	46.1	36.9	39.5	42.1
Shank link	38.9	42.1	45.3	34.7	37.4	40.0

and standard deviations for the torso are given for a young, healthy male population in Table 3.9 (from Snyder, Chaffin, and Schultz, 1972). Other torso link dimensions can be estimated for various male and female populations by multiplying the stature estimates in Table 3.10 by the link proportion constants given in Figure 3.8.

TABLE 3.9
Torso Link Dimensions for Young Males in the United States (Snyder, Chaffin, and Schutz, 1972)

Link Centers	Mean(cm)	SD(cm)
L5/S1 to L4/L5 disc centers[a]	3.66	0.23
L4/L5 to L3/L4 disc centers	3.63	0.79
L3/L4 to L2/L3 disc centers	3.86	0.81
L2/L3 to L1/L2 disc centers	3.63	0.25
T12/L1 to T8/T9 disc centers	11.28	0.74
T8/T9 to T4/T5 disc centers	9.47	0.68
T4/T5 to C7/T1 disc centers	8.92	0.25
C7/T1 to C6/C7 disc centers	1.93	0.10
C6/C7 to C5/C6 disc centers	1.80	0.10
C5/C6 to C4/C5 disc centers	2.78	0.13
C4/C5 to C3/C4 disc centers	1.78	0.10
C3/C4 to C2/C3 disc centers	1.82	0.13
C7/T1 disc to shoulder center[b]	16.43	1.40

[a] Disc centers (vertebral interspaces). Located at the center of intervertebral discs and determined by the intersection of a perpendicular plane drawn through the center of the adjoining vertebral bodies, and a horizontal plane equally dividing the intervertebral space.
[b] Shoulder center (proximal head of the humerus). The center of mass of the proximal (glenohumeral) head of the humerus, as determined by the intersection of lines drawn perpendicular to the shaft and at 90° across the head of the humerus in an anterior–posterior view of the subject's radiograph.

TABLE 3.10
Population Stature Data (Webb Associates, 1978)

Definition: The vertical distance from the standing surface to the top of the head. The subject stands erect and looks straight ahead.

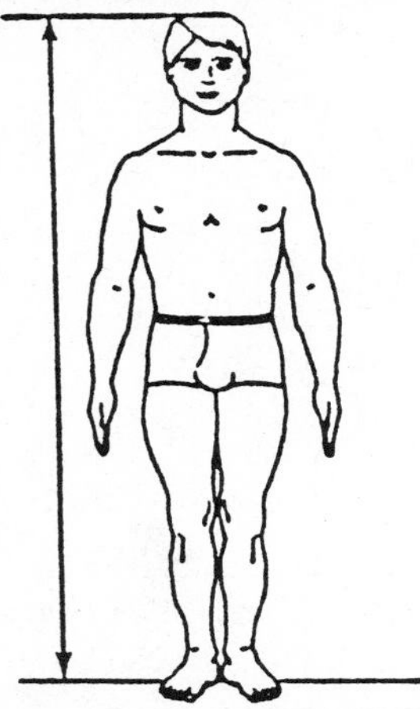

Sample and Reference	Survey Date	No. of Subjects	Age Range	Descriptive Statistics[a]			
				$\bar{X}$	SD	5%ile	95%ile
Females							
USAF Women	1968	1905	18–56	162.1	6.0	152.4	172.1
U.S. HEW civilians	1971–1974	6822	18–74	161.8	6.4	151.1	172.2
British civilians	1957	4995	18–55+	160.1	6.6	149.5	171.2
Swedish civilians	1968	215	20–49	164.7	6.1	154.6	174.7
Japanese civilians	1967–1968 1972–1973	1622	25–39	153.2	4.8	145.3	161.1
Males							
USAF flying personnel	1967	2420	21–50	177.2	6.2	167.2	187.7
U.S. HEW civilians	1971–1974	6823	18–74	175.3	7.1	163.6	187.0
NASA astronauts	Dates vary	60	28–43	176.4	4.7	167.4	182.8
RAF flying personnel	1970–1971	2000	18–45	177.4	6.2	167.3	187.8
Italian military	1960	1342	18–59	170.8	6.2	160.2	180.8
French fliers	1973	65	27–32	175.6	5.3	166.9	184.6
German AF	1975	1004	Not reported	176.7	6.2	166.8	187.1
Japanese civilians	1967–1968 1972–1973	1870	25–39	165.3	5.8	155.8	174.8

[a] Data given in centimeters.

TABLE 3.11
Estimates of Mass Distributions (kg) for Various Male and Female Percentiles[a]

	Male			Female		
Link	5%	50%	95%	5%	50%	95%
Hand	.4	0.4	0.6	0.3	0.4	0.5
Forearm	0.9	1.2	1.6	0.7	1.0	1.4
Upper arm	1.6	2.1	2.8	1.3	1.7	2.5
Head, neck, and trunk	33.0	43.4	56.8	27.2	35.8	52.1
Arms, head, and torso above L5/S1 disc[b]	27.2	35.8	46.8	22.4	29.5	42.9
Upper leg	5.7	7.4	9.7	4.7	6.2	8.9
Lower leg	2.6	3.4	4.5	2.2	2.8	4.2
Foot	.7	1.0	1.4	.7	.9	1.3
Body weight[c]	57.1	75.2	98.3	47.1	62.1	90.1

[a] Estimates are from Dempster (1955) as corrected for fluid loss by Clauser et al. (1969).
[b] Based on Morris, et al. (1961).
[c] Based on National Health Survey, *Weight, Height and Selected Body Dimensions of Adults*, PHS Pub 1000, Series 11, No. 8, (1965).

Clearly the torso dimensions are subject to large errors due to the infinite variety of torso shapes that are feasible with such a multi-segmented link system. A first attempt at describing the dimensional complexities of the torso for various extreme postures was performed by Chaffin, Schultz, and Snyder (1972). Many other studies will be necessary to provide a means to easily depict the dimensional qualities of the torso. Until then, simple one-, two-, or three-link models will have to suffice for biomechanical analysis.

3.2.2 Segment Weight Data

Body-segment weights have been correlated with total body weight, and have produced correlation coefficients from approximately 0.8 to 0.9 (Webb Associates, 1978). Resulting proportionality factors were given in Table 3.5. These factors have a standard error estimate of less than approximately 5% of the mean weight of a segment.

Table 3.11 presents body-segment weights by percentiles using population total body weights obtained in a United States civilian population survey. Appendix B presents a similar set of data on additional body segments for select population body weights (Webb Associates, 1978).

3.2.3 Segment Mass-Center Location Data

Living subject segmental mass-center location data are relatively rare. Cadaver data are still the primary source. In this regard, the data from eight

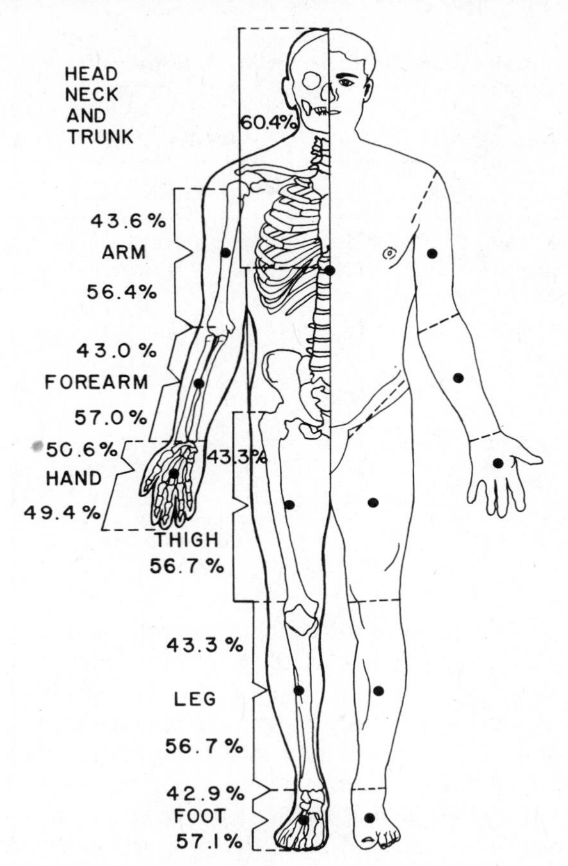

FIGURE 3.9 Link boundaries and mass center locations as percentage of link lengths (Dempster, 1955).

male cadavers dissected by Dempster (1955) are most often quoted in biomechanical studies. His mass-center locations were expressed as a percentage of the segment link length. Figure 3.9 displays these results. A detailed description of the mass-center locations according to Dempster is presented in Appendix B.

Further cadaver dissections were performed for this purpose by Clauser, McConville, and Young (1969). These resulted in small modifications of the mass-center locations according to Dempster, displayed in Table 3.6. Application of the average of these values to the link length values presented in Section 3.2.2 provides the mass-center locations displayed in Table 3.12.

3.2.4 Segment Moment of Inertia and Radius of Gyration Data

Once again, segment moment of inertia values have been obtained primarily from the more precise cadaver studies. In particular, Dempster's detailed

TABLE 3.12
Distances to Segment Mass Centers (cm)

Dimension	Male			Female			Percent Link Length Dempster (1955)[a]
	5%	50%	95%	5%	50%	95%	
Wrist-to-hand CM	6.7	7.0	7.4	6.1	6.4	6.7	—
Elbow-to-lower arm CM	11.0	11.7	12.3	9.9	10.4	11.0	43.0%
Shoulder-to-upper arm CM	12.5	13.2	14.0	11.6	12.1	12.5	43.6%
Hip-to-trunk, neck, head CM[b]	18.1	19.3	22.5	16.7	17.9	19.1	39.6%
Knee-to-upper leg CM	23.0	24.8	26.1	22.2	23.2	24.2	56.7%
Ankle-to-lower leg CM	23.0	23.2	24.9	19.3	20.6	22.1	56.7%
Heel-to-foot CM	10.6	11.4	12.3	9.4	10.3	11.1	42.9%

[a] All dimensions are based on the Dempster percentage of link length estimates.
[b] When in erect posture, measured from hip to top of head.

investigation in 1955 is most often quoted, with additional values from Becker (1972) and Chandler et al. (1975). These studies have provided average estimates of the radius of gyration K for body segments when turning around proximal and distal joint centers, as illustrated in Table 3.7. The radius of gyration values, when computed from the three principal axes of rotation at the segment's center of mass, were shown by Chandler et al. (1975) to be proportional to the length of the segment. Braune and Fischer in 1889 had suggested this, and used the proportionality constant of 0.3. Thus the radius of gyration can be estimated as

$$K = CL \tag{3.8}$$

where

- K is radius of gyration (cm)
- C is a proportionality constant with an average value of 0.3, or with values given in Table 3.13 for the three principle axes of rotation about the segment centers of mass
- L is link length value (cm)

The segment moments of inertia about the center of mass I_{cm} are computed as follows:

$$I_{cm} = MK^2 = MC^2L^2 = 0.09\,ML^2 \tag{3.9}$$

Values for the moments of inertia of 5, 50, and 95 percentile males are

TABLE 3.13
Radius of Gyration (K) as a Ratio (C) of Segment Length L

Segment	Motion Axis	Link	C
Head	x	Head Length	.32
	y		.31
	z		.34
Torso	x	Torso length	.43
	y	(Suprasternale height	.35
	z	–trochanterion height)	.21
Upper arm	x	Acromion–radiale	.26
	y		.25
	z		.10
Hand	x	Hand breadth	.50
	y		.46
	z		.27
Thigh	x	Trochanterion height	.28
	y	–fibular height	.28
	z		.12
Shank	x	Fibular height	.28
	y		.28
	z		.08
Foot	x	Foot length	.26
	y		.25
	z		.12

Motion axes to allow the motion are defined as:
x = sagittal plane motion
y = frontal plane motion
z = horizontal plane motion (around long axis)

included in Appendix B (from Webb Associates, 1978) for reference. They were derived from the multiplication of Chandler's data in Table 3.13 by the segment mass and length data given in preceding subsections.

3.3 SUMMARY OF ANTHROPOMETRY IN OCCUPATIONAL BIOMECHANICS

It should be clear that the field of anthropometry provides not only reliable methods of measuring parameters related to the human size and form, but also appropriate data from various cadaver and living subject surveys. Of particular interest to occupational biomechanics are the anthropometric data describing the physical properties of the body segments. Though more limited than the whole-body data, they are sufficient to allow quantitative biomechanical analyses of industrial tasks to be performed. As will be seen in the second part of the book, these analyses are providing new insights into the cause and control of mechanical trauma in the workplace today.

The contributions of anthropometry to occupational biomechanics will not have been properly summarized without acknowledgment that anthropometrists have also provided new methods and data on human strength and joint range-of-motion. These two particular human attributes are so important to the occupational biomechanist that a separate chapter in this text (Chapter 4) has been devoted to discussion of them.

REFERENCES

Becker, E. B., *Measurement of Mass Distribution Parameters of Anatomical Segments,* SAE Paper 720964, Society of Auto Engineers, Detroit, 1972.

Braune, W. and O. Fischer, *The Center of Gravity of the Human Body as Related to the German Infantryman,* Leipzig, Germany (ATI 138, 452, Available from U.S. National Technical Information Office), 1889.

Chaffin, D. B., R. K. Schutz and R. G. Snyder, *A Prediction Model of Human Volitional Mobility,* SAE Paper 720002, Society of Auto Engineering, Detroit, 1972.

Chandler, R. F., C. E. Clauser, J. P. McConville, H. M. Reynolds, and J. W. Young, *Investigation of Inertial Properties of the Human Body,* AMRL-TR-74-137, Aerospace Medical Research Laboratories, Ohio, 1975.

Clauser, C. E., J. T. McConville, and J. W. Young, *Weight, Volume and Center of Mass of Segments of the Human Body,* AMRL-TR-69-70, Aerospace Medical Research Laboratories, Ohio, 1969.

Cleaveland, G. H., *The Determination of the Center of Gravity of Segments of the Human Body,* unpublished Doctoral Dissertation, University of California, Los Angeles, 1955.

Dempster, W. T., *Space Requirements of the Seated Operator,* WADC-TR-55-159, Aerospace Medical Research Laboratories, Ohio, 1955.

Drillis, R. and R. Contini, *Body Segment Parameters,* BP174-945, Tech. Rep. No. 1166.03, School of Engineering and Science, New York University, New York, 1966.

Frankel, V. H. and M. Nordin, *Basic Biomechanics of the Skeletal System,* Lea and Febiger, Philadelphia, 1980, pp. 121.

LeVeau, B., *Biomechanics of Human Motion,* 2nd ed., Saunders, Philadelphia, 1977, pp. 205–215.

Miller, D. I. and R. C. Nelson, *Biomechanics of Sport,* Lea and Febiger, Philadelphia, 1976, pp. 48–53 and 88–110.

Morris, J. M., D. B. Lucas and B. Bressler, "Role of the Trunk in Stability of the Spine," *J. Bone Jt. Surg.* **43A**(June), 327–339, 1961.

Plagenhoef, S. C., "Methods for Obtaining Kinetic Data to Analyze Human Motions," *Res. Q. Amer. Assoc. Health Phys. Ed.,* **37,** 103–112, 1966.

Roebuck, J. A., K. H. E. Kroemer, and W. G. Thomson, *Engineering Anthropometry Methods,* Wiley-Interscience, New York, 1975, pp. 52–70 and 173–186.

Snyder, R. G., D. B. Chaffin, and R. K. Schutz, *Link System of the Human Torso,* HSRI Report 71-112, Highway Safety Research Institute, University of Michigan, Ann Arbor, Michigan, and AMRL-TR-71-88, Aerospace Medical Research Laboratories, Ohio, 1972.

U.S. HEW-Public Health Service, Vital and Health Statistics—Weight and Height of Adults, Series 11 (211), National Center for Health Statistics, Hyattsville, Md, 1979, pp 28.

Webb Associates, *Anthropometric Source Book,* Vol. I, NASA 1024, National Aeronautics and Space Administration, Washington, D.C., 1978, pp. IV-1 to IV-76.

4

MECHANICAL WORK-CAPACITY EVALUATION

4.1 INTRODUCTION

The preceding anthropometric chapter presented measurement methods and data detailing biomechanical properties of various body segments. This chapter describes two important biomechanical properties of the intact musculoskeletal system: joint motion and muscle strength. These two properties define an indiviual's ability to perform mechanical tasks, such as reaching to an object or exerting manual forces on a control or object. Joint motion and muscle strength data also allow biomechanical models to be developed that can be used to predict both the population's capability to obtain various working postures, and the manual forces that can be produced in a given posture.

The chapter is divided into three sections. The first discusses methods used to measure joint motion and includes data describing the population's normal joint motions. The second describes various methods of acquiring and using human strength data. The third is a brief discussion of the limitations in the available data.

4.2 JOINT MOTION—METHODS AND DATA

Joint *motion* is often referred to as joint *mobility* and joint *flexibility*. The distinction is vague, and the term *motion* appears to be more commonly used today. Though joint motion could be defined relative to a specific absolute posture, it is generally conceded that body segments rotate about a joint; thus, the range of motion (or maximum angular deviation) available at the joint is the best means to express joint mobility. This measure is quite acceptable for primarily single-axis joint rotation found when flexing or extending the elbow or knee. For more complex joints, such as at the hips and shoulders, motion must be expressed with reference to a particular plane of motion; i.e., in the sagittal, frontal (coronal), or horizontal (transverse) planes. Because of this complication, a set of definitions has been developed for all major joints that assumes that motion is measured with the person beginning in the *standard anatomical position* depicted in Figure 4.1. This position requires the person to stand erect, face forward, and hold the arms down at the side with palms facing forward (i.e., the thumbs point away from body). The resulting traditional terms are illustrated in Figure 4.1.

Unfortunately, even with such diagrams the terminology can be ambiguous. For instance, when the arm is moved outward and up over the head, the shoulder joint is said to be in flexion if the motion occurs in the sagittal plane and in abduction if in the frontal plane. Thus the direction or plane of motion of a segment can affect the definition of the angle at a joint. This confounding of motions and joint angles has resulted in a new system of joint mobility terminology adovcated by Roebuck (1968). It is explained in detail in the text by Roebuck, Kroemer, and Thomson (1975) and combines, in the motion terminology, information regarding the principle plane of motion, direction of motion from resting position (i.e., in, out, left, right), and type of motion (i.e., translation, rotation, or torsion). To obtain a simple terminology, the terms were modified, shortened, and combined, as in Table 4.1.

Though this more systematic terminology has not been widely used, it does provide a consistent language in which joint motion can be defined in the future. For the present, however, the reader will have to accept the fact that the classic motion terminology, even though it is neither complete nor consistent in various texts, is the standard in the literature. Thus, care must be exercised when referring to a particular range-of-motion datum.

In this context, if one is to present values for others to use in the future, it would be prudent to present:

1. A definition that is complete and conforms to other investigator's use of terms as closely as possible.
2. A drawing—or drawings—that illustrate:
   a. Body-segment positions at the extremes of the motion, along with the general body posture.
   b. Body-segment axis positions used to measure the joint angle of concern.
3. A description of the population measured.

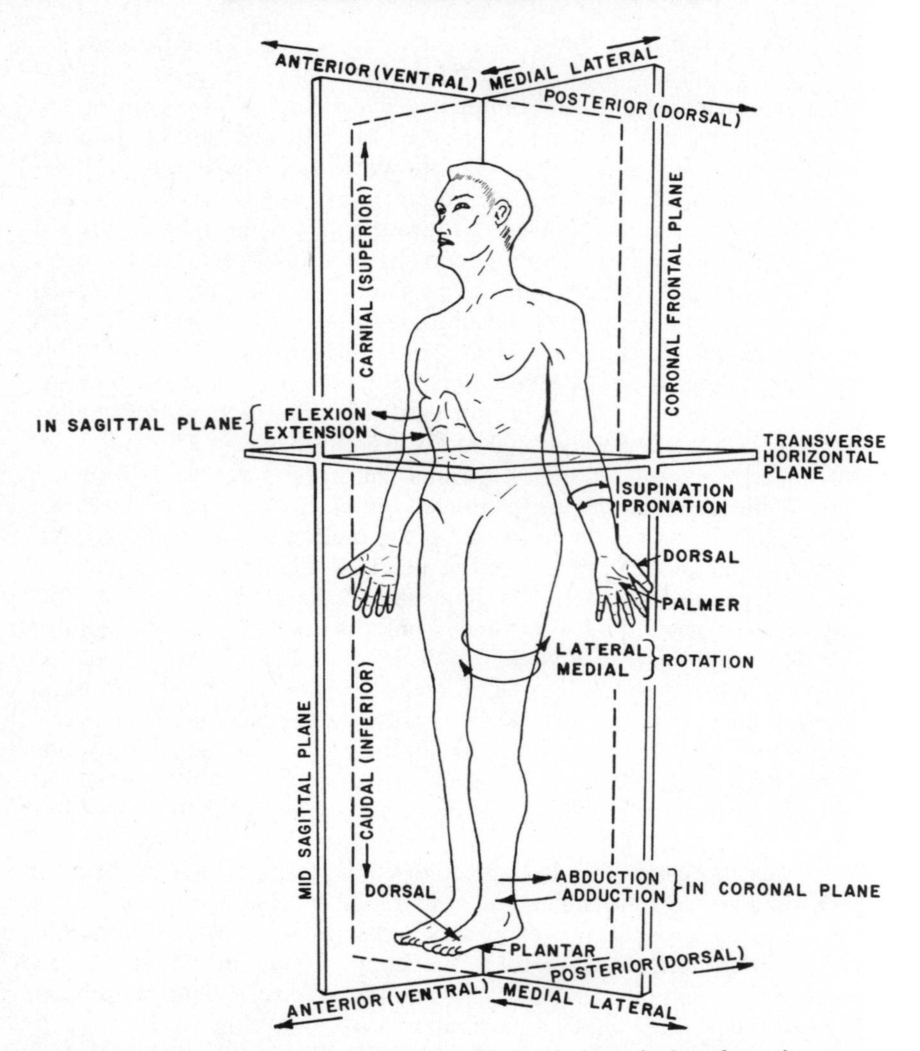

FIGURE 4.1 Standard anatomical position with classical terminology for major movements shown.

4. A summary of the data, including both mean and variance statistics.
5. A description of the measurement method.

This latter requirement is critical in that a deviation in a measured angle of as much as $\pm 15°$ can easily occur because of undetected shifts in the measurement reference points used to define extreme positions of the body segment. Such shifts have been reported when comparing data obtained from photographs with data from actual measurements. This concern is amplified

TABLE 4.1
New Movement Terms, Triplanar System (from Roebuck, Kroemer, and Thomson, 1975)

(Plane + Direction + Type = Motion)

Plane	Direction	Type	Motion
Front(al)	-e-	Vection	Frontevection
Sag(ittal)	-in-	Rotation	Frontinvection
Trans(versal)	posi-	Torsion	Sagevection
	negi-		Saginvection
			Transevection
			Transinvection
			Fronterotation
			Frontinrotation
			Sagerotation
			Saginrotation

in the following discussion in which measurement methods of joint motions are discussed.

4.2.1 Measurement Methods of Joint Motion

The various devices that have been developed to measure joint motions seek to satisfy the following criteria:

1. Accuracy—the measurement device reports the true deviation of the joint angle.
2. Repeatability—when measurements are repeated, the data are consistent.
3. Ease of use—the measurement device is easy to use with minimum time in preparation and in recording of results.
4. Cost—the measurement device should be inexpensive.
5. Measurement flexibility—the measurement system can be used to measure many different types of motions.

The accuracy and repeatability criteria are often dependent on each other. If a measuring device attached to a body segment shifts its position during various tests, then both repeatability and accuracy will be lost. If the device is well attached, reducing the repeatability error, inaccuracy can still exist if care is not taken to obtain a good estimate of the center-of-rotation of a joint and to align the device appropriately. Ease of use and cost are also somewhat interdependent. An inexpensive system may be easy to use, but limited to only certain types of motions. Then, measurement flexibility is lost, and accuracy and repeatability may also be sacrificed. To meet these various criteria, several different systems have been developed. Some of them are described briefly here.

Goniometry. A goniometer is a protractor with two reference arms. A typical manual goniometer is shown in Figure 4.2 from Roebuck, Kroemer, and Thomson (1975). The reference arms are aligned with the long axes of the adjoining body segments, with the center of the protractor over the estimated joint center. The protractor is then taped into position and the subject is asked to move through the entire range of motion, while the angles are visually measured and then recorded. By including a rotational transducer where the reference arms are attached, an electronic readout can be obtained, as described by Adrian (1968) and Ricci (1967), for example. This electrogoniometer is sometimes referred to as the *ELGON* (i.e., ELectronic GONiometer).

Clearly, the advantage of a goniometer is the cost and its ease of use. But such a device is limited to simple planar motions of body segments to which the reference arms can be easily aligned and attached. Because of the soft tissue bulk, particularly around the hips and shoulders, and the three-dimensional motion of these and other joints, the use of simple goniometers for such joint motion measurement has been limited in anthropometric studies.

Flexometer. The disadvantages of goniometry are reduced by attaching a gravity-sensitive measurement device to the midsection of the body segment to be moved, and by measuring the motion deviation relative to the gravity vector rather than an adjoining body segment. This is possible with a device known as a *flexometer* developed by J. R. Leighton (1955). The device is shown in Figure 4.3. It has the advantage of ease of use and relatively low cost and avoids the attachment problems at the shoulder and hips. Obviously, error is still possible, however, if the adjoining body segment moves or if the device shifts due to muscle contraction. It also requires time (a few seconds) for the pendulum to stabilize before a reading is recorded, though a newer version has reduced this problem. Despite these limitations, the device has been used in population surveys, as reported by Roebuck (1968) and Harris and Harris (1968), data from which will be presented later in this section.

Spatial Imaging. The advent of the digital computer has made it possible to easily record the location of reference points in space obtained from photographs or from electronic video systems. By attaching light-reflective markers (or light-emitting diodes) to the body and having the person demonstrate extreme postures, angular deviation data can be obtained for the various body joints, even in three dimensions. This technique is referred to as *photogrammetry.*

Once the camera(s) are setup and calibrated (a procedure that is quite time consuming), photogrammetric range-of-motion data can easily be ob-

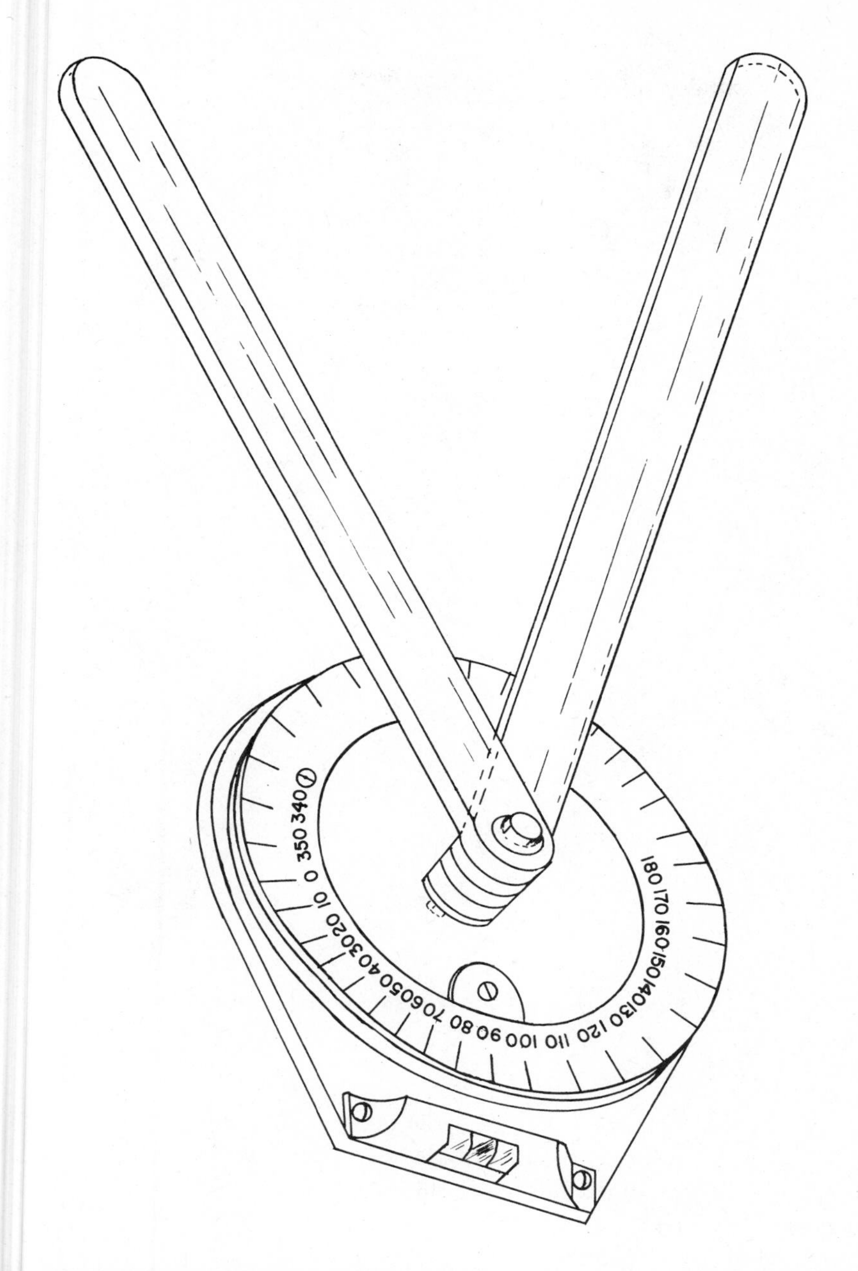

FIGURE 4.2 Bubble level goniometer for joint motion measurements (Roebuck, 1968).

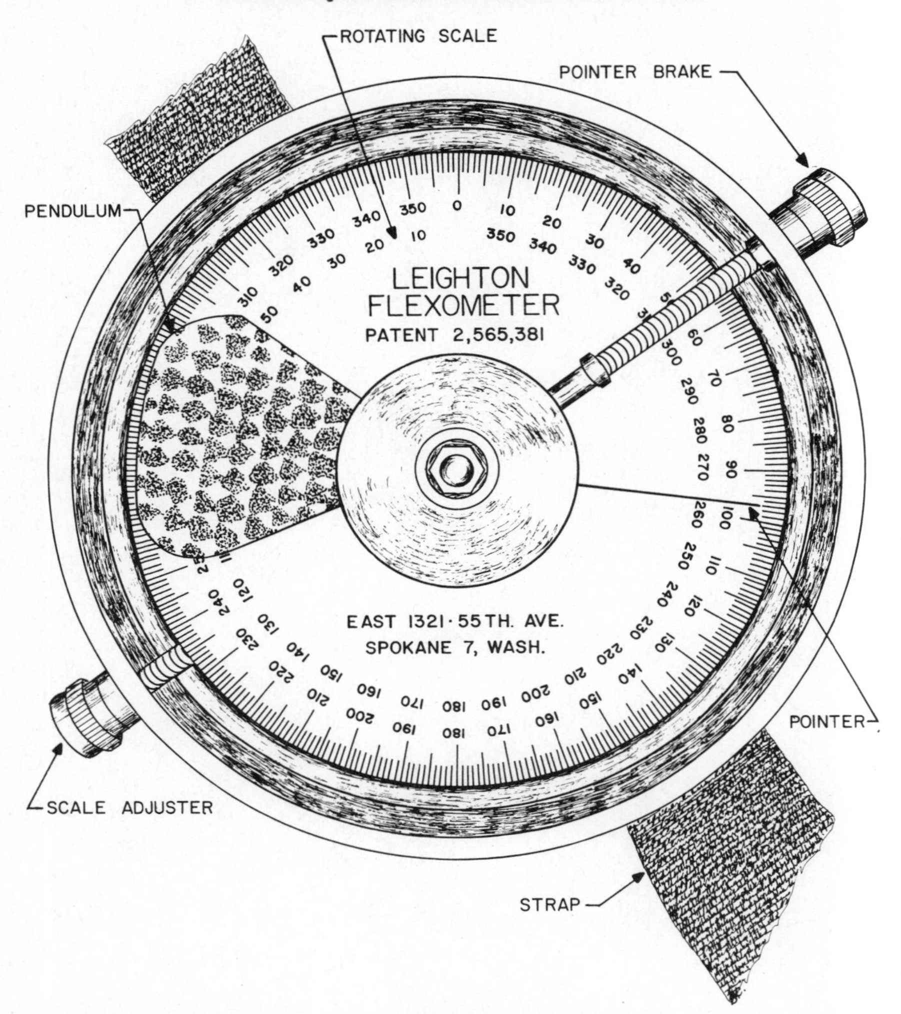

FIGURE 4.3 Original flexometer by J. R. Leighton was improved to include pendulum damping and easier-to-read scale (Leighton, 1957).

tained. Clearly, a large cost is involved in (1) equipment to obtain the spatial images, (2) transducing the data to digital signals that must be calibrated to accurately represent the reference markers in space, and (3) computing time necessary to derive the joint-angle deviations. Because of these requirements, photogrammetric range-of-motion data have only recently been reported. Ayoub (1972) believes such systems are well warranted, however, particularly when dynamic motion analysis is also being sought. Because of its potential, a more detailed discussion of photogrammetry is given in Chapter V.

4.2.2 Normal Ranges of Joint Motion

Since the early part of the century, studies of joint mobility have been published by Gilliland (1921), Glanville and Kreezer (1937), and the American Academy of Orthopaedic Surgeons (1965), to name a few. These represent attempts to describe normal motions of the joints from sample data representative of specific populations. Because the measurement techniques used have varied, direct comparisons are difficult to make.

For general reference, the values determined by Barter, Emanuel and Truett (1957) appear to provide representative baseline values for young healthy males. The measurement postures used are illustrated in Figure 4.4. The data were obtained from photographs that were manually scaled to determine joint angles. Table 4.2 presents the resulting values as depicted by Laubach (Webb Associates, 1978).

Many factors can affect the utility of such norms, such as those in work design situations. Some of the more relevant factors are discussed in the following section.

4.2.3 Factors Effecting Range-of-Motion Data

Age, gender, body stature, and body weight have been studied to determine their possible effect on joint mobility.

Age. Age has been found to have a complex effect on joint motion. It appears that in the 10 to 16-year age group, joint motion decreases by about 10% from that of the first decade. From about 16 to age 70, however, there is no further significant systemic change in asymptomatic populations, according to Salter and Darcas (1953). Considering that various musculoskeletal conditions are prevalent in the population with aging, however, this result poses a dilemma for widespread industrial application. Until larger, cross-sectional surveys of particularly older popoulations are performed, the values by Salter and Darcas will have to suffice as representative of a normal healthy population.

Gender. A study of 100 men and 100 women in the 20 to 50-year age group by Sinelkinoff and Grigorowitsch (1931) indicated that women generally had greater joint mobility than men. Their mean values in degrees and the present ratio between the two is given in Table 4.3.

Anthropometric Dimensions. Various studies of the possible effect of body stature and weight have been reviewed by Laubach (1969) and Roebuck, Kroemer, and Thomson (1975). In general, it is concluded that these anthropometric factors are not well correlated to joint motion. Body fat content, however, was found to be predictive of decreased joint motion by Laubach (1969). It was not clear, however, whether this was caused by the extra tissue bulk around a joint, or whether the more obese subjects simply performed less stretching exercises as part of their normal daily activities.

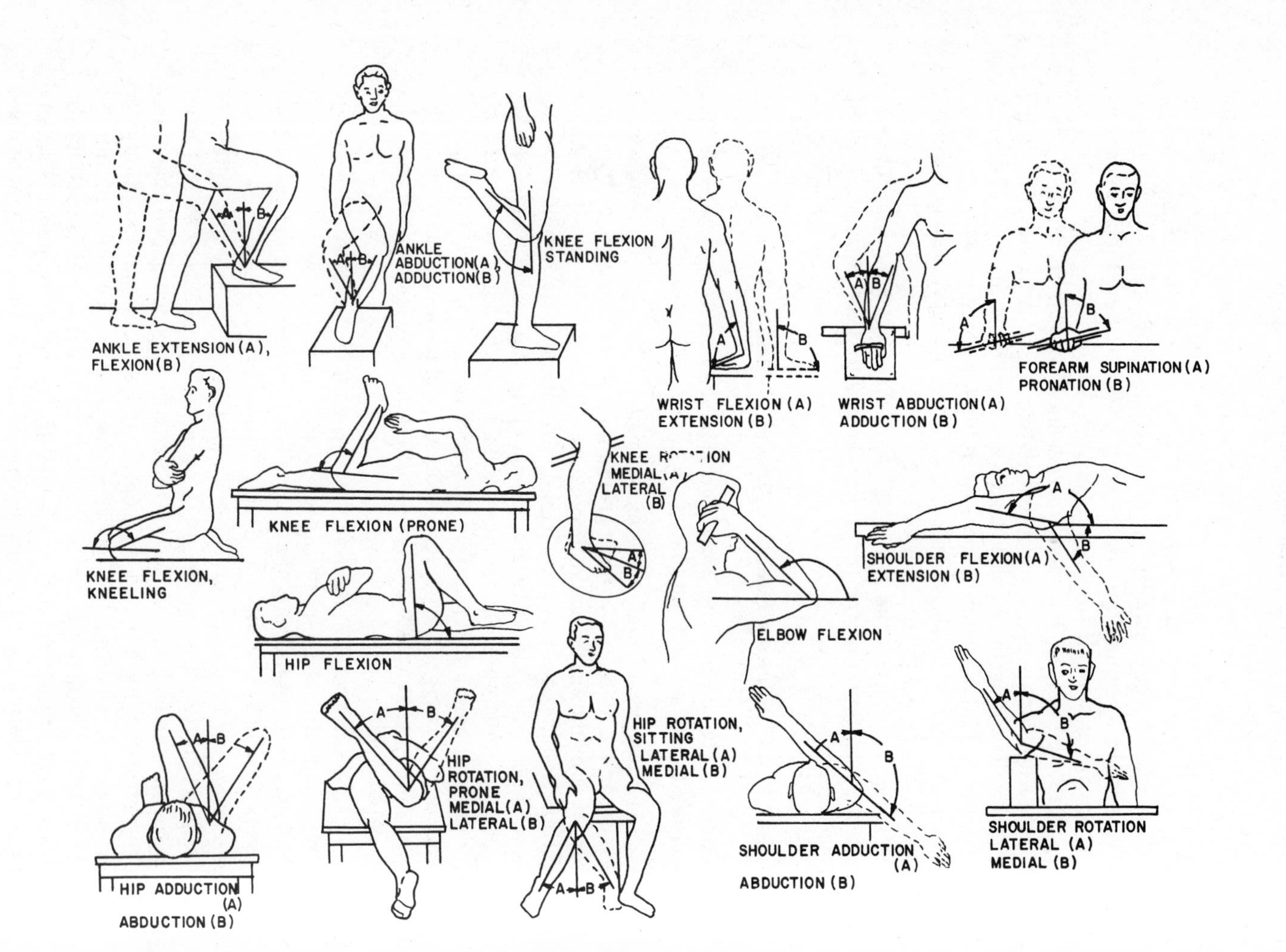
ANKLE EXTENSION (A),
FLEXION (B)
ANKLE
ABDUCTION (A),
ADDUCTION (B)
KNEE FLEXION
STANDING
WRIST FLEXION (A)
EXTENSION (B)
WRIST ABDUCTION (A)
ADDUCTION (B)
FOREARM SUPINATION (A)
PRONATION (B)
KNEE FLEXION,
KNEELING
KNEE FLEXION (PRONE)
KNEE
MEDIAL
LATERAL
(B)
SHOULDER FLEXION (A)
EXTENSION (B)
HIP FLEXION
ELBOW FLEXION
HIP ADDUCTION
(A)
ABDUCTION (B)
HIP
ROTATION,
PRONE
MEDIAL (A)
LATERAL (B)
HIP ROTATION,
SITTING
LATERAL (A)
MEDIAL (B)
SHOULDER ADDUCTION
(A)
ABDUCTION (B)
SHOULDER ROTATION
LATERAL (A)
MEDIAL (B)

TABLE 4.2

Range of Joint Mobility Values Corresponding to Postures in Figure 4.4 from Barter, Emmanuel, and Truett, 1957 Study as Depicted by Laubach (Webb Associates, 1978)[a]

Movement	Mean	SD	5 %ile	95 %ile
Shoulder flexion	188	12	168	208
Shoulder extension	61	14	38	84
Shoulder abduction	134	17	106	162
Shoulder adduction	48	9	33	63
Shoulder medial rotation	97	22	61	133
Shoulder lateral rotation	34	13	13	55
Elbow flexion	142	10	126	159
Forearm supination	113	22	77	149
Forearm pronation	77	24	37	117
Wrist flexion	90	12	70	110
Wrist extension	99	13	78	120
Wrist abduction	27	9	12	42
Wrist adduction	47	7	35	59
Hip flexion	113	13	92	134
Hip abduction	53	12	33	73
Hip adduction	31	12	11	51
Hip medial rotation (prone)	39	10	23	56
Hip lateral rotation (prone)	34	10	18	51
Hip medial rotation (sitting)	31	9	16	46
Hip lateral rotation (sitting)	30	9	15	45
Knee flexion, voluntary (prone)	125	10	109	142
Knee flexion, forearm (prone)	144	9	129	159
Knee flexion, voluntary (standing)	113	13	92	134
Knee flexion forced (kneeling)	159	9	144	174
Knee medial rotation (sitting)	35	12	15	55
Knee lateral rotation (sitting)	43	12	23	63
Ankle flexion	35	7	23	47
Ankle extension	38	12	18	58
Foot inversion	24	9	9	39
Foot eversion	23	7	11	35

[a] Measurement technique was photography. Subjects were college-age males. Data are in angular degrees.

In this latter regard, Leighton reported studies of college baseball and basketball player, swimmers, and shot and discus throwers that disclosed increased joint mobility directly related to the specific motor skill requirements of a sport (Leighton, 1957). In general, it appears that people who practice any routine or competitive sports in which joint motion is stressed will be

TABLE 4.3
Differences in Joint Mobility Between Men and Women based on the Snelkinoff and Grigorowitsch (1931) Study as presented by Laubach (Webb Associates, 1978)

	Men's Mean	Women's Mean	% Ratio[a]
Shoulder abduction (rearward)	59.8[b]	61.4	103%
Elbow flexion–extension	142.1	149.9	105%
Wrist flexion–extension	141.4	154.0	109%
Wrist adduction–abduction	62.2	72.7	117%
Hip flexion (with extended knee)	83.5	86.8	104%
Hip flexion (with bent knee)	117.9	121.0	103%
Knee flexion–extension	140.5	140.1	100%
Ankle flexion–extension	62.6	66.9	107%

[a] Percentage ratio obtained by dividing the women's reported mean value by the men's reported mean value; e.g., 61.4 divided by 59.8 = 103%.
[b] Mean values reported in angular degrees.

more flexible than others, and this flexibility can be maintained through one's lifetime by such activities.

Two-joint Muscle Effect. As discussed in Chapter 2, muscles can contribute to a limitation in joint mobility because of their inherent length-tension properties. When the muscle is extremely shortened, it cannot produce a useful amount of tension. If gravity or tissue compression causes resistance to extreme rotation of a body segment, then joint motion can be limited by the diminished force capability of the muscles. When a muscle acts about only one joint, the described limitation to joint motion is not great. Unfortunately, many joints are spanned by two-joint muscles, and thus the mobility at one joint depends to some degree on the position of an adjacent joint.

A recent pilot study of this effect was conducted by Laubach (Webb Associates, 1978). He found the following major effects:

1. There is a major decrement (up to about 47°) in shoulder flexion when the elbow is flexed.
2. Elbow flexion was decreased in some subjects by concomitant shoulder flexion, but the results were not consistent with all subjects.
3. Hip flexion is decreased by knee flexion, but the extent of the effect is not known as the leg weight varied in different test positions chosen in the experiments.
4. Knee flexion is markedly reduced (up to about 34°) when the hip is also flexed.
5. Ankle dorsiflexion (flexion) is reduced (by about 3°) when the knee is also flexed.

Laubach concluded from these results that more research should be conducted to improve the understanding of the effect of two-joint muscles on joint motions (Webb Associates, 1978). Until such research is done, a person referring to mobility data such as that presented in Table 4.2 should be cognizant of these potential limitations in the data.

4.3 MUSCLE STRENGTH EVALUATION

The topic of muscular strength has been of great interest to orthopaedic and rehabilitation medicine, as well as to exercise physiology for many decades. From earlier studies it has been recognized that (1) the maximum force producing capability (i.e., strength) varies considerably between people and between tasks, (2) static (or isometric) strengths are not necessarily correlated with dynamic strengths, and (3) like anthropometric measurements, several different measurement criteria must be met to develop and use strength data successfully.

Since many jobs in industry require exertions that approach or exceed individual strengths, occupational biomechanics expertise has been sought in the last few decades to either assist in redesigning the jobs to reduce or eliminate such exertions, or to assure that only people with sufficient strength to safely perform such jobs are so employed. Further, in the process of studying the biomechanics involved when people perform various jobs in industry, it has become evident that the limitations to safe job performance are often the result of muscular force insufficiency (i.e., lack of muscular strength).

This section describes human strength, its measurement, and normal data obtained from different populations. In the second part of the book, applications of strength data to both the design of industrial tasks and in selection and placement of workers will be presented.

4.3.1 Muscular Strength Definition

The term *muscular strength* (muscle strength) is chosen to separate it from the strength of bones, ligaments, and tendons. In this context, strength is the maximum force that a muscle can develop under prescribed conditions. Since it requires the muscle to be activated voluntarily by a person, some authorities refer to muscular strength as producing *maximum voluntary exertion levels*. In this regard, the measured strength values, even with well-motivated subjects, are probably below the physiological tolerance of the muscle-tendon-bone system, thus providing a safety factor against unusual exertion requirements. The extent of this "safety factor" is not well known, but it could be as great as 30% (Hettinger, 1961). From gross epidemiology, however, when a job requires exertions above a person's volitional strength,

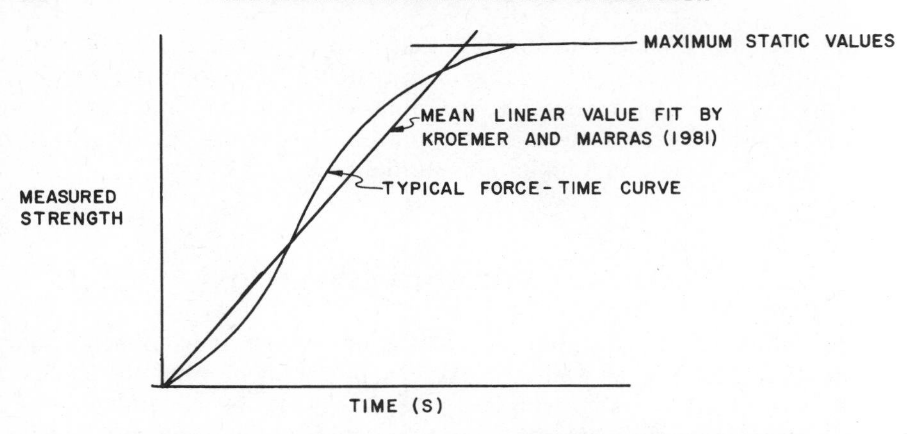

FIGURE 4.5 Typical force–time relationship (Miller and Nelson, 1976).

the person appears to be at a greatly increased risk of injury (Chaffin et al, 1978, and Keyserling et al, 1980).

Clearly, such muscular exertions can result in body-segment motions, in which case the resulting force measured is referred to as *dynamic strength*. On the other hand, the individual may be restrained against motion, so that a *static strength* (sometimes referred to as the *isometric strength*) is obtained. In either case, since the skeletal muscles operate to produce or attempt to produce body segment rotations, the output measures can be either (1) force-produced at some point on the body, or (2) moment- or torque-produced about a given joint. In whole-body exertions—for example, lifting boxes or pushing carts—force measures are generally used. When a specific body action is involved (e.g., turning a screwdriver or manipulating a machine control), moment or torque measures are often used. The latter are also important in modelling human strength, as they depict muscle moment capabilities at various joints.

Whether the exertion is dynamic or static, or measured as a force or moment, the length of time of the exertion is critical. At the initiation of an exertion, time is required for the neural and muscular components to produce maximal output. This is shown by Miller and Nelson (1976) in Figure 4.5. An empirical model of the muscle force-to-time relationship was devised by Kroemer and Marras (1981). They fit a linear line to the force-time relationship and determined the slope of the line for various single joint actions—for example, knee extension and flexion, finger flexion, and elbow flexion. This showed that the maximum strength values were reached in less than 1 sec, with some subjects displaying maximum efforts in approximately one-third sec. Kroemer and Marras also concluded that if the subjects were purposely attempting to perform at prescribed submaximal levels, additional force-buildup time was necessary, presumably to sense the level of effort and consciously adjust to it.

As one would expect, instructions to a person greatly affect strength performance. A study reported by Caldwell et al. (1974) disclosed that when subjects were told to *increase* their exertion level to the maximum and *hold* it, only 3% reached maximum levels in 1 sec, with 18% requiring more than 4 sec. When told to *jerk* (i.e., to *apply the force as rapidly as possible*), 38% of the subjects took 1 sec, while none took as long as 4 sec.

The resulting strength values from tests based on such diverse instructions also vary. If muscle exertions that can impart a momentum to a body segment are used to move large body masses, peak strength values will be higher when *jerk* instructions are given than when the subjects are told to *increase and hold* (Kroemer and Howard, 1970). When smaller muscle actions (i.e., grip strength) are measured, the influence appears to be small (Caldwell et al., 1974).

As an exertion becomes more complex, such as when lifting a heavy object, many muscle actions need to be coordinated, and additional time is needed to build up the necessary forces. Practice in such tasks will, of course, shorten the force build-up time. For these reasons, it has been recommended in an *Ergonomics Guide for the Assessment of Human Static Strength*, developed by the American Industrial Hygiene Association (AIHA), that as much as 4 sec be permitted before recording strength values with naive subjects (Chaffin, 1975). Shorter time delays can be used if the exertions involve a limited set of muscles and the instructions specify a fast contraction.

Similarly, the length of the exertion period must be considered. If the exertion period is too long, then muscle fatigue develops and strength is then decreased. The AIHA *Ergonomics Guide* on the subject indicates that static exertions can be maintained for approximately 4 to 6 sec, during which the instantaneous maximum and 3-sec mean value can be recorded (Chaffin, 1975). Kroemer (1970) recommends that the total exertion period be less than 10 sec to avoid fatigue effects. Under these conditions, however, if the same muscle groups are retested, a minimum 2-min rest period is also recommended.

4.3.2 Static and Dynamic Strength-Testing Methods

Static strengths have been obtained by a variety of researchers, and standardized procedures have evolved. These procedures specify the following (Chaffin, 1975):

1. Exertion duration of 4 to 6 sec.
2. A measuring device that can:
   a. Record peak and 3-sec time averaged exertion levels.
   b. Be applied to a person without creating discomfort due to localized pressure.

   c. Be easily adjusted for measurement of different types of exertions and for measurements of individuals of varied anthropometry.
3. Adequate rest between exertions—from 30 sec to 2 min.
4. Carefully stated instructions to
   a. Avoid coercion of subject.
   b. Inform subject of potential risks involved.
   c. Allow subject to control pace/rest if extra time is needed.
   d. Inform subject of future reporting of data and potential consequences.
5. Provide positive, general verbal feedback to the subject, but during testing restrict specific performance values from the subject to avoid uncontrolled competition when groups of people are tested together.
6. Minimize all environmental distractions—for example, unusual temperatures, noise or lighting, and spectators.
7. Standardization of test postures and body supports and restraints.
8. Complete reporting of test conditions, subject biographic data, and resulting statistics.

By following such procedures it has been shown that static strength values can be obtained that are:

1. Safe—risk of injury appears to be small (possibly less than 1:1000 people tested) and of a comparatively minor nature (i.e., muscle strain).
2. Reliable—test/retest values are highly correlated (r = .9), with coefficients of variation of less than 10% (Keyserling, 1980).
3. Practical—devices are marketed below $2000 (U.S.) and five different measurements can be obtained in less than 15 min by a trained technician.

Postural Effect on Strength. One of the major variables affecting a person's strength (both static and dynamic) is posture. This results from two factors—one biomechanical and one physiological. The biomechanical factor can be appreciated by realizing that skeletal muscles act about joints to cause rotation of adjacent body segments. The muscles accomplish this by producing moments at the joint. In other words, the force of the muscle acts through finite moment arms. As the two body segments rotate—that is, the joint angle changes—the moment arms vary. Figure 4.6 illustrates this for the elbow at three different angles. Similar angle effects exist for the other major elbow flexor muscles. Thus, the muscle force F_M required to produce a constant moment at the elbow will vary inversely as the moment arm M. On the other hand, if the force producing capability of the muscle were constant over the range of motion, then the resulting maximum strength

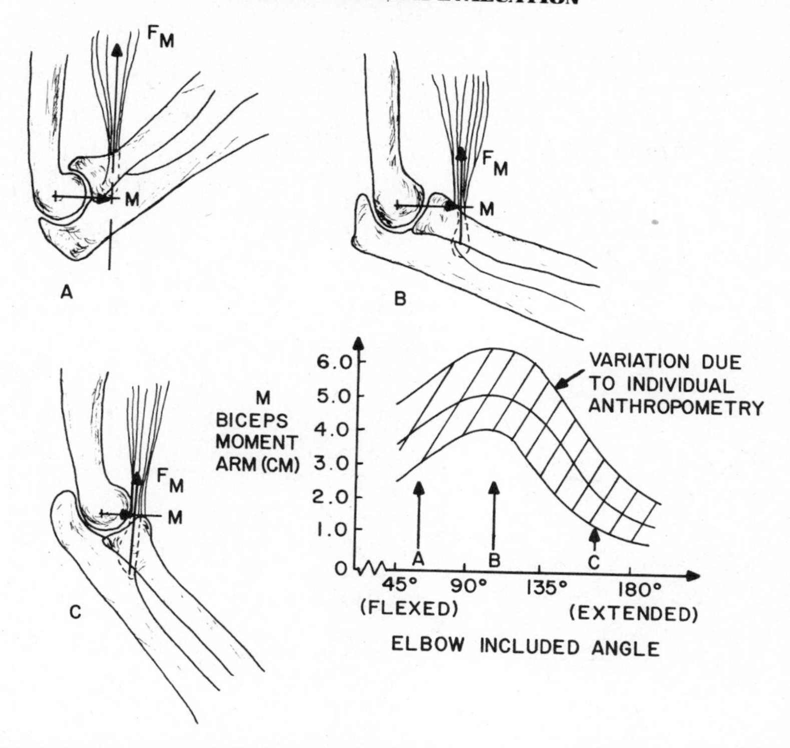

FIGURE 4.6 Length variation in the moment arm of the biceps brachii at different elbow flexion angles.

moment in elbow flexion would occur when M is greatest (i.e., about 90°). This latter presumption is not the case, however, since the phsyiological length-tension relationship alters the maximum F_M throughout the range of motion. As discussed in Chapter 2, each muscle possesses a length-tension property. The orientation of the muscle fibers, the position of the muscle relative to the joint, and other factors will collectively alter the effect of the length-tension relationship of any given muscle on the maximum F_M values. Thus, quantitative models of F_M for various muscles, as well as the individual muscle moment arms at each joint, have not been developed at this time.

What is documented is the result of these two processes (i.e., the *muscular moment strength* at a joint). In fact, Clarke (1966) published a book describing many different muscle strengths as a function of joint angle. Other researchers have contributed to these data (Elkins, et al., 1951, and Schanne, 1972). Later in this chapter, normative strength values will be summarized for specific muscle actions.

In summary, it should be clear from the preceding that static strength values can be obtained, but that care is necessary. Particular concern must

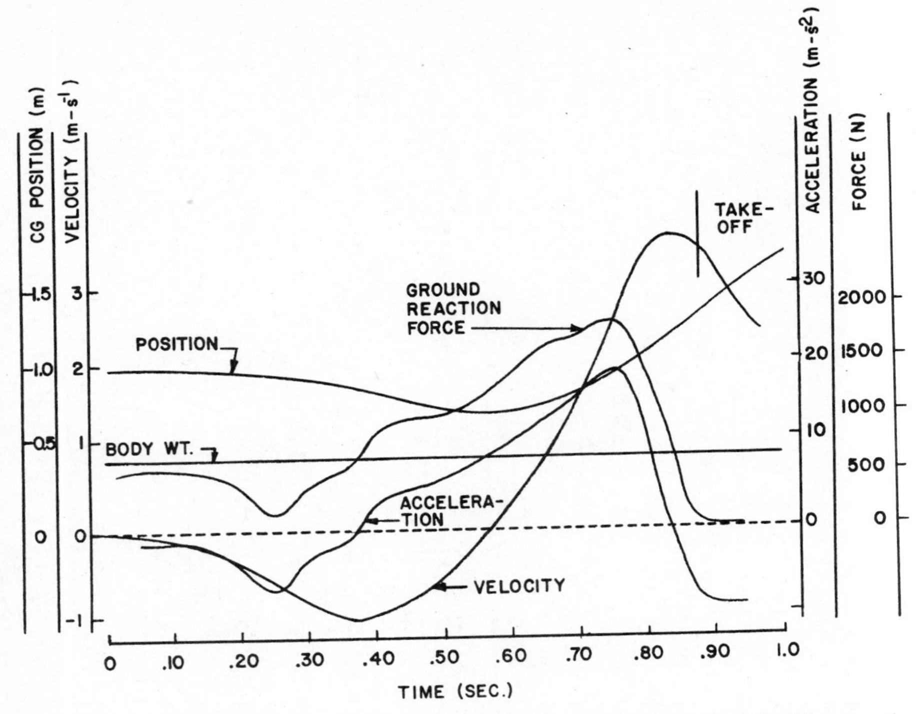

FIGURE 4.7 Vertical position velocity and acceleration of the whole-body CG while executing a vertical jump take-off (Miller and Nelson, 1976).

be shown with regard to instructions and positioning of the subjects. When obtained properly, static muscular strength values can provide important information on the biomechanical capabilities of workers to perform strenuous jobs.

Dynamic Strength Measurement. This is much more complicated than static strength measurement. By definition, dynamic strength requires motion. Body motions have two effects on muscular force capability. First, from a biomechanical standpoint, the motion of body segments may require significant muscle force simply to accelerate the body mass and overcome inertia. Of course, there will also be a deceleration phase of any motion, resulting in momentum, which will assist the muscles in producing additional effective force output. When large body segments are accelerated and decelerated quickly, the effective force output can vary considerably over time. This is illustrated in Figure 4.7 by Miller and Nelson (1976) for a person performing a vertical jump. The ground reaction force is the force output of the person. As can be seen, this force varies in proportion to the vertical acceleration of the whole-body mass, with a peak instantaneous force much greater than the average. In reviewing the literature, and performing his own experiments, Kroemer (1970) concludes that peak strength values are highly variable due to the dynamics of the task, and thus will be difficult to measure and predict without careful specification of the dynamic task parameters.

The second factor effecting dynamic strength is due to the physiology of muscle—namely, its force-velocity property described in Chapter 2. As muscle is rapidly required to shorten and produce force, the maximum force F_M will be reduced as a function of the velocity of muscle shortening. Asmussen, Hansen, and Lammert (1965) measured both static and dynamic strengths in elbow and shoulder exertions. The total distance and velocity of the arm motions were expressed as a ratio of arm length. Rapid arm motions (60% of arm length per second) resulted in dynamic peak strengths that were 75 to 80% of static strengths. They reported an average correlation between the static and slow dynamic strengths of $r = 0.80$ at velocities of 15% of arm length per second. Other researchers have reported similar or higher correlations (Carlson, 1970; Berger and Higgenbotham, 1970). Stothart (1970), however, reports correlations of about $r = 0.70$ between static and dynamic strengths at the beginning of an elbow-flexion motion, but much lower correlations toward the ending deceleration phase.

Unfortunately, muscle velocity-of-shortening effects are not well documented for isolated human muscles. Further, the fact that the moment arms of the muscles acting about a joint vary with the joint angle, and that acceleration effects are not easily observed without sophisticated measurement equipment, combine to cause a great deal of complexity in the measurement and prediction of dynamic strength. This complexity has led some researchers to develop pragmatic solutions to the question of how much a person can lift, push, or pull, as discussed next.

One approach to the problem of predicting a person's dynamic lifting strength, for instance, has been to assume that the maximum load to be lifted will be moved slowly. This assumption then allows one to use the more easily measured static strengths to represent dynamic lifting capacity. From the earlier discussions, however, it is still necessary to carefully define the postures requiring most strength when performing such a dynamic activity. If such postures are not well chosen, the static strengths will either over- or underpredict dynamic lifting capability. Studies by Garg et al. (1980) have shown that the maximum load lifted from the floor to a table can be underpredicted by the static strength data by as much as 54% even with reasonable care in choosing the static-strength test postures. This is particularly true if it is assumed that the load is moved vertically from its initial position, and the static strength tests are then performed with a vertical lift type of test beginning in this posture. Often, when a load is lifted, it is both vertically lifted and pulled in toward the body. Thus the load vector is not solely vertical until the peak acceleration occurs, which is about 0.4 sec into the lift (Konz and Desai, 1976). Figure 4.8 shows this effect. When static lifting strength tests were administered by Garg and his colleagues (1980) with the

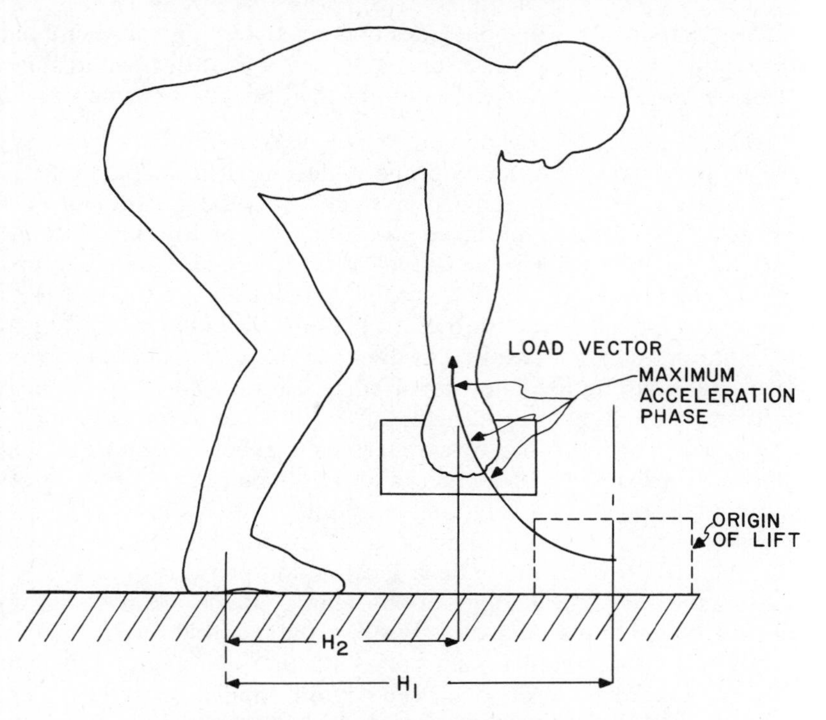

FIGURE 4.8 Example of load lifting trajectory.

horizontal distance at the closer H_2 location shown in Figure 4.8 than at the initial box location H_1, the static values overpredicted the maximum loads that the subjects chose to lift by about 20%. This overprediction of the maximum load lifted could be explained by the fact that the load was accelerated to some degree during the dynamic lifts, thus adding an unknown inertia load to the dead weight on which the subjects were basing their estimate of maximum dynamic lifting capability. Such an inertia effect (between 10 and 20%) is consistent with that found by Park and Chaffin (1974) in a study of dynamic lifting performed by workers in industry.

In the context of a specific test of dynamic lifting capability, Pytel and Kamon (1981) developed a single dynamic strength test using a modified Mini-Gym. This device, which will be illustrated in Chapter 5, has a handle connected via a rope to a winch, which rotates at a preselected speed when the rope is pulled. The force applied to the handle is measured by a load cell inserted between the handle and connecting rope. Pytel and Kamon have used the device to measure lifting strengths at two different lift speeds. A preliminary study of 10 men and 10 women of college age disclosed that the speed selected for the lift test (0.73 and 0.97 m/sec) had a large effect on dynamic lifting strength values—that is, the slower speed lifts averaged about 50% higher than the faster lifts. The dynamic lifting strengths measured at the slower speeds were well correlated ($r = 0.9$) with the maximum loads that the subjects chose to lift in a tote box from the floor to chest height, though the measured dynamic strength values generally overpredicted the chosen tote box loads by as much as 50%.

Psychophysical Strength Method. Dynamic strength capability, whether of a well-defined muscle group or on a whole-body basis, is difficult to measure and predict. This difficulty has motivated a second practical approach for the estimation of dynamic load handling capability. It has been referred to as the *psychophysical* method of setting work capacity norms (Snook, 1978). The psychophysical approach requires a laboratory simulation of a task of interest, in which the subjects are allowed to adjust the load after each attempted performance. The subjects continue to make such adjustments until they subjectively believe the load to be their maximum. The amount of load added or subtracted is not known by the subjects until after a test session is complete. In fact, the initial load set by the experimenter is not known by the subject. Thus, the subject is choosing an abstract load. The experiments are accomplished in sessions lasting approximately 30 to 45 min to assure that the subjects have chosen loads that they believe they can handle throughout a workday.

Snook (1978) has reported that acceptable test/retest values (within 15% of each other) can be expected with well-motivated and trained subjects. The technique has been used to estimate the amount of acceptable load that can be lifted, push, and pulled (Emmanuel et al., 1956; Switzer, 1962; Snook and Irvine, 1967; Snook et al., 1970; and Ayoub et al., 1980). Some values from the use of this technique are reported later in this chapter. Until more comprehensive dynamic strength testing methods and biomechanical models are developed, psychophysical limits, based on simulations of specific tasks of interest, may be the most accurate method of determining a person's acceptable performance limit for a given task. Clearly it is an expensive method that require extensive cooperation and motivation on the part of the subject. It is also not clear how these psychophysical limits relate to the more physiologically determined strength capability of a person. More discussion of this point is presented later in this chapter.

4.3.3 Normal Muscle Strength Values

Static (isometric) strength values have been reported from studies of various populations performing different types of exertions. It should be clear from the preceding discussion that many individual and task variables effect the resulting strength values. This section first illustrates the extent of variation in the static strengths of various population strata when performing different exertions. Both localized (segmental) muscle strengths and whole-body

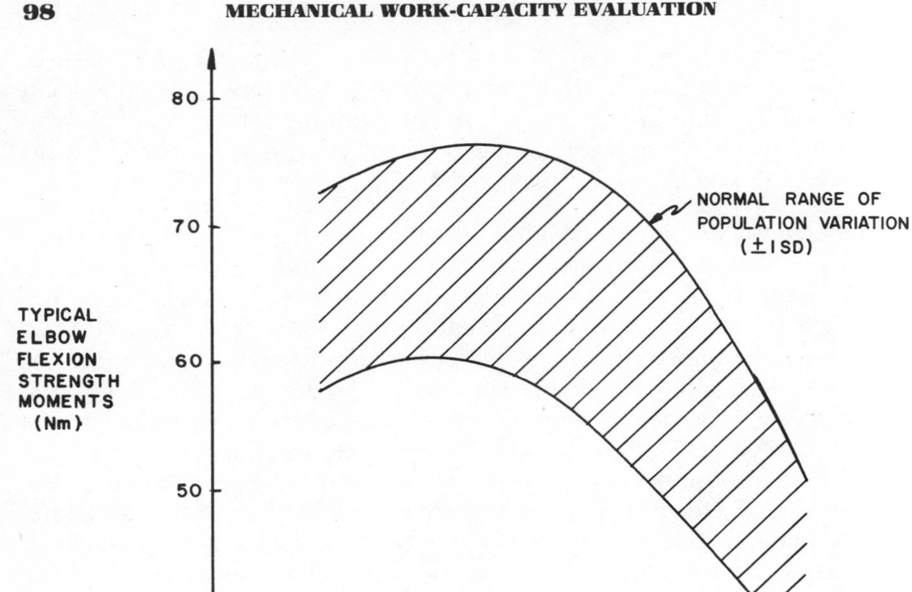

FIGURE 4.9 Variation in elbow flexion strength moments as a function of elbow angles (Schanne, 1972).

static strengths are discussed, followed by a discussion of dynamic values obtained by the psychophysical method.

Localized Static Muscle Strengths. When a person attempts to rotate a restrained body segment about a joint with maximal exertion, a localized muscle exertion is required. In such a case the resulting strength values are expressed as *muscle strength moments*. Strength moments for the trunk and upper extremity have been categorized by Schanne (1972) for 10 men and 10 women of college age. Based on the preceding discussion of postural effects on strength, Schanne developed strength-moment prediction equations based on the angle of the joints during specific muscle exertions. A graphical example of these equations is given in Figure 4.9 for the elbow flexor strength moment as a function of the elbow angle. Schanne's work also showed that because of the presence of the muscles that span two joints, the prediction of strength at one joint would require consideration of the angle of joints adjacent to the primary joint. Thus the prediction of the elbow flexion strength moment was based on both elbow and shoulder angles. This procedure was repeated, and resulted in 19 strength-moment prediction equations depicting the major muscle actions of the body. The resulting prediction equations for a given individual's strengths accounted or approximately 90% of the strength variation found when the subjects assumed various postures, though a large variation in the strength moments existed between individuals. A similar method was utilized by Burggraaf (1972) to depict leg strengths.

Stobbe (1982) measured strength moments of 16 muscle functions in 25 men and 22 women employed in three different industries. His values were obtained with carefully defined body restraints and postures, usually near the center-of-joint range of motions. The device used for the measurements is depicted in Chapter 5. Table 4.4 shows the resulting muscle strength moment values. These disclose a large variation in population strength moments.

Other types of localized muscle strengths have been measured for various population groups. A recent study by Laubach (Webb Associates, 1978) measured total arm strengths in 76 positions for 55 U.S. Air Force men seated in an aircraft pilot's seat. Seat back angles were systematically varied. As expected, arm strengths were lower when the elbow was extended and the exertion direction was perpendicular to the axis of the forearm. If the arm was flexed or abducted above the shoulder, strength values were also reduced. Seat back angle was shown to be important as it effected the arm postures and thereby the exertion levels. Similar arm strength studies were conducted by Hunsicker (1957) and Kroemer (1975) for seated operators, and by Rohmert (1966) for a standing operator, and the results were similar.

Whole-body Static Muscle Strengths. When the body is not restrained during an exertion, such as by a seat, shoulder harness, or table top, many more muscles will be required than considered in the localized strength situations just presented. When many muscles are involved, such as in lifting a load or pushing or pulling on a cart, "the exertion" can be considered to be a *whole-body exertion* to distinguish it from the previous examples. In such exertions, posture remains one of the major factors effecting static strength values, as it effects the magnitude of moments at each joint caused by a load acting on the hands, as well as the muscle moment strengths available at each joint. Biomechanical models of these effects will be presented in Chapter 6.

Static, whole-body muscle strengths of workers have been measured by various researchers at the University of Michigan's Center for Ergonomics. The measurements were obtained by asking the volunteers to lift, push, or pull on a handle attached to a load cell and electronic readout device (see Chapter 5 for details of the measuring device). The handle was positioned

TABLE 4.4

Static Muscle Strength Moment Data (N m) for 25 Men and 22 Women Employed in Manual Jobs in Industry (Stobbe, 1982)

		Male (%ile)			Female (%ile)		
Muscle Function	Joint Angles[a]	5	50	95	5	50	95
Elbow function	90° Included to arm (arm at side)	42	77	111	16	41	55
Elbow extension	70° Included to arm (arm at side)	31	46	67	9	27	39
Medial humeral (shoulder) rotation	90° Vertical shoulder (abducted)	28	52	83	9	21	33
Lateral humeral (shoulder) rotation	5° Vertical shoulder (at side)	23	33	51	13	19	28
Shoulder horizontal flexion	90° Vertical shoulder (abducted)	44	92	119	12	40	60
Shoulder horizontal extension	90° Vertical shoulder (abducted)	43	67	103	19	33	57
Shoulder vertical adduction	90° Vertical shoulder (abducted)	35	67	115	13	30	54
Shoulder vertical abduction	90° Vertical shoulder (abducted)	43	71	101	15	37	57
Ankle extension (plantar flexion)	90° Included to shank	69	126	237	31	81	131
Knee extension	120° Included to thigh (seated)	84	168	318	52	106	219
Knee flexion	135° Included to thigh (seated)	58	100	157	22	62	104
Hip extension	100° Included to torso (seated)	94	190	419	38	97	180
Hip flexion	110° Included to torso (seated)	118	185	342	57	126	177
Torso extension	100° Included to thigh (seated)	164	234	503	71	184	348
Torso flexion	100° Included to thigh (seated)	89	143	216	49	75	161
Torso lateral flexion	Sitting erect	95	159	261	50	94	162

[a] Joint angles are achieved by a fixture described in Chapter 5, with complete statistics and procedures in reference by Stobbe (1982).

in various locations relative to where the person stood. The person was allowed to select the body posture that he or she believed would produce the largest exertion level, but was not allowed to reposition the feet. Repeated tests were performed to assure that the values were consistent (with a coefficient of variation less than 10%). Subjects were recruited from a diverse set of industries, but all were employed in manual jobs. Table 4.5 depicts the mean and standard deviations of these tests. On inspection of these gross strength values it is obvious that they are highly dependent upon the handle positions. This underscores the complexity of the strength producing process and the need to study the process with sophisticated biomechanical models.

Dynamic Strength. As dynamic strengths are even more complex than static strengths, normative population values have not been published for a large variety of activities and worker populations. In recent years, the psychophysical studies of lifting, pushing, and pulling conducted by both Liberty Mutual Insurance Company and Texas Technological University have produced results that show the relative range of load people subjectively believe they can handle for eight hours. An analysis of these studies for load lifting by Ayoub et al. (1980) produced the mean values given in Table 4.6. In these studies, three different sized boxes (of varying sagittal plane depth) were lifted. The boxes were lifted once a minute from varying heights, relative to the person's knuckle, shoulder, and overhead reach when standing erect. The people chose their own speed of lifting. Approximately 100 men and 100 women engaged in a variety of work and leisure time activities were included in the various studies quoted by Ayoub.

As in the static lifting strength tests, one must be impressed with the effect of postures on these dynamic values as dictated by the box size and load movement heights. The studies also indicate that when the frequency of such lifts was increased from one every minute to as many as 12 each minute, the lifting limits decreased in a linear fashion, presumably to avoid fatigue. Snook (1978) showed that when lifts were performed less frequently than one every minute, load lifting values increased by approximately 40%. Postural differences discussed earlier, however, make it difficult to compare these psychophysical limits directly with static strength limits (Garg et al., 1980). The 40% correction would raise the psychophysical limits in Table 4.6 to be more equivalent to the isometric lifting strengths displayed in Table 4.5.

Collectively, these studies indicate that muscle strengths are complex. To explain such strengths one must carefully consider body postures and methods of restraining the body. Such factors are often dependent upon workspace geometry, as well as work methods. The occupational biomechanist must be particularly cognizant of the effect of such factors on strength performance. Specific discussion of the design of workspaces, work seats, and materials handling as they affect strength performance will be

TABLE 4.5
Static Strengths Demonstrated by Workers when Lifting, Pushing, and Pulling with Both Hands on a Handle Placed at Different Locations Relative to the Midpoint Between the Ankles on Floor

	Handle Location (cm)[a]		Male Strengths (N)			Female Strengths (N)		
Test Description	Vertical	Horizontal	Sample Size	Mean	SD	Sample Size	Mean	SD
Lift—leg partial squat[b]	38	0	673	903	325	165	427	187
Lift—torso stooped over[b]	38	38	1141	480	205	246	271	125
Lift—arms flexed[b]	114	38	1276	383	125	234	214	93
Lift—shoulder high & arms out	152	51	309	227	71	35	129	36
Lift—shoulder high & arms flexed	152	38	119	529	222	20	240	84
Lift—shoulder high & arms close	152	25	309	538	156	35	285	102
Lift—floor level-close (squat)	15	25	309	890	245	35	547	182
Lift—floor level-out (stoop)	15	38	170	320	125	20	200	71
Push down—waist level	118	38	309	432	93	35	325	71
Pull down—above shoulders	178	33	309	605	102	35	449	107
Pull in—shoulder level-arms out	157	33	309	311	80	35	244	53
Pull in—shoulder level-arms in	140	0	205	253	62	52	209	62
Push-out—waist level-stand erect	101	35	54	311	195	27	226	76
Push-out—chest level-stand erect	124	25	309	303	76	35	214	49
Push-out—shoulder level-lean forward	140	64	205	418	178	52	276	120

[a] Handle locations are measured in mid-sagittal plane, vertical from floor and horizontal from mid-point between ankles.
[b] These three tests are defined in Figure 4.11.
Note: 4.45 N = 1 lbf or 9.81 N = 1 kgf = 1 Kp.

TABLE 4.6
Psychophysical Limits for Load Lifting (Ayoub et al., 1980)[a]

Height of Lift (cm)	Sagittal Plane Box Dimensions (cm)	Male Mean (N)	Female Mean (N)
Floor to knuckle height when erect	30.5	296	194
	45.7	261	171
	61.0	236	152
Knuckle to shoulder height when erect	30.5	263	141
	45.7	233	129
	61.0	205	127
Shoulder to reach height when erect	30.5	221	120
	45.7	204	110
	61.0	195	112

[a] The values represent acceptable lifting limits (N) based on lifting frequency of once per minute sustained for eight hours.

presented in the second part of this text. It should also be apparent that individual attributes (e.g., gender, age, stature, and body weight) can effect a person's strength capability. A discussion of these variables follows.

4.3.4 Personal Factors Effecting Strength

It should be evident from inspection of the preceding normal values that the strengths of the healthy adult population vary greatly, with the strongest being 6 to 8 times stronger than the weakest. Many personal factors effect strength performance. Discussion of some of the more important factors follows.

Gender. Of the many factors that could be listed, gender accounts for the largest difference in mean strengths for any easily identified population strata. Women, on average, are weaker than men. The traditional difference often quoted is that a women's mean strength is approximately two-thirds that of a man's (Roebuck, Kroemer, and Thomson, 1975). What should be realized, however, is that this is an average of many different muscle strengths. Laubach (Webb Associates, 1978) has compared various strengths between the two sexes, and concluded that some of the female strength values average as low as 35% of male strengths, while others average closer to 85%. Figure 4.10 depicts these results for four different types of static strengths as well as the psychophysically measured dynamic strengths quoted earlier.

In general, female average strengths compare more favorably with male strengths for lower extremity static efforts and various dynamic lifting, pushing, and pulling activities. Muscular exertions involving flexion, abduction, and rotation of the arm about the shoulder appear to be particularly difficult

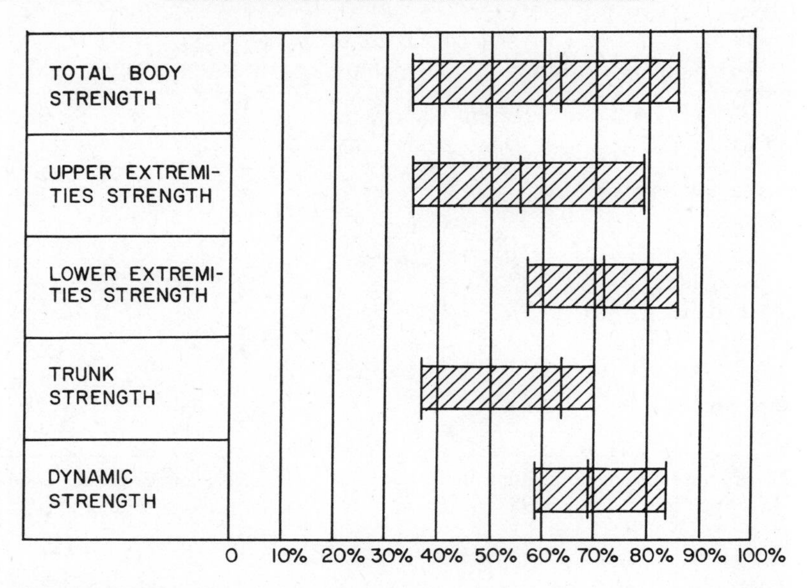

FIGURE 4.10 The range and average mean percentage differences in muscle strength characteristics between women and men (from Laubach, in Webb Associates, 1978).

for a woman relative to a man, possibly because of the smaller muscle moment arms provided by the shoulder/thoracic skeletal frame of the average woman (Stobbe, 1982).

In considering gender effects, however, it is most important to realize that a great deal of the variation in population strengths are *not* explained by gender alone. This is obvious from the static lifting strength distributions shown in Figure 4.11 from Chaffin et al. (1978). There is a large overlap between the strength distributions for men and women, with many women being stronger than many men.

Age. Several studies of strength in different age groups have been conducted for many years (Asmussen and Heeboll-Nielsen, 1962; Cathcart, 1927; Chaffin et al., 1977, to name a few). In general, muscle strength appears to be greatest in the late 20's and early 30's, with a general decline thereafter. The average population strength at age 40 appears to be approximately 5% less, and at age 60, 20% less than in the late 20's (Roebuck, Kroemer, and Thomson, 1975; Hertzberg, 1972).

Anthropometry. Body weight and stature have been studied to determine their influence on strength. From one study it appears that both total body weight and lean body mass (body mass corrected for fat) of 77 young men correlated with muscle static strengths, resulting in r values of approximately 0.3 to 0.5 (Laubach and McConville, 1969). Stature was not as well correlated in this same study, an observation later confirmed in women by Nordgren, 1972.

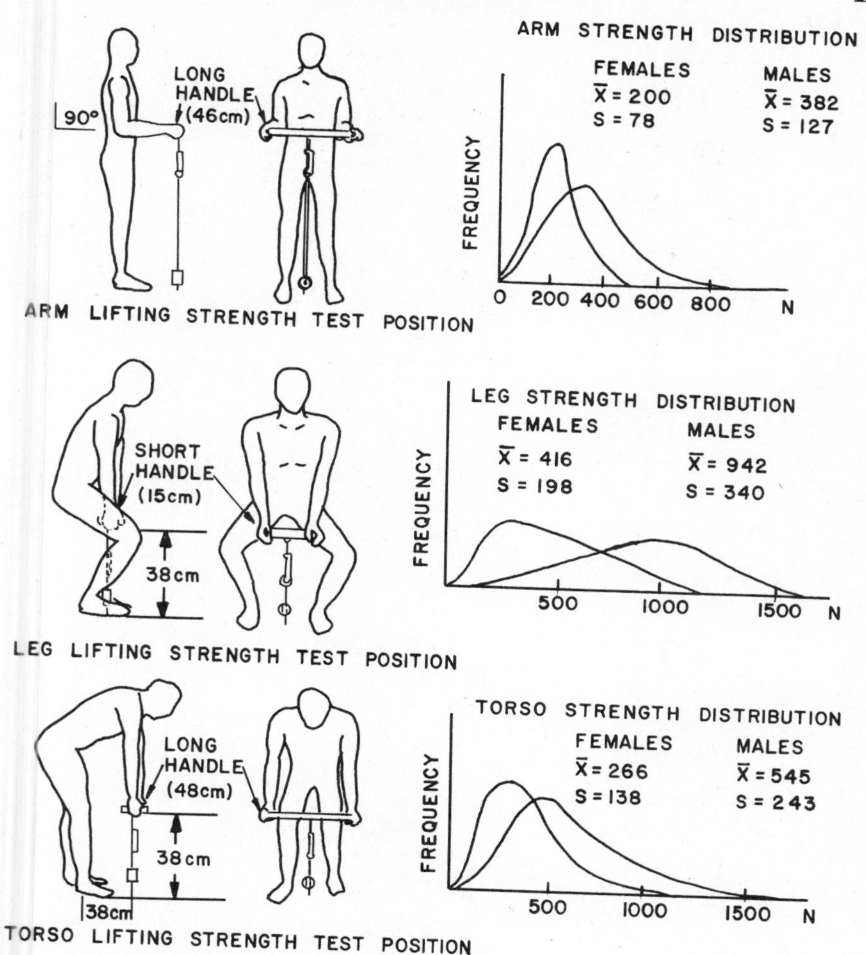

FIGURE 4.11 Static strength test postures and results for 443 male and 108 female workers employed in four different industries (Chaffin, Herrin, and Keyserling, 1978).

Herrin (in Chaffin et al., 1977) used a multiple correlation approach and showed that the three static strengths depicted in Figure 4.11 could be predicted by combined consideration of gender, age, body weight, and stature. Table 4.7 provides the mean values obtained by such an approach. Unfortunately, only about 36% of the total variance in the strength values for the worker population was explained by the combined consideration of all four variables.

Strength Intercorrelations. Anthropometric variables—for example, body segment length and circumference, and somatotype ratings—have been combined with localized muscle strengths—for example, shoulder and arm

TABLE 4.7

Average Static Lifting Strength Predictions (N) for Tests Depicted in Figure 4.11 with Worker Population Stratified by Age, Gender, Body Weight, and Stature (from Chaffin, Herrin, Keyserling and Foulke, 1977)

Age (yr)	Stature (cm)	Torso Lift Strengths				Arm Lift Strengths				Leg Lift Strengths			
		Male Body Weight		Female Body Weight		Male Body Weight		Female Body Weight		Male Body Weight		Female Body Weight	
		445 N	890 N	445 N	890 N	445 N	890 N	445 N	890 N	445 N	890 N	445 N	890 N
20 yr	152 cm	335	573	224	351	323	393	183	249	497	1010	367	586
	183 cm	394	691 (max)	283	468	342	430 (max)	197	286	834	1098 (max)	410	674
50 yr	152 cm	251	405	140 (min)	183	287	322	144 (min)	178	675	780	252 (min)	357
	183 cm	310	523	198	300	306	359	162	215	696	822	273	399

Note: 4.45 N = 1 lbf or 9.81 N = 1 kgf = 1 kp.

strengths—to predict static lifting strengths (Mital and Ayoub, 1980). The variance explained by using combinations of up to nine anthropometric and strength variables was about 62% of the static lifting strengths of 73 male and 73 female industrial workers. A similar correlation was obtained by Herrin using the three static lifting tests illustrated in Figure 4.11 to predict lifting strengths in various positions required in the jobs performed by the 594 men and women workers tested. Stobbe (1982) studied the intercorrelations between various static strengths exhibited by 30 men and 30 women employed in manual jobs in three different industries. He found that by using seven whole-body standardized strength tests he could account for approximately 84% of the variance measured in the 16 different localized muscle strengths listed in Table 4.4. These localized muscle strengths were chosen to represent major muscle moment strengths of the body. Mital and Ayoub (1980) also reported that by combining anthropometric and static strength measures, approximately 90% of the variance in dynamic lifting capability (as determined by psychophysical experiments) could be explained.

In summary, it appears that anthropometric descriptors alone are not well enough correlated with strength to be of practical value. Thus, if one believes it is necessary for a person to possess a certain amount of strength to perform a given task safely, a measurement of that person's size, shape, or weight alone will not suffice. More recent studies of strength intercorrelations, however, raise the possibility that the strength required to safely perform a job can be measured in a number of different ways with good predictability. A discussion of the use of such testing as part of pre-employment medical examinations is given in Chapter 13.

4.4 SUMMARY AND LIMITATIONS OF MECHANICAL WORK CAPACITY DATA

New methods of measuring both joint mobility and strength have been developed in the last two decades. These methods have provided accurate and repeatable data. Furthermore, the methods can be used by trained technicians following published procedures.

The population data resulting from the use of these new methods are still rather limited, however. For this reason, the concept of normative joint motion and strength data must be carefully applied, especially in industries where manual exertion is sometimes performed in extremely awkward postures and worker physical attributes vary greatly. The extrapolation of existing data to older workers, women, and physically limited individuals for the purpose of job design and personnel job placement is not simple. Measuring the mobility or strength at one joint only as a predictor of a person's general capability to reach, lift, or move objects is not valid. Because mobility and strength are so individual, multiple measurements will be necessary under most situations to accurately characterize a person's mechanical work capacity. The single best set of measurements for this characterization is not known, and in fact may not exist. What is clear, however, is that the definition of one's mechanical work capacity is highly dependent upon the task to be performed. Thus, any set of tests to define the mechanical work capacity of an individual will also depend upon the specific manual attributes of a task expected to be performed by that person.

Lastly, it must be realized that joint motions and strengths are not fixed human attributes, though clearly genetic endowment is important. Unfortunately, adequate longitudinal studies have not been done to define the extent of modification in motion and strength that can be attributed to specific types of exercises for adults. It is not unreasonable, however, for stretching exercises to increase joint motion by as much as 5 to 10%, and for isometric strength training exercises to increase adult strength by 30 to 40% in a period of four to six months. Whether these induced changes are prophylactic when a person performs heavy physical work is not yet proven by epidemiological data, though anecdotal evidence certainly supports the possibility.

From a practical standpoint, the preceding suggests that the occupational biomechanist will need to be involved in sophisticated evaluations of both the biomechanical requirements of jobs and the mechanical work capacity of workers. Simple measurements of an individual's capacity to perform manual exertions will be limited to situations in which well-documented job strength and reach demands are known. Other than in such cases, multiple joint mobility and standardized strength tests will be necessary to define a person's mechanical work capacity for a prospective job. Finally, it must be conceded that well-motivated individuals can volitionally improve their performance on such tests through specific training programs. Thus, retesting will be necessary if such a training effort on the part of an individual is evident.

REFERENCES

Adrian, M. J., "An Introduction to Electrogoniometry," *Kinesiological Review*, Phys. Ed. Div., Amer. Assoc. for Health, Phys. Ed. and Recreation, Wash., D.C., 1968, pp. 12–18.

American Academy of Orthopaedic Surgeons, *Joint Motion—Method of Measuring and Recording*, Amer. Acad. of Ortho. Surg., Chicago, 1965.

Asmussen, E., O. Hansen, and O. Lammert, "The Relation Between Isometric and Dynamic Muscle Strength in Man," *Communications of the Testing and Observation Institute*, No. 20, Danish Nat'l Assoc. for Infantile Paralysis, Hellerup, Denmark, 1965.

Asmussen, E. and K. Heeboll-Nielsen, "Isometric Muscle Strength in Relation to Age in Men and Women," *Ergonomics*, **5**(1), 167–169 (1962).

Ayoub, M. M., "Human Movement Recording for Biochemical Analysis," *Int. J. Prod. Res.* **10**(a), 35–51 (1972).

Ayoub, M. M., A. Mital, G. M. Bakken, S. S. Asfour, and N. J. Bethea, "Development of Strength and Capacity Norms for Manual Materials Handling . . . ," *Human Factors*, **22**(3), 271–283 (1980).

Barter, J. T., I. Emanuel, and B. Truett, *A Statistical Evaluation of Joint Range Data,* WADC-TR-57-311, Wright-Patterson Air Force Base, Ohio, 1957.

Berger, R. A. and R. B. Higgenbotham, "Prediction of Dynamic Strength from Static Strength in Hip and Knee Extension," *Amer. Corrective Therapy J.*, **24,** 118–120 (1970).

Burggraaf, J. D., *An Isometric Biochemical Model for Sagittal Plane Leg Extension,* unpublished Ms. thesis, Center for Ergonomics, University of Michigan, Ann Arbor, Michigan, 1972.

Caldwell, L. S., D. B. Chaffin, F. N. Dukes Du Bobos, K. H. E. Kroemer, L. L. Laubach, S. H. Snook, and D. E. Wasserman, "A Proposed Standard Procedure for Static Muscle Strength Testing," *Amer. Ind. Hyg. J.*, **35,** 201–206 (1974).

Cathcart, E. P., "The Physique of Women in Industry," *Proc. Roy. Inst. Gr. Brit.*, **25,** 185–199 (1927).

Carlson, B. R., "Relationship Between Isometric and Isotonic Strength," *Arch. Phys. Med. Rehabil.*, **51,** 176–179 (1970).

Chaffin, D. B., "Ergonomics Guide for the Assessment of Human Strength," *Amer. Ind. Hyg. J.*, **36,** 505–510 (1975).

Chaffin, D. B., G. D. Herrin, W. M. Keyserling, and J. A. Foulke, *Preemployment Strength Testing,* Technical Report No. 77-163 of National Institute for Occupational Safety and Health, Cincinnati, 1977.

Chaffin, D. B., G. D. Herrin, and W. M. Keyserling, "Preemployment Strength Testing," *J. Occup. Med.*, **20**(6), 403–408 (1978).

Clarke, H. H., *Muscular Strength and Endurance in Man,* Prentice-Hall, Englewood Cliffs, 1966.

Elkins, E. C., U. M. Leden, and K. G. Wakim, "Objective Recording of the Strength of Normal Muscles," *Arch. Phys. Med.*, **32,** 639–647 (1951).

Emmanuel, I., J. W. Chaffee, and J. Wing, *A Study of Human Weight Lifting Capabilities for Loading Ammunition . . .*, Tech. Report, WADC-TR-56-367, Wright-Patterson Air Force Base, 1956.

Garg, A., A. Mital, and S. S. Asfour, "A Comparison of Isometric Strength and Dynamic Lifting Capability," *Ergonomics,* 23, 13–27 (1980).

Gilliland, A. R., "Norms for Amplitude of Voluntary Movement," *J. Amer. Med. Assoc.*, **77,** 1357–1359 (1921).

Glanville, A. D. and G. Kreezer, "The Maximum Amplitude and Velocity of Joint Movement in Normal Male Adults," *Human Biol.*, **9**(2), 197–211 (1937).

Harris, M. L. and C. W. Harris, "A Factor Analytical Study of Flexibility," Paper presented at National Convention, Amer. Assoc. Health, Phys. Ed. and Recreation, St. Louis, 1968.

Hertzberg, H. T. E., "Engineering Anthropology," in H. P. Van Cott and R. G. Kinkade, Eds., *Human Engineering Guide to Equipment Design* (revised edition), McGraw-Hill, New York, 1972.

Hettinger, T., *Physiology of Strength,* Thomas, Springfield, 1961, pp. 13.

Hunsicker, P. A. and G. Greey, "Studies in Human Strength," *Research Quarterly,* **28,** 109–122 (1957).

Keyserling, W. M., G. D. Herrin, D. B. Chaffin, T. J. Armstrong, and M. L. Foss, "Establishing an Industrial Strength Testing Program," *Amer. Ind. Hyg. J.*, **41,** 730–736 (1980).

Konz, S. and C. Desai, "Lifting Forces for Nine Lifting Heights," *Aggressologie,* **17,** 55–59 (1976).

Kroemer, K. H. E., *Human Force Capabilities for Operating Aircraft Controls at 1, 3 and 5 G's,* Tech. Report, AMRL-TR-73-54, Wright-Patterson Air Force Base, Ohio, 1975.

Kroemer, K. H. E., "Human Strength: Terminology, Measurement and Interpretation of Data," *Human Factors,* **12**(3), 297–313 (1970).

Kroemer, K. H. E. and J. M. Howard, "Toward Standardization of Muscle Strength Testing," *J. Med. and Science in Sports,* **2**(4), 224–230 (1970).

Kroemer, K. H. E. and W. S. Marras, "Evaluation of Maximal and Submaximal Static Muscle Exertions," *Human Factors* **23**(6), 643–653 (1981).

Laubach, L. L., "Body Composition in Relation to Muscle Strength and Range of Motion," *J. Sports Med. and Physical Fitness,* **9**(5), 89–97 (1969).

Laubach, L. L. and J. T. McConville, "The Relationship of Strength to Body Size and Typology," *Med. and Science in Sports,* **1,** 189–194 (1969).

Leighton, J. R., "An Instrument and Technique for Measurement of Range of Joint Mobility," *Arch. Phys. Med. Rehabil.,* **36,** 571–78 (1955).

Leighton, J. R., "Flexibility Characteristics of Four Specialized Skill Groups of College Athletes," *Arch. Phys. Med. Rehabil.,* **38,** 24–28 (1957).

Miller, D. I. and R. C. Nelson, *Biomechanics of Sport,* Lea and Febiger, Philadelphia, 1976, pp. 28–35, 57.

Mital, A. and M. M. Ayoub, "Modeling of Isometric Strength and Lifting Capacity," *Human Factors,* **22**(3), 285–290 (1980).

Nordgren, B., "Anthropometric Measures and Muscle Strength in Young Women," *Scand. J. Rehab. Med.,* **4,** 165–169 (1972).

Park, K. S. and D. B. Chaffin, "A Biochemical Evaluation of Two Methods of Manual Load Lifting," *AIIE Trans,* **6**(2), 105–113 (1974).

Pytel, J. L. and E. Kamon, "Dynamic Strength as a Predictor for Maximal and Acceptable Lifting," *Ergonomics,* **24**(9), 663–672 (1981).

Ricci, B., *Physiological Basis of Human Performance,* Lea & Febiger, Philadelphia, 1967, pp. 70–73.

Roebuck, J. A., "A System of Notation and Measurement for Space Suit Mobility Evaluation," *Human Factors,* **10**(1), 70–94 (1968).

Roebuck, J. A., K. H. E. Kroemer and W. G. Thomson, *Engineering Anthropometry Methods,* Wiley-Interscience, New York, 1975, pp. 77–128.

Rohmert, W., *Maximal Forces of Men Within the Reach Envelope of the Arms and Legs* (in German) Research Report No. 1616, State of Northrhine–Westfalia, Westdeutscher Verlag Koeln-Opladen, 1966.

Salter, N., and H. D. Darcus, "The Amplitude of Forearm and of Humeral Rotation," *J. of Anatomy,* **87,** 407–418 (1953).

Schanne, F. T., *Three Dimensional Hand Force Capability Model for a Seated Person,* unpublished Ph.D. dissertation, University of Michigan, Ann Arbor, Michigan, 1972.

Snelkinoff, E., and M. Grigorowitsch, "The Movement of Joints as a Secondary Sex and Constitutional Characteristic," Zeitschrift fur Konstitutionslehre, **15**(6), 679–693 (1931).

Snook, S. H., "The Design of Manual Handling Tasks," *Ergonomics,* **21**(12), 963–985 (1978).

Snook, S. H. and C. H. Irvine, "Maximal Acceptable Weight of Lift," *Amer. Ind. Hyg. Assoc. J.,* **28,** 322–329 (1967).

Snook, S. H., C. H. Irvine, and S. F. Bass, "Maximal Weights and Work Loads Acceptable to Male Industrial Workers," *Amer. Ind. Hyg. Assoc. J.,* **31,**79–586 (1970).

Stothart, J. P., *A Biochemical Analysis of Static and Dynamic Muscular Contraction,* unpublished Ph.D. dissertation, Pennsylvania State University, University Park, Pennsylvania, 1970.

Stobbe, T. J., *The Development of a Practical Strength Testing Program in Industry,* unpublished Ph.D. dissertation, University of Michigan, Ann Arobr, Michigan, 1982.

Switzer, S. A., *Weight Lifting Capabilities of a Selected Sample of Human Males,* Tech. Report, WADC-TR-62-57, Wright-Patterson Air Force Base, Ohio, 1962.

Webb Associates, *Anthropometric Source Book,* Vol I, NASA Ref. 1024, Nat'l Aero. Space Admin., Chapters VI & VII, 1978.

5

BIOINSTRUMENTATION FOR OCCUPATIONAL BIOMECHANICS

5.1 INTRODUCTION AND MEASUREMENT SYSTEM CRITERIA

It should be clear from the preceding chapters that occupational biomechanics, both in research and application, is highly empirical. Furthermore, data acquisition in this field requires complex measurement systems. Since the required measurements are of human functions—that is, biomechanical parameters are to be estimated—the measurement systems should meet certain biomedical criteria (Chaffin, 1982; Brand and Crowninshield, 1981). Some general criteria follow:

1. The measurements should accurately estimate a specific, definable, human motor function—that is, they should provide well-correlated and unbiased estimates of the function of interest.
2. The data should be repeatable under prescribed conditions.
3. The measurements should provide estimates of the limits of function—that is, distinguish between normal and abnormal.
4. The measurement system should not alter the function being estimated.
5. The measurement system should be safe to use.
6. The measurement system should be practical—that is, easy to setup and use, insensitive to outside influences, and inexpensive.

What follows is a description of several different methods of acquiring biomechanical data that are relevant to the evaluation of human kinematics and kinetics in occupational activities. The presentation is in four parts. First is a description of the various methods available to estimate human motion parameters. This is followed by a discussion of the use of electromyography in industry. Next is a description of volitional force measurement systems. The measurement of abdominal and intradiscal pressure is also presented, followed by a brief discussion of a force platform system.

5.2 HUMAN MOTION ANALYSIS SYSTEMS

5.2.1 Basis for Measurement of Human Motion

Biomechanical studies must begin with a kinematic description of a person's posture and movements. In a static analysis, the joint center locations in space must be determined, and with movement, the velocity and acceleration vectors of each body segment center of mass must be estimated. Providing such data can be relatively simple or quite complicated, depending on the evaluation to be made. If a static, planar evaluation of an activity is to be performed, then the use of goniometers strapped to the person at various joints (described in Section 4.2.1) can be quite sufficient to quantify postures of interest. As an alternative, reflective markers can be placed over estimated joint centers and a photograph taken and combined with anthropometric data on the subject.

In contrast to these rather simple methods, three-dimensional static analyses and two- or three-dimensional dynamic analyses require significantly greater measurement sophistication. In a three-dimensional analysis, all joint and mass-center locations must be resolved with respect to a single reference axis system. If motion is involved, then the movement of one body segment relative to all others must be considered, or the motion of each segment relative to an absolute (stationary) coordinate system must be known.

This latter requirement is illustrated by considering a person lifting a weight from about floor level to a shelf. A five-link kinematic model can be used for such an analysis, as developed by Fisher (1967) and Muth et al. (1978). The kinematic linkage is described in Figure 5.1. It assumes a conventional axis system (X horizontal, Y vertical) with the origin stationary at the ankle joint.

The instantaneous linear acceleration components ax and ay for each joint center j can be estimated by a sequential kinematic analysis, beginning at the stationary ankle joint $j = 1$:

$$ax_1(t) = 0 \text{ (ankle stationary)}$$

$$ay_1(t) = 0 \tag{5.1}$$

$$ax_{j+1}(t) = -L_i[\dot{\theta}_j^2(t) \cos \theta_j(t) + \ddot{\theta}_j(t) \sin \theta_j(t)] + ax_j(t)$$

$$ay_{j+1}(t) = -L_i[\dot{\theta}_j^2(t) \sin \theta_j(t) - \ddot{\theta}_j(t) \cos \theta_j(t)] + ay_j(t) \tag{5.2}$$

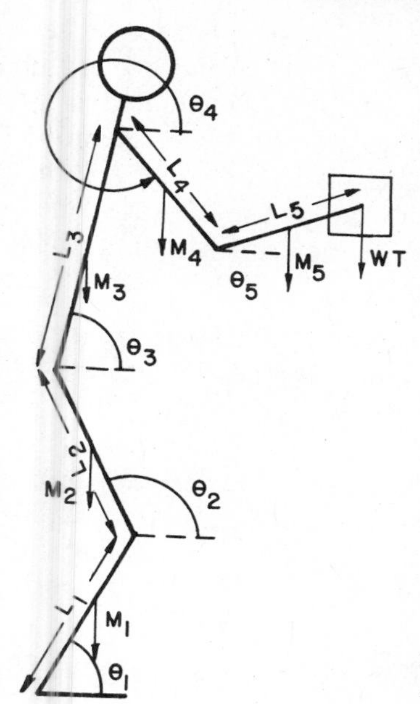

FIGURE 5.1 Five link model of worker lifting (Muth, Ayoub, and Gruver, 1978).

where

- L_i are link lengths
- $\theta_j(t)$ is the angular position of joint j relative to a stationary Cartesian axis system
- $\dot{\theta}_j(t)$ is the angular velocity of i segments at joint j
- $\ddot{\theta}_j(t)$ is the angular acceleration of i segments at joint j

These same equations of motion are used to estimate the instantaneous linear accelerations of the body segment centers of mass by substituting the distance from the joint j to the mass center of segment i for the values of L_i in the equation.

Thus, by measuring angular changes at joints it is possible to generate estimates of necessary linear acceleration components. These linear acceleration values are then used to compute the inertial forces and moments resulting from the motion, as described in detail in Chapter 6 on biomechanical models. Such joint angular data can be obtained by methods to be described. It should be mentioned first, however, that instantaneous linear and angular accelerations can be measured directly with the use of accelerometers strapped to the body segments. This method is now presented.

Accelerometer Measurement of Body Motion. Small, lightweight accelerometers are commercially available to measure the linear acceleration of an object. They measure the deformation (strain) in a small beam or electro-active crystal caused by the inertial force of a given mass attached to the beam or crystal. By carefully locating six linear accelerometers in orthogonal orientations, both linear and angular accelerations of a body segment can be measured directly. Numerical integration of the acceleration data is then performed to estimate velocity and displacement values. If angular data only are required, five accelerometers can be used, as described by Morris (1973). Accelerometers are also used to measure vibration; they will be described in Chapter 12.

The advantages and disadvantages of the use of accelerometer systems have been summarized by Chao (1978). In essence, the advantages are as follows:

1. The method provides a direct means of measuring the desired parameter—acceleration—and thus avoids errors induced by numerical differentiation required when using displacement data. This is particularly important in the analysis of fast motions—for example, in sports and impact studies.
2. Analog data are produced, allowing direct and convenient handling and filtering of the data.
3. Total instrumentation cost is reasonable.
4. The measurements can be obtained without being able to see the body segments during complex motions.

Though these advantages are impressive, the disadvantages far outweigh them for occupational biomechanics investigations. They are as follows:

1. The use of six or more accelerometers attached at each of several joints induces a complexity in the analysis that requires a highly trained biomechanics specialist to assure that the raw data are properly obtained and analyzed.
2. Instrument signal drift or small noise levels can add significantly to the errors in the displacement estimates so important in a multiple-linkage analysis.
3. The attachment of accelerometers and associated wires and preamplifiers can interfere with motions.
4. Establishing an initial set of conditions for each accelerometer is difficult because of their different orientations with respect to gravity.

It must be concluded that direct acceleration measurements are probably best reserved for simple (two- or three-link) analysis of high-speed motions studied in a laboratory or clinical environment. Thus, alternative methods have been developed for studies of manual activities in industry. Two common methods are now described.

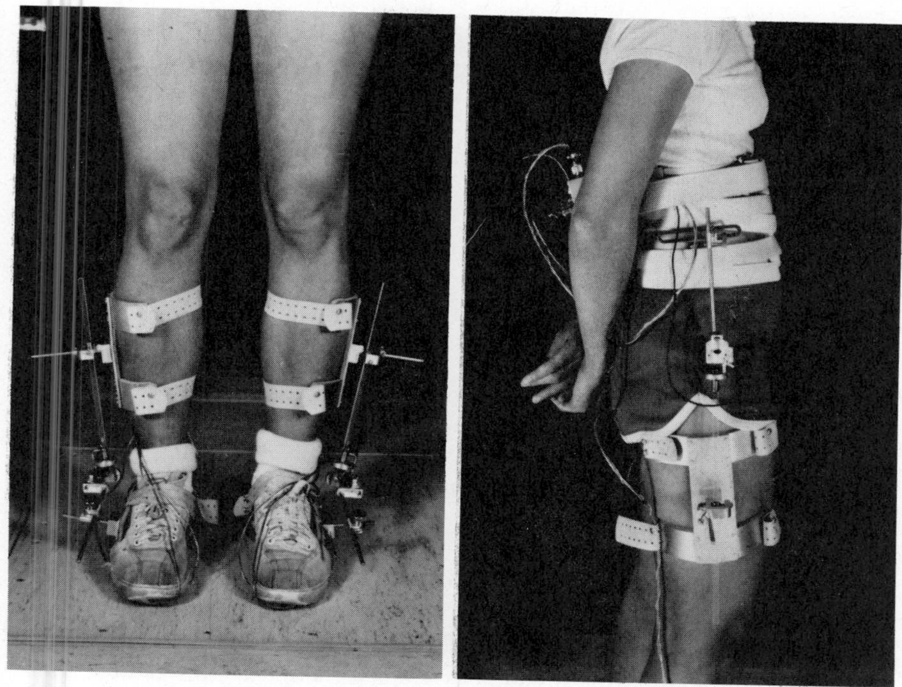

FIGURE 5.2 Electromechanical goniometer system used by Moss Rehabilitation Hospital (photo courtesy of E. Chao).

Goniometric Methods. The analysis of motion at a simple hinge joint (e.g., elbow or knee) has been accomplished for many years by attaching goniometers over the joints, as described in Chapter 4 (Figure 4.2). When a rotary electrical transducer or potentiometer is attached to the arms of the goniometer, the device is referred to as an electrogoniometer (Elgon). Goniometers and Elgons have provided the means to estimate joint range of motions in many studies. As stated earlier in this chapter, however, it is necessary when performing a biomechanical analysis of a task to measure multiple joint angles and often motions during a prescribed activity. Furthermore, because joints move about multiple axes, complex measurement systems are necessary. Such systems need to measure the joint rotations about three axes and be self-aligning—that is, translate as the joint center moves.

Goniometer devices to measure such motions have been developed. For multiple-linkage motion studies they require that an exterior, lightweight skeleton be attached to the person's body. Figure 5.2 displays one such system. By allowing the attachment arms of the goniometer (which are strapped to the person) to shorten and lengthen to compensate for small joint-center translation, the angular measurement devices at the joints of the exoskeleton are kept aligned over the joint centers. By using three orthogonally positioned rotary transducers at each joint, motion about specific axes are recorded directly.

The advantages and disadvantages in using a goniometric approach have also been summarized by Chao (1978). The main advantages are as follows:

1. It is easy to use and is relatively inexpensive.
2. It provides direct measurement of body positions that are easily understandable.
3. It avoids the requirement to be able to see a segment's position during complex motions.
4. It results in repeatable measures of joint angles.

The disadvantages in use of a goniometric system, however, are considerable:

1. Positioning of the exoskeletal joints or goniometers over human joints is difficult and requires care to minimize alignment errors.
2. Interaction (cross-talk) of the signals obtained from the three transducers at a joint are inevitable, though a theoretical procedure, along with a self-aligning linkage system can minimize the errors (Chao, 1980).
3. Soft-tissue motion can induce errors by shifting the exoskeleton relative to the human skeleton.
4. Abnormal joint geometries and motions confound alignment of the transducer axes with the joint centers.
5. The resulting joint-angle data are transformed into an inertial (gravity-dependent) reference axis system for biomechanical analyses.
6. The joint-angle displacement data must be numerically differentiated to provide velocity and acceleration data, which can be in error because of noise in the displacement signals.
7. The attachment of an exoskeleton can alter normal motion patterns.

The use of an exoskeleton-goniometer system for human motion studies has been valuable in clinical settings—for example, in comparing the effect of a specific treatment on human movement patterns involving two or three links (body segments). It is not clear, however, that such a method can be used to study whole-body motions typical of industrial tasks, especially in a plant environment, though the potential exists. With the advent of improved transducers and telemetry systems, such measurements will undoubtedly become more useful in occupational biomechanics in the future.

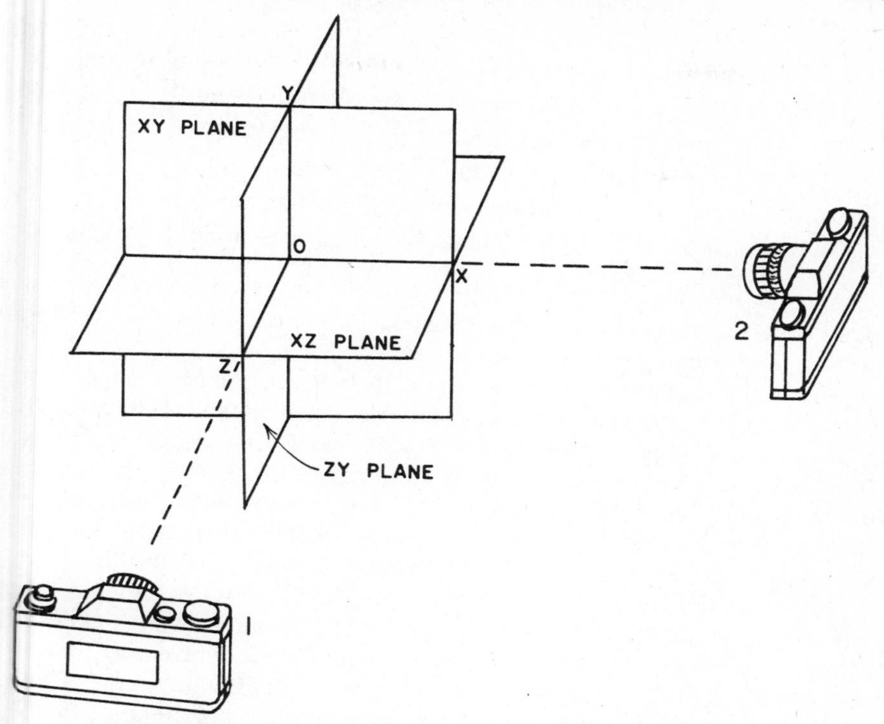

FIGURE 5.3 Three-dimensional reference system (Martin and Pongratz, 1974).

Photogrammetric Methods. These methods are the ones most often used in occupational biomechanics. One camera can be used when a motion is performed in one plane, while for three-dimensional analysis two cameras are used. Figure 5.3 depicts an orthogonal arrangement for two cameras. When the cameras are positioned with perpendicular optical axes, accurate three-dimensional coordinates of a point in space can be obtained by application of the following formula (Martin and Pongratz, 1974):

$$x = \frac{XYx\left(1 - \frac{ZYz}{Dz}\right)}{1 - \frac{(XYx)(ZYz)}{(Dz)(Dx)}} \tag{5.3}$$

$$y = ZYy\left(1 - \frac{x}{Dx}\right) \tag{5.4}$$

$$z = ZYz\left(1 - \frac{x}{Dx}\right) \tag{5.5}$$

where

x,y,z are the actual values of coordinates of a given point in space
XYx is the x coordinate measured in the XY film plane
ZYy is the y coordinate measured in the ZY film plane
ZYz is the z coordinate measured in the ZY film plane
Dx is the distance from film plane 2 to the origin along the x axis
Dy is the distance from film plane 1 to the origin along the y axis

A non-perpendicular photographic system, using cameras in the same plane but separated by a given distance has been advocated by Ayoub, Ayoub, and Ramsey (1970).

The above techniques assume optically perfect cameras. Since optical distortion exists, various techniques have been devised to reduce errors from those sources. One method involves the use of a third camera in an orthogonal arrangement (along the y axis in Figure 5.3). This provides a means of checking the estimates, by using the film data from the cameras in pairs—that is, cameras 1 and 2, 2 and 3, 1 and 3—in the equations to yield three separate estimates of the x, y, and z coordinates. This overdetermined system is then resolved by using the mean of the three estimates. Such a system was used by Chaffin, Schutz and Snyder, (1972), with an estimated mean error of approximately 1.0 cm. A second technique for correcting for optical distortion has been suggested by Andriacchi et al. (1979). A grid of known calibration points is used to define a "correction" polynomial, fit to the points. Once the coefficients in the polynomial are determined for the calibration grid, the polynomial is used to "correct" the estimates obtained in equations 5.3, 5.4, and 5.5. A cluster of 10, 19, and 29 calibration points was used to test this technique, and resulted in an average resolution of 0.57 cm for 10 points, 0.56 cm for 19 points, and 0.45 cm for 29 points, which the authors believe would be sufficient for human motion analysis. Further discussion of distortion correction in photogrammetric methods is given by Miller and Nelson (1976).

Another requirement of a photogrammetric system is its ability to accurately locate a moving target. This can be accomplished by using a strobe light set close to the optical axes of the cameras. Reflective markers are taped either directly over the joint centers on the skin or on tight-fitting clothing. The result is a set of discrete points on a film frame that, when joined, form a linkage movement diagram, as illustrated in Figure 5.4 for a lifting activity. The strobe rate depends on the velocity of motion. In high-velocity motions of the arm, such as when throwing a ball, a strobe rate of 80–100 flashes/sec may be required to have enough data to accurately estimate peak accelerations. In handling heavy loads wherein slower movements occur, too fast a strobe rate causes the reference points to overlap. Thus, an analysis of lifting may only require a strobe rate of 20–40 flashes/sec. In general, the rule is to use the highest strobe rate that allows points to be easily distinguished during a specific motion.

SAMPLING RATE: 20 Hz

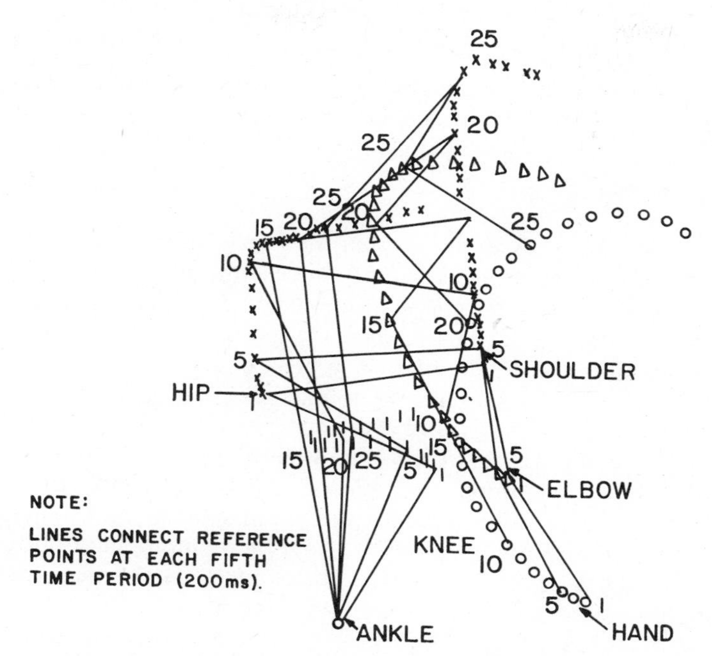

FIGURE 5.4 Typical sagittal plane motion diagram of lifting heavy load from floor to waist high (76 cm) table (Freivalds, Chaffin, and Garg, 1983).

The problem of overlapping reference points is eliminated by using motion pictures. In doing so, however, a stationary set of at least three reference points must be included in each film frame. The coordinates of each body point are then measured relative to these stationary points in each film frame analysis. This procedure avoids any dependence on the film alignment capability of the film transport mechanisms in the camera and projector.

The estimation of the angular velocity $\dot{\theta}_j$ and acceleration $\ddot{\theta}_j$ values can be accomplished by a number of methods, given accurate locations of joint centers for given time intervals. A graphical method is described by Plagenhoef (1971). Pearson (1961) and Fisher (1967) used a Taylor series expansion of the angular data at each joint. This yields the following estimates of velocity and acceleration:

$$\dot{\theta}_j(t) = \frac{\theta_j(t + \Delta t) - \theta_j(t - \Delta t)}{2t}$$

$$\ddot{\theta}_j(t) = \frac{\theta_j(t + \Delta t) + \theta_j(t - \Delta t) - 2\theta_j(t)}{\Delta t^2}$$

where

$\dot{\theta}_j(t)$ is the angular velocity of joint j at t interval of time

$\ddot{\theta}_j(t)$ is the angular acceleration of joint j at t interval of time

$\theta_j(t + \Delta t)$ is the angular displacement of joint j at t plus small Δt interval, with Δt dependent on the strobe or movie film rate

$\theta_j(t - \Delta t)$ is the angular displacement of joint j at t minus a small Δt interval

Use of a Taylor series expansion as the basis for the above equations provides some data-smoothing capability since additional terms can be added to the series. This reduces the sometimes large acceleration effects when an error in a particular coordinate value is used in the computations. Other mathematical smoothing techniques have been devised, however. Some require the statistical fitting of a polynomial function to the displacement curves using a mean least-squared error criterion. The first and second derivatives of the resulting polynomial prediction equation represent the velocity and acceleration functions (Jackson, 1979). A slightly different approach has been to fit spline functions to the displacement data (Zernicke et al., 1976; Wood and Jennings, 1979). A method employing Fourier transforms of the data has been advocated by Hatze (1981). All of these techniques require computerized analysis of the data, and all depend on the user deciding how much data smoothing is necessary. Peak acceleration will be underestimated if the smoothing is too great. Too little smoothing, on the other hand, and small errors in the displacement data, will generate large errors in the acceleration estimates. A good general rule is to measure repeated trials of any motion with little data smoothing. If the acceleration peak values occur at approximately the same times in the motions, are of the same relative magnitudes, and the accelerations are regular when plotted—that is, have very little variance from period to period—then the data are probably providing good motion estimates and can be used in subsequent biomechanical analyses.

Video Spot Locator Systems. It should be clear from the preceding that if a fast sampling rate can be obtained without the extra cost of film data coding, then a major improvement in motion analysis is possible. This has led to the development of various video systems for locating contrasting spots against a given background. Though reflective markers can be used along with a strobe light for providing point images for traditional television-based video systems, contemporary systems rely on rapidly pulsed light emitting diodes (LED's) attached to the person. The light rays from these flashing diodes are detected by a camera with a dual-axis photosensitive diode instead of a film. When a single spot of light hits the photosensitive diode, it outputs the spot's two-dimensional location to a digital computer.

Video spot locator systems that are commercially available can track up to 30 LED's flashed at more than 300 Hz. Woltring and Marsolais (1980)

have evaluated the utility of such systems and believe that with careful calibration of the cameras to reduce optical and diode signal distortion errors, the systems are quite accurate for normal motion studies, but not for sports activities.

Chao (1978) has provided a thorough critique of both photographic and videographic systems. The advantages of these methods are as follows:

1. They produce estimates of a motion in an absolute reference frame,and thus are easily used in biomechanical studies.
2. Joint centers can be accurately located by projecting the intersection of the long axes of segments when several (three or more) body reference markers are placed on each segment (rather than over an estimated joint center).
3. Body reference markers will be smaller and lower in mass than either an exoskeleton–goniometric or accelerometer-type of system; thus, interference with motion will be minimal.
4. Since multiple camera systems allow the determination of motions in three dimensions over time, energy expenditure can easily be determined for diverse activities.
5. The resulting data directly produce stick diagrams of postures and motions that are important in describing specific motor behaviors.

The limitations to such systems can also be summarized. This is perhaps best done separately for each system, as suggested by Chao (1978). First, the disadvantages of the *photogrammetric* systems:

1. The film data must be manually decoded, which can be quite time consuming and prone to error.
2. Use of two cameras requires additional care in maintaining alignments and additional time in film-data reduction.
3. If still cameras are used, reference points cluster together if the sampling rate is too low, causing difficulty in maintaining synchronization of the reference points.
4. Reference points can be obscured from the camera by other body segments.
5. Additional calculations that may produce cumulative errors are required to estimate relative motions.

The use of *videographic* (T.V. and optoelectronic) systems also has disadvantages that must be recognized:

1. To acquire high resolution and fast sampling with multiple cameras, complex and expensive ($25,000 minimum) electronic equipment is necessary.

2. Lighting conditions will need to be controlled (dim lights with no point light sources in background.
3. Large amounts of data that need to be stored in a digital computer are quickly acquired, usually requiring direct data transfer when high sampling rates are used.
4. It uses a relatively new technology requiring advanced expertise in computers and electronics.

In summary, various motion measurement systems that are useful when performing biomechanical studies have been developed. Several different criteria have been presented to assist in the selection of a system to meet the data needs of a particular biomechanical analysis. It should be recognized that new measurement systems using completely different techniques are technically feasible. This discussion has briefly described a few commonly used systems. It can be expected that our fascination with human motion in sports, dance, and work, coupled with new sensing and data analysis systems will lead to major improvements in existing measurement systems. This should be anticipated when contemplating the development of a human kinematic study in the future.

5.3 APPLIED ELECTROMYOGRAPHY

5.3.1 Theory of the Use of Electromyography in Occupational Biomechanics

All living cells are surrounded by a membrane. Because the membrane is selectively permeable to ions and also actively transports ions, a *resting membrane potential* exists. This potential is typically on the order of 70–90 millivolts (mV) for an axon or muscle fiber, the outside being positive relative to the inside. A depolarization of a cell to a threshold value initiates an *action potential* as the membrane permeability changes. Having been generated in a motor nerve neuron, the action potential propagates along the cell's axon down to the so-called motor end plate (see Chapter 2). The impulse is then transmitted by chemical transmitters to the muscle fibers. An action potential propagating down the motor neuron normally activates all its branches, and these in turn activate all the muscle fibers of the motor unit. When the muscle fiber membrane is depolarized, the depolarization propagates in both directions along the fiber, causing a wave of contraction that produces a brief twitch and subsequent relaxation. During the twitch, a small electrical potential is generated. Since all muscle fibers of a motor unit do not contract simultaneously, a complex *motor unit potential* results from superimposition of several muscle fiber action potentials. By placing an electrode either within the muscle-indwelling or intramuscular electrode, or on the skin-surface electrode, the potential can be recorded. Such a meas-

TABLE 5.1
Some Common Factors Affecting Ratio of Myoelectric Activity to Muscle Force Levels

Measurement Factors	*General Effect of Ratio (EMG/Load)*
Large electrode	Increase
Close electrode proximity to muscle	Increase
High Electrode impedance	Decrease
Surface electrode (vs. indwelling)	Varied
Bipolar electrode (vs. monopolar)	Varied
Bipolar electrode spacing increase	Varied
Physiological Factors	
Prolonged contraction (fatigue)	Increase
High muscle temperature	Decrease
Specific muscle tested	Varied
Highly strength-trained muscle	Decrease
Muscle length	Varied
High speed of shortening	Increase

urement is called *electromyography,* often abbreviated to EMG. Each muscle has a number of motor units. A tension increase can occur either by an increase in stimulation rate for a given motor unit or by recruitment of additional motor units. Thus, the recorded EMG voltage, called *myoelectric activity,* is usually the sum of several motor unit potentials. Because of the different characteristics of electrodes, the signals picked up vary considerably in their characteristics. Surface electrodes record the algebraic sum of all motor unit potentials reaching the electrode site. They are useful particularly for recording the activity of superficial muscles and in situations where a more general recording is sufficient. For detailed analysis of individual muscle activities, indwelling electrodes must be used. Although needle electrodes are the most common in clinical electromyography, fine-wire electrodes are usually used in kinesiology. They require a hypodermic needle for insertion, but are quite comfortable following needle withdrawal. A discussion of the advantages and disadvantages of different electrodes is given by Basmajian (1974).

The reason for the recording and processing of myoelectric signals in kinesiology is to predict muscle tension. The relationship of EMG activity to muscle force is dependent on several factors, some of which appear in Table 5.1. The relationship appears to be monotonic in the sense that an increase in tension is paralleled by an increase in myoelectric activity, but non-linear under many circumstances. EMG signals processed by means of so-called full-wave rectification and low-pass filtering techniques have been extensively used to study EMG force relationships such as Fig 5.5. While reasonably reproducible relationships are found under isometric conditions,

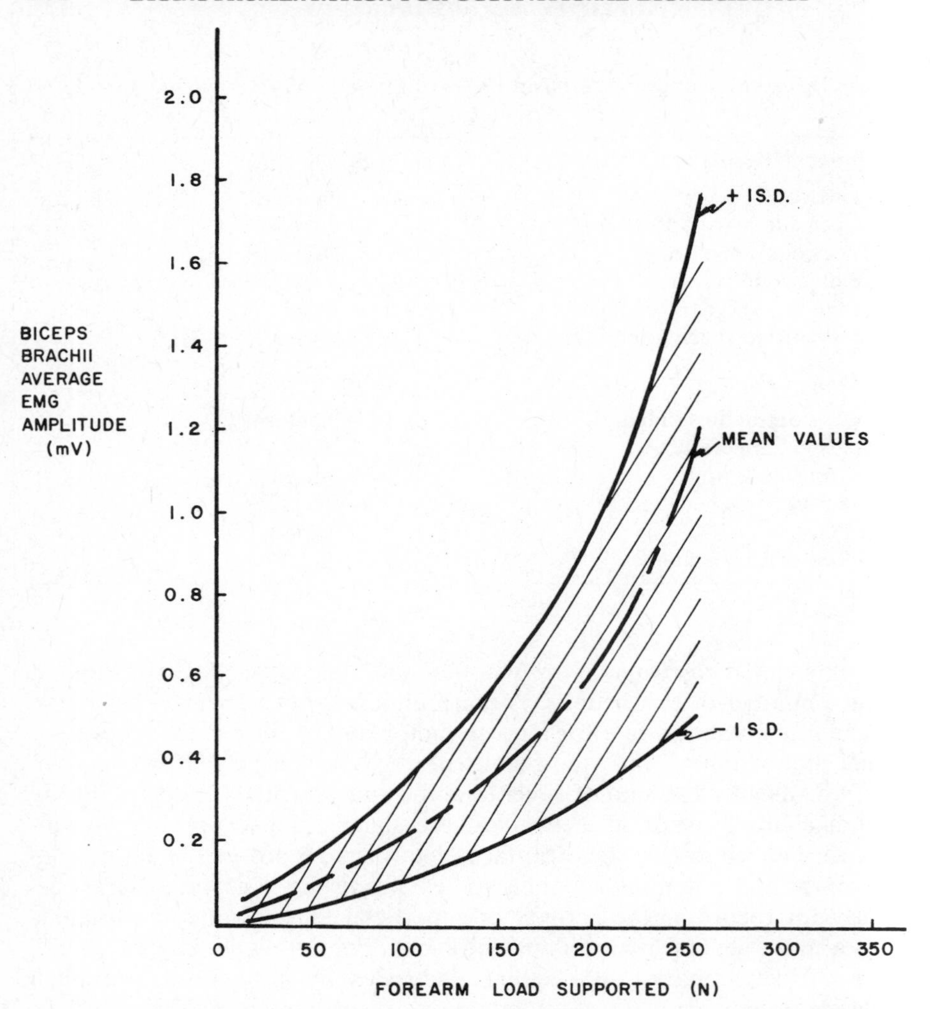

FIGURE 5.5 Non-linear relationship of mean and standard deviations of averaged EMG potentials from biceps brachii (surface electrodes) at various loads and for repeated tests with different people (Zuniga and Simmons, 1969).

although not linear over the entire force range, there is great uncertainty about the validity of the relationship under dynamic conditions. The fact that so many factors affect the relationships of the electric activity of a muscle and the force provided by that muscle indicates that care is necessary in predicting muscle contraction levels from EMG data. One procedure to minimize errors in these estimates follows. First, the electrodes are attached to a person in a way that achieves a low electrical impedance between the electrodes. Often this may require abrading of the epidermal layer to remove dry, high-resistance skin cells, and the use of biocompatible electrode paste (AgCl salt) between the skin and electrode. Dymond (1978) presents an excellent discussion of electrode considerations, and should be consulted for details on such measurements. Once a reasonably low impedance is achieved relative to the input impedance of the amplifier (a 1 : 10 ratio is a desirable minimum), the subject is requested to perform a measured graded exertion of the muscle or muscle group being monitored. At each exertion level the myoelectric activity, usually the RMS values, is recorded along with the external moment produced by the exertion. By following this procedure, many of the factors that affect the EMG/muscle force ratio are avoided, since the procedure results in a person- and activity-specific "calibration curve." This curve can then be used to estimate the load moments or forces created when the person performs a more complex task. Of course, the precision of the resulting moment or force estimate depends on how well the calibration experiments simulate the specific postural requirements of the motions being studied on the job. Sometimes, subtle changes in postures or motion dynamics can result in different muscles being used to execute a motion, and hence in an altered EMG/muscle force relationship (Grieve and Pheasant, 1976).

Despite this qualifier, myoelectric activity measurements have been used extensively to estimate the relative magnitude and temporal relationships required of various muscles during a number of common and sport activities. Such measurements have also been used in evaluating the physical requirements of occupational tasks as indicated in Table 5.2.

In addition to using the EMG to evaluate relative muscle activity, it has also been found to be useful to evaluate local muscle fatigue. Such an estimate of muscle fatigue relies on changes in the spectral characteristics of the myoelectric signal. A comparison of the electric activity from the biceps brachii during a fatiguing static contraction is shown in Figure 5.6. It shows both an amplitude increase and frequency decrease with fatigue. Under rested conditions, the mean frequency of the myoelectric signal may be twice that found when a muscle is fatigued (Stulen and DeLuca, 1979; Lindstrom, Kadefors, and Petersen, 1977). The exact frequency varies with many of the same factors listed in Table 5.1 that affect the myoelectric amplitude. Nevertheless, fatigue-induced myoelectric frequency shifts toward lower frequencies and concomitant amplitude increases have been useful in many occupational biomechanical studies to understand the length of time a muscle group can be statically or frequently contracted to a certain level (Chaffin, 1973; Herberts, Kadefors and Broman, 1980; Hagberg, 1981).

5.3.2 EMG Measurement System

Given the preceding EMG signal characteristics, criteria for a measuring system can be stated as follows:

TABLE 5.2
Some Examples of EMG Studies of Occupational Activities

Source	Muscles and Activity
Tichauer et al. (1972)	Forearm and biceps muscles during forearm torsion at various elbow angles
Tichauer, Miller and Nathan (1973)	Low-back and thigh muscles during sustained load carrying.
Tichauer (1966)	Hand and arm muscles during hand-tool gripping
Morris, Benner and Lucas (1962)	Back muscles during various standing postures.
Grieve and Pheasant (1976)	Leg, low-back, and biceps muscles during lifting.
Andersson, Ortengren, and Herberts (1977)	Back muscles during lifting in various postures.
Ortengren et al. (1975)	Shoulder and back muscles in assembly line work.
Herberts, Kadefors, and Broman (1980)	Shoulder muscles during arm abduction.
Norman, Winter, and Pierrynowski (1979)	Leg and back muscles during load carrying.
Andersson, Ortengren, and Schultz (1980)	Back muscles during load lifting in seated postures.
Sweetman et al. (1976)	Back muscles in bending.
Armstrong, Chaffin, and Foulke (1979)	Finger/flexor muscles in grasping objects.
Chaffin, Lee, and Freivalds (1980)	Muscle strength estimation.

1. High amplifier input impedance to assure constant system gain with varied electrode impedance. It should be at least 10 times the electrode/skin impedance; for surface electrodes, 1 MΩ is usually adequate, sometimes larger for indwelling.
2. High amplifier common-mode signal rejection to reduce electrical noise from 60- or 50-Hz electrical devices and input lead capacitance, noise rejection should be above 80 dB, preferably 90 dB.
3. Use of a small portable preamplifier worn on the subject to reduce electrode input lead length and resulting noise.
4. Amplifier band width—for surface electrodes, 10 to 1000 Hz is suggested; for indwelling electrodes, 20 to 2000 Hz.
5. Variable step gain capable of amplifying signals of from 3 μV to 2 mV.
6. Subject protection through fault-current limiting circuit at input.

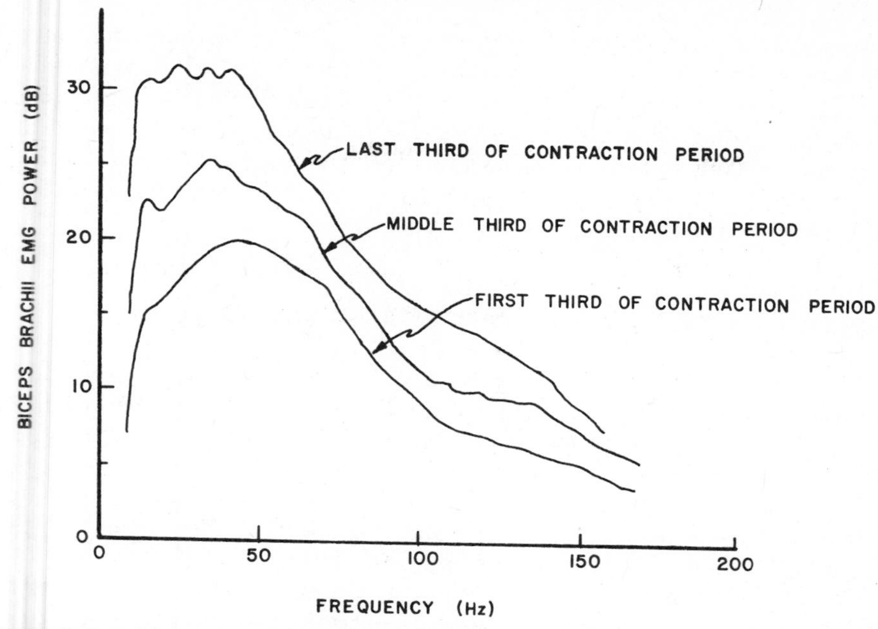

FIGURE 5.6 Monopolar electrode mean EMG spectra averaged from four subjects holding 90 N load to exhaustion on wrist with horizontal forearm (Stern, 1971).

7. Output levels compatible with typical recording devices or A-D computer input devices (±1 V).

An example of a measuring system meeting these criteria is the one developed for industrial use by Foulke, Goldstein, and Armstrong (1981). It was designed to use a commercially available AC voltmeter. This system provides both an RMS voltage estimate, as well as an instantaneous signal output compatible with various other displays and recorders. The authors report excellent results in several studies of manual work in industrial plants (Armstrong et al., 1982; Armstrong and Chaffin, 1979).

For additional analysis, the instantaneous EMG signal is recorded on a portable instrumentation-quality tape recorder. The resulting tape can then be evaluated by either analog or digital processing. Often this involves choosing muscle contraction periods during the performance of the job when the RMS level is relatively constant—that is, when a static exertion is performed for several seconds. The myoelectric signals obtained during the periods allow a power spectra to be determined—that is, the voltage at given frequencies are estimated. This provides statistics, such as power percent in given frequency bands and the mean frequency of the myoelectric signal, that are necessary to predict muscle fatigue states.

In summary, the use of electromyography in occupational biomechanics is relatively new. It has become possible with the advent of safe, portable, and reliable measuring equipment, as well as the development of procedures for gathering and analyzing EMG data in a plant environment. Such measurements are becoming indispensible in understanding how specific muscles collaborate to execute a variety of motor functions in industry, as well as estimating localized muscle fatigue.

5.4 MUSCLE-STRENGTH MEASURING SYSTEMS

The external force or moment capability of a person performing a maximum volitional exertion was referred to in Chapter 4 as a person's *muscle strength*. It should be obvious from the description of the measurement methods employed that the strength of a particular muscle is not measured per se, but rather the methods provide estimates of the capacity of groups of muscles to perform a specific function—for example, flexing the forearm at the elbow, rotating the arm inward, gripping an object.

Since various methods for strength testing were presented in Chapter 4, the discussion here will be limited to a description of the equipment necessary to make such measurements in an occupational setting. Specifically, two types of static strength-testing devices will be described—one for localized (regional) muscle function testing, one for whole-body exertions. A simple device for measuring dynamic strengths will also be presented.

5.4.1 Localized Static Strength-Measurement Systems

The following static strength-measurement system criteria have been synthesized from the recommendations of Kroemer (1970), Caldwell et al (1974), and Chaffin (1975):

1. The measuring device should be coupled to a body segment in order to
   a. Provide a known and repeatable location for the force application.
   b. Disperse the contact force over a sufficient area to avoid local discomfort.
   c. Measure the force in a given direction.
2. The device should easily accommodate the measurement of different muscle exertions for individuals of varied anthropometry.
3. The measuring device should be attached to a fixture that is both structurally capable of resisting high force loads, far greater than expected for extra safety, and that provides necessary bracing of body segments against expected reactive forces.
4. The system read-out device should provide:

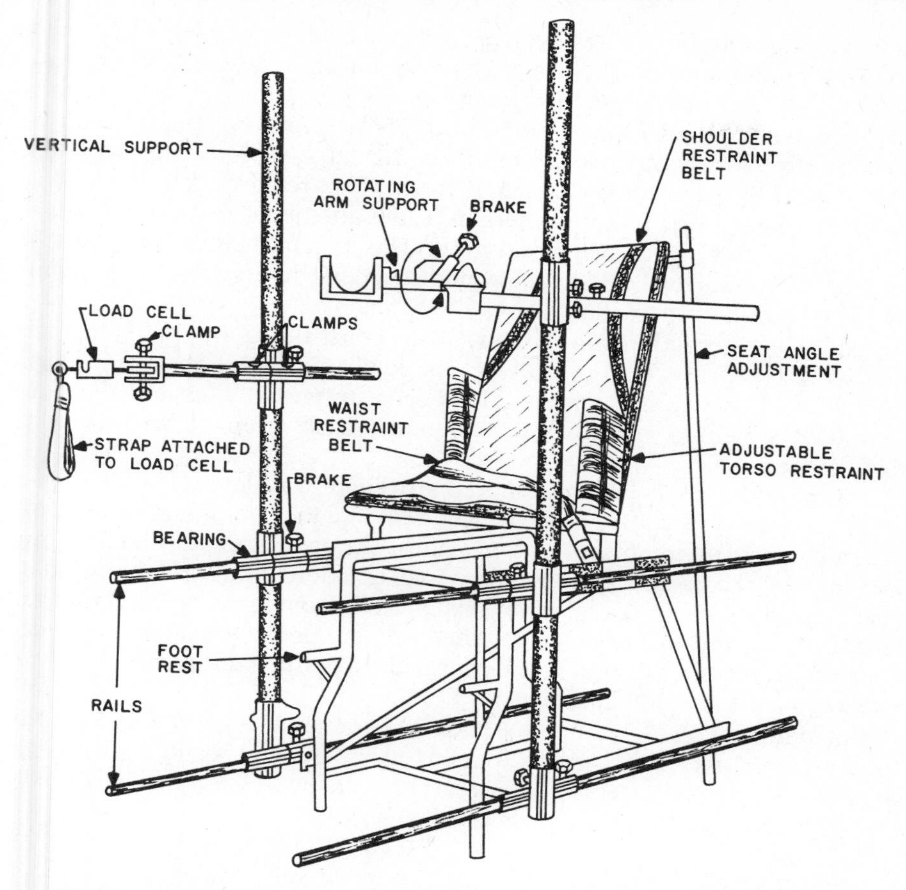

FIGURE 5.7 T. Stobbe strength test fixture used to test specific muscle actions.

a. Peak signal display.
b. Time-weighted average display of the exertion over a three-sec period.
c. Compatible output to common instrumentation recorders and A-D computer interfaces.

The last criterion implies the use of some type of electronic force-measuring system. Before recognition of such a criterion, mechanical, pneumatic, and hydraulic devices were used. Some typical examples of these are described by Roebuck, Kroemer, and Thomson (1975). A modern system devised by Stobbe (1982) for measuring specific muscle strengths is depicted in Figure 5.7. This is an upgraded version of a system advocated by Wasserman et al (1974). It includes a method of restraining various body segments against reactive forces by using a seated posture with adjustable belts and padded braces. An electronic load cell is used to measure the exertion forces. The load cell is attached to an arm with five degrees of freedom so as to be easily moved and locked in various locations necessary to obtain specific strength measurements. In this particular system, a single-axis load cell is used to measure the exertion force. This requires that the cell be carefully aligned in a direction normal to the attempted motion in order to measure the total force of the exertion. This requirement is alleviated by use of a three-axis load cell, which would cost considerably more and require more complex instrumentation.

A similar strength-testing fixture was constructed by Backlund and Nordgren (1968). Instead of moving one load cell to test various strength functions, they used a group of 12 load cells strategically located about the seat. They reported measuring 33 different muscle-strength functions with this device.

A similar device for testing children's strength was devised by Owings et al (1975). This system uses strain gauges built into the restraining seat fixture. When the child attempts an exertion against various restraints, simultaneous measures of the exertion force and various reaction forces are obtained. By processing the signals in real-time through a minicomputer, the direction and magnitude of the exertion force is determined, and the experimenter can detect whether the exertion was performed as instructed by comparing various reactive-force combinations.

For simultaneous force measurements, a computer is essential to resolve the force vectors. If a single load cell is used, amplifiers and signal conditioning systems that meet the display criteria stated earlier are commercially available. One such system is shown in Figure 5.8.

5.4.2 Whole-Body Static Strength-Measuring System

A biomechanical analysis of a job will often reveal two or three manual tasks that require a large amount of strength. These tasks can be simulated by setting up a strength-test fixture to allow workers to assume the postures required in the performance of the job tasks. Because such testing often involves many simultaneous muscle actions, it is referred to as *whole-body strength testing*. The method for such testing was described in Chapter 4 (Section 4.3).

The strength-measurement criteria presented earlier in this chapter apply to whole-body exertions as well as localized muscle exertions. Because several different types of exertions may be evaluated for an anthropometrically heterogeneous population, the test fixture must be easily adjusted from one test to another. In practice this means that:

1. The vertical location of the load cell from the floor can easily be adjusted to simulate objects being lifted or moved on the job at various heights.

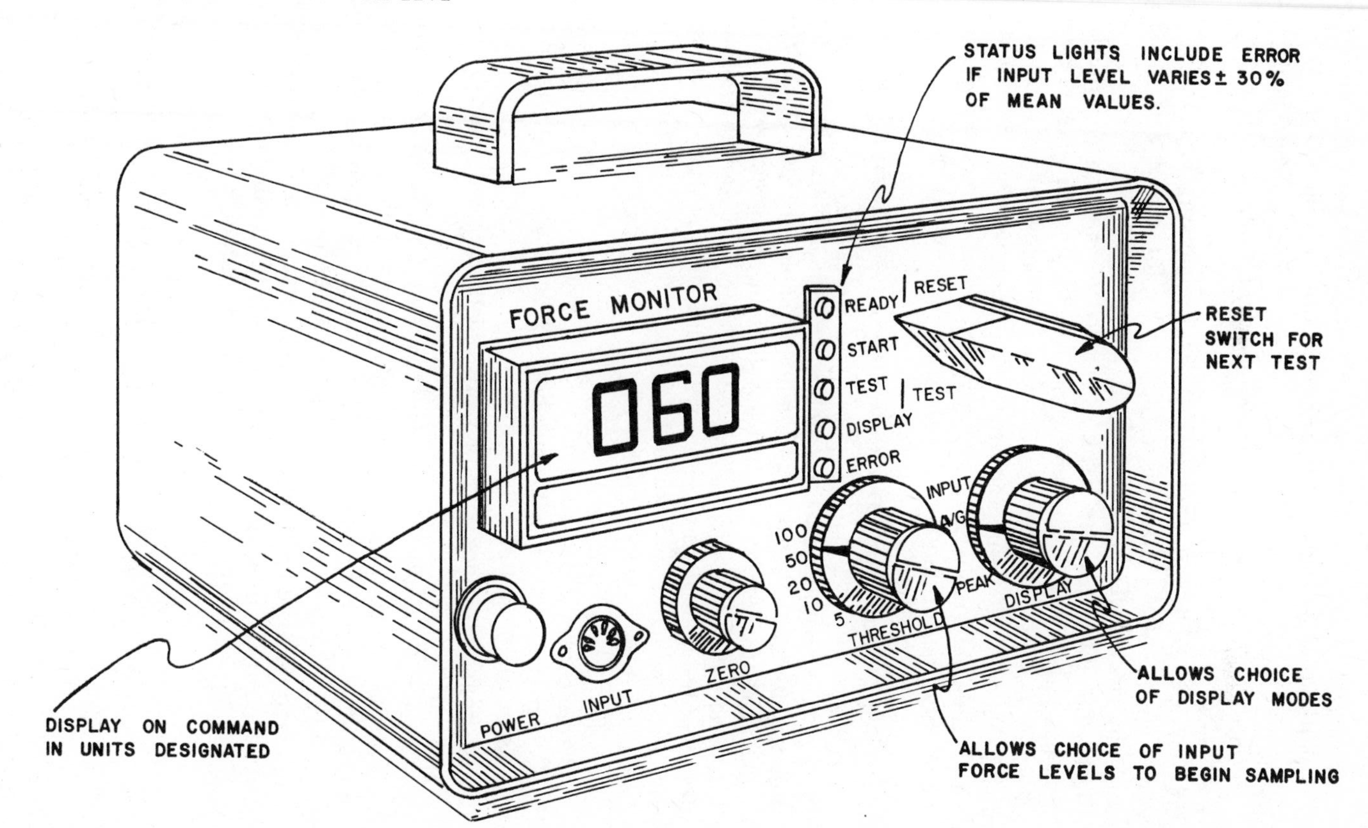

FIGURE 5.8 Strength monitoring and display system (courtesy of Prototype Design and Fabricating Co., Ann Arbor, Michigan).

2. The load call can be positioned to measure varied force directions—that is, pushes, pulls, lifts, pull-down forces, and so on.
3. The load cell can be attached to various handles, straps, tools, boxes, and so on required in the performance of a job.

Of course, even though the load cell fixturing must be adjustable, it must also be capable of withstanding high reactive forces—that is, it must be structurally rigid under high loads.

Figure 5.9 illustrates one fixture designed to meet these criteria. This system has been used to measure whole-body static strengths of over 5000 workers in various industries, as described in Chapter 4 (Section 4.3.3). It appears that the use of such a fixture, combined with appropriate data acquisition and analysis systems, can provide a practical method of measuring static strength performance in a variety of work situations.

5.4.3 Whole-Body Dynamic Strength-Measuring System

As was the case with whole-body static exertions, a biomechanical job evaluation may reveal the need to assess the dynamic strength of a person performing a specific task. A measuring system for performing such an analysis of lifting activities has been suggested by Pytel and Kamon (1981). Figure 5.10 illustrates their fixture, along with three test postures used in a pilot study (see comments in Section 4.3.2 for details of the pilot study). The components in this system consist of:

1. An electronic load cell and velocity transducer connected to a strip chart recorder to determine both velocity and instantaneous forces over the period of a motion.
2. A constant-velocity motor with adjustable speed control.

Such a device allows the experimenter to preset the velocity of the movement based on the motion-distance/time requirements of the task being simulated. Correlations with both static whole-body strengths and lifting capacity as determined by a psychophysical method appear to be good, though based on a limited pilot study. With further development and evaluation in various industrial settings, dynamic strength-testing systems should become of practical value in evaluating worker strength performance.

5.5 INTRADISCAL PRESSURE MEASUREMENT

5.5.1. Measurement Development Technique

The intervertebral disc consists of two parts: the nucleus pulposus and the annulus fibrosus (Figure 5.11). The nucleus (center), which is surrounded

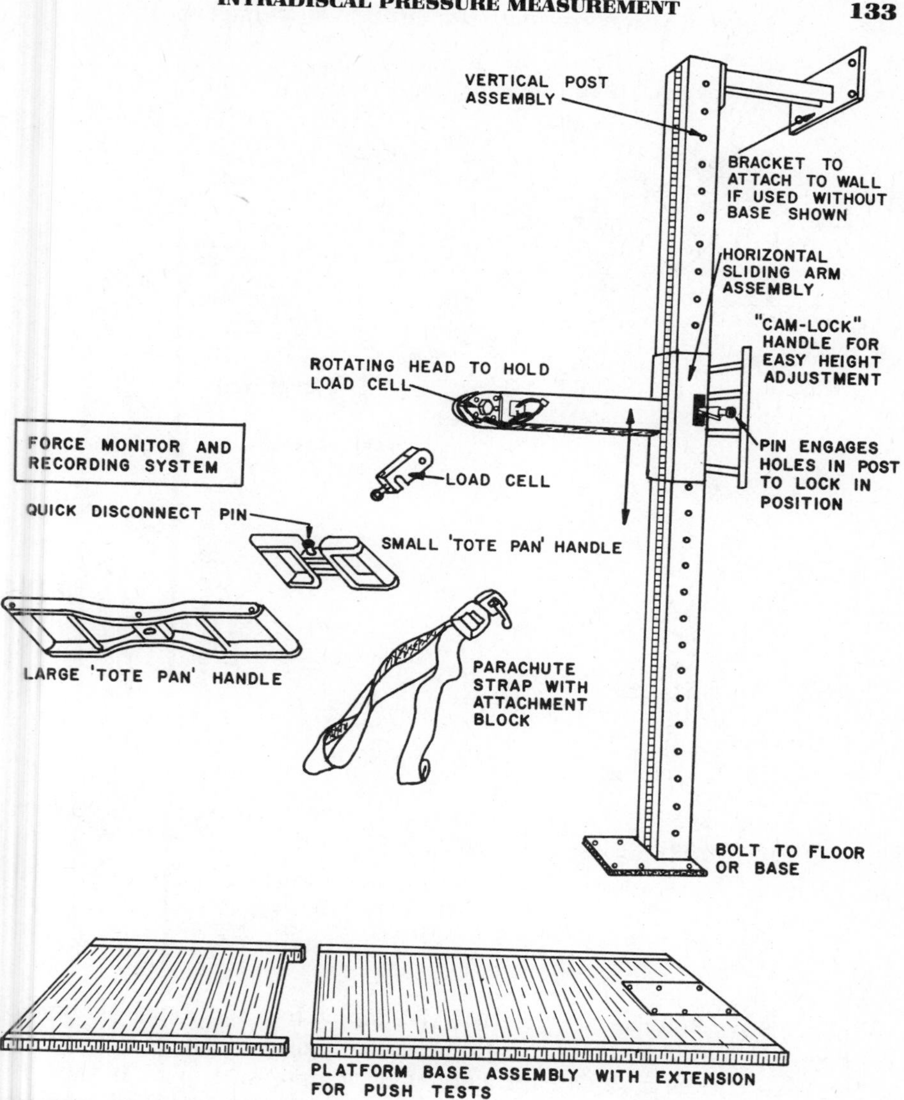

FIGURE 5.9 Static strength test fixture with two handle sizes and adjustable strap shown (courtesy of Prototype Design and Fabrication Co., Ann Arbor, Michigan).

by the annulus, is a watery gel with a water content of more than 80% in young people. The surrounding annulus consists of fibrous lamellae, with the fibers angled at approximately 20° to each other in adjacent sheets. The fibrous lamellae are more numerous and stronger in the anterior and lateral aspects than in the posterior. The integral structure allows the disc to distribute loads and transform compression loads into radially directed tensile

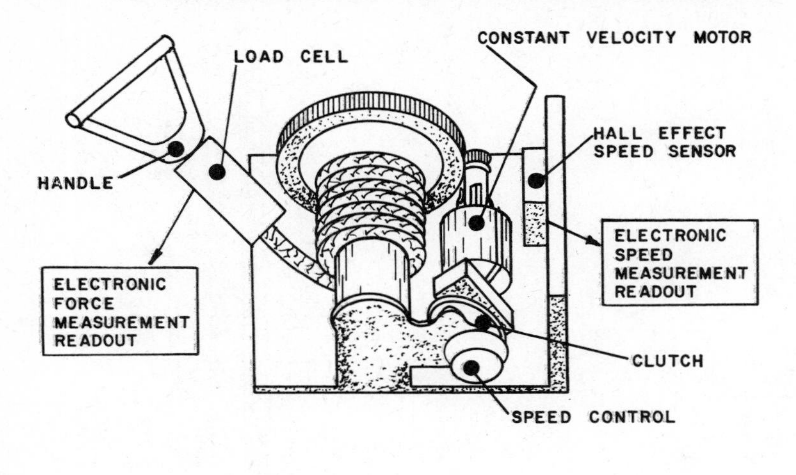

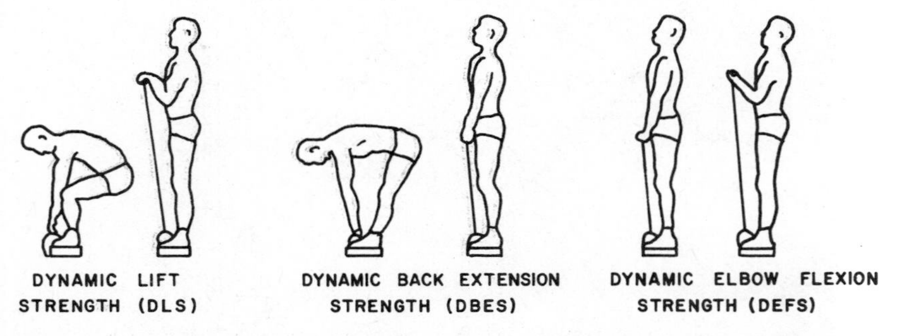

FIGURE 5.10 (Top) A schematic of major components in dynamic strength tester. (Bottom) The starting and ending postures in three tests used in evaluation of tester (Pytel and Kamon, 1981).

forces that the annular structure is well-adapted to resist. It should be mentioned here that in addition to carrying radially distributed stresses, the annulus also carries a significant amount of compressive load. It should be further noted that the fluid content is considerably reduced with increasing age.

The high fluid content of the disc nucleus has already been noted at the beginning of the century. Petter (1933) found that the central substance bulged when the disc was sectioned transversely, and attempted to measure the disc expansion. Following attempts at pressure measurement in intact discs, Nachemson (1960) presented experimental evidence that the nucleus pulposus behaved hydrostatically in all but the most severely degenerated discs. He also found that the pressure, which increased linearly with increasing compressive load, was about 30–50% higher than the applied load per unit of area. *In vitro* studies in the 60's and 70's (Berkson et al, 1979;

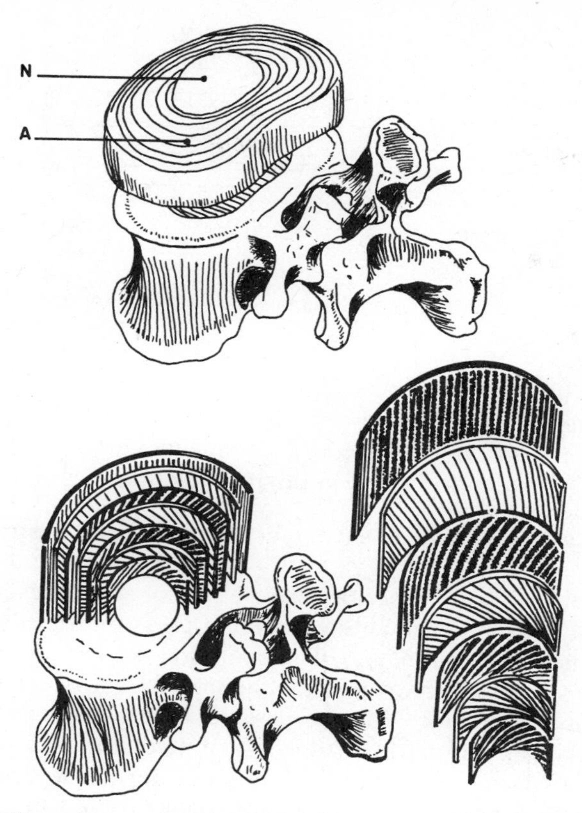

FIGURE 5.11 Fibers of the annulus fibrosus (A) enclosing the nucleus pulposus (N) (Kapandji, 1974).

Nachemson et al, 1979; Schultz et al, 1979) have confirmed the hydrostatic properties of the healthy disc and also that a no-load pressure exists within the disc, probably caused by ligamentous pre-stress. Pressure changes *in vitro* have been found to be linearly related to compressive loads, and the major portion of an intradiscal pressure change under *in vitro* conditions is due to a change in compressive load. Other modes of loading, such as flexion, extension, lateral bending, and torsion, are also accompanied by pressure changes, however.

The *in vitro* experiments have been followed by *in vivo* measurements of disc pressures in different positions and when various maneuvers and occupational activities are performed. The results of these studies will not be given here. Some appear in various chapters of this book. For reference, the interested reader is referred to Andersson (1982). The *in vitro* measurements were performed using a membrane-covered liquid-filled needle connected to a pressure transducer. A similar method was used for the first *in*

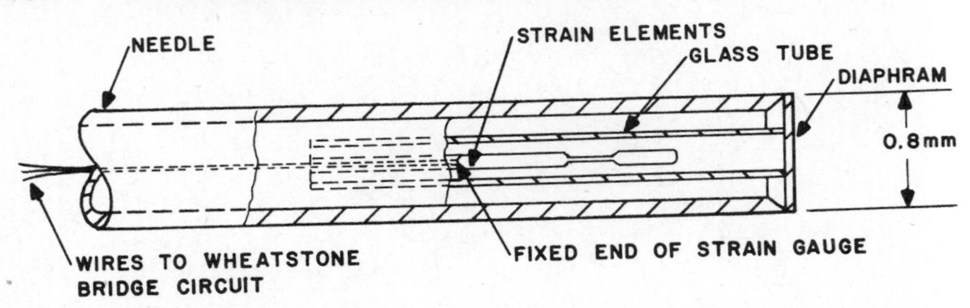

FIGURE 5.12 Intra-discal pressure transducer on end of needle as developed by Nachemson and Elfstrom (1970).

vivo measurements (Nachemson and Morris, 1964). In subsequent studies, the membrane needle was replaced by a semiconductor strain-gauge transducer needle, a system described in the next section.

5.5.2 Intradiscal Pressure-Measurement System

The measurement system described here is the one first described by Nachemson and Elfstrom (1970). It consists of a pressure transducer attached to the tip of a needle (Figure 5.12). The operation of the transducer is based on the piezoresistive effect of a semiconductor strain gauge embedded in rigid resin in an elastic tube. When a uni-axial load is applied to this "strain tube" in the axial direction, the load is transmitted to the gauge through the rigid resin. The strain tube is mounted in the center of the transducer, with one end fixed to a pressure-sensitive diaphragm. When the diaphragm is displaced due to pressure, the electrical resistance of the gauges will change. The transducer is connected as a Wheatstone bridge. A change in the electrical resistance of the gauges due to pressure changes produces an out-of-balance current from the bridge that is then amplified. The output signal can be read either on a panel meter or fed to a recording system.

The needle is inserted via a lateral approach into the center of the disc. At first, a guiding needle with a mandrin is inserted. The insertion procedure is followed by TV fluoroscopy, and the final position checked by roentgenogram. The mandrin of the guiding needle is then withdrawn and the transducer needle inserted to its full length. Following this, the needle is withdrawn about 5 mm to insure that it is not in contact with the tissue of the nucleus pulposus. In that position, the Wheatstone bridge is zero-balanced and the transducer needle is again inserted into the nucleus.

5.5.3 Applications and Limitations in Occupational Biomechanics

Intradiscal pressure measurement is the most direct and reliable method to assess loads on the spine. From pressure readings and knowledge of the disc area, the load can be calculated quite readily. The measurement technique

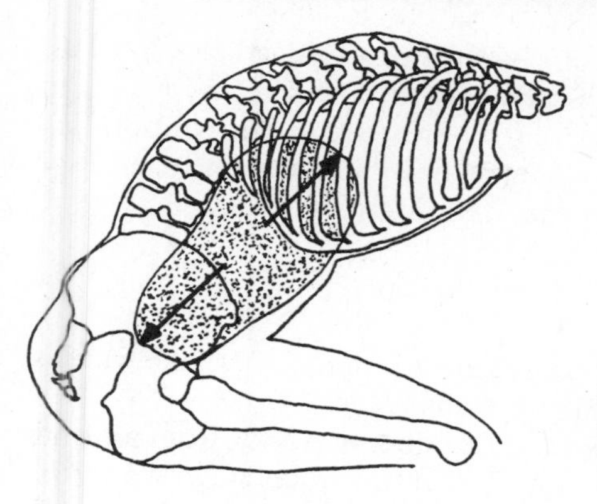

FIGURE 5.13 Diagram to suggest how abdominal fluid acts to support the trunk while lifting (Bartelink, 1957).

has been used for laboratory measurements of many different types of vocational activities and postures, as summarized by Andersson (1982). Disc-pressure measurements have also been used to validate biomechanical models in which the loads on the lumbar spine were predicted (Schultz, et al, 1982). In these experiments, several tasks were performed under well-controlled conditions imposing compressional loads on the spine up to about 2400 N, corresponding to intradiscal pressures up to 1.6 MPa. The predicted load and measured quantities correlated well.

The main disadvantage of the method is that it is invasive and can only be used *in vivo* under strict laboratory conditions, and even then with several restrictions. The needle does not permit large and rapid movements, and reliable measurements of hydrostatic pressures require comparatively healthy discs.

5.6 INTRA-ABDOMINAL (INTRAGASTRIC) MEASUREMENTS

5.6.1 Measurement Development

The idea that pressures within the trunk might assist with its mechanical efficiency was suggested in the 1920's (Keith, 1923). The underlying theory is that when, for example, a weight is lifted, a flexion moment develops about the spine. This moment is counter-balanced by the posterior back muscles. The pressure in the trunk cavities assist in this respect, producing an extension moment. The muscle contraction force needed for equilibrium can be reduced; as a result the stress on the vertebral column is reduced (Figure 5.13). Davis in 1956 found that the intragastric pressure did increase when the trunk moment was increased, and his findings were later confirmed by Bartelink (1957). Morris et al (1961), using a mathematical model of forces acting on the spine, concluded that the load on the spine was reduced by 30% on the lumbosacral disc because of the support from the pressures within the trunk. Subsequent studies have been conducted at somewhat lower load-reduction levels, as discussed below. Early measurement techniques using rubber balloons have been replaced either by pressure-sensitive radio pills that transmit intestinal pressures by telemetry, or by strain gauge transducer catheters that are either swallowed or inserted intrarectally.

5.6.2 Measurement System

The two types of measurement systems used most frequently are the wireless radio pill and the wire-connected pressure transducer.

The pressure-sensitive *radio pill* is available commerically from a firm in Surrey, United Kingdom. It transmits at the central frequency of approximately 400 kHz. Pressure variation alters the frequency of transmission; for a pressure change of 10 kPa it is between 10 and 40 kHz. The pill consists of a transistor oscillator, the frequency of which is controlled by a diaphragm-operated variable inductor. A silver battery is used as a power source. The pill signals are detected by a unidirectional antenna placed on the subject's abdomen after the pill has been swallowed.

The catheter-mounted *pressure transducers* are, for example, silicone beams with different resistors on each side of the beam. A miniature diaphragm is mounted on the beam and pressure variations cause this diaphragm to deflect the beam. The transducer elements convert mechanical position to electrical signals. The catheter can either be introduced orally, about 45 to 55 cm, to place the transducer well inside the gut, or rectally 12 to 15 cm, to place it well inside the abdominal cavity.

The output signal from both systems is usually graphically displayed on an ink recorder or sampled on magnetic tape. The two measurement methods have recently been tested simultaneously (Nordin et al, 1983). Both systems show good correspondence in the wave form of the curves generated, i.e. in terms of start of decrease, increase of intra-abdominal pressure in both time and shape. Poorer agreement was found in recorded peak difference—that is, in highest and lowest pressure response to a given task. Thus, measurement data cannot be directly compared. While the radio pill has advantages in being less invasive and quite easy to swallow, it is presently too expensive to be disposable and very sensitive to temperature changes. So, calibration is essential. The catheter transducers give excellent readings, but are somewhat uncomfortable to be used in industrial settings.

5.6.3 Applications and Limitations in Occupational Biomechanics

Following the early studies, which were aimed at establishing the relationship between trunk moments and trunk pressures, pressure responses to lifting have been studied extensively in the laboratory. Typically, the pressure response can be divided into an initial peak, a lower sustained pressure while the load is raised, and a further peak associated with load placement.

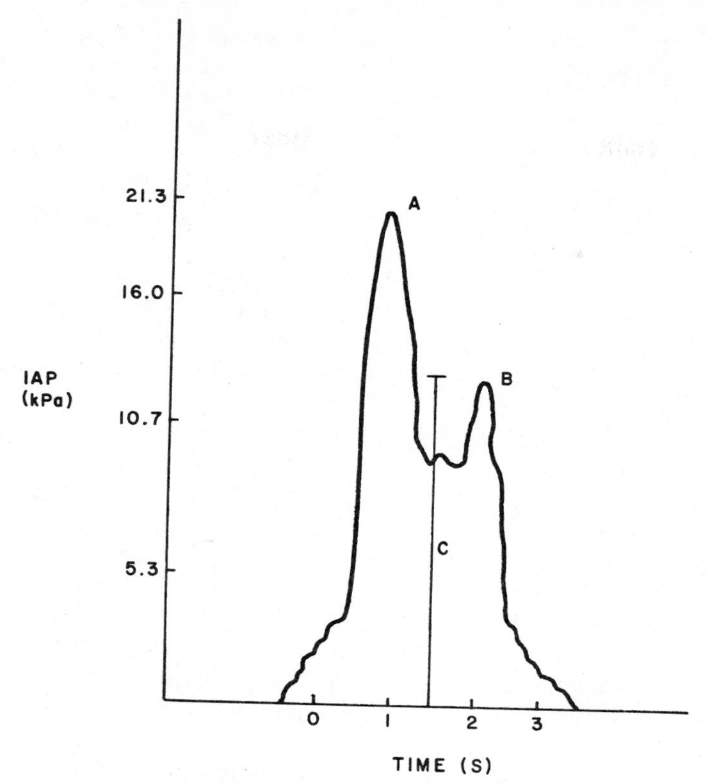

FIGURE 5.14 Intra-abdominal pressure record of subject lifting load of 250 N from ground level to 1.11 m using full stoop (back flexed) technique. First peak pressure A = 21 kpa (160 mmHg); second peak pressure B = 12 kpa; mean pressure C = 12.5 kpa. Pressure quotient (mean pressure × time of lift) is 22.5 kpa-s.

Various ways of analyzing this response curve have been used, including measuring the magnitude of the peaks, the mean pressure time, and the pressure-time coefficient (Figure 5.14). The peaks are believed to be associated with the extra moment required to accelerate the load and the trunk at the start and end of the handling maneuver; they are not encountered in static activities. Trunk pressures and other movements and maneuvers such as pulling and pushing, skiing, and walking have also been studied (Davis and Troup, 1964; Eie, 1966; and Grillner et al., 1978).

The relationship between the trunk moment and the increasing pressures found in the laboratory prompted the use of intragastric pressure measurements to assess load during work. Stubbs (1973) and Davis and Stubbs (1978b) used the radio pill to measure directly at the work site. They found some indication that workers in occupations with peak pressures within the trunk of 100 mm of mercury or more were more likely to report back injuries.

Based on that assumption, a series of observations were made on 200 young men during highly controlled physical activities. These included pushing, pulling and lifting with one or both hands while standing, sitting, and kneeling, and with the hands in 27 different standard positions within the reach envelope. The results have been published in three papers to provide guidelines for safe load levels for manual work (Davis and Stubbs, 1977a, 1977b, 1978a). The validity and practical use of these diagrams are now being investigated. From the extensive data obtained in the survey, coefficients of variance to determine the reliability of the method were calculated. The mean coefficient of variance for all activities was 32.3%. This means that if in a population an average individual obtains a pressure of 59 mm Hg for a given activity, the one SE range for a male population would be about ± 16 mm Hg (Davis, 1981).

The effect of age on the pressure has also been studied. The capacity of a lift which produces a standard pressure at different ages was estimated in a general industrial male worker population of 200. A highly significant decline in capacity after 50 years of age was found. Stubbs (1980) also made a summary of studies of trunk pressure measured in construction workers and other industrial workers on the job. Hamberg et al. (1978) studied the effect of systematic strengthening exercises for the abdominal muscles on the intra-abdominal pressure. They found that the exercises resulted in stronger abdominal and back muscles, but no increase in intra-abdominal pressure was found when lifting before and after the exercise period.

Intra-abdominal-pressure recording is a safe and simple measurement method causing little discomfort to the subject investigated. One single value or curve emerges, simplifying the data handling and evaluation procedures. The main obstacle to using the technique to estimate the load on the spine is the uncertainty about the relationship of pressure and spine compression. There is considerable ambiguity in the studies made so far. The pressure apparently rises in parallel with the load increase over a considerable force range when the load is static and the position is symmetrical. Asymmetry in load and posture influences the relationship, however. Moreover, the pressure response is not continuous but divided during the lift into an initial response and a sustained response. The relationship of the initial response and the trunk moment is not understood. There is also a lack of knowledge of the factors that control the intra-abdominal pressure. The situation is further complicated by the fact that there is uncertainty about how well intragastric and intra-intestinal pressure measurements actually reflect the true intraperitoneal pressure. The sustained pressures measured in various load situations are far too small to give adequate load relief and protect the vertebral bodies from crush fractures. Perhaps the pressures within the trunk are greater than the measurements indicate.

When a lift is initiated, we assume the intra-abdominal pressure will yield its major support to the spine when the trunk moment is at its maximum. This does not seem to be the case. It is also uncertain to what degree voluntary and reflex-triggered contractions of the abdominal muscles relieve the spine by increasing the intra-abdominal pressure. When this relief occurs is uncertain.

The role of intra-abdominal pressure measurements in investigations of work force need to be further clarified. It seems that asymmetrical loading of the trunk can create large stresses on some component structures without simultaneous intra-abdominal-pressure increases. This casts some doubt on the use of intra-abdominal pressure to develop safe levels of manual handling, but it is an interesting approach well worth further exploration.

5.7 FORCE PLATFORM SYSTEM

The forces exerted by the foot against the ground during walking or climbing on and off of ladders are of major interest in occupational biomechanics. By understanding how shoe soles, floor materials, ladders, and so on affect the foot forces, improved designs to reduce the large number of slip- and trip-related injuries that occur each year in industry, as well as to decrease the chronic biomechanical stresses that lead to lower extremity and back disorders, can be specified.

To acquire data regarding foot-ground forces, platforms that are supported by sensitive force transducers have been developed in the last 100 years. A modern platform typically measures about 40 × 60 cm. Such a device is referred to as a *force plate* or, more typically, a *force platform*. Mounted level with a floor it permits the measurement of three orthogonal forces and moments exerted on it during normal walking, running, or jumping.

A typical platform is shown in Figure 5.15. The force transducers can be linear variable-differential transformers, resistance strain gages, or piezoelectric crystals. The type of force transducer chosen must take into account the need for a mechanically stiff system that will not deform under dynamic loads of 5 kN or more, while still being sensitive to force changes of less than 1 N. Thanks to the development of stable, solid-state amplifiers that maintain calibration over long periods, piezoelectric crystal transducers appear to be most often used to meet measurement demands today. Such devices can provide a system that will tolerate loads of 5 kN, have a sensitivity of less than 0.2 N, and have a resonate frequency above 1 kHz.

There are several important uses for a force platform today. Dividing the horizontal force components (Y and X) by the vertical component (Z), when the foot is on the platform, enables an estimate of the instantaneous coefficient of friction necessary to stop the foot from sliding to be made. Some typical values from using this procedure are given in Section 6.4.3. Another application is in the determination of the dynamic effects of load lifting, discussed in Chapter 6. In general, when used with biomechanical models of dynamic activities, the force platform becomes a powerful tool in un-

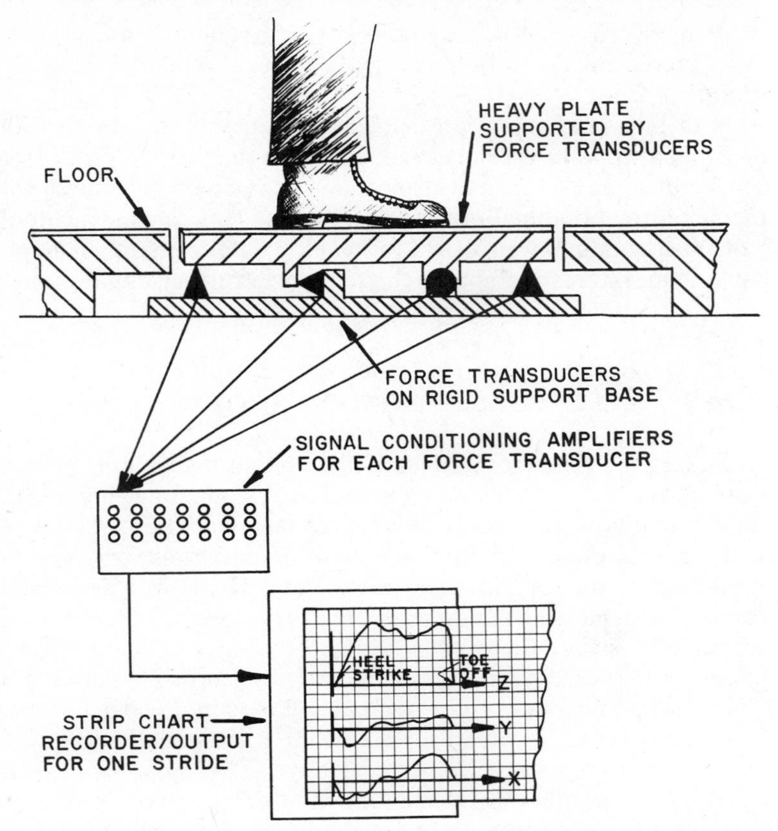

FIGURE 5.15 Diagram of major components comprising a modern force platform system, with illustration of three force components resulting from a single foot contact during normal stride.

derstanding the dynamic kinetic problems associated with complex human movements.

REFERENCES

Andersson, G. B. J., Ortengren, and P. Herberts, "Quantitative Electromyographic Studies of Back Muscle Activity Related to Posture and Loading," *Orthop. Clinic. North Amer.*, **8,** 85–96 (1977).

Andersson, G. B. J., "Measurements of Load on the Lumbar Spine," *Symposium On Idiopathic Low Back Pain,* Mosby, St. louis, 1982, pp. 220–251.

Andersson, G. B. J., R. Ortengren, A. Schultz: Analysis and Measurement of the Loads on the Lumbar Spine During Work at a Table. *J. Biomech.*, **13,** 513–520 (1980).

Andriacchi, T. P., S. J. Hampton, A. B. Schultz, and J. O. Galante, "Three-dimensional Coordinate Data Processing in Human Motion Analysis," *J. of Biomechanical Engineering,* **101,** 279–282 (1979).

Armstrong, T. J. and D. B. Chaffin, "Carpal Tunnel Syndrome and Selected Personal Attributes," *J. Occup. Med.*, **21**(7), 481–486 (1979).

Armstrong, T. J., D. B. Chaffin, and J. A. Foulke, "A Methodology for Documenting Hand Positions and Forces during Manual Work," *J. Biomech.*, **12**, 131–133 (1979).

Armstrong, T. J., J. A. Foulke, B. S. Joseph, and S. A. Golstein, "Investigation of Cumulative Trauma Disorders in a Poultry Factory," *J. Amer. Ind. Hyg. Assoc.*, **43**(2), 103–116 (1982).

Ayoub, M. A., M. M. Ayoub, and J. D. Ramsey, "A Stereometric System for Measuring Human Motion," *Human Factors*, **12**(6), 523–535 (1970).

Backlund, L. and L. Nordgren, "A New Method for Testing Isometric Muscle Strength under Standardized Conditions," *Scand. J. Clin. Lab. Invest.*, **21**, 33–41 (1968).

Bartelink, D. L., "The Role of Abdominal Pressure in Relieving the Pressure on the Lumbar Intervertebral Discs," *J. Bone Joint Surg.*, **39B**, 718–725 (1957).

Basmajian, J. V., *Muscles Alive: Their Functions Revealed by Electromyography*, 3rd ed., Williams and Wilkins, Baltimore, 1974.

Berkson, M. H., A. Nachemson, and A. B. Schultz, "Mechanical Properties of Human Lumbar Spine Motion Segments. Part II: Responses in Compression and Shear; Influence of Gross Morphology," *J. Biomech Engineering*, **101**, 53–57 (1979).

Brand, R. A., and D. Crowninshield, "Comment on Criteria for Patient Evaluation Tools," *J. Biomech.*, **14**(9), 655(1981).

Caldwell, L. S., D. B. Chaffin, F. N. Dukes-Dobos, K. H. E. Kromer, L. L. Laubach, S. H. Snook, and D. E. Waserman, "A Proposed Standard Procedure for Static Muscle Strength Testing," *AIHA J.*, **35**, 201–212 (1974).

Chaffin, D. B., "Functional Assessment for Heavy Physical Labor" in M. H. Alderman and M. J. Hanley, *Clinical Medicine for the Occupational Physician*, Dekker, New York, 1982, pp. 187–192.

Chaffin, D. B., "Ergonomics Guide for the Assessment of Human Static Strength," *AIHA J.*, **35**, 505–510 (1975).

Chaffin, D. B., "Localized Muscle Fatigue—Definition and Measurement," *J. Occup. Med.*, **15**(4), 346–354 (1973).

Chaffin, D. B., M. Lee, and F. Freivalds, "Muscle Strength Assessment for EMG Analysis," *Med. Sci in Sports and Exercise*, **12**(3), 205–211 (1980).

Chaffin, D. B., R. K. Schutz, and R. G. Snyder, "A Prediction Model of Human Volitional Mobility," *SAE Paper*, No. 720002, SAE, Inc. New York (1972).

Chao, E. Y., "Experimental Methods for Biomechanical Measurements of Joint Kinematics," in B. N. Feinberg and D. G. Fleming, Eds., *CRC Handbook for Engineering in Medicine and Biology*, Vol. 1, CRC Press Inc., Cleveland, Ohio, 1978, pp. 385–411.

Davis, P. R., "The Use of Intra-Abdominal Pressure in Evaluating Stresses on the Lumbar Spine," *Spine*, **6**, 90–92 (1981).

Davis, P. R. and D. A. Stubbs, "Safe Levels of Manual Forces for Young Males," *Applied Ergonomics*, **8**(pt 1), 141–150 (1977a); **8**(2), 219–228 (1977b); **9**(3), 33–37 (1978a).

Davis, P. R. and D. A. Stubbs, "A Method of Establishing Safe Handling Forces in Working Situations," *Safety in Materials Handling*, U.S. Department of Health, Education and Welfare, (1978b), pp. 34–38.

Davis, P. R. and J. D. G. Troup, "Pressures in the Trunk Cavities When Pulling, Pushing, and Lifting," *Ergonomics*, **7**, 465–474 (1964).

Dymond, A. M., "Instrumentation for Bioelectric Measurements," in B. N. Feinberg, D. G. Fleming, A. Burstein, and E. Bahnink, Eds., *CRC Handbook of Engineering in Medicine and Biology*, Vol. 1, CRC Press Inc., Cleveland, Ohio, 1978, pp. 3–40.

Eie, N., "Load Capacity of the Low Back," *J. Oslo City Hosp.*, **16**, 73 (1966).

Fisher, B. O., *Analysis of Spinal Stresses During Lifting*. Unpublished M.S. thesis, Industrial Engineering, University of Michigan, Ann Arbor, Michigan, 1967.

Foulke, J. A., S. A. Goldstein, and T. J. Armstrong, ''An EMG Preamplifier System for Biomechanical Studies,'' *J. Biomech.,* **14**(6), 437–438 (1981).

Freivalds, A., D. B. Chaffin, and A. Garg, ''A Dynamic Biomechanical Evaluation of Lifting Maximum Loads,'' *J. Biomech.* (in press).

Grieve, D. W. and S. T. Pheasant, ''Myoelectric Activity, Posture and Isometric Torque in Man,'' *Electromyogr. and Clin. Neurphysiol.,* **16,** 3–21 (1976).

Grillner, S., J. Nilsson, and A. Thorstensson, ''Intra-Abdominal Pressure Changes During Natural Movements in Man,'' *Acta Physiol. Scand.* **103,** 275–283 (1978).

Hagberg, M., ''Work Load and Fatigue in Repetitive Arm Elevations,'' *Ergonomics,* **24**(7), 543–555 (1981).

Hamberg, J., et al., ''Abdominal Muscle Activity and Intra-Abdominal Pressure When Lifting with Different Technique Before and After Training of the Abdominal Muscles,'' *Report to the Work Environment Fund* (D77/322) (in Swedish), Stockholm (1978).

Hatze, H., ''The Use of Optimally Regularized Fourier Series for Estimating Higher-Order Derivatives of Noisy Biomechanical Data,'' *J. Biomech.,* **14,** 13–18 (1981).

Herberts, P., R. Kadefors, and H. Broman, ''Arm Positioning in Manual Tasks: An Electromyographic Study of Localized Muscle Fatigue,'' *Ergonomics,* **23**(7), 655–665 (1980).

Jackson, K. M., ''Fitting of Mathematical Functions to Biomechanical Data,'' *IEEE Tr., Biomed. Eng.,* **28,** 122–124 (1979).

Keith, A., ''Man's Posture: Its Evolution and Disorders. Lecturer IV: The Adaptations of the Abdomen and Its Viscera to the Orthograde Posture,'' *Br. Med. J.,* **1,** 587–590 (1923).

Kroemer, K. H. E., ''Human Strength: Terminology, Measurement and Interpretation,'' *Human Factors,* **12**(3), 297–313 (1970).

Lindstrom, L., R. Kadefors, and I. Petersen, ''An Electromyographic Index for Localized Muscle Fatigue,'' *J. Appl. Physiol.,* **43,** 750–754 (1977).

Martin, T. P. and M. B. Pongratz, ''Validation of a Mathematical Model for Correction of Photographic Perspective Error,'' in R. C. Nelson and C. A. Morehouse, Eds., *Biomechanics IV,* University Park Press, Baltimore, Md., 1974, pp. 469–475.

Miller, D. I. and R. C. Nelson, *Biomechanics of Sport,* Lea and Febiger, Philadelphia, 1976, pp. 119–161.

Morris, J. M., D. B. Lucas, and B. Bresler, ''Role of the Trunk in Stability of the Spine,'' *J. Bone Joint Surg.,* **43A:** 327 (1961).

Morris, J. R. W., ''Accelerometry—A Technique for the Measurement of Human Body Movement,'' *J. Biomech.,* **6,** 729–736 (1973).

Morris, J. M., G. Benner, and D. B. Lucas, ''An Electromyographic Study of the Intrinsic Muscles of the Back in Man,'' *J. Anat. London,* **96**(4), 509–520 (1962).

Muth, M. B., M. A. Ayoub, and W. A. Gruver, ''A Nonlinear Programming Model for the Design and Evaluation of Lifting Tasks,'' in C. G. Drury, Ed., *Safety in Manual Materials Handling,* Nat'l Inst. for Occup. Safety and Health, Cincinnati, 1978. pp. 96–109.

Nachemson, A., Lumbar Intradiscal Pressure, *Acta Orthop Scand.,* **43**(suppl), 1–104 (1960).

Nachemson, A. and G. Elfstrom, Intravital Dynamic Pressure Measurements in Lumbar Disc. A Study of Common Movements, Maneuvers and Exercises,'' *Scand. J. Rehabil. Med.,* **1**(suppl), 1–40 (1970).

Nachemson, A. and J. Morris, ''In Vivo Measurements of Intradiscal Pressure,'' *J. Bone Joint Surg.,* **46A,** 1077–1092 (1964).

Nachemson, A., A. B. Schultz, and M. H. Berkson, ''Mechanical Properties of Human Lumbar Spine Motion Segments. Influences of Age, Sex, Disc Level, and Degeneration,'' *Spine,* **4,** 1–8 (1979).

Nordin, M., G. Elfstrom, and P. Dahlquist: Intra-Abdominal Pressure Measurements Using a Wireless Radio Pressure Pill and Two Wire-Connected Pressure Transducers: A Comparison," *Scand. J. Work Environ. Health* Tech Report, Stockholm Sweden (1983).

Norman, R. W., D. A. Winter, and M. R. Pierrynowski, "The Utility of Combining EMG and Mechanical Work Rate Data in Load Carriage Studies," *Proceedings of 4th Congress of Int. Soc. Electrophysiol. Kinesiol.*, ISEK, Sahlgren Hosp., Goteborg, Sweden, 1979, pp. 148–149.

Ortengren, R., B. J. G. Andersson, H. Broman, R. Magnusson, and I. Petersen, "Vocational Electromyography: Studies of Localized Muscle Fatigue at the Assembly Line," *Ergonomics,* **18,** 157–174 (1975).

Owings, C. L., D. B. Chaffin, R. G. Snyder, and R. H. Norcutt, *Strength Characteristics of U.S. Children for Product Safety Design,* Consumer Product Safety Commission Tech. Report #011903-F, 1975.

Pearson, J. R. and D. R. McGinley, *Dynamic Analysis of the Upper Extremity for Planar Motions,* University of Michigan Tech. Report 04468, Office of Research Administration, Ann Arbor, Michigan, 1961.

Petter, C. K., "Methods of Measuring the Pressure of the Intervertebral Disc," *J. Bone Joint Surg.,* **15,** 365 (1933).

Plagenhoef, S., *Patterns of Human Motion,* Prentice-Hall, Englewood Cliffs, New Jersey, 1971, pp. 28–31.

Pytel, J. L. and E. Kamon, "Dynamic Strength Test as a Predictor for Maximal and Acceptable Life," *Ergonomics,* **24**(9), 663–672 (1981).

Roebuck, J. A., K. H. E. Kroemer, and W. G. Thomson, *Engineering Anthropometry Methods,* Wiley-Interscience, New York, 1975, pp. 108–128.

Stern, M. M., *A Model that Relates Low Frequency EMG Power to Flucturations in the Number of Active Motor Units,* Unpublished Ph.D. Dissertation, University of Michigan, Ann Arbor, 1971.

Schultz, A. B., D. N. Warwick, M. H. Berkson, and A. Nachemson, "Mechanical Properties of Human Lumbar Spine Motion Segments. Part I: Responses in Flexion, Extension, Lateral Bending, and Torsion," *J. Biomech. Engineering,* **101,** 46–52 (1979).

Schultz, A. B., G. B. J. Andersson, R. Ortengren, A. Nachemson, and K. Haderspeck, "Loads on the Lumbar Spine: Validation of a Biomechanical Analysis by Measurements of Intradiscal Pressures and Myoelectric Signals," *J. Bone Joint Surg.,* **64A,** 713–720 (1982).

Stobbe, T., *The Development of a Practical Strength Testing Program for Industry,* Unpublished Ph.D. Dissertation, University of Michigan, Ann Arbor, 1982.

Stubbs, D. A., "Manual Handling in the Construction Industry," *Construction Industry Training Board Report* (1973).

Stubbs, D. A., "Trunk Stresses in Construction and Other Industrial Workers," *Spine,* **6:**83–89 (1981).

Stulen, F. B. and C. J. DeLuca, "Median Frequency of the Myoelectric Signal as a Measure of Localized Muscle Fatigue," *Proceedings of 4th Congress of Int. Soc. Electrophysiol. Kinesiology,* ISEK, Sahlgren Hosp., Goteborg, Sweden, 1979, pp. 92–93.

Sweetman, B. J., W. J. Jayasingehe, C. S. Moore, and J. A. D. Anderson, "Monitoring Work Factors Relating to Back Pain," *Postgraduate Med. J.,* **52**(Sup. 7), 151–155 (1976).

Tichauer, E. R., "Some Aspects of Stress on Forearm and Hand in Industry," *J. Occup. Med.,* **8**(2), 63–71 (1966).

Tichauer, E. R., H. Gage, and L. B. Harrison, "The Use of Biomechanical Profiles in Objective Work Measurement," *J. Ind. Eng.,* **4,** 20–27 (1972).

Tichauer, E. R., M. Miller, and I. M. Nathan, "Lordosimetry: A New Technique for the Measurement of Postural Response to Materials Handling," *Am. Ind. Hyg. J.,* **34,** 1–12 (1973).

Wasserman, D. E., T. Germann, D. V. Goulding, and F. Pizzor, *An Instrument for Testing Isometric Strength and Endurance,* NIOSH Tech. Rep. 74-109, Nat. Inst. for Occup. Safety and Health, Cincinnati, 1974.

Woltering, J. P. and E. B. Marsolais, "Optoelectric (Selspot) Gait Measurement in Two- and Three-Dimensional Space—A Preliminary Report," *Bulletin of Prosthetics Research,* BP R-10, **17**(2), 46–52 (1980).

Wood, G. A. and L. S. Jennings, "On the Use of Spline Functions for Data Smoothing," *J. Biomech.,* **12,** 477–479 (1979).

Zernicke, R. F., G. Caldwell, and E. M. Roberts, "Fitting Biomechanical Data with Cubic Spline Functions," *Res. Q.,* **47**(1), 9–19, (1976).

Zuniga, E. N. and D. G. Simons, "Nonlinear Relationship Between Averaged Electromyogram Potential and Muscle Tension in Normal Subjects," *Arch. Phys. Med. Rehab.,* **50,** 613–620 (1969).

6

OCCUPATIONAL BIOMECHANICAL MODELS

6.1 WHY MODEL?

The preceding chapters have described various methods of measuring biomechanical properties. For many of these properties, survey data have been systematically gathered, yielding population norms and performance limits. Given these measuring techniques and data, why is it necessary to develop biomechanical models?

Several different answers can be given. First, consider the more general question: Why model any system? The response is that models are representations that we can understand, even though such representations may require gross simplifications and assumptions. By comparing a model's behavior with the actual behavior of the system, we obtain further insight into how components of the system function and are coordinated to achieve desired outcomes. Each time a model doesn't predict a system's behavior correctly, we can rationally change certain parts of the model, thus gaining insight into the complex nature of the real system. It goes without saying that the human biomechanical system is very complex. Though we can measure many of its basic properties, internal forces can rarely be measured directly; we have hardly begun to understand how different properties interact and are coordinated to produce desired motor activities. Further, we have just begun to understand why, under certain biomechanical loads, some people are injured, while under supposedly similar conditions no injury occurs to others.

In this context one might claim that biomechanical models are strictly academic, whose purpose is the englargment of our understanding of musculoskeletal functions. This is certainly one motivation. However, occupational biomechanical models deal with the evaluation of very real situations (i.e., they allow one to determine the maximum allowable magnitude for a load held in various postures, the appropriate size of tools, the least stressful configuration of workplaces and seats, etc.). In these situations not only is the human highly variable, but the external loads and postural requirements of different industrial tasks also vary greatly. Thus, it is often not possible, because of time and cost constraints, to set up a laboratory simulation of a task and gather adequate performance capacity data from representative volunteers. It is often necessary to have normal human performance data available very early in the design of a job in order to consider a variety of alternative job conditions, work methods, and personnel stereotypes. Biomechanical models can sometimes help to rationally interpolate and extrapolate limited musculoskeletal capacity data from different sources to quickly provide specific design guides.

Finally, it must be conceded that there are work situations in which a biomechanical model is the only means to predict potentially hazardous loading conditions on certain musculoskeletal components—for example, when one picks up a heavy load. In one posture the load may pose no particular hazard to the low back. In a slightly different posture the combined effects of the load and body weight on the low back exceed limits that most authorities agree are hazardous. Only a model of the biomechanical properties of a person performing a specific task can yield such insight.

For these reasons and others, this chapter describes various biomechanical models appropriate to the analysis of work situations. It begins with elementary, static models of isolated body segments. These are expanded into three-dimensional, whole-body models. Dynamic models are discussed for analysis of sagittal plane motions, such as lifting and pushing or pulling loads. Special models are developed for analysis of the low back and hand and wrist—two body regions that account for the majority of serious work-related injuries and illnesses in industry today. A description of a model to predict worker strength capabilities is also included. The final section speculates about the need for new models for occupational biomechanics in the future.

6.2 PLANAR STATIC BIOMECHANICAL MODELS

6.2.1 Single-Body Segment Static Model

Consider an anthropometrically average-sized man holding a load in both hands at about waist height in front of his body. The load is equally balanced between both hands, and the forearms are horizontal. For this example, the load is a 10-kg mass. One question is: What rotational moments and forces are acting at the man's elbow?

This problem can be analyzed as a problem in *kinetics*—that is, a problem in which external forces act on a mass to impart motion. In this particular case, since our hypothetical man is *holding* the load, a *static* analysis is appropriate. A *dynamic kinetic* analysis will be described later for a person *moving* the load.

In solving this particular static problem, one must first determine the magnitude of the external force acting on the stationary mass. Clearly, if the man released the mass it would accelerate downward due to the pull of the gravitational force—that is, the mass of the load is attracted by the mass of the earth. The gravitational attraction creates the *weight* of the mass, and its magnitude is proportional to the mass by the factor g in the equation

$$\text{Weight} = Mg \tag{6.1}$$

where

Weight is measured in Newtons (N)
mass M is measured in kilograms (kg)
g is the gravitational acceleration (m-s^{-2})

The gravitational acceleration is usually considered to be constant—9.8067 m-s^{-2}. Strictly speaking this is only true at sea level. It would be slightly smaller on a high mountain, but the difference is too small to consider for practical problems in occupational biomechanics. Using an approximate value for g, we can now compute one external force acting on the mass being held:

$$\text{Weight} = 10 \text{ kg} \times 9.8 \text{ m-s}^{-2} = 98 \text{ N (downward)} \tag{6.2}$$

The designation (downward) is not necessary when dealing with weight, but it leads to a distinction between mass, which has magnitude only, and weight, which is a force. Forces are vector quantities, and as such have four characteristics:

1. Magnitude
2. Direction
3. Line of action
4. Point of application

For our example, the weight-vector characteristics are easy to specify. The magnitude is 98 N, the direction is downward, the line of action is vertical, and the point of application is at the center of mass of the load.

Since the load in our example is not moving, some other external force must be acting to uniquely cancel the weight vector. Of course, this second

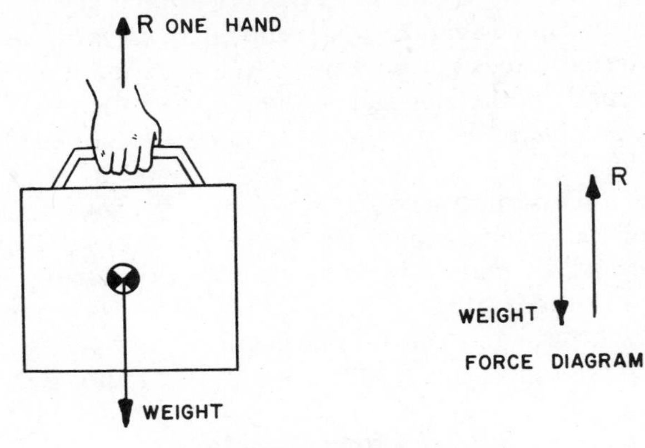

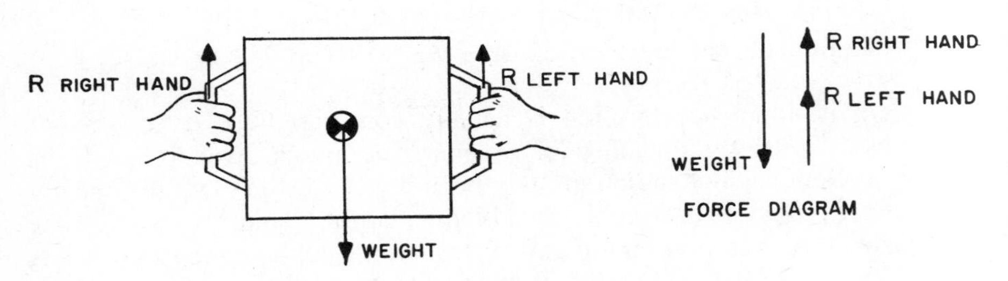

FIGURE 6.1 Load and force diagrams when holding load with one and two hands.

force is provided by the hands acting on the load. In some manner, now to be described, upward lifting forces provided by the hands place the load in physical *equilibrium*. In fact, since the object is not moving, it is said to be in *static equilibrium*. This means that the additive effect of all external forces acting on a mass is zero. This statement, in fact, is often referred to as the *first condition for equilibrium*. In other words, in holding the load, gravity creates a weight force to which the hand-lifting force is a reaction. The hand *reactive force R* must have an equal but opposite direction, and it must act along the same line of action, otherwise the object would rotate. This system of forces is illustrated in Figure 6.1 for cases in which the load is equally divided between the two hands and one hand is used to lift the box from the top. In the one-handed lift, the reactive upward hand force is simply equal to the downward weight. Since the force acts along the same line of action as the weight, no rotation occurs. When two hands support the load, the reactive hand force is divided between the two hands, and can be estimated as follows, assuming symmetric loading:

$$\begin{aligned} \sum \text{forces} &= 0 \\ -98 \text{ N} + 2R \text{ hand} &= 0 \\ R \text{ hand} &= \frac{98 \text{ N}}{2} \\ &= 49 \text{ N} \end{aligned} \tag{6.3}$$

At this point it must be made clear that the center of mass is located exactly between the two hands, and thus the weight is equally divided. If this were not so, then further analysis of the moments of the forces would be necessary to determine the relative weight on each hand. Such situations will be considered later in this chapter.

The concept of a *free-body diagram* of forces acting on a body is often helpful, and is included for this case. In a free-body diagram, force vectors are scaled in the drawing to indicate magnitude. They can then be arranged graphically to determine the magnitude of an unknown force quantity. In the simple case just discussed, the solution is obvious. When forces are not parallel, however, a graphical approach becomes more useful, as will be seen later in the chapter.

What we have accomplished so far is to determine how much force is acting on each hand when holding a 10-kg mass. We can now estimate the elbow forces and moments, assuming a *planar static analysis*—that is, with the forces acting in a single plane. First, consider the forearm and hand to be one body segment. For a numerical solution, the length, weight, and center-of-mass location will be assumed to represent an average male. The line of action of the weight will be considered to act through the center of mass of the hand, which approximates the center of grip when using a power-type of grip on a handle. The force system is illustrated in Figure 6.2. Note that in this system the reactive force R_E is assumed to act at the elbow. It acts to maintain force equilibrium against the combined weight forces of the load and forearm–hand segment. Thus:

$$\begin{aligned} \sum \text{forces}_E &= 0 \\ -49 \text{ N} - 15.8 \text{ N} + R \text{ elbow} &= 0 \\ R \text{ elbow} &= 64.8 \text{ N (upward)} \end{aligned} \tag{6.4}$$

In this particular case, the single elbow-reactive force, which is created by ligaments and muscle actions at the elbow, is not acting through the line of actions of the external forces (weights). Thus, the reactive force of 64.8 N is sufficient to keep the forearm and hand from moving in a line (*trans-*

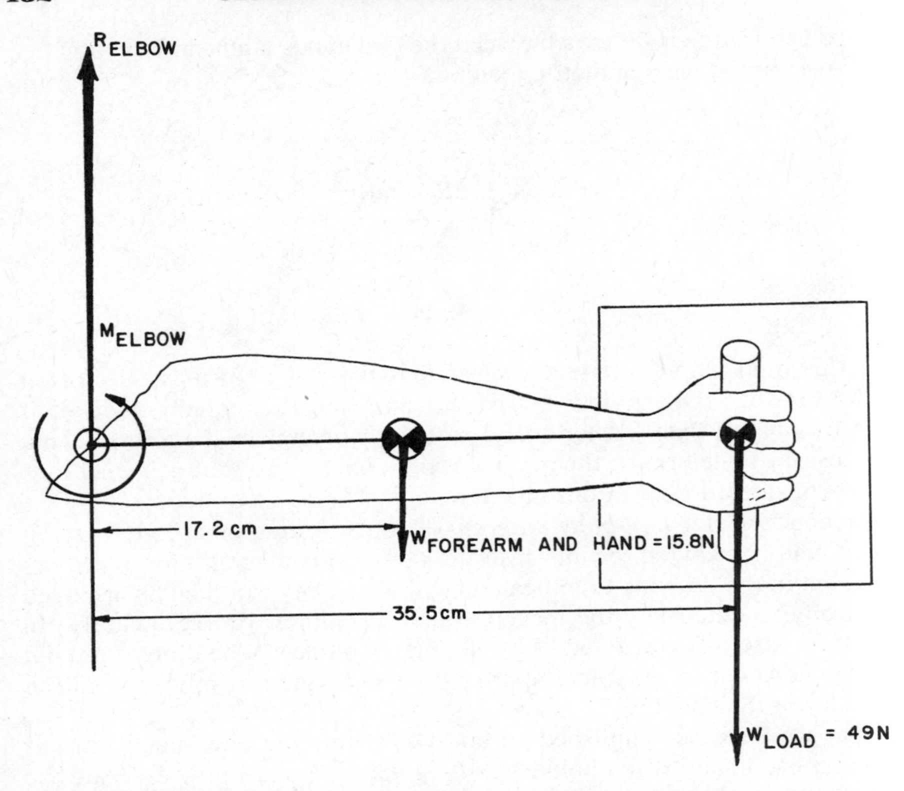

FIGURE 6.2 Free body diagram of forearm and hand in horizontal position holding load.

lational motion), but it is not able to stop the segment from *rotational motion*. In this case, the body segment and load weights, acting a distance away from the supporting elbow-reactive force create a *moment*—that is, a tendency to rotate. The magnitude of a moment is simply the product of a force and the perpendicular distance (i.e., its *lever* or *moment arm* distance) from its line of action to the point of rotation. In our example, with the forearm–hand segment held horizontal, its weight has a moment arm from the elbow of 17.2 cm, based on average male anthropometry. Moments, like forces, are vectors, and thus direction about a point of rotation as well as magnitude must be considered. The combined moment effects of the two weights shown in Figure 6.2 are counteracted by an equal but opposite-in-direction reactive moment M_E at the elbow, otherwise the segment will rotate. The *second condition of equilibrium* now can be stated:

$$\sum \text{moments} = 0$$

which yields, assuming downward forces to be negative,

$$\begin{aligned} 17.2 \text{ cm } (-15.8 \text{ N}) + 35.5 \text{ cm } (-49 \text{ N}) + M_E &= 0 \\ (-271.8 \text{ N cm}) + (-1739.5 \text{ N cm}) + M_E &= 0 \end{aligned} \tag{6.5}$$

$$M_E = 2011.3 \text{ N cm or } 20.113 \text{ N m (counterclockwise)}$$

We now have the answer to the question posed earlier regarding the external forces and moments at the elbow during the holding of a 10-kg mass. In the process we have determined that the load creates two types of effects—one, a force effect that moves objects along the line of action of the force; the second, a moment effect that tends to rotate segments about supporting joints. The reactive forces are important at a joint for they indicate the magnitude of the necessary tensile forces in ligaments and muscles holding a joint together and shearing and compressive forces acting on joint-contact surfaces. The reactive moment at a joint are also important in that they represent the strength moment required of specific muscle actions to maintain posture or impart motions.

In the example, if the forearm-hand segment were not horizontal, but held at some other angle, the moment at the elbow would be reduced. Figure 6.3 depicts the forearm–arm segment at various rotation angles measured relative to the horizontal. As can be seen, the greatest moment is at the horizontal, when the moment arm d is equal to the distance from the elbow joint to the center of mass (17.2 cm). As the arm is lowered or raised, the moment arm d decreases as a function of the cosine of the angle θ. When the arm is hanging vertically, $d = 0$. In this orientation there is no moment requirement and the elbow flexor and extensor muscles can relax. Numerically, the elbow moment equation for an average male is

$$\begin{aligned} M_E &= dW \\ &= \cos\theta \ 17.2 \ (15.8) \end{aligned} \tag{6.6}$$

For the case where a load is held in the hand, the equation for an average man becomes

$$M_E = \cos\theta \ [17.2 \ (15.8) + 35.5 \text{ (load)}] \tag{6.7}$$

Thus, the load, like the weight of the arm, has an additive effect on the elbow moment, with its maximum moment value when the arm is horizontal and a minimum (zero) effect when the arm is vertical. It is worth noting that because objects are held in the hands at the end of the forearm, when the forearm is horizontal the moment arm for these loads, relative to that of the forearm weight, can become large (35.5 cm vs. 17.2 cm). The large load

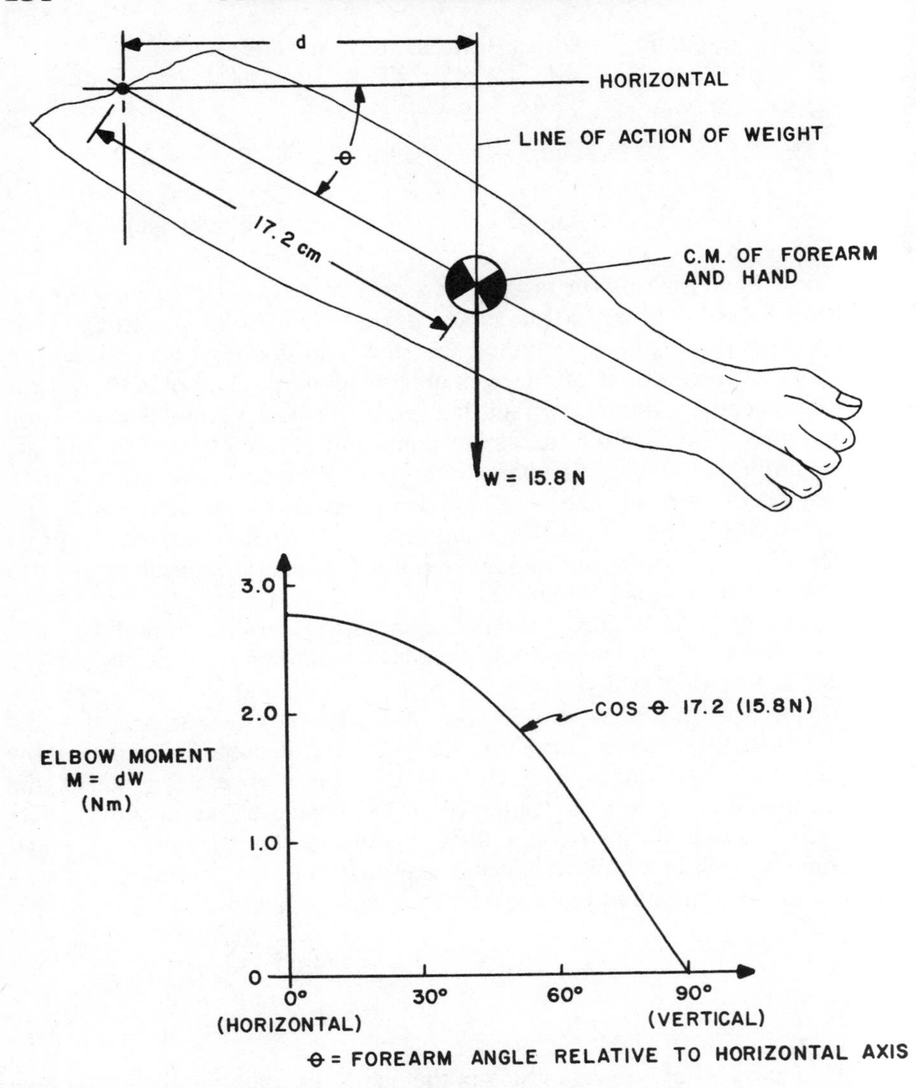

FIGURE 6.3 Elbow moment due to forearm–hand weight held in various postures (average male anthropometry).

moment arm, when multiplied by the magnitude of the load, results in high elbow load moment values. Thus, one finds that it may not be difficult to pick an object up from a table and carry it with the forearms near vertical, but that it may be impossible to lift the object to place it on a shelf. Later in this chapter a discussion of this postural effect in load lifting will be presented, further expanding on the moment concept.

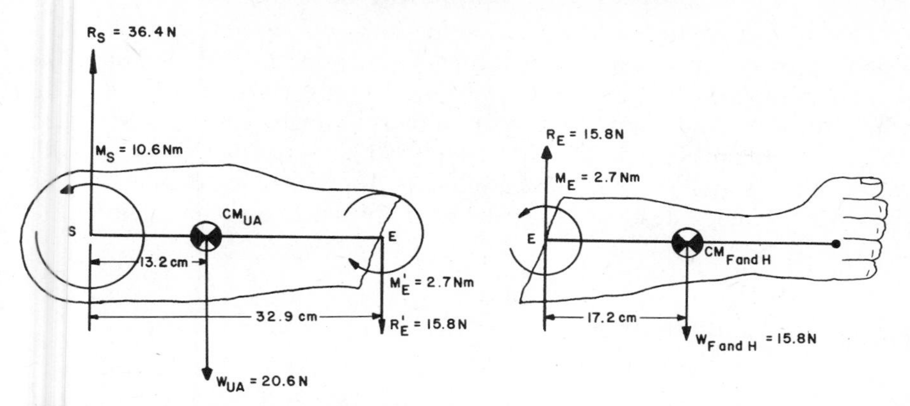

FIGURE 6.4 Two-link static planar analysis of upper arm and forearm–hand segments held horizontal.

6.2.2 Two-Body Segment Static Model

In the preceding cases the forearm and hand were considered as one body segment. The mass center location for the combined segment was given. When body segments retain a constant orientation to each other, it is sometimes easiest in an analysis to first estimate the center of mass of the combined segments and simply add the external load moments, as done in the preceding case. Often, however, an analysis will require body segment postures to be systematically varied, as well as varying the external loads. In this situation it is best to treat each body segment as a separate link in a *kinetic chain*. The analysis starts at the point of application of external load (often the hands) and proceeds in sequence, solving the equilibrium equations for each body segment, until reaching the segment that supports the body (usually the feet).

What follows is an analysis for the upper arm and forearm–hand segments treated as two separate links. At first, no external load is included. Later in the analysis a load will be added to the hand. First consider the case of the arm and forearm–hand held horizontal, as in Figure 6.4. From the preceding analysis we found the *elbow reactive force and moment* necessary to maintain static equilibrium. For an average man, the values were:

$$
\begin{aligned}
R_E &= W_{F\&H} \\
&= 15.8 \text{ N} \\
&= 15.8 \text{ N (upward)} \\
M_E &= (ECM_{F\&H})(W_{F\&H}) \\
&= 17.2 \text{ cm } (15.8 \text{ N}) \\
&= 2.7 \text{ N m (counterclockwise)}
\end{aligned}
$$

To carry the analysis to the adjoining link—the upper arm—all of the external forces and moments operating on the arm must be considered. In this case it means that the weight W_{UA} of the upper arm and the resultant elbow force R_E and moment M_E caused by the weight of the forearm–hand link must be considered. The static equilibrium equations at the shoulder result in a reactive force R_S and a moment M_S for the average man (a line over the variables designates a distance between two points identified by the variables):

$$\sum F_S = 0$$

$$-W_{UA} - R_E + R_S = 0$$

$$R_S = W_{UA} + R_E \tag{6.8}$$

$$= 20.6 \text{ N} + 15.8 \text{ N}$$

$$= 36.4 \text{ N (upward)}$$

$$\sum M_S = 0$$

$$(\overline{SCM}_{UA})(-W_{UA}) + (\overline{SE})(-R_E) - M_E + \text{M}_S = 0 \tag{6.9}$$

$$\begin{aligned} M_S &= (\overline{SCM}_{UA})(W_{UA}) + (\overline{SE})(R_E) + (M_E) \\ &= 13.2 \text{ cm } (20.6 \text{ N}) + (32.9 \text{ cm})(15.8 \text{ N}) + 2.7 \text{ N m} \\ &= 2.7 \text{ N m} + 5.20 \text{ N m} + 2.7 \text{ N m} \\ &= 10.6 \text{ N m (counterclockwise)} \end{aligned}$$

Intuitively, we know that it is much more strenuous to hold the entire arm in a horizontal position than just the forearm–hand segment. One reason for this increased exertion level should now be clear. By including the upper arm as part of the link system, the upper arm weight created a moment at the shoulder of 2.7 N m, and the forearm–hand weight created a moment of 5.2 N m. Both of these moments add to the elbow moment of 2.7 N m to create the high shoulder moment. As we shall see, if an extra load exists in the hand, the shoulder moment becomes very large when the arm is horizontal, due to the large moment arms provided by the arm-forearm links.

As an example of this, consider the hand load of 49 N used in the previous one-link analysis. Recall from equations 6.4 and 6.5 that this created an elbow reactive force and moment of

$$R_E = 64.8 \text{ N}$$

$$M_E = 20.1 \text{ N m}$$

The static equilibrium equations for the shoulder are not altered in form by the inclusion of this additional load, since the load effect is already included in the elbow analysis. The values for an average man holding a 49-N load in the hand with the horizontal arm posture shown in Figure 6.4 are

$$
\begin{aligned}
R_S &= W_{UA} + R_E \\
&= 20.6 \text{ N} + 64.8 \text{ N} \\
&= 85.4 \text{ N (upwards)} \\
M_S &= (\overline{SCM}_{UA})(W_{UA}) + (\overline{SE})(R_E) + (M_E) \\
&= 2.7 + 21.3 + 20.1 \\
&= 44.1 \text{ N m (counterclockwise)}
\end{aligned}
$$

Thus, a 49 N load in the hand quadruples the moment requirement on the shoulder compared with the unloaded arm. In fact, the shoulder flexion strength moment data discussed in Chapter 4 show that a shoulder moment of 44.1 N m may only be achievable by approximately 50% of women and 95% of men.

If a different arm posture than the horizontal were to be considered, the preceding analysis would be altered to consider the angles of both segments. Figure 6.5 displays a more general configuration of the arm for planar static analysis. The static equilibrium equations for this form now include two angles for the shoulder and elbow, measured relative to the horizontal axis:

$$\sum F_E = 0$$

$$-L_H - W_{F\&H} + R_E = 0 \tag{6.10}$$

$$R_E = L_H + W_{F\&H}$$

$$\sum M_E = 0$$

$$\cos \theta_E[\overline{EH}(-L_H) + \overline{ECM}_{F\&H}(-W_{F\&H})] + M_E = 0 \tag{6.11}$$

$$M_E = \cos \theta_E[\overline{EH}(L_H) + \overline{ECM}_{F\&H}(W_{F\&H})]$$

$$\sum F_S = 0$$

$$-W_{UA} - R_E + R_S = 0 \qquad \text{(same as eq. 6.8)}$$

$$R_S = W_{UA} + R_E$$

$$\sum M_S = 0$$

$$\cos \theta_S[\overline{SE}(-R_E) + \overline{SCM}_{UA}(-W_{UA})] - M_E + M_S = 0$$

$$M_S = \cos \theta_S[\overline{SE}(R_E) + \overline{SCM}_{UA}(W_{UA})] + M_E \tag{6.12}$$

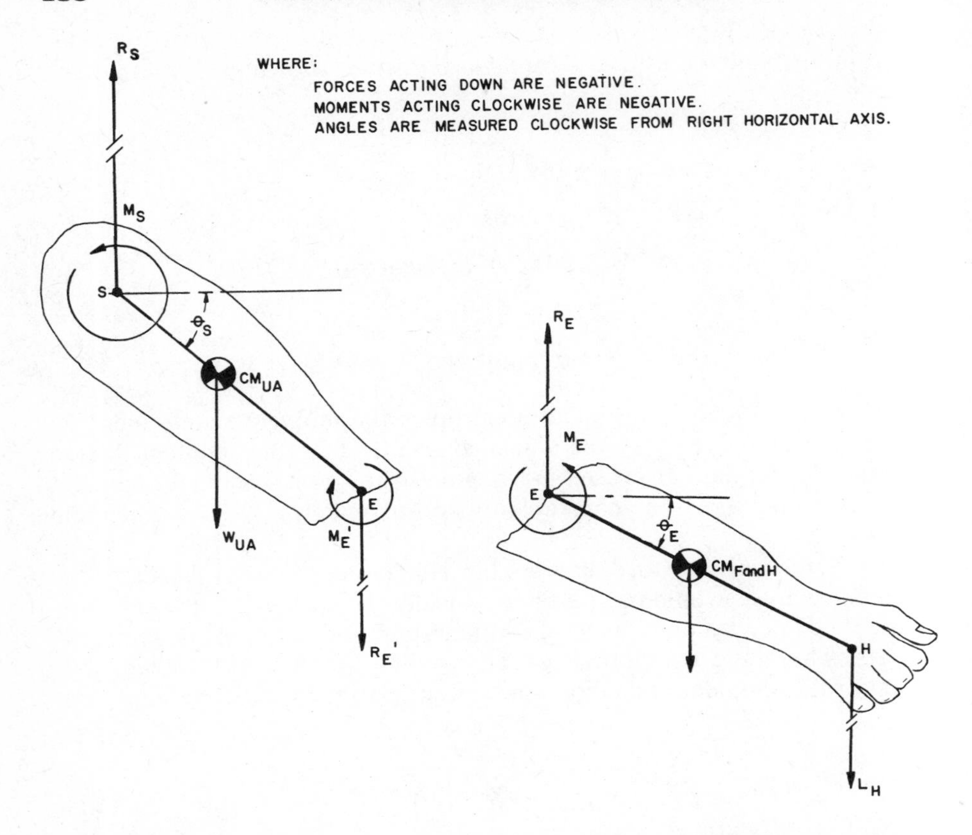

FIGURE 6.5 Two-link static planar analysis of upper arm and forearm–hand segments including angle effect.

It should be noted that altering the posture of the arm will have a great effect on the moments at the elbow and shoulder, but will have no effect on the external reactive forces since they have remained as a *parallel force system*. A change in the direction of the load acting on the hands, however, will have an effect on both the reactive moments and forces at the joints. This case is considered next.

6.2.3 Static Planar Model of Nonparallel Forces

The preceding models have all considered gravity as the source of external forces and hence have assumed *parallel force systems*. If a person is pushing or pulling on a load, however, this is not the case. This complexity can be considered in planar static models by resolving the forces into two orthogonal components. A Cartesian coordinate system is usually used, with one axis vertical and the other horizontal in a planar analysis. To accomplish this,

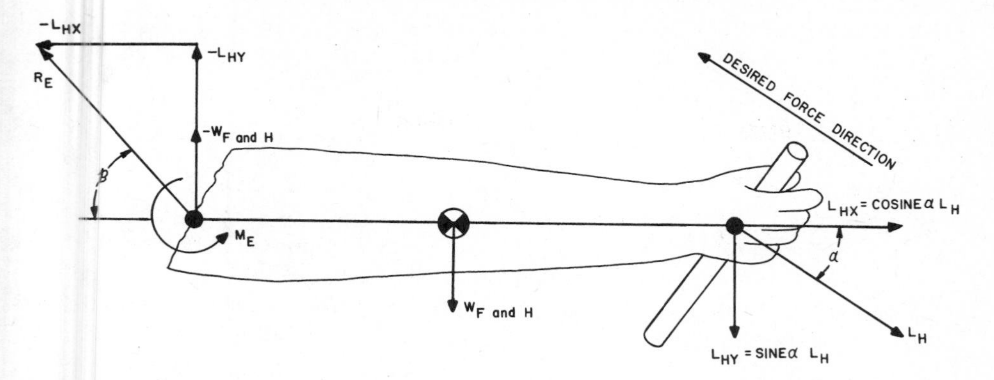

FIGURE 6.6 Illustration of non-parallel coplanar concurrent force system requiring resolution of load force on hands and composition of reactive forces at elbow.

recall that any force can be treated as the vector sum of any two other forces. In essence, a given force of interest is considered to be the *resultant* of a pair of *component* forces acting perpendicular to each other and with a common point of application for all three. In this case, these forces are said to be *coplanar* (in one plane), nonparallel and *concurrent* (have a common point of application). Figure 6.6 shows such a force system for a load on the hand resulting from the act of pulling on a handle. The two components of the load are shown acting along the horizontal (X) and vertical (Y) axes. The magnitude of each is found by application of the cosine law, which yields:

$$L_{HX} = L_H \cos\alpha \tag{6.13}$$

$$L_{HY} = L_H \sin\alpha \tag{6.14}$$

Once a non-parallel force is resolved into components that are either parallel or perpendicular to other forces in the system—for example, body segment weights—it becomes much easier to combine the forces into a *composite* force system by simply adding or subtracting the colinear components. This is demonstrated in Figure 6.6, which shows the reactive force at the elbow composed of the two vertical forces $W_{F\&H}$ amd L_{HY}, and the horizontal force L_{HX}. The magnitude of the elbow reactive force can be found by application of the Pythagorean theorem of right triangles. This states that the magnitude of the hypotenuse is equal to the square root of the sum of the squared lengths of the sides of a right triangle. In this case, the hypotenuse is the desired reactive force magnitude R_E and the sides of the triangle are the values of L_{HX} and the sum of L_{HY} and $W_{F\&H}$. Thus:

$$R_E = \sqrt{L_{HX}^2 + (L_{HY} + W_{F\&H})^2} \tag{6.15}$$

The angle β of R_E relative to the horizontal can be found by application again of the cosine equations (6.13 and 6.14) or by using the tangent relationship:

$$\text{cotan } \beta = \frac{L_{HX}}{L_{HY} + W_{F\&H}} \tag{6.16}$$

In a multisegment analysis, the reactive force components are used in the analysis of forces acting at adjacent joints. The composite reactive force is sometimes computed at the end of the linkage to determine the total effect (magnitude and direction) of all external forces acting on the body. At each joint in a kinetic linkage the individual reactive force components are necessary, as they indicate the external forces acting to create compression, shear, and tensile forces in bone and ligament structure at a joint.

The elbow moment equation must also be expanded in a nonparallel force system analysis. For the system shown in Figure 6.6, the equation becomes:

$$\sum M_E = 0$$

$$\overline{ECM}_{F\&H}(-W_{F\&H}) + \overline{EH}_X(-L_{HY}) + \overline{EH}_Y(-L_{HX}) + M_E = 0$$

In other words, the two components of the load are each multiplied by their respective moment arms. In this particular case, however, the horizontal component L_{HX} has a line of action through the elbow, and thus its moment arm EH_Y is zero. Ignoring this term yields

$$M_E = \overline{ECM}_{F\&H}(W_{F\&H}) + \overline{EH}_X(L_{HY}) \tag{6.17}$$

Of course, if the arm were not horizontal while pulling on the handle, then both load components could have finite moment arms that would have to be computed by the cosine law.

6.2.4 Planar Static Analysis of Internal Forces

The preceding discussion has considered only external forces acting on a body segment under gravity or the pull force of a control handle. These forces produce what we have referred to as *load moments* at various body joints. These load moments can be compared to muscle strength moments, and thus provide a means to evaluate the demand for different strengths when one performs a specific manual task in industry. This particuar strength evaluation method will be discussed in detail in Section 6.5.3 of this chapter.

Besides providing insight into the strength moment requirements of a task, the preceding modeling techniques are the basis for estimating the forces on various musculoskeletal tissues. To analyze these internal forces it is necessary to know the force line of action and point of application of muscles

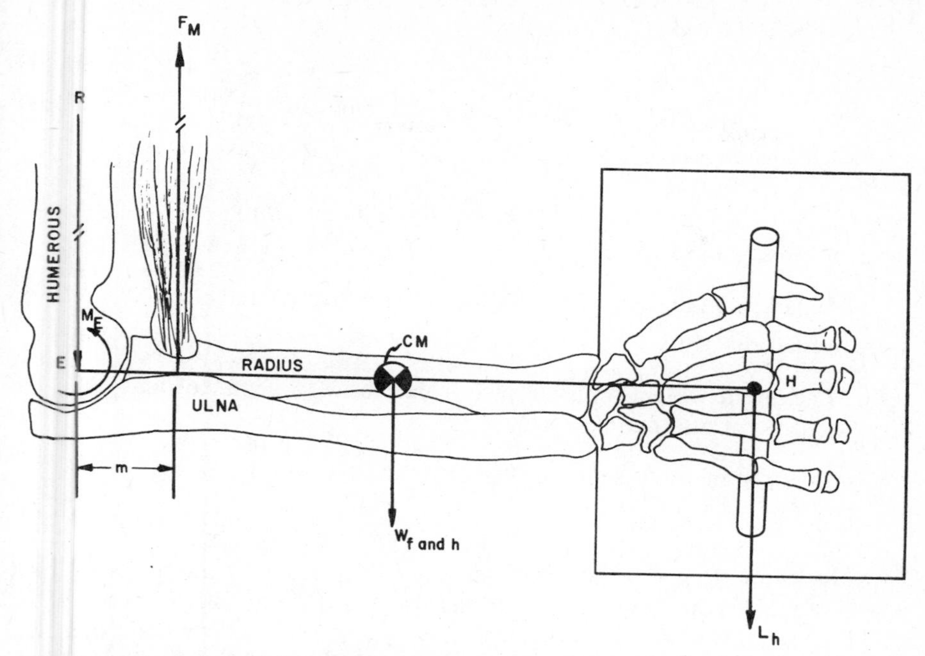

FIGURE 6.7 Analysis of internal muscle force F_M and joint reaction force R created by weight of forearm–hand $W_{F\&H}$ and load L_H held horizontal.

within the musculoskeletal structure. For this reason, kinesiological data is of importance in biomechanical modeling as well as anthropometric and mechanical capacity data.

The following example for elbow flexion will demonstrate such a method of analysis. In this case, consider the elbow to be at 90°, with the forearm–hand segment held horizontal. The primary muscle action is assumed to be provided by the biceps brachii. Its point of force application where its distal tendon inserts on the radius is assumed to be 5 cm from the elbow center of rotation, with a vertical line of action. Figure 6.7 illustrates this system. Note that the reactive force R necessary to stabilize the point of rotation against translation—the elbow center—acts through the center. Therefore, it does not contribute a moment effect. The resulting moment equation can be expressed as follows:

$$\sum M_E = 0$$

$$-M_{load} - M_W + M_{FM} = 0$$

$$M_{FM} = M_{load} + M_W$$

which can also be expressed as:

$$m(F_m) = \overline{EH}(L_H) + \overline{ECM}(W_{F\&H}) \tag{6.18}$$

F_m is the force of muscle required to stabilize arm

m muscle moment arm

$\overline{EH}$ load moment arm from elbow to hand grip center

L_H load on hand

$\overline{ECM}$ segment weight moment arm from elbow to center of mass of forearm and hand

$W_{F\&H}$ weight of body segment (forearm and hand)

From the example given earlier (Figure 6.2), the reactive moment M_E for an average male arm holding a 49 N load was found to be 20.1 N m = (17.2 × 15.8) + (35.5 × 49). In this new analysis it is assumed that the muscle force F_m acting though its moment arm m is capable of exerting a muscle strength moment $m(F_m)$ of equivalent magnitude. Thus, the moment equilibrium conditions result in

$$m(F_m) = 20.1 \text{ N m}$$

and with m = 5 cm (0.05 m):

$$F_m = \frac{20.1}{.05} = 402 \text{ N (upward)}$$

Thus, an external load of 49 N in the hand creates an internal muscle force requirement almost 10 times greater (402 N). This mechanical disadvantage of muscles—that their moment arms are often small compared with those of external loads—can require muscles all over the body to be highly stressed while performing tasks in certain postures, even though the external loads appear to be relatively low.

The analysis of internal forces in this case would not be complete without also determining the joint reactive force R_E acting on the distal trochlea of the humerus in the trochlear notch of the ulna. This is easily accomplished in the parallel force system of Figure 6.2 by simply summing all forces, including both the external forces and the internal muscle force. The direction (or sign) of R_E depends on the solution in such a case.

$$\sum F = 0$$

$$R_E + F_m - W_{f\&h} - L_h = 0$$

$$-R_E = -F_m + W_{f\&h} + L_h \tag{6.19}$$

$$R_E = -402 + 15.8 + 49$$

$$= -337.2 \text{ N (downward)}$$

This indicates that the high forces created by muscles during a voluntary exertion impose equally large forces on adjoining skeletal structures. Later, when low back modeling is discussed, this concept of muscles loading skeletal structures will be used to explain how a 200 N load in the hand can create a compression load on the L_5/S_1 disc of more than 4000 N.

One of the most important assumptions in the preceding estimation of internal forces is that a singular muscle action accounts for the internal force. More often, several muscles share the moment requirement at a joint. In fact, subtle changes in posture may result in one muscle in a group contributing substantially or contributing little to the required action. Such is the case at the elbow, as described by MacConaill and Basmajian (1969). There are three primary flexor muscles at the elbow—the biceps brachii, the brachialis, and the brachioradialis. Electromyographic studies have shown that, in very general terms, the brachioradialis is highly active in fast forearm motions but not nearly as active during slow flexion. Further, the biceps brachii is highly active along with the brachialis during elbow flexion with the forearm-hand supine (palm up), but much less active if the forearm–hand is prone (palm down). In fact, this differential effect of the biceps led Tichauer (1967) to propose that the biceps be considered a major outward rotator of the forearm, as well as an elbow flexor muscle.

From a biomechanical modeling perspective, the preceding discussion presents major problems. First is the problem of *muscle force redundancy*. That is, the presence of more than one muscle action requires that several force lines of action and points of application be considered simultaneously in the analysis. This makes the trigonometry more difficult. More importantly, however, it also means that there may be more muscle forces of unknown magnitude in the system than there are independent equations, and thus a unique solution is not possible. The problem in this case is said to be *statically indeterminate*. This limitation in modeling can be demonstrated in our present example by assuming that two muscles flex the elbow—the brachialis, which appears to be the primary flexor under most conditions, and the biceps brachii. Figure 6.8 illustrates a reasonable configuration for these muscles. The external loads are assumed to remain as

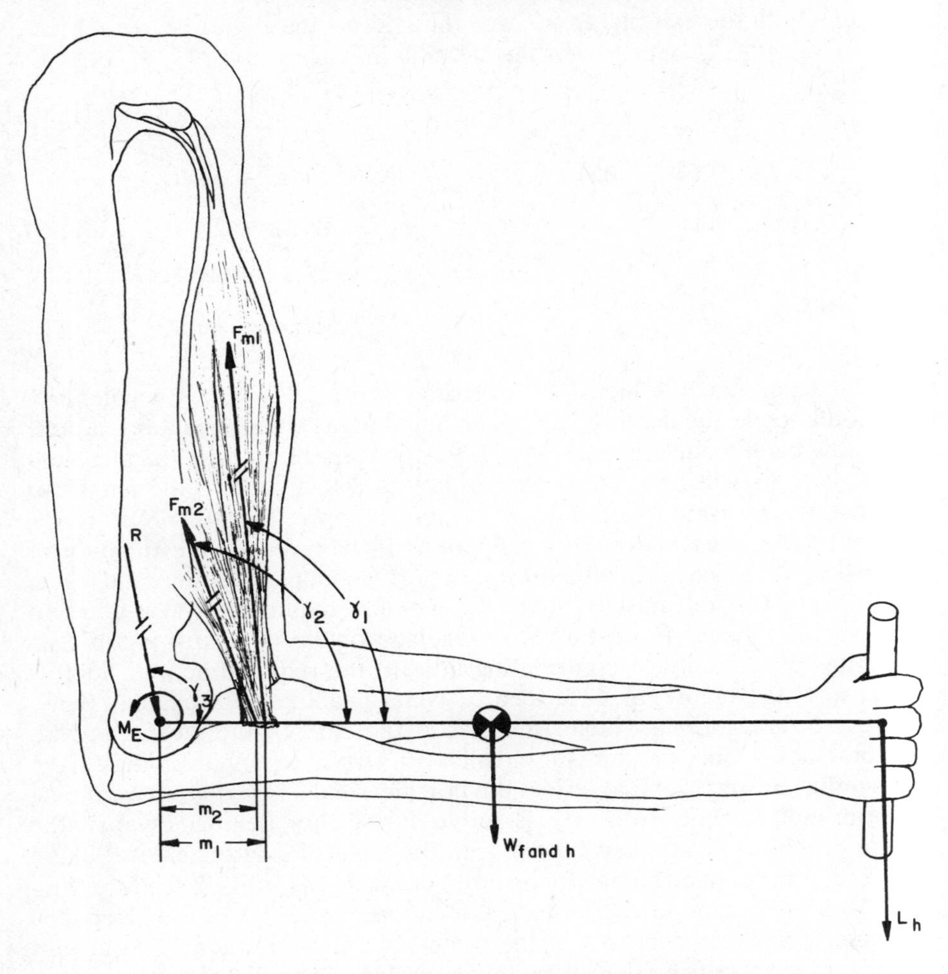

FIGURE 6.8 Two-muscle system for elbow flexion.

before—average male anthropometry and a load held with the forearm horizontal. The moment equilibrium equation (equation 6.18) is expanded to include the actions of both muscles, along with their lines of action and their insertion points of force application, yielding

$$\sin \gamma_1 m_1 F_{m_1} + \sin \gamma_2 m_2 F_{m_2} - M_E = 0 \tag{6.20}$$

where

γ_1 and γ_2 are muscle line-of-action angles known from anatomy
m_2 and m_2 are muscle points-of-force-application distances known from anatomy

M_E is the known reactive moment at elbow due to external loads
F_{m_1} and F_{m_2} are muscle forces to be estimated

Likewise, the force equilibrium equations are expanded to include both muscle forces and to consider the fact that they are no longer parallel. This gives two independent equations for forces in the X (horizontal) and Y (vertical) directions. The equations are

$$\sum F_X = 0$$

$$\cos\gamma_3 R_E + \cos\gamma_1 F_{m_1} + \cos f_2 F_{m_2} = 0 \tag{6.21}$$

and

$$\sum F_Y = 0$$

$$\sin\gamma_3 R + \sin\gamma_1 F_{m_1} + \sin\gamma_2 F_{m2} - W_{f\&h} - L_h = 0 \tag{6.22}$$

In the force equations, two additional variables of unknown magnitude have been included: the reactive force R_E and its angle of action γ_3. Thus the problem has a total of four unknowns and only three independent equations. A unique solution is not possible without more information.

Several different approaches have been used to solve the problem. First, an assumption can be made that each muscle exerts a force in proportion to its cross-sectional area. Thus cross-section data, available in various anatomy books, can be consulted to provide the additional information necessary to form a fourth independent equation:

$$F_{m_1} = KF_{m_2} \tag{6.23}$$

where K is the ratio of two muscle's cross-sectional areas. The values of K can also be estimated by using EMG amplitudes that approximate the relative force levels within muscles for a given activity.

Another approach is to perform a three-dimensional (non-coplanar) analysis, as discussed later in this chapter. In the process of setting up the system in three dimensions, of course, the muscle lines of action would have to be described in relation to three axes. This additional information would provide three force-equilibrium equations and three moment-equilibrium equations, while only introducing one more unknown (the angle for the reactive force R_E in relation to the third axis).

A completely different approach is to assume that the body is attempting to minimize or maximize some particular function. An example could be to minimize the total muscle exertion—that is, minimize $F_m = F_{m_1} + F_{m_2}$. Various mathematical methods have been developed to allow consideration of this and other optimization criteria (see Crowninshield, 1978; Crowninshield and Brand, 1981; Seireg and Arvikar, 1975). Obviously, the results obtained depend on the selection of appropriate biomechanical objective functions. Unfortunately, at this time it is not clear which particular biomechanical objective is dominant during the performance of a manual task. In one situation the body's learned motor responses may cause a minimization of individual muscle forces or some combination of muscle forces. In another situation the body may be limiting certain forces on ligaments or contacting joint surfaces. Clearly, past performance of the task alters muscle coordination and enhances overall motor skill by subtle alterations in individual muscle actions. Also, wide variations in individual motor behavior patterns are known to exist. For all these reasons, the modeling of multiple muscle and joint forces during the performance of various manual exertions remains a major research activity in biomechanics. For most of the remainder of this text, when internal forces are discussed, single muscle equivalent models will be used. This assumes that multiple muscle actions can be considered to exert a composite effect with a known single line of action and point of application.

6.2.5 Multiple-link Co-planar Static Modeling

The discussion in Section 6.2.2 describes a static two-link analysis of the external forces acting on the upper extremity. In that case a load held in the hands was combined with body segment weights to produce load moments and reactive forces at the elbow and shoulder joints. This was accomplished by using the fact that the reactive forces and moments acting at the elbow can be added as vectors to the weight of the upper arm to estimate the effects at the shoulder. Thus, the solution procedure is first to solve the equilibrium equations for the load moments and reactive forces at the joint adjacent to the application of an external load—for example, at the elbow for our preceding examples. Then the resulting values are used to solve the equilibrium conditions for the next adjacent joint—for example, the shoulder. The procedure is continued in sequence until all the load moments and joint reactive forces are determined at each joint in the kinetic linkage system.

One co-planar, multiple-linkage static model is depicted in Figure 6.9 for symmetric sagittal plane activities while lifting. In this system the forces are considered to act in parallel, producing only one force equilibrium condition in the vertical direction. The reactive forces at each joint can be expressed as follows:

$$\sum R_j = 0$$
$$\sum R_j = R_{j-1} + W_L \qquad (6.24)$$

where

R_j are the reactive forces at each j joint

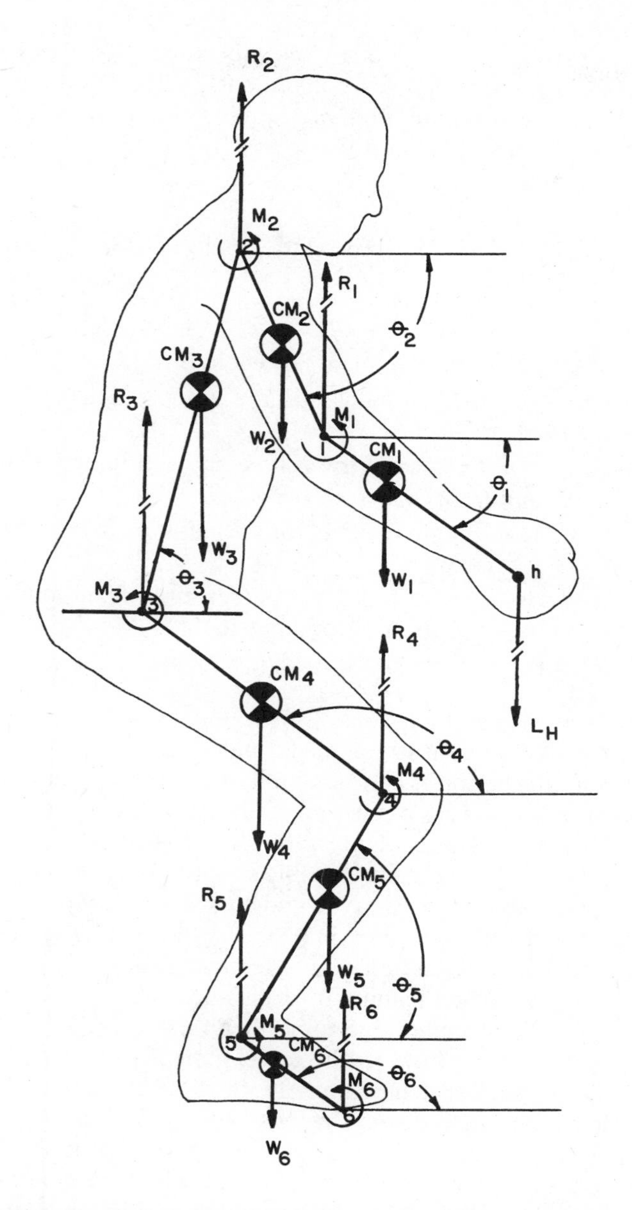

FIGURE 6.9 Six-link coplanar parallel static force system for analysis of sagittal plane lifting showing reactive forces and moments at major joints.

$R_{j\text{-}1}$ are the reactive forces at the previous adjacent joint in the solution
W_L are the weights of each L link

Because the links are not parallel, the joint angles θ_j with respect to a reference axis—horizontal right is convenient—are considered when estimating the moment arms. The moment equilibrium conditions can then be expressed in the following general form:

$$\sum M_j = 0$$

$$M_j = M_{j\text{-}1} + [\overline{jCM_L}\,(\cos\theta_j)W_L] + [\overline{j\,j\text{-}1}\,(\cos\theta_j)R_{j\text{-}1}] \qquad (6.25)$$

where

M_j are the reactive load moments at each joint j
$\overline{jCM_L}$ are the distances from joint j to the Center of Mass of link L (from anthropometric data)
θ_j are the postural angles of the L links at each joint j with respect to the horizontal right axis
W_L are the body segment Weights for each link L (from anthropometric data)
$\overline{j\,j\text{-}1}$ are the body segment link lengths measured from joint j to the adjoining joint j-1 (from anthropometric data)
$R_{j\text{-}1}$ are the reactive forces at the adjacent joints j-1

The reactive forces and moments for three different postures are shown in Figure 6.10. Average anthropometric male data were used in the computations. From inspection it can be readily seen that posture has no effect on the external ractive forces but has a very large effect on the load moments. Since the skeletal muscles respond to the load moments, it is possible even with simple static models to gain valuable insight into what postures require specific muscle groups to be active, and to what relative magnitude each muscle group must contract. For instance, when standing erect with the arms held out in front of the body, only a small moment is produced in the low back/hip region. When leaning forward and reaching out, however, the combination of the relatively large torso–head mass and large moment arm for the arm masses contribute to the development of a large load moment at the low back/hip region. Unfortunately, in practice, extended reaches are sometimes necessary in industry because of the layout of machines and workplaces. If not carefully evaluated, from a biomechanics perspective, extended reaches can cause low back pain and other musculoskeletal problems, even though a load in the hands does not exist or is relatively light in weight.

In this context, the addition of a load to the hands will increase the load moments at each joint in proportion to the moment arm distances that the

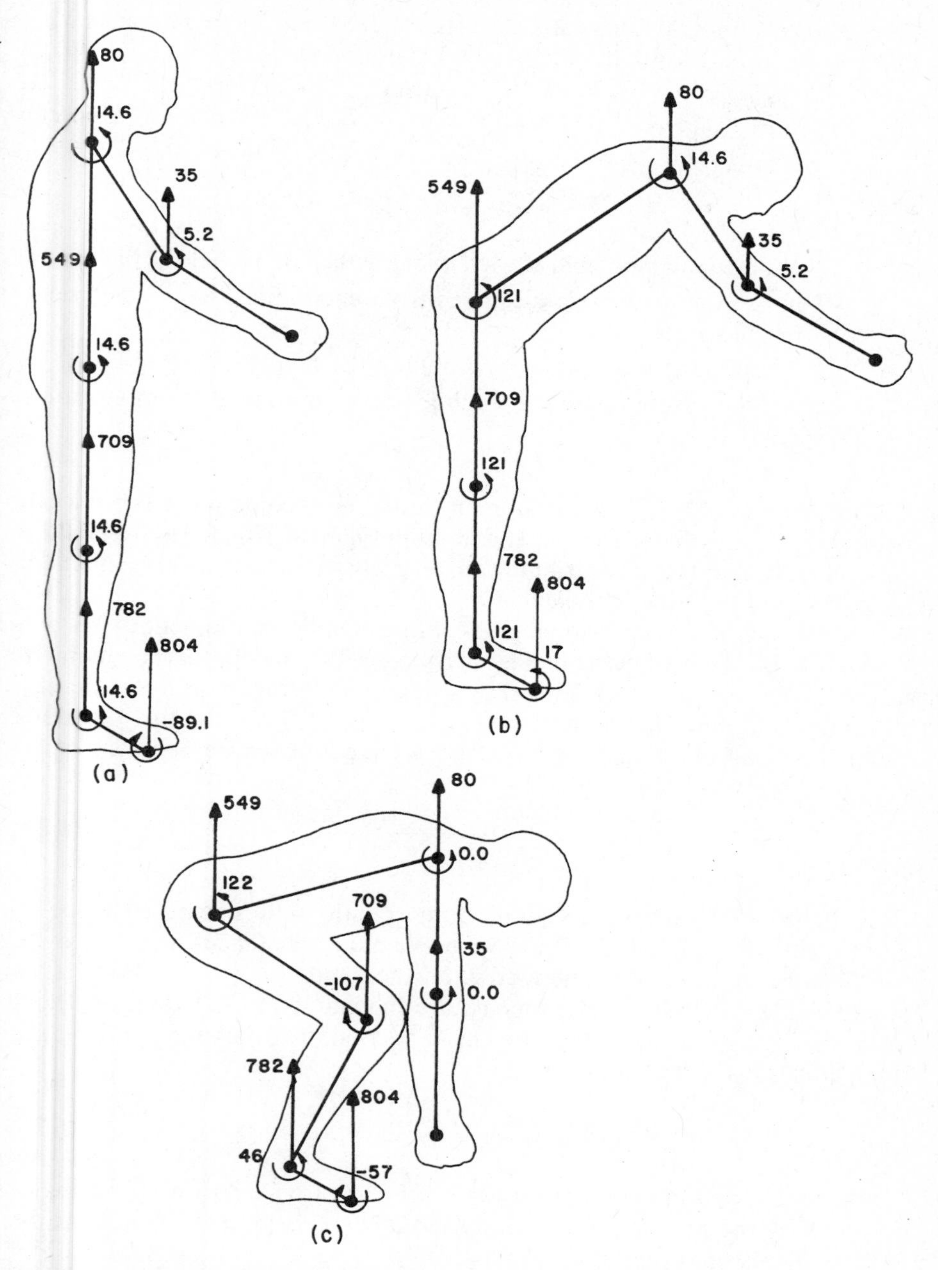

FIGURE 6.10 Reactive moments (N m) and forces (N) for three different postures using average male anthropometric data (no load in hands).

load is away from each joint. If a scale drawing is made of the posture of interest, these moment arms can be scaled, and thus the moment values with no load (equation 6.25) can be expanded as follows:

$$M_{j/L} = M_{j/L=0} + \overline{j\text{-}h}(L) \tag{6.26}$$

where

$M_{J/L}$ are the load moments at each joint j with a load L held in the hands
$\overline{j\text{-}h}$ are the moment arm distances from each joint j to the load held in the hands
L is the magnitude of the load held in the hands
$M_{J/L=0}$ are the load moments at each joint j with no load (from equation 6.25)

As can be imagined, the $\overline{j\text{-}h}$ moment arm can become quite large when leaning forward with the arms extended, as in Figure 6.10b. In such a posture the load in the hands can contribute considerably to the already large moment at the hips while lifting an object.

Another method of determining how much additional moment exists at a joint due to a load held in the hands is based on the concept that the moment arm distances $\overline{j\text{-}h}$ can be estimated by using the difference between the moments at each joint M_j with no load in the hands ($L = 0$) and the M_j values with a 1 N load in the hands ($L = 1$ N). This can be expressed as:

$$\overline{j\text{-}h} = \frac{(M_{j/L=1} - M_{j/L=0})}{(L = 1\text{ N})} \tag{6.27}$$

This requires that the static equilibrium equations be computed twice to solve for the two M_j values. Once the moment arm distances $\overline{j\text{-}h}$ are known, each can be multiplied by the load L on the hands to give the $M_{j/L}$ values due to the load. These load moments are then added to the values without load $M_{j/L=0}$ to yield the total moment at each joint. Combining and simplifying equations 6.26 and 6.27 yields:

$$M_{j/L} = M_{j/L=0} + (M_{j/L=0} - M_{j/L=1})L \tag{6.28}$$

When the solution is set up in this form it provides an easy means of determining the relative effects of varied loads L held in the hands on the load moments at various joints in a given posture. If varied postures are of interest while holding a constant load in the hands, then the best method is to include the load along with the forearm–hand weights in computing the elbow-load moment $M_{j=1}$ and reactive force $R_{j=1}$ for each posture of interest. This is exactly what was done in the earlier examples of the single link model with a hand load (see Figure 6.2 and equation 6.7). Once the elbow moments and reactive forces are computed including the hand load, the effect of the hand load is automatically included in the sequential computations for the remaining joint moments and forces, as shown in equation 6.25.

6.3 STATIC THREE-DIMENSIONAL MODELING

Two-dimensional coplanar models are very useful in the evaluation of many occupational tasks. In some cases a person will use only one arm when lifting, pushing, or pulling an object, while the other arm is used to counterbalance or stabilize the rest of the body. In such a situation the external forces acting on the body must be treated in three dimensions, and the forces are considered to be *noncoplanar*. This results in six independent equilibrium equations in reference to three orthogonal axes at each joint:

$$\sum F_X = 0 \qquad \sum F_Y = 0 \qquad \sum F_Z = 0$$

$$\sum M_X = 0 \qquad \sum M_Y = 0 \qquad \sum M_Z = 0$$

Because there now are six equations, six unknowns can be considered in various problems. It should be realized at this point that in the preceding coplanar models the moments were estimated about a center of rotation, and a clockwise/counterclockwise designation was acceptable for its directional effect. In a noncoplanar analysis it is necessary to designate the moment directions with respect to one of the reference axes. In these cases a right-hand rule is usually adopted in which the line of action of the moment is defined as operating along a line perpendicular to the plane in which the force and moment arm exist. The direction along such a line is defined by allowing the fingers of the right hand to point in the direction of the force line of action that causes the moment. The thumb then indicates the direction of the moment. Figure 6.11 depicts this rule. Keep in mind that the rule is useful when adding noncoplanar moments, in which a consistent direction of action must be considered to allow the necessary vector operations. In the physical sense, a moment operates in the plane of the force and its moment arm that creates the moment, as was the case in the coplanar models illustrated earlier.

Because it is difficult to geometrically depict and solve three-dimensional force and moment systems, a basic understanding of vector algebra is necessary. It is beyond the scope of this text to present such knowledge in a comprehensive form, but a simple example of its application is given to demonstrate the solution methods that are possible. Further basic treatment of the topic can be found in the text by Shames (1967), and additional discussion of the topic is given in Miller and Nelson (1976).

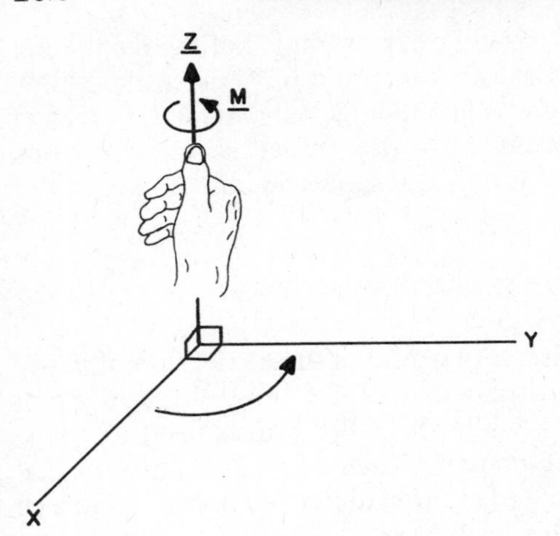

FIGURE 6.11 Right-handed coordinate reference frame (Miller and Nelson, 1976).

Consider a person reaching out with the right arm to push a load upward. The load is located above and to the right of the shoulder, as in Figure 6.12. The load magnitude is 100 N. The arm weight will at first be considered insignificant. The question raised is: What are the reactive forces and moments operating at the shoulder?

To solve this problem, the external force vector direction operating on the hand must be specified in relation to the shoulder. To accomplish this, an orthogonal-axis reference system is defined, with its origin through the center of rotation of the shoulder. Then, two points are specified on the force vector relative to this reference system. In the example, a point D will be defined as the hand grip center, with X, Y, and Z coordinates of 25, 10, and 12 cm, respectively. A second reference point must be chosen on the force vector. For this point, assume that the elbow moves into alignment with the force, so that the elbow represents a convenient second point on the line of action of the force vector and is measured and found to have coordinates of 25, -15, and 12 cm from the origin of the reference axes at the shoulder.

The first step in the solution is to express the force direction in unit vector form. *Unit vectors,* designated i along the X axis, j along the Y axis, and k along the Z axis, represent orthogonal components of the vector in a given scale. The force vector direction from D to E in the example can be specified D (DX, DY, DZ) to E (EX, EY, EZ), which in unit vector form becomes:

$$\text{DE} = (\text{E}X - \text{D}X)i + (\text{E}Y - \text{D}Y)j + (\text{E}Z - \text{D}Z)k \qquad (6.29)$$

Using the coordinates given, the vector direction can be expressed:

$$\begin{aligned} \text{DE} &= (25 - 25)i + (-15 - 10)j + (12 - 12)k \\ &= 0i - 25j + 0k \\ &= -25j \end{aligned}$$

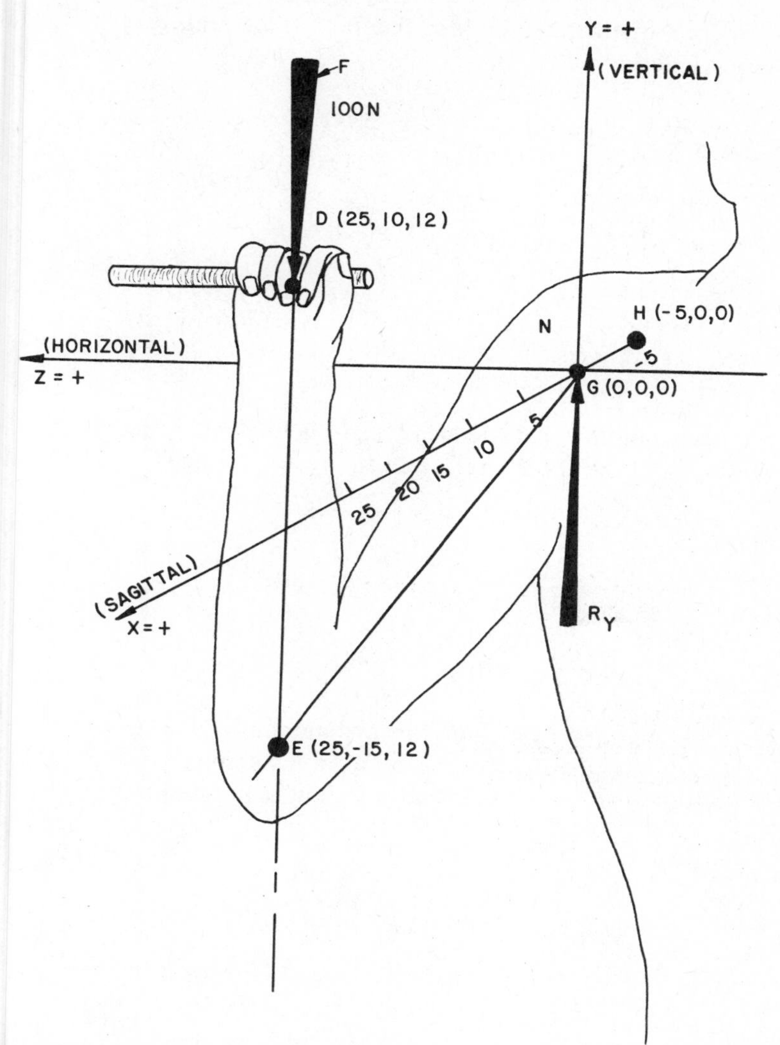

FIGURE 6.12 Non-coplanar analysis at right arm pushing load upward (see text for analysis of moments and forces at shoulder).

This vector direction is converted to a unit vector U by dividing by its magnitude, which can be found by application of the Pythagorean theorem:

$$\begin{aligned} U &= \frac{0i - 25J + 0k}{\sqrt{0^2 - 25^2 + 0^2}} \\ &= \frac{1}{\sqrt{25^2}}(-25j) \end{aligned} \tag{6.30}$$

When multiplied by the magnitude of the force 100 N, this gives the external force vector F acting on the hands:

$$
\begin{aligned}
F &= \frac{100}{\sqrt{25^2}}(-25j) \\
&= \frac{100}{25}(-25j) \qquad (6.31) \\
&= \frac{-2500}{25} \text{ or } -100 \text{ N in } Y \text{ axis (downward)}
\end{aligned}
$$

With the assumption that the arm weight is negligible, the static equilibrium conditions in the three axes yield the reactive forces at the shoulder:

$$
\begin{aligned}
R_X &= 0 \\
R_Y &= 100 \text{ N (upward)} \\
R_Z &= 0
\end{aligned}
$$

If the arm weight were included in a static analysis, it would simply be added to the components of the external forces acting in the Y vertical direction. Assuming an average male arm weight of approximately 36 N, then

$$
\begin{aligned}
F_Y &= 0 \\
-100 \text{ N} - 36 \text{ N} + R_Y &= 0 \\
R_Y &= 136 \text{ N (upward)}
\end{aligned}
$$

The estimation of the moments at the shoulder requires additional use of vector algebra. The first step is to decide on the axis about which the moment should be estimated. Recall, moments can be independently estimated about three orthogonal axes. To accomplish this, an axis of interest is specified by locating two reference points on it. In the example, a reference axis system was defined through the shoulder, so the reference points can be used to estimate the three moment components. Thus, one point of reference is the intersection of the reference axes. This will be designated G(0,0,0). To compute the moment M_X about the X axis, a second point on the axis is arbitrarily chosen. It is designated H(-5,0,0). The direction of the axis can now be expressed in unit vector form (as was done for the force vector). Its direction is

$$\begin{aligned} \text{GH} &= (-5 - 0)i + (0 - 0)j + (0 - 0)k \\ &= -5i \end{aligned}$$

which as a unit vector designated N becomes

$$\begin{aligned} N &= \frac{-5i + 0j + 0k}{5^2 + 0^2 + 0^2} \\ &= \frac{1}{25}(-5i) \end{aligned}$$

At this point in the solution, both the force and one axis of rotation are designated as unit vectors F and N. The moment of the force F and axis N is the triple scalar product of $r(F \cdot N)$, where r represents any position vector running from a point on the rotation axis N to a point on the force vector F. Once again, r is designated as a unit vector. It can be found by subtracting the coordinates of any two pairs of points on F and N. For the example, the unit vector value of r from point G on the N axis to point D on the force vector is

$$\begin{aligned} r &= (25 - 0)i + (10 - 0)j + (12 - 0)k \\ &= 25i + 10j + 12k \end{aligned}$$

The moment M_N about the N axis is now computed:

$$\begin{aligned} M_N &= r(F \cdot N) \\ &= \frac{1}{\sqrt{25}} \times \frac{100}{\sqrt{25^2}} \begin{vmatrix} 25 & 10 & 12 \\ 0 & -25 & 0 \\ -5 & 0 & 0 \end{vmatrix} \end{aligned} \tag{6.32}$$

where the matrix column values are i, j, k (left to right) coefficients and the rows are r, F, N (top to bottom) coefficients. The determinant of the matrix is found by multiplying each of the r coefficients in the top row by the sum of the cross-products of the coefficients in the F and N rows that are not in each r coefficient's column. Thus:

$$M_N = \frac{1}{5} \times \frac{100}{25} [25[(-25 \times 0) - (0 \times 0)] - 10 [(0 \times 0) - (-5 \times 0)]$$

$$+ 12[(0 \times 0) - (25 \times -25)]]$$

$$= \frac{100}{125} [25(0 - 0) - 10(0 + 0) + 12(0 - 125)]$$

$$= 0.8(-1500)$$

$$= -1200 \text{ N cm}$$

$$= = -12 \text{ N m}$$

Therefore, about the N axis (which coincides in the example with the X reference axis) the 100 N hand force creates a moment of 12 N m. By the right-hand rule, the negative sign for the moment allows it to be designated as operating backward (away from the viewer in Figure 6.12) along its reference axis N. The moments about the other two Y and Z axes would be estimated in a similar fashion.

It is hoped that this example illustrates the method of solution using vector algebra to add, subtract, and multiply the force and distance values in three dimensions. It should be obvious that the computations become quite tedious in a three-dimensional analysis. It is for this reason that the solutions must rely on computerized models. Such a model for three-dimensional static strength evaluations of jobs has been developed by Garg and Chaffin (1976). The linkage system used in this model is illustrated in Figure 6.13. The model allows the user to specify anthropometry, body postures, and hand loads of interest. The outputs from the model are the reactive forces and moments at each of the joints of the linkage for the designated input values. The model also allows the user to compare the joint load moments with muscle strength moments to predict the strength exertion requirements of a job relative to normal populations. This latter aspect will be discussed later in this chapter.

In summary, vector algebra provides a mathematical method of computing joint load moments and reactive forces in three dimensions. The computations are complex, but computer programs have been written to provide the analysis capability when it is found that a task cannot be evaluated with the simpler coplanar models.

6.4 DYNAMIC BIOMECHANICAL MODELS

The introduction of human motion into biomechanical models introduces two types of complexity. First, the motion must be described in a *kinematic* fashion. This means that the direction of motion of a body segment, as well as its velocity and acceleration–deceleration profile, must be measured over the period of the motion. Chapter 5 describes the instrumentation available to provide these measurements.

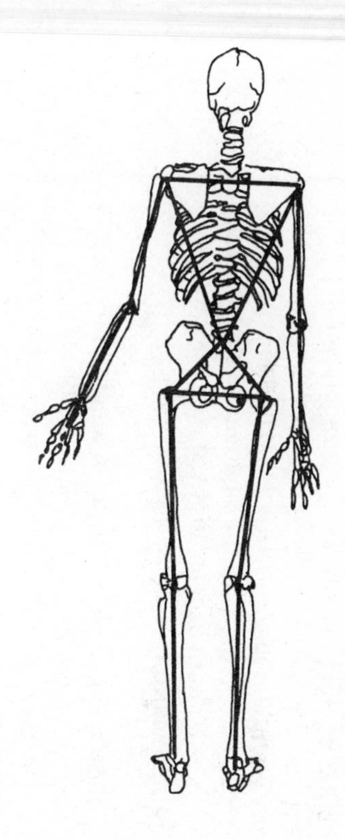

FIGURE 6.13 Linkage representation used in the biomechanical model.

The second complexity regarding human motion biomechanics is in modeling of the *kinetic* forces and moments of complex linkage systems during movement. As a human motion is executed—even a simple one—complex inertial forces are created by the changes in motion velocity and direction. These changes result in accelerations and decelerations of the body segments which, by application of Newton's second law ($F = ma$), create inertial forces.

What follows is a description of a single-body segment in motion. The results of the analysis are then generalized to a biodynamic coplanar model of a person lifting a weight, followed by a similar model of a person pushing and pulling a cart.

6.4.1 Single-Segment Dynamic Biomechanical Model

As a body segment is pivoted about a joint center, inertial forces in addition to those due to gravity alone act on it. These additional inertial forces can be depicted in a coplanar motion as two orthogonal forces. One of these is a force tangent to the arc of the segment motion and acting at the segment's center of mass. The magnitude of this tangential force F_t is simply $F_t = ma$, where a is the instantaneous linear acceleration or deceleration of the segment's mass center. Because the object is rotating about a known joint center, the value of F_t can also be expressed as

$$F_t = m\ddot{\theta}r \tag{6.33}$$

where

F_t is the tangential force acting at its mass center
m is the segment mass
$\ddot{\theta}$ is the instantaneous angular acceleration of the segment
r is the distance from the joint center to the mass center

The second force acting as a result of a segment's motion is referred to as the *centrifugal force of rotation*. It is caused by the particles of mass attempting to move in a straight line rather than an arc about the joint center. The greater the square of the velocity of rotation about the center of rotation, the greater the centrifugal force pulling away from the center. Also, if the radius of the arc of rotation is large, so will be the centrifugal force. The centrifugal force acts at the center of mass and along the radius of the arc of motion. For this reason it is sometimes referred to as the *radial force*, expressed by

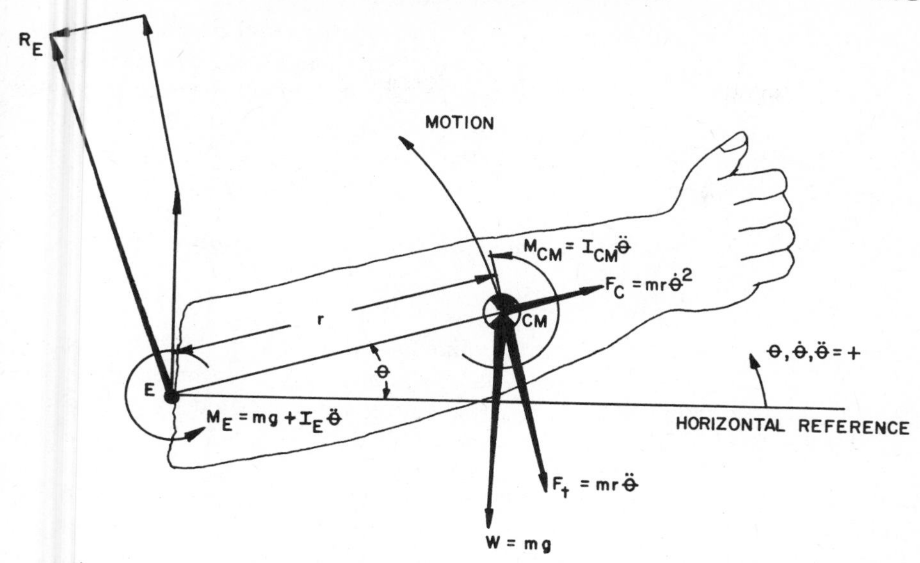

FIGURE 6.14 Single-segment coplanar analysis at elbow of forearm–hand flexion motion.

$$F_C = mr\dot{\theta}^2 \tag{6.34}$$

where

F_C is the centrifugal or radial force of rotation away from joint center
r is the distance from the joint center to the mass center
$\dot{\theta}$ is the instantaneous angular velocity of the segment

It should be noted that because the centrifugal force acts along the radius of rotation, its line of action also passes through the joint center. It therefore does not contribute to the moment at the joint, but does create a reactive force necessary to hold the structure (especially at the joint) together during high-velocity motions. This reactive force is referred to as the *centripedal force*, and acts to counteract the centrifugal force of rotation.

The preceding can be illustrated for a simple flexion at the elbow as shown in Figure 6.14. The positive (counterclockwise) acceleration creates a tangential force acting downward, which is added to the weight of the segment to create a composite reactive force at the elbow. Also contributing to this composite reactive force at the elbow is the centripedal force caused by the angular velocity of the motion about the elbow.

The moment at the elbow is also affected by the movement. For the example in Figure 6.14, the moment at the elbow is comprised of the static effect of gravity ($mg \cos \theta r$) and a dynamic inertial effect. The latter has two components—an instantaneous acceleration effect at the mass center due to the tangential rotation force ($rF_t = mr\ddot{\theta}r$) and a rotational acceleration effect due to the mass distribution (shape) about the center of mass. Because body segment mass is distributed in a cylindrical-like shape, it resists rotation of its long axis about its center of mass. The effect of the segment rotation about its center of mass can be found by multiplying its moment of inertia I_{CM} at the center of mass by the instantaneous rotational acceleration ($\ddot{\theta}$). Thus, the sum of the moments at the elbow can be expressed by

$$M_E = M \text{ static} + M \text{ dynamic tangential} + M \text{ dynamic rotational} \tag{6.35}$$

$$M_E = mg \cos \theta r + mr\ddot{\theta}r + I_{CM}\ddot{\theta}$$

If appropriate moment of inertia values are available in the literature, or can be measured for a segment about a joint center j, rather than at the center of mass, then the two dynamic terms are simplified by use of the parallel axis theorem ($I_j = I_{CM} + mr^2$), giving

$$M_j = M_j \text{ static} + I_j\ddot{\theta} \tag{6.36}$$

If inertial values are not available about a joint center, the radius of gyration k_L of a body segment may be referenced. The k_L's can be used to estimate the moment of inertia I_{CM} at the center of mass, where

$$I_{CM} = mk^2 \tag{6.37}$$

The values of I_{CM} are then used to compute the instantaneous moments at the joint center M_j during a motion by substitution in the parallel axis theorem, giving

$$I_j = mk^2 + mr^2 = m(k^2 + r^2) \tag{6.38}$$

which can be used in equation 6.36 to compute the M_j values.

Amis, et al. (1980) used this procedure to estimate moments about the elbow during high-speed (maximum) flexion. They recorded the forearm–hand segment rotation about the elbow with a high-speed cine camera at 200 frames per second. Their experiments disclosed that typical angular velocities $\dot{\theta}_E$ reached about 15 rad/sec, with average angular accelerations of about 200 rad/sec^2 during the first two-thirds of the motion, and approximately -550 rad/sec^2 (deceleration) in the last third of the movement. Figure 6.15 displays these results. The resulting dynamic elbow moments varied from an average of about 14 N m during the acceleration phase to 35 N m during the deceleration phase. Interestingly, when these dynamic moments were compared with isometric strength values obtained at various elbow angles, the latter decelerations caused elbow moments approaching the triceps extension strength limits.

The experiment by Amis and the accompanying biomechanical analysis indicate the hazard of performing high-speed movements. It appears, even

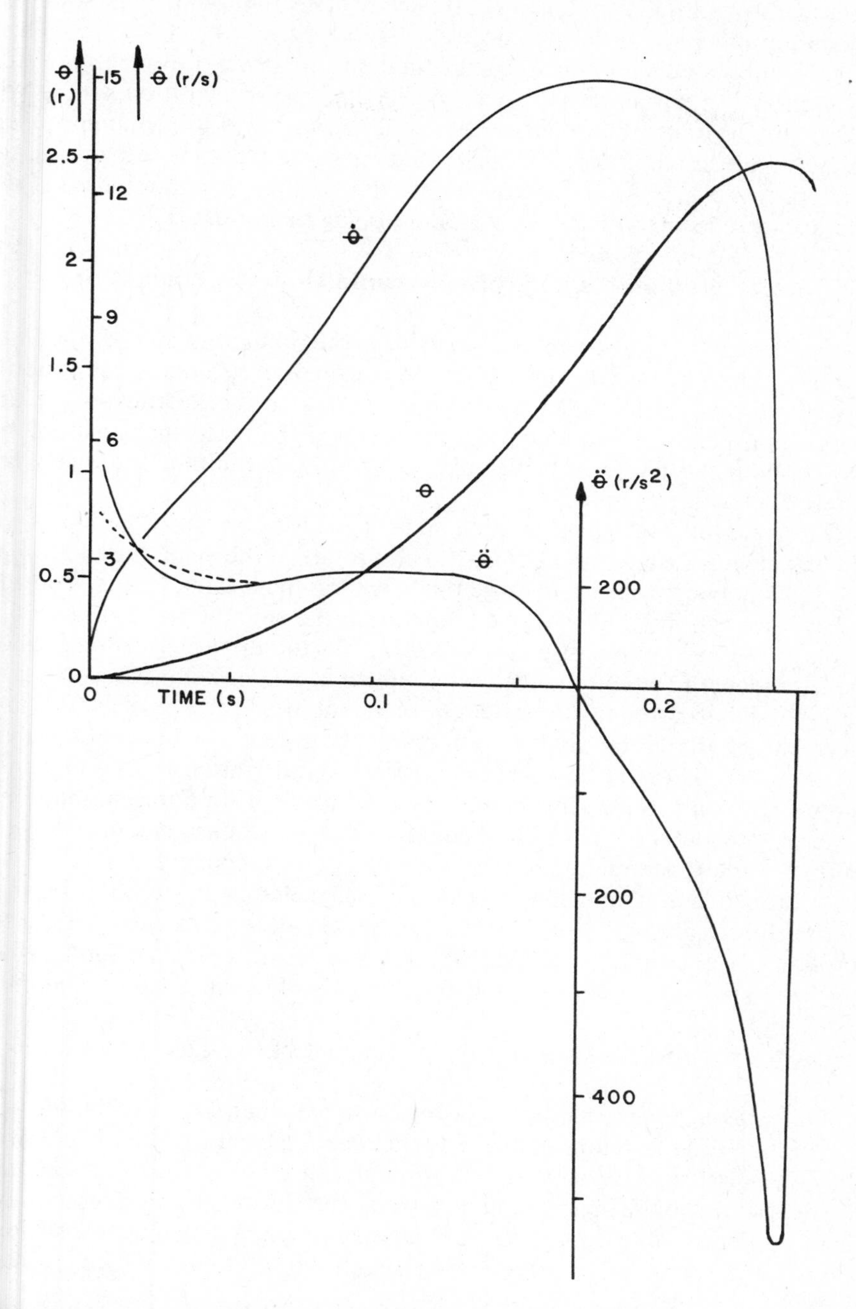

FIGURE 6.15 Kinematic relations for typical high speed forearm flexion (Amis et al., 1980).

without the additional mass of an object held in the hand, that volitional dynamic moments can be imparted to a segment that are large enough to possibly subluxate a joint or strain an antagonistic muscle attempting to stop the motion. As a mass is added to the hands, the acceleration profile will necessarily be lower, but the moments could be larger. Until dynamic muscle strength limits are better defined, it would appear to be prudent to encourage workers to develop smooth movements that reduce accelerations and decelerations, especially if heavy loads are being manipulated.

6.4.2 Multiple-Segment Biodynamic Model of Load Lifting

As was done in the static coplanar analysis earlier in this chapter, the addition of more than one link in the system is considered by adding the reactive forces and moments caused by an adjacent link to the kinetics of the link in question. Thus, a sequential set of analyses are performed, usually beginning with the forearm–hand link, and continuing until a stationary link (often the foot) is reached.

One complicating aspect of such an analysis for two or more links is that the links can translate (move linearly) and rotate at the same time. A simple two-link analysis will demonstrate this effect on the inertial forces. Consider the arm as two links being raised quickly in the sagittal plane. Figure 6.16 displays the motion posture and variables. As the upper arm rotates about the shoulder, the elbow is accelerated upward. The inertial forces acting on the elbow, therefore, must consider *both* the rotational effects of the forearm-hand link at the elbow and the effects of the forearm–hand motion about the shoulder. These latter forces are indicated in Figure 6.16 as a tangential and centrifugal force acting at the center of mass of the forearm–hand link relative to a radius R from the shoulder. These are designated $m_1R\ddot{\theta}_2$ and $m_1r\ddot{\theta}_2$ for the tangential and centrifugal forces, respectively.

A French mathematician, Coriolis, disclosed that in addition to these inertial forces acting on the distal segment, the difference between the two rotational accelerations ($\ddot{\theta}_1$ and $\ddot{\theta}_2$) causes a separate inertial force, which has been shown by Plagenhoef (1971) for a two-link motion to be equal to

$$2\dot{\theta}_1\dot{\theta}_2 r_1 m_1 (\text{Coriolis Force for distal segment of two-link motion}) \quad (6.39)$$

Thus, if one segment rotates faster or slower than its adjacent segment, a Coriolis Force develops and affects the reactive forces at the joint between the two segments. Because it acts through the joint, however, it exerts no moment at the joint. In a three-link system, the Coriolis Force becomes even more complicated to compute, so computer programs have been developed to estimate the reactive forces and moments in multilink coplanar whole-body motion studies (Plagenhoef, 1971).

If a kinematic analysis of a person's movement is completed, as described in Chapter 5, the instantaneous linear accelerations of the segment mass

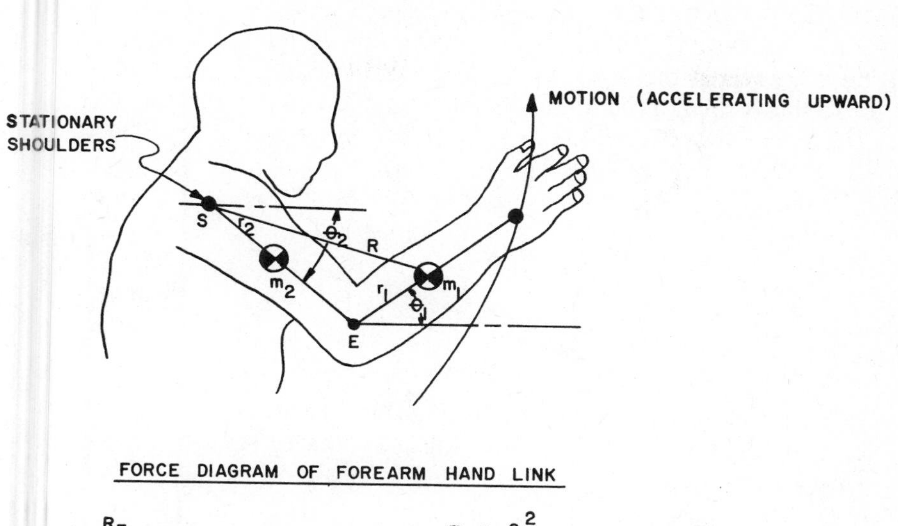

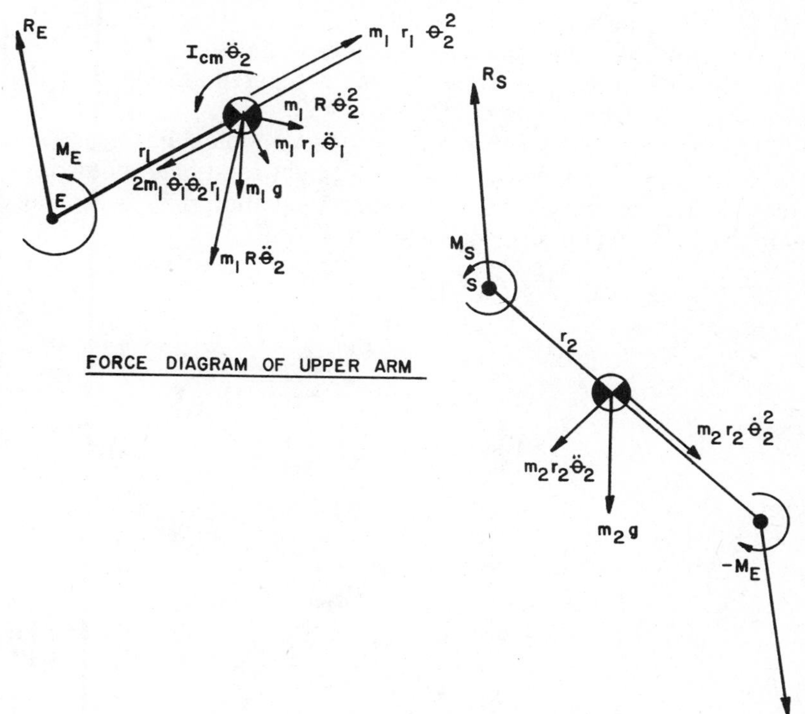

FIGURE 6.16 Two-segment dynamic coplanar analysis showing mass–acceleration forces.

centers are obtained. These linear accelerations can be used to directly estimate the dynamic effects in a multisegmental motion and thus avoid the need to consider a Coriolis Force analysis. For example, consider a person lifting a load from the floor, as in Figure 6.17. The instantaneous reactive force in the vertical direction is computed by adding the dynamic effects of ma_y to the gravity effect of segment weights mg at each joint. Thus,

$$\sum F_{jy} = 0$$

$$m_L g + m_L a_{Ly} + R_{(j\text{-}1)y} - R_{jy} = 0 \tag{6.40}$$

$$R_{jy} = m_L g + m_L a_{ly} + R_{(j\text{-}1)y}$$

where

R_{jy} is the reactive force at joint j in the y direction
$R_{(j\text{-}1)y}$ is the reactive force at the adjacent joint (j-1) in the y direction
m_L is the mass of segment L
a_{Ly} is the measured or computed instantaneous acceleration of segment L in the y direction at the center of mass

The horizontal reactive forces R_{jx} are computed in a similar manner, but the segment weights are excluded.

The moments at the joints are computed by adding the linear acceleration effects to the static effects, plus the segment rotational moment of inertia effect. Thus, the static moment of inertia equation 6.25 is expanded to include the dynamic effects, yielding

$$\sum M_j = 0$$

$$\begin{aligned} M_j = {} & M_{j\text{-}1} + \overline{jCM}_L(\cos\theta_j)m_{LG} + \overline{jCM}_1(\cos\theta_j)m_L a_{Ly} \\ & + \overline{jCM}_L(\sin\theta_j)m_L a_{Lx} + \overline{j\,j\text{-}1}(\cos\theta_j)R_{(j\text{-}1)y} \\ & + \overline{j\,j\text{-}1}(\sin\theta_j)(R_{(j\text{-}1)y}) + I_L\ddot{\theta}_j \end{aligned} \tag{6.41}$$

where

M_j is the load moment at each joint j
$M_{j\text{-}1}$ is the load moment at the adjacent joint j-1
θ_j are the postural angles of each joint j relative to the horizontal right axis
m_L is the mass of link L
$\overline{j\,j\text{-}1}$ are the body segment link lengths
$\overline{jCM}_L$ is the distance from joint j to the center of mass of link L
a_{Ly} or x is the instantaneous linear acceleration x or y components of link L at the centers of mass

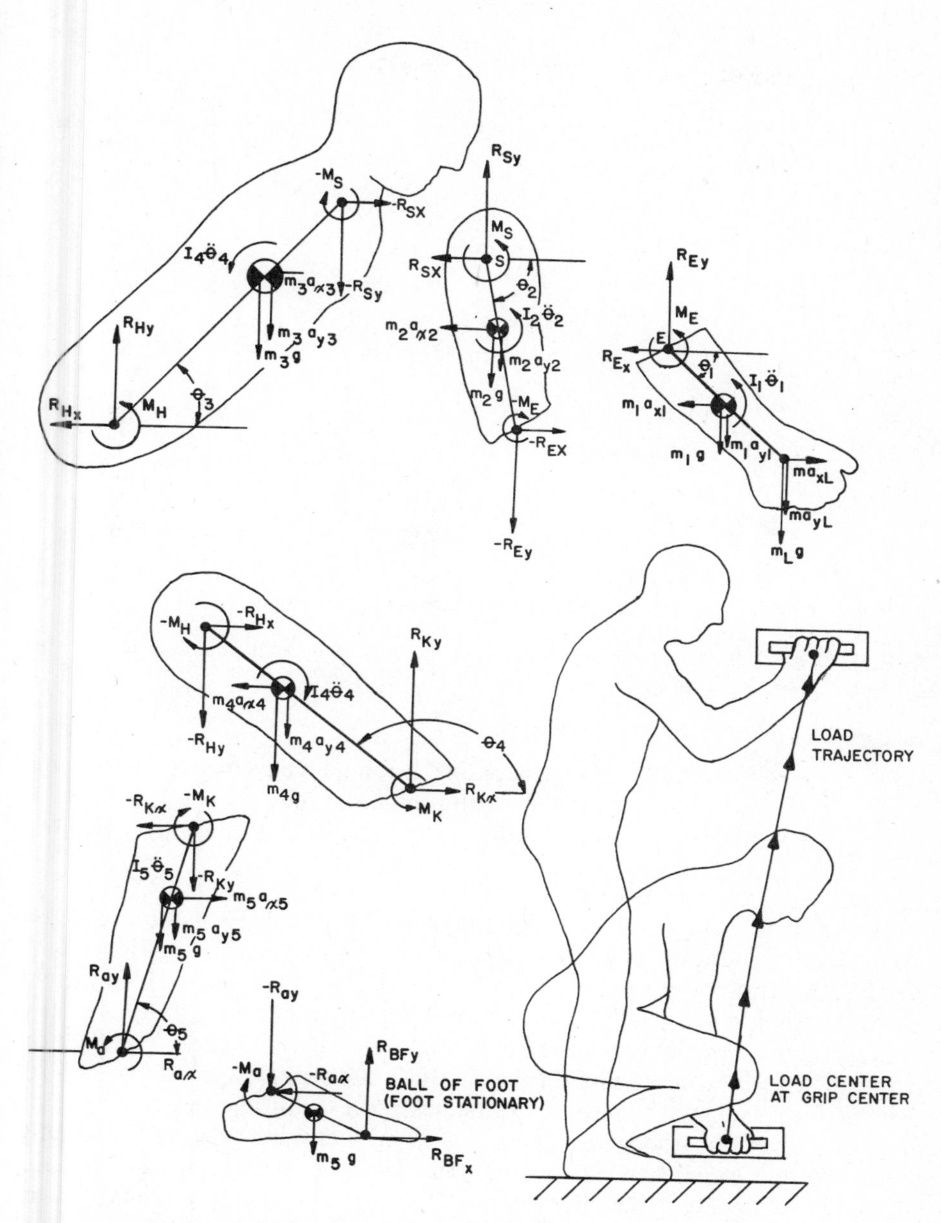

FIGURE 6.17 Dynamic coplanar analysis of box lifting (see text for explanation).

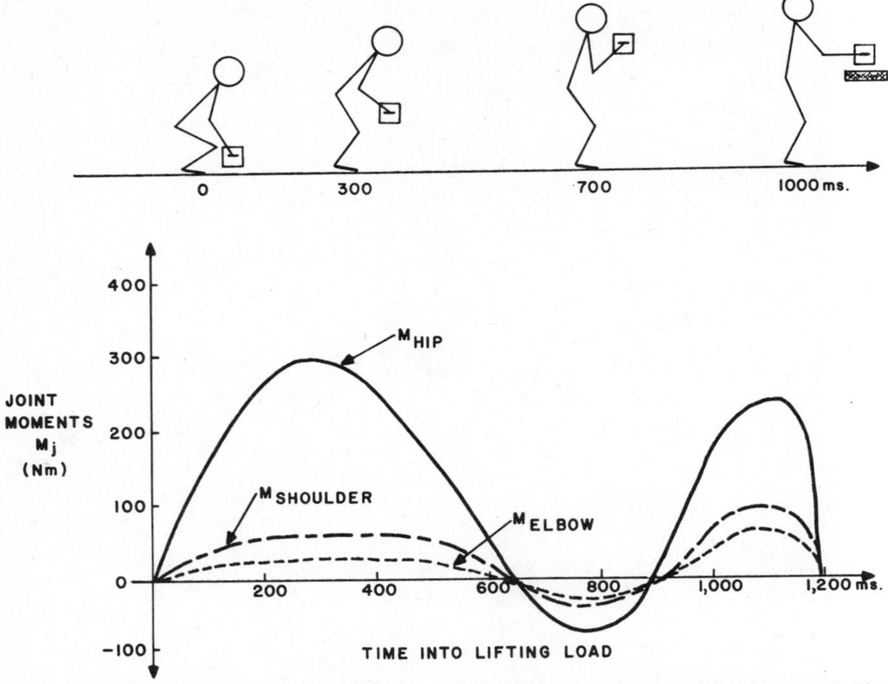

FIGURE 6.18 Typical moments at joints during lifting of 18 kg load from floor to shelf 108 cm above floor and 38 cm in front of ball of foot (Fischer, 1967).

I_L is the moment of inertia of link L about an axis through the center of mass normal to the sagittal plane of motion

$\ddot{\theta}_j$ is the angular acceleration of the segment about joint j relative to the horizontal

This type of model has been used by Fisher (1967) and El-Bassoussi (1974) to determine the joint moments during lifting of various loads. Figure 6.18 shows a typical moment profile of a person lifting a load with a dead weight of approximately 20% of the person's psychophysical lifting capacity (Fisher, 1967). Inspection of these moment profiles indicates that motion imparts large inertial effects during the first 200 to 400 msec of the lift. As will be discussed later regarding low back stresses in lifting, the inertial loads in lifting from the floor can combine with an awkward posture (especially if the object is large) to produce large internal forces in the low back region.

The need for dynamic biomechanical models has been summarized by Plagenhoef (1971) as follows:

The ability to interpret the forces and moments obtained (at each joint) allows a more critical analysis of motion. Timing of the relative accelerations of each

body part and the magnitude of the moments of force indicate the motion's efficiency. The contribution of each body segment to the whole motion may also be found.

6.4.3 Coplanar Biomechanical Models of Foot Slip Potential While Pushing a Cart

Another example of the insight provided by a dynamic analysis is provided when one studies the act of pushing a heavy cart. The inertia of the mass of the cart and/or the rolling resistance of the cart wheels creates a hand load that can become quite high relative to a person's isometric push strength. From the biodynamic standpoint, one of the major questions of concern in such an act is: What are the peak normal and shearing forces at the foot/floor contact point? An answer to this question would assist in designing the type of flooring or shoe material necessary to keep the person's foot from slipping when required to perform such an act, or it would provide insight into the maximum dynamic push forces that one could expect, given a certain floor/shoe condition and a certain pushing posture.

The discussion of foot slipping cannot proceed without introducing the concept of the *coefficient of friction*. Simply stated, the coefficient of friction is the ratio of maximum force acting parallel to a contact surface that resists motion of one body on the other, divided by the normal force acting on the contact surface. Figure 6.19 illustrates this concept during a cart-pushing task.

The contact surface is assumed to be at the floor and the ball of the foot—that is, the person is learning forward. The normal forces F_N and F_R are considered to be vertical and equal in magnitude, and the shearing and friction forces F_S and F_F are considered to act horizontally and are equal in magnitude, provided slipping doesn't occur. In this case the values of these variables can be determined by the preceding models. In a simple static case, the normal forces are equal to body weight mg, and the horizontal forces are equal to the push force created by the cart's resistance to move. It will be shown that even in such a simple case it is possible for a person to slip, however. This occurs when the maximum value of the force of friction is less than the shear force created by the pushing force—that is, F_S is greater than $F_{F\max}$.

The value of $F_{F\max}$ is dependent upon the contact surface conditions and the normal forces. In general, $F_{F\max}$ values are proportional to the normal forces, and hence define the coefficient of friction:

$$F_{F\max} = \mu F_R \tag{6.42}$$

where

μ is the coefficient of friction for the given contact surfaces

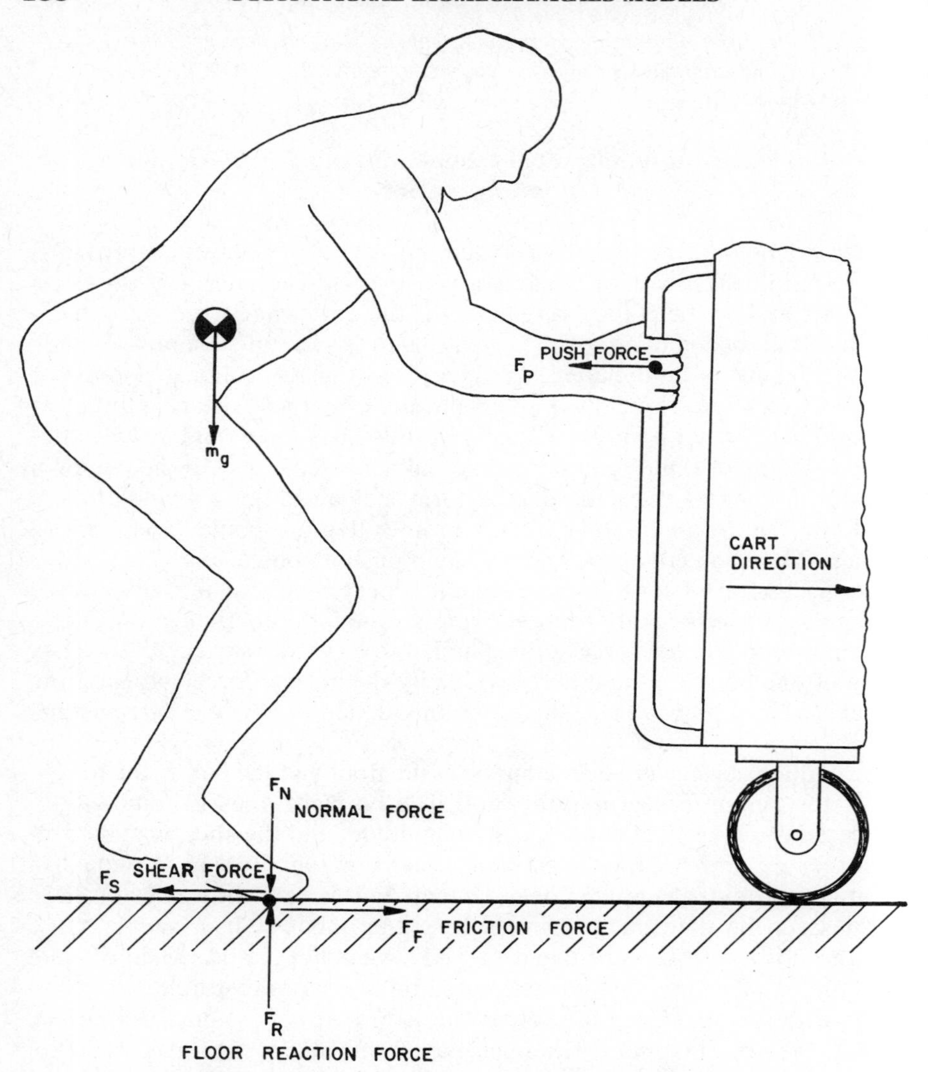

FIGURE 6.19 Foot/floor forces during cart pushing and static external loads.

F_R is the normal reactive forces pressing the surfaces together

$F_{F\max}$ is the force of friction (maximum) necessary to resist slip motion

Values for μ under static conditions will normally range from 0.2 when smooth surfaces are wet, to 0.7 when rough and dry. Thus, the limiting $F_{F\max}$ values for a person of light body weight (445 N) could be as low as 90 N in this static analysis. Since muscular push strengths when a person is well braced against slipping can easily exceed 90 N, as shown by Kroemer

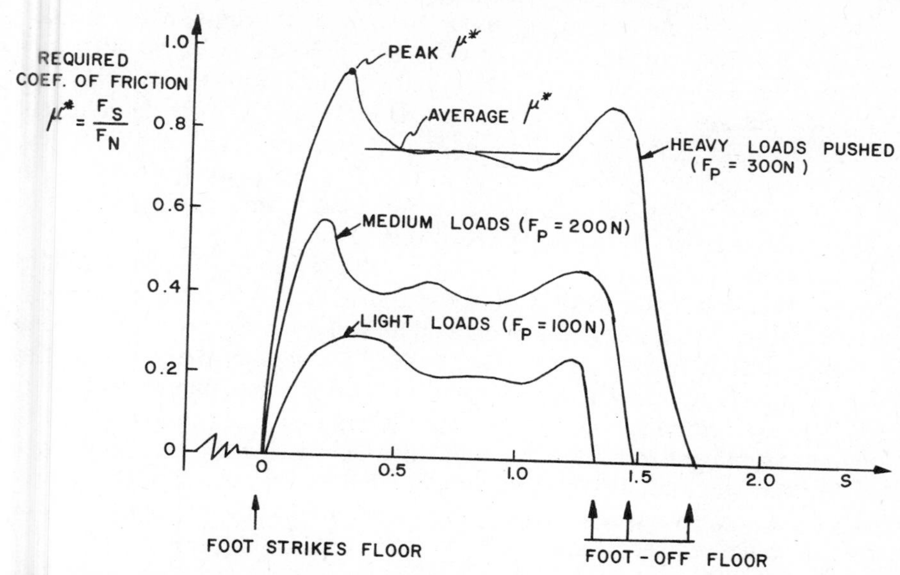

FIGURE 6.20 Required dynamic coefficients-of-friction μ^* during cart pushing with different hand loads (Lee, 1982).

and Robinson (1971) and Ayoub and McDaniel (1974), the potential for foot slip is very great in such acts in industry.

An even more revealing insight into this hazard is produced when a biodynamic analysis is performed. Such an analysis was recently completed by Lee (1982), in which subjects pushed carts across a floor mounted force platform, while hand push forces and body segment motions were measured. The force platform also provided measurements of the foot shear and normal forces, and confirmed the biodynamic predictions obtained with a biomechanical model similar to that depicted in Figure 6.17, but with the hand loads horizontal instead of vertical. What occurs biodynamically during the time the foot is in contact with the floor is graphically illustrated in Figure 6.20 during a typical pushing task. In essence, immediately after the foot strikes the floor, the shearing force relative to the normal force, which can be thought of as the required coefficient of friction μ^* necessary to prevent slipping, is large until the forward momentum of the body carries the body center of mass over the foot. At this central phase of the motion the normal force is higher than at the beginning of the stride, and thus the required μ^* is reduced. Toward the end of the stride, the normal forces are lower, even though the shear force may remain high (if a constant push force on the handle is required), and thus the required μ^* to prevent slipping increases until the other foot makes floor contact. Inspection of the typical values in Figure 6.20 discloses very high peak values for the required μ^* to prevent slipping, especially when pushing heavy loads.

Dynamic coefficient of friction values obtained for various shoe/floor surfaces are not known. They may be different than published static values (possibly because of non-linear boundary layer effects at the contacting surfaces). Despite this, it is clear with reference to even the average values of μ^* in Figure 6.20, that pushing (and pulling) of carts can be hazardous unless high-friction walking surfaces and slip-resistant shoes are provided. This latter point will be discussed further in connection with manual materials handling activities in the second part of this text.

From the preceding it should be realized that when a person is handling a submaximal load, the motion can be—and often is—performed quickly. In such cases the inertial effects may result in sudden peak forces on muscles and ligaments that are already stressed to their physiological limits, resulting in strain and sprain types of injuries. In such cases, biodynamic models are necessary, even though computationally difficult to use. As the load handled is increased, slower movements result, thus decreasing the inertial effects. In this case, static analyses can provide excellent insights into the source of stress on various tissues. In this regard, various static analyses have been used to evaluate alleged hazardous physical work conditions. A few of the resulting models are described in the following section.

6.5 SPECIAL PURPOSE BIOMECHANICAL MODELS OF OCCUPATIONAL TASKS

Statistical analyses (epidemiology) of musculoskeletal injury patterns in industry have indicated that certain physical activities produce more than a random number of injuries to a particular anatomical region. In particular, the large number of workers who are disabled because of low back pain associated with manual labor has resulted in concerted efforts to relate the biomechanical aspects of such labor to resulting tissue injury. In this regard, the acts of lifting, pushing, and pulling heavy loads in various postures have been biomechanically modeled to predict resulting forces and moments on, for example, the L5/S1 spinal disc. This modeling effort will be described as representative of one special purpose biomechanical model of occupational tasks. Since the low back has also been shown to become painful when a person maintains an awkward sitting posture for too long a period, biomechanical modeling of sitting postures will also be described.

The wrist is an anatomical region that has also been shown to be injured by frequent hand exertions, especially when performed in extreme postures. The modeling of such acts is presented in this section. Lastly, the prediction of muscle strengths in different postures is important as it provides the basis for guidelines regarding the expected number of people who can perform a given manual task. Biomechanical modeling of muscle strengths is described as another example of special purpose biomechanical models of manual activities in industry.

6.5.1 Low Back Biomechanical Models

The preceding static and dynamic models of load lifting have shown that the moments at the hip joint can become quite large, especially when a load that cannot be held close to the body is lifted. Since the lumbar spine is anatomically close to the hip joints, a similar effect occurs about the joints of the lumbar spine, which in flexion and extension can be considered to be near the center of the spinal discs. In fact, Tichauer (1971) proposed that the load moment about the lumbosacral disc (L5/S1) should be used as the basis for setting limits for lifting and carrying loads of various sizes to avoid muscle fatigue in the lumbar extensor (erector spinae) muscle group.

From a biomechanical perspective, the fact that large moments are created at the lumbar spine when lifting heavy loads raises the question of the nature of the internal forces that must be present to stabilize the spine while incurring such large load moments. A simple static sagittal-plane model of the lumbar spine during lifting was proposed by Morris, Lucas, and Bressler (1961). This model assumed that two types of internal forces acted to resist the external load moment. One is the extensor erector spinae muscles that exert their force approximately 5 cm posterior to the centers of rotation in the spinal discs. The second stabilizing force was assumed to be caused by abdominal pressure acting in front of the spinal column, pushing the upper torso into extension, thus resisting the load moment acting on the lumbar spine. What resulted from application of this type of model was a realization that large compression forces developed in the spinal column, acting to compress the discs during a load lifting act. These compression forces were later confirmed by Nachemson and Elfstrom (1970) in a series of experiments in which the pressure within the center portion of the discs was measured (by inserting a needle attached to a pressure transducer) in volunteers who performed various lifting maneuvers. Interestingly, separate biomechanical tests of cadaver spinal segments showed that the compression forces predicted in the Morris, Lucas, and Bressler model of lifting were sufficient to create micro-fractures of the cartilage endplates between the spinal vertebral bodies and the intervertebral discs, according to Armstrong (1965), Morris et al (1961), Nachemson (1971), and Sonoda (1962).

Thus, it appeared reasonable to develop a more complete model to estimate the compressive forces created at various disc endplate interfaces during specific physical tasks. It was hoped that such a model would provide an important contribution toward the further understanding of low-back injury mechanisms related to specific physical activities, especially lifting activities.

The lumbosacral disc was chosen for particular attention in this model, based on statistics regarding back disorders. These showed that between 85 and 95% of all disc herniations occur with relatively equal frequencies at the L4/L5 and L5/S1 levels, (Krusen et al, 1965; Smith et al, 1944; Armstrong, 1965). Since the L5/S1 disc incurs the greatest moment in lifting activities because it most often has a large moment arm relative to any load in the hands, it was chosen to represent lumbar stresses during lifting.

To perform an analysis of the lumbar spinal stresses, the following concepts were included by Chaffin (1975) to refine the Morris, Lucas, and Bressler model. First, the geometry of an average erect spinal column and pelvis was developed from the dimensions of Fick (1904), Lanier (1939) and Chaffin, Schutz, and Snyder (1972). The dimensions of this average male column were then proportionally scaled, based on hip-to-shoulder anthropometric data, to permit the study of smaller or larger individuals. This resulted in a hip joint to L5/S1 disc link that was approximately 20% of the hip-to-shoulder link length. Next, the superior surface of the sacrum was estimated to be at an angle of 40° from the horizontal when standing erect (Thieme 1950). These data provided the basis for a kinematic torso model with two links—a pelvic–sacral link and a lumbar–thoracic link. Figure 6.21 shows the linkage.

The motion of the two links when stooping or squatting to lift an object was determined from empirical data of Dempster (1955). This showed an approximate relationship between the pelvic angle and the two external measured angles at the hip θ'_H and θ'_T. The relationship is graphed in Figure 6.22. It shows that as the torso is inclined forward in flexion, the pelvis contributes to the motion after approximately 20 to 30° by rotating at the rate of 2° for each 3° of torso inclination. Conversely, if the thigh is flexed, the pelvis rotates counterclockwise after about 10 to 15° at the rate of 1° for each 3° of thigh flexion. If both thigh and hip flexion occur simultaneously, it is assumed that the angle β is the difference between the two contributing effects.

The relevant forces are now considered, as shown in Figure 6.23. The analysis of the forces acting on the L5/S1 disc begins with the estimation of the moments at the hip M_H. These can be computed by application of the segmental approach (summing moments for each body segment) in the linkage system described in equation 6.25. For simplification, assume that the composite center-of-mass location of the body segments above the hip is known. This allows the moments at the hip to be expressed as:

$$M_H = M_{\text{body weight}} + M_{\text{load}}$$

$$M_H = b'mg_{\text{body weight}} + h'mg_{\text{load}} \tag{6.43}$$

The hip moment is then used to predict the expected abdominal pressure P_A. The abdominal pressure is created when the diaphragm and abdominal wall muscles are tensed in a reflex fashion. An average value can be estimated from the data of Morris, Lucas, and Bresler (1961). Fisher (1967) correlated these data with the hip moments M_H at various included angles between the thigh and torso—that is, the sum of the absolute values of θ'_H and θ'_T. The resulting empirical prediction equation was:

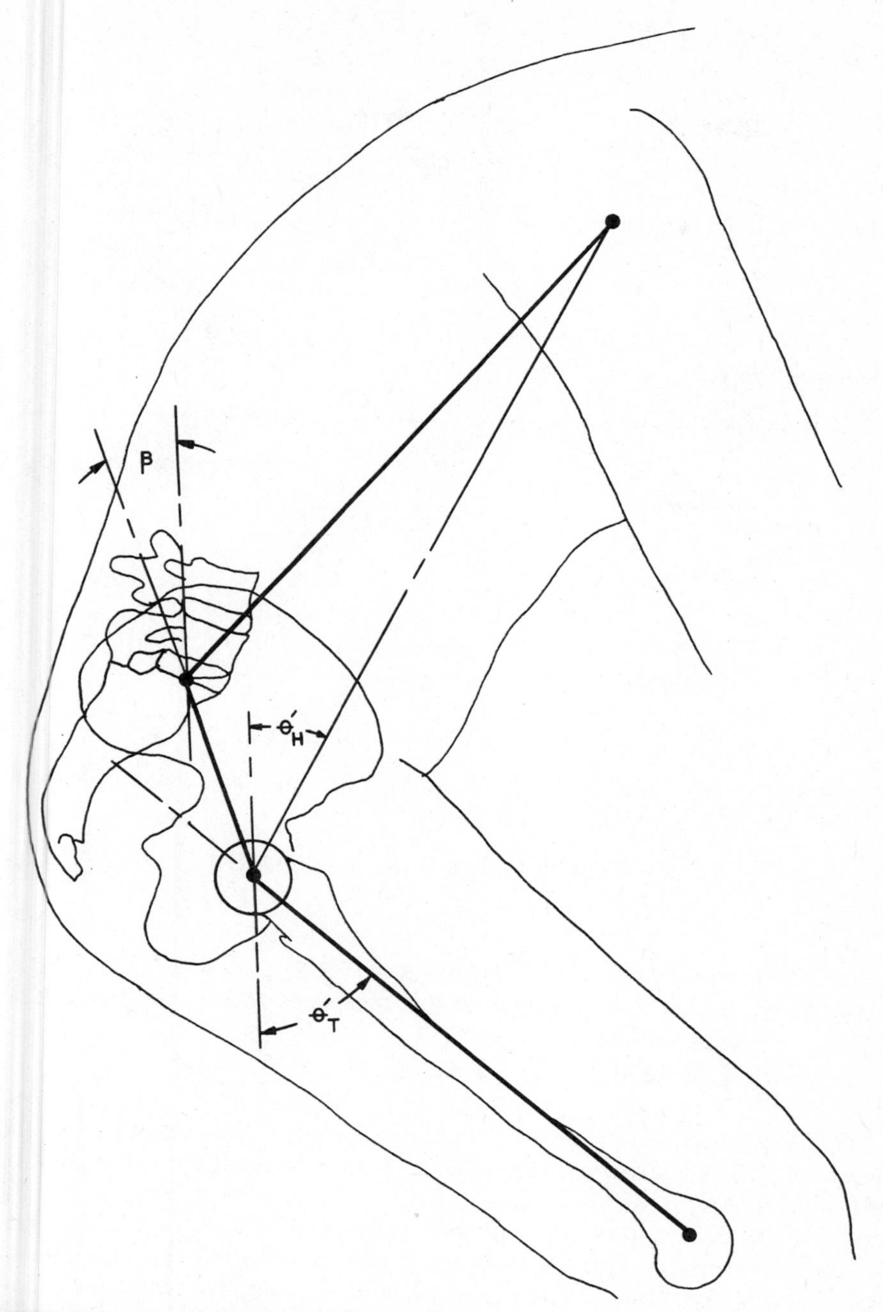

FIGURE 6.21 Torso linkage system used to model forces on superior sacral surface of L5/S1 disc.

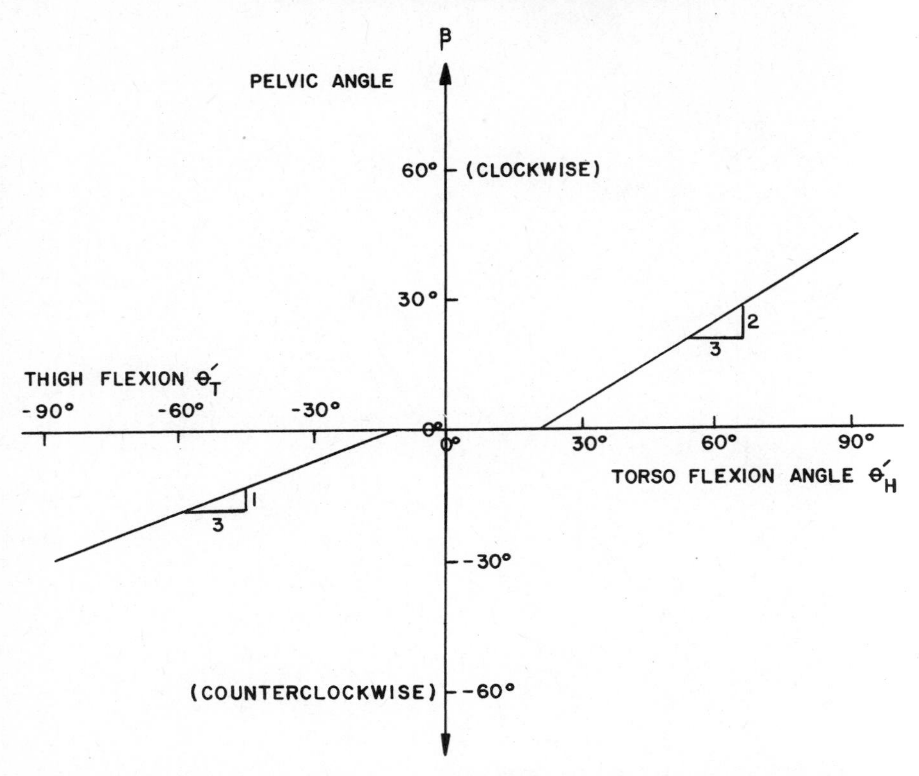

FIGURE 6.22 Pelvic angle β changes as a function of torso flexion θ'_H and thigh flexion θ'_T measured from the vertical axis (adapted from Dempster, 1955).

$$P_A = 10^{-4}[43 - 0.36(\theta'_H + \theta'_T)][M_H^{1.8}] \tag{6.44}$$

where

P_A is the abdominal pressure (mm of Hg)
θ'_H and θ'_T are the hip and thigh angles measured relative to the vertical axis (degrees)
M_H is the load moment at the hip (N m)

This expression was found by Fisher to have a simple correlation coefficient of 0.73. The error was attributed to (1) not knowing the exact position of the trunk during each test, thus causing an unknown variation in both the assumed angles and moments, and (2) not being able to assess the time rate of force application. This latter factor is important in that experiments by Asmussen (1968) disclosed that the method of lifting—quick jerk or slow sustained pull—can significantly affect the abdominal pressure reflex, with a peak pressure about 20% higher than that incurred in a more sustained

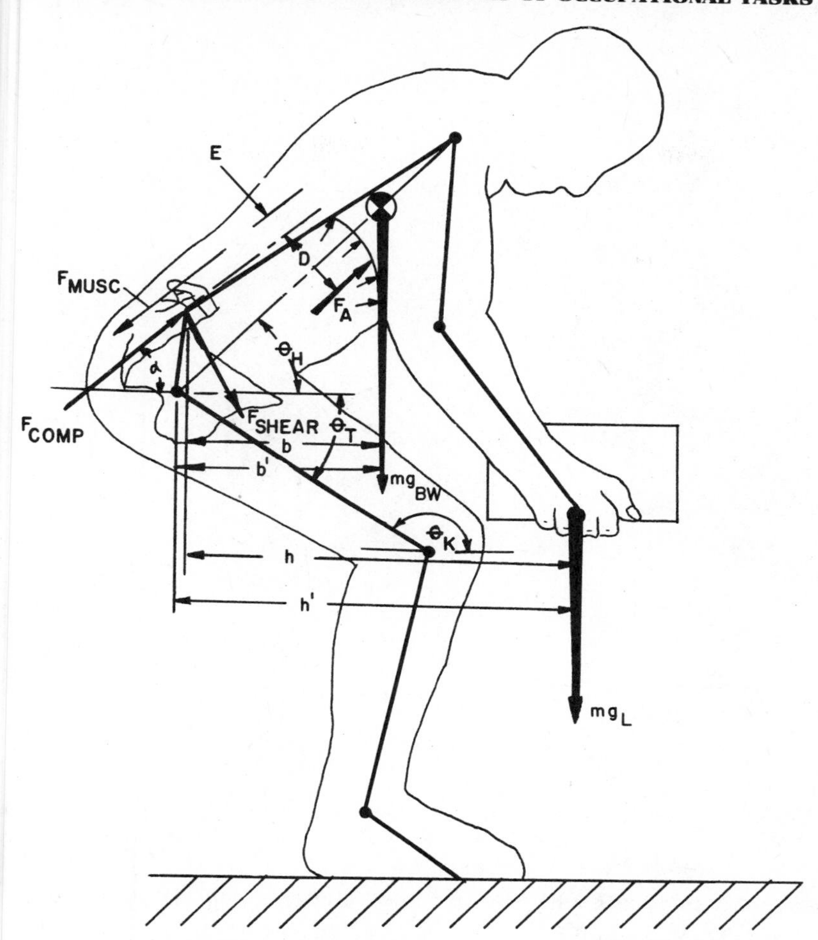

FIGURE 6.23 Simple low-back model of lifting as adapted by Chaffin (1975) for static coplanar lifting analyses.

exertion. A limit to the value of P_A has been suggested by Davis and Stubbs (1978) to be 90 mm Hg, although values of 150 mm Hg are not uncommon in those who regularly lift weights.

The amount of force F_A created by the abdominal pressure can be estimated by assuming the following two conditions, according to Morris, Lucas, and Bressler (1961): (1) an average diaphragm area of 465 cm^2 upon which the abdominal pressure can act, and (2) a line of action of the force that acts parallel to the line of action of the compression forces on the lower lumbar spine. The moment arm D of the force F_A has been assumed by Chaffin (1975) to vary as the sine of the angle at the hips θ'_H, with the erect

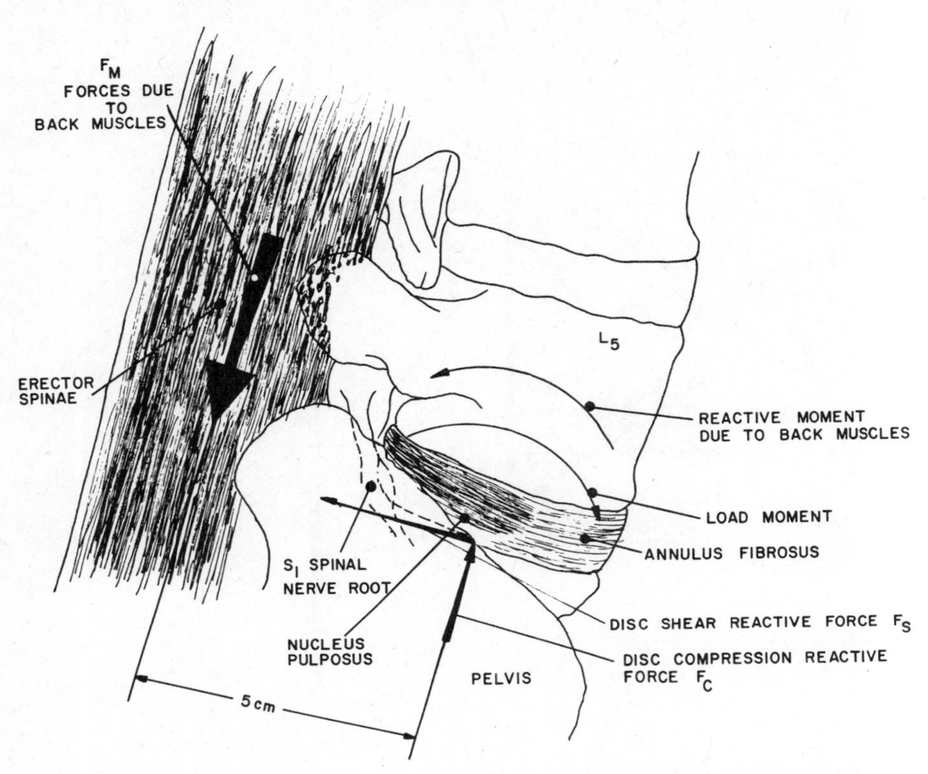

FIGURE 6.24 Forces and moments on spine in simplified spinal model of load lifting.

position having a moment arm of about 7 cm, increasing to about 15 cm when stooped over at $\theta'_H = 90°$ from the vertical, or $\theta_H = 0$ from the horizontal.

Additional compressive forces on the lumbar spine due to an antagonistic action of the abdominal muscles are assumed to be negligible, since Bartelink (1957) showed that the rectus abdominis, which could mechanically cause a spinal compressive force, was relatively inactive during lifting activities. The abdominal pressure was attributed, therefore, to the oblique and transverse muscles, which are not well positioned to directly assist or hinder in sagittal-plane flexion or extension of the trunk. This assumption of no antagonism was also utilized by Schultz (1981) and Lee (1982) in more sophisticated torso models.

The line of action of the erector spinae muscles of the lower lumbar back are assumed in this simple model to act parallel to the normal force of compression on the L5/S1 disc, and with a moment arm E of 5.0 cm, as illustrated in Figure 6.24. This is an average muscle moment arm, based on values published by Bartelink (1957), Perey (1957), and Thieme (1950).

Given the preceding information, the model becomes statically determinate with three unknowns and three coplanar equilibrium equations. The unknown magnitudes are (1) the magnitude of the erector spinae muscle force, (2) the magnitude of the reactive compression force at the top of the sacrum, and (3) the magnitude of the reactive shearing force across the top of the sacrum. The moment equation for this system is

$$\sum M_{\mathrm{L5/S1}} = 0$$

$$b(mg_{bw}) + h(mg_{load}) - D(F_{\mathrm{A}}) - E(F_m) = 0 \qquad (6.45)$$

Since the compression and shear forces are assumed to act at the disc center of rotation, and thus are not considered in the moment equation, the equation can be rearranged to solve for the unknown muscle force F_M:

$$F_M = \frac{b(mg_{bw}) + h(mg_{load}) - D(F_A)}{E}$$

where

F_M is the effective erector spinae muscle force necessary to stabilize the spine
b, h, D, and E are the moment arms of relevant forces (see Figure 6.23)
mg_{bw} is the weight of the body segments above the L5/S1 level
F_A is the effective force due to abdominal pressure acting at the center of the diaphragm ($F_A = P_A$ 465 cm^2)
mg_{load} is the weight of the load in the hands

The forces acting parallel to the disc compression force (as shown in Figure 6.23) can be expressed by:

$$\sum F_{\mathrm{COMP}} = 0$$

$$\sin \alpha mg_{bw} + \sin \alpha mg_{\mathrm{load}} - F_A + F_M - F_C = 0 \qquad (6.46)$$

Since all values are known except the reactive force of compression F_C, the equation can be solved for this unknown. Similarly, the reactive shear force across the L5/S1 disc can be solved by the third equilibrium equation:

$$\sum F_{\mathrm{SHEAR}} = 0$$

$$\cos \alpha mg_{bw} + \cos \alpha mg_{\mathrm{load}} - F_S = 0 \qquad (6.47)$$

Note in this equation that the internal muscle and abdominal forces are considered to be normal to the shear forces and act only in compression. If anatomical and kinesiological data that provide the line of action for additional muscles and their relative force contributions are considered, it becomes possible to expand the three equilibrium equations to better estimate the relative compression and shearing forces. Such a model will be discussed later in this chapter. For now, this simple model assumes that the reactive shearing forces are produced by the lumbar facet joints, which was anatomically shown to be possible in the lumbar spine by Kraus (1973).

Though highly simplistic, the preceding model of the low back is of assistance in predicting the relative forces on the low back when lifting various loads in front of the body. A few examples of the insight one can gain from such a model follow:

Lifting 450 N Load. Assume for this case the posture displayed earlier in Figure 6.23. The kinematic and anthropometric data are

$$h = 30 \text{ cm} \qquad \theta_H = 30°$$

$$b = 20 \text{ cm} \qquad \theta_T = 40°$$

$$mg_{bw} = 350 \text{ N (above L5/S1 level)}$$

The moment at the hip for an average anthropometric male in this posture is found to be 200 N m. This value, combined in equation 6.44 with these hip and thigh angles yields an abdominal pressure prediction of

$$\begin{aligned} P_A &= 10^{-4}[43 - 0.36(70°)][200^{1.8}] \\ &= 24 \text{ mmHg} \\ &= 0.32 \text{ N/cm}^2 \end{aligned}$$

which, when multiplied by the diaphragm area of 465 cm^2, produces an estimate of the abdominal force F_A:

$$\begin{aligned} F_A &= 0.32(465) \\ &= 148 \text{ N} \end{aligned}$$

Using this force value in the moment equilibrium equation 6.45, the erector spinae effective force is calculated:

$$\begin{aligned} F_M &= \frac{(20 \times 350) + (30 \times 450) - (11 \times 148)}{5} \\ &= \frac{7000 + 13500 - 1628}{5} \\ &= 3774 \text{ N} \end{aligned}$$

For comparison, this value would be within the 2200 to 5500 N limit estimated by Farfan (1973) as the normal range of strength capability of the erector spinae muscles. From a muscular limit, therefore, exerting 450 N of lifting force in the posture illustrated in Figure 6.23 appears feasible for many people, though some would have difficulty. This indeed appears to be the case, based on the isometric lifting stengths. These indicate that approximately 90% of men and 50% of women could perform such an exertion. Thus, some muscle strains could occur if a weak person attempted such a lift.

The compression force on the L5/S1 disc is computed by equation 6.46. This first involves estimating the angle α from the pelvic rotation data described in Figure 6.22. For this example α was estimated to be about 60°—that is, 50° from horizontal when erect, which is a normal lordotic angle, plus 10° of backward rotation due to the 40° thigh angle. The resulting compression force F_C is

$$
\begin{aligned}
F_C &= \sin 60°(350) + \sin 60°(450) - 148 + 3774 \\
&= 303 + 390 - 148 + 3774 \\
&= 4319 \text{ N}
\end{aligned}
$$

This value is not excessive compared with vertebral/cartilage endplate crush values from the cadaver spine studies of Evans & Lissner (1965), Sonoda (1962), and Perey (1957). In these studies, isolated cadaver spines of various ages were carefully placed in combined compression and moment loading conditions until the specimens failed. Follow-up microscopic inspection of the failed specimens disclosed that the cartilage endplates most often failed. The compression failure values from these studies are summarized in Figure 6.25. Based on these results, the National Institute for Occupational Safety and Health (NIOSH, 1981) has recommended that predicted L5/S1 compression values above 3400 N be considered potentially hazardous for some workers. If the values are greater than 6400, the job is hazardous to most workers. The example of lifting a 450 N load produces a compression force estimate above the lower limit, and thus care would be recommended in selecting, training, and monitoring workers who elect to perform such lifting.

It appears that chronic low-back pain is often the result of degeneration of the disc (Armstrong, 1965; Rowe, 1971; Morris, et al, 1961). It is believed that repeated microfractures and bone scarring of the cartilage endplate leads to a weakening of the annulus fibrosus of the disc, which thus protrudes into the spinal canal, as depicted in extreme form in Figure 6.26. Since pain nerve fibers are not known to exist in the cartilage endplate per se, damage to this structure could occur frequently during heavy lifting but would only be realized later when the disc, and annulus fibrosus in particular, has failed to maintain structural integrity. Such failure would then cause nerve root

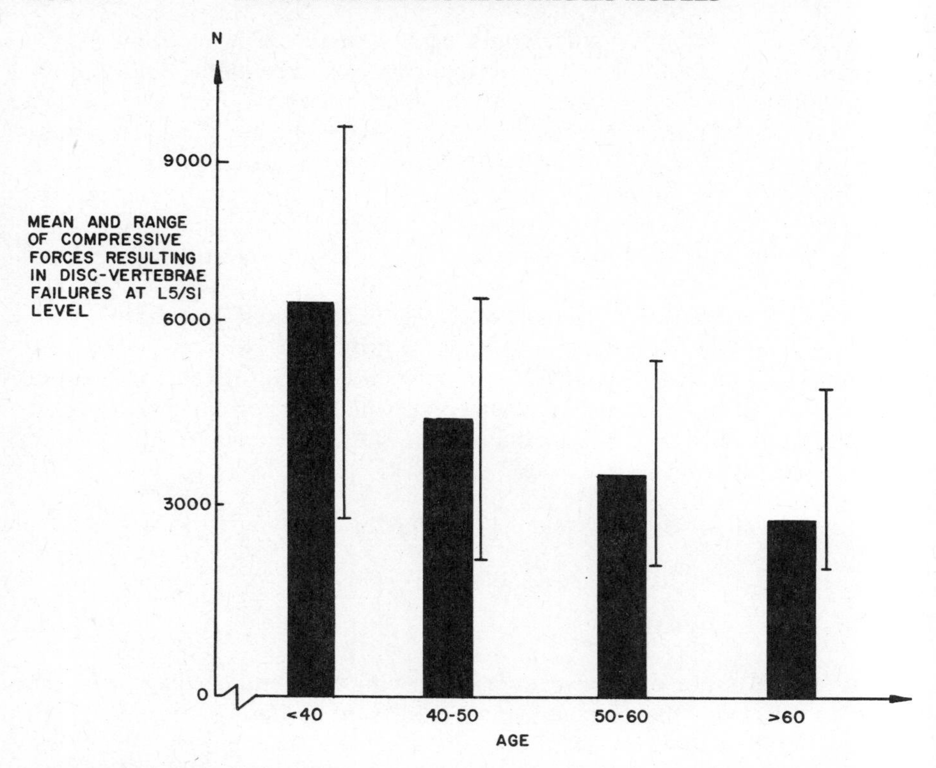

FIGURE 6.25 Composite of compression failure values obtained from cadavers of different ages (see text).

compression or distortion of the ligaments around the disc or at the posterior facet joints, which are believed to be innervated by pain receptors.

In short, large spinal-disc compression forces can be produced by muscular exertion, especially when lifting. The resulting values, if repeated, are believed to increase the risk of disc degeneration and accompanying chronic low-back symptoms.

The disc shear forces in the example of lifting 450 N can also be estimated by equation 6.47:

$$
\begin{aligned}
F_S &= \cos 60°(350) + \cos 60°(450) \\
&= 175 + 225 \\
&= 400 \text{ n}
\end{aligned}
$$

Such shear forces, whether incurred in flexion, extension, or torso twisting activities, are resisted primarily by the posterior facet joints in the lumbar spine, as well as the annulus fibrosus of the disc. The strength of the facet joints has been examined by Fiorini and McCommond (1976). They ac-

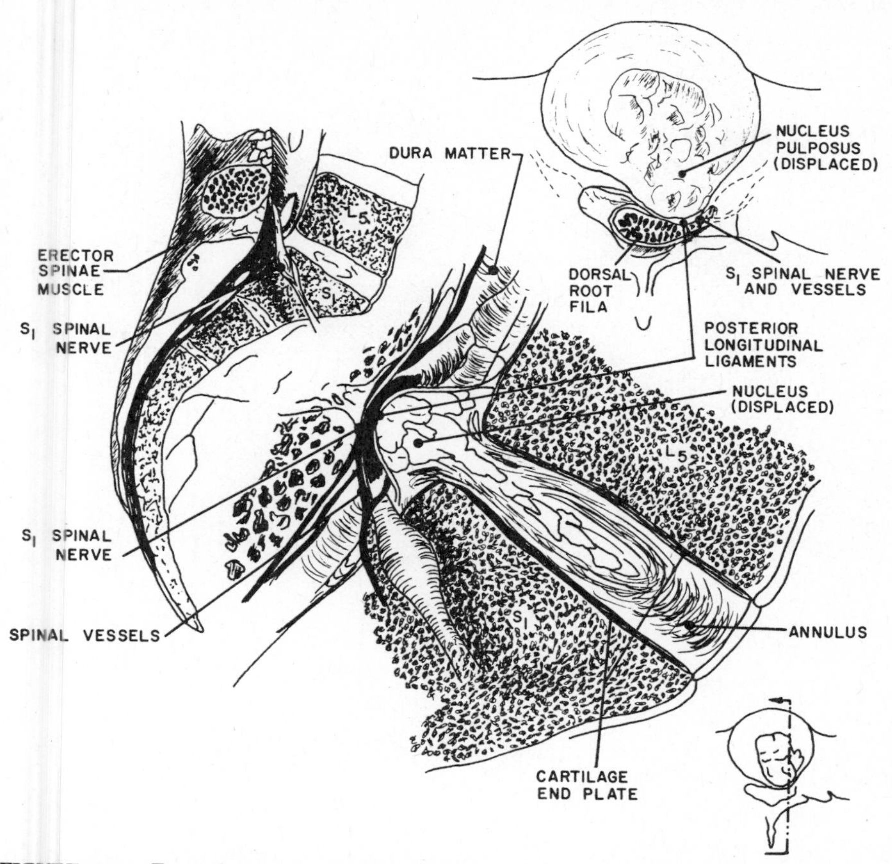

FIGURE 6.26 Example of L5/S1 disc herniation resulting in spinal nerve root compression.

knowledge that under normal physiological conditions the facets can resist shear forces, though the limit to such forces is not well documented (Schultz et al, 1979). Fiorini and McCommand (1976) propose that if a disc space is narrowed by degeneration, abnormally high stresses can result in the facet joints. Because pain receptors are abundant in the facet joint and associated ligaments, such misalignment could cause low back pain.

It appears that maintenance of the biomechanical and geometric integrity of the spinal discs is necessary if one is to avoid increased risk of low back pain. For this reason, biomechanical models of spinal discs along with experimental studies have been developed in the last decade. Several different approaches have been taken to relate the external moments and forces acting on the disc to internal tissue stresses and resulting strains. Generally, these models treat the disc as a mesh of tensile elements, which deform because of (1) the intradiscal pressure resulting from the high compression forces,

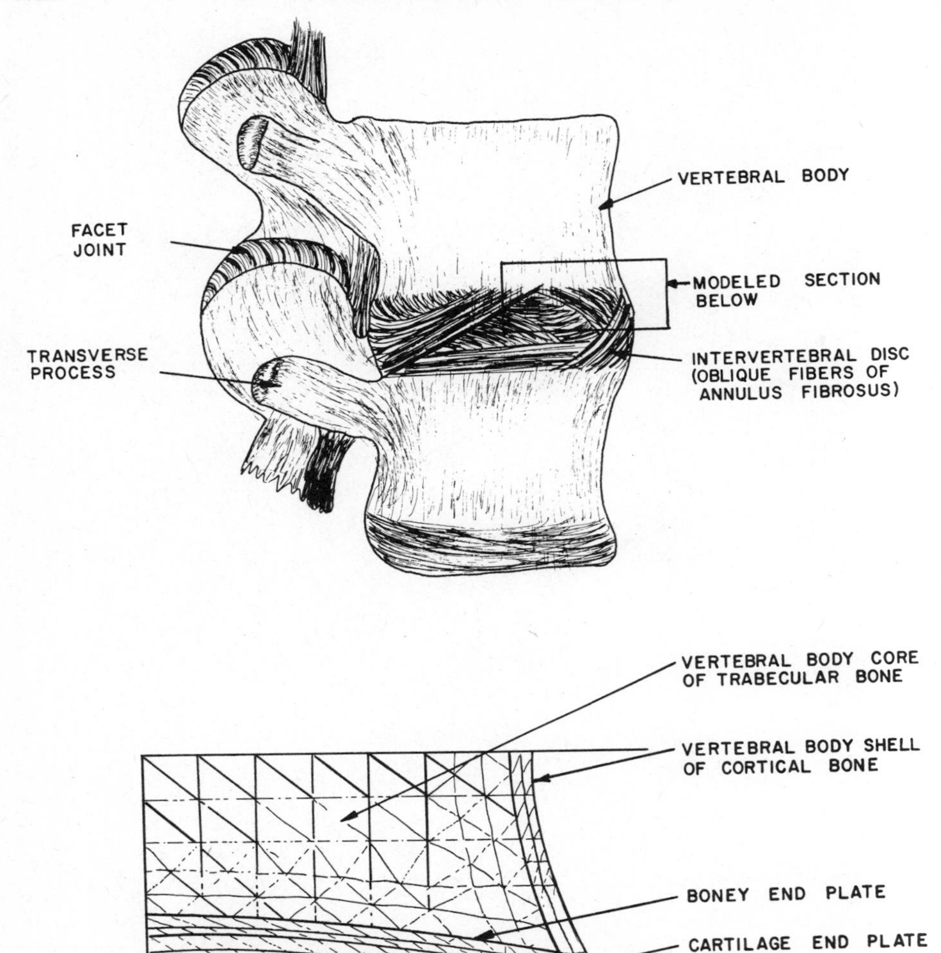

FIGURE 6.27 Finite element model of vertebral body and disc (Kulak, Belytschko, and Schultz, 1976).

or (2) the shearing or torsional loads acting on the structure. Because the disc is defined as a complex set of mechanical elements, the modeling method is referred to as finite element modeling. Figure 6.27 shows a typical geometry for such an element mesh of the disc by Kulak, Belytschko, Schultz, and Galante (1976). As the exterior forces are altered, the mesh will deform as a function of the geometry and material stiffness properties assigned to each element. The deformation (strain) in the mesh is tabulated and compared with cadaver-disc experimental results under similar loading conditions. Such models provide insight into how gross external forces and spinal geometries combine to cause overstrain and injury to specific elements of the disc. Obviously, simplifications in the material properties and structural geometries are necessary in such models to allow them to be computationally practical and interpretable. Despite these simplifications, Schultz and Andersson (1981) believe that planar-motion finite element models that have been developed, though difficult to compute, are reasonably predictive of (1) the disc narrowing that occurs due to compression force loadings, and (2) the radial and tangential stresses in the fibers of the annulus for given intradiscal pressures. Clearly, this type of internal structural modeling is an important research area in the search to improve the etiology of most low back pain, which is now designated as *idiopathic*—without obvious tissue pathology.

Because the exact source of low back pain often cannot be diagnosed, it does not mean that the present state of biomechanical knowledge isn't currently valuable in understanding and reducing the incidence and severity of low back pain, especially in industry. The preceding example of the application of a simple sagittal-plane model of the low back provides a basis for understanding how a load held in various postures can create potentially harmful compressive disc forces. Another example of the use of such a simple model follows to further demonstrate its utility.

Lifting Variable Loads in Three Different Postures

In this example, first consider an anthropometrically average male holding a load of varying magnitude in the same posture as depicted earlier in Figure 6.23—that is, with an H distance of 30 cm from the L5/S1 disc. If the load is incrementally increased in magnitude from no load to over 500 N, the predicted compression force on the L5/S1 disc (F_C in equation 6.46) would increase as in Figure 6.28. The slight nonlinearity in the F_C response is due to the nonlinear abdominal pressure effect (equation 6.44).

If the load is moved closer or further from the torso (H is varied along with the torso and arm postures), the F_C response is also greatly affected. This is particularly informative when the F_C values are compared with the suggested NIOSH Action Limit, where it is believed some workers would be at risk. This limit is also shown in Figure 6.28 along with the Maximum Permissible Limit. This analysis shows that a potential spinal hazard exists for some workers when they lift a 500 N load close to the body ($H = 20$ cm) or in lifting a 75 N load further away from the body ($H = 50$ cm).

Lifting a Large Object from Floor. If a small object is lifted from the floor, a person can squat down with the feet beside the object, and thus

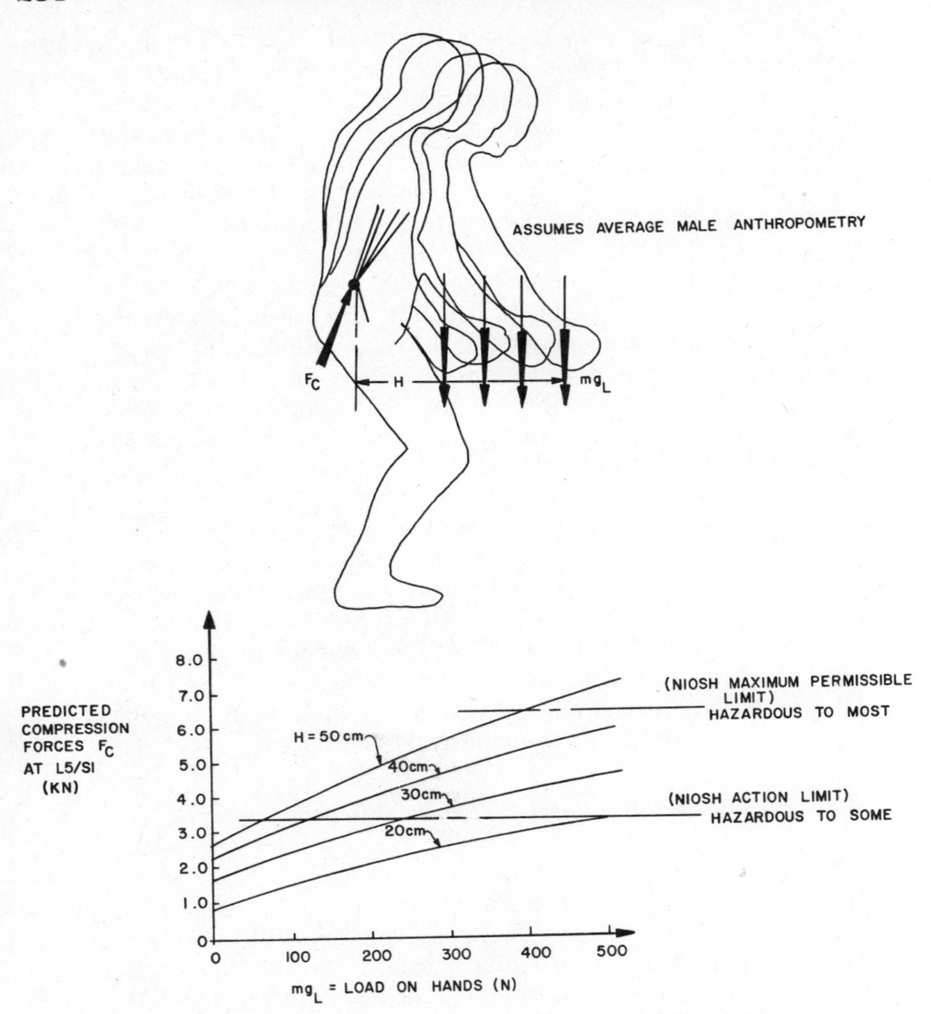

FIGURE 6.28 Predicted L5/S1 disc compression forces for varying loads lifted in four different positions from body.

maintain the object close to the torso while lifting, minimizing H and the resulting force of compression on the L5/S1 disc. When an object is too large to straddle, however, it must be lifted around the knees when one attempts to squat down and lift it. In such a situation, stooping over the object may result in less compression on the L5/S1 disc than the squat type of lift. A biomechanical comparison of the two methods was reported by Park and Chaffin (1974), based on cineographic analysis of workers handling loads in a plant. Typical lifting postures are shown in Figure 6.29. Also shown are

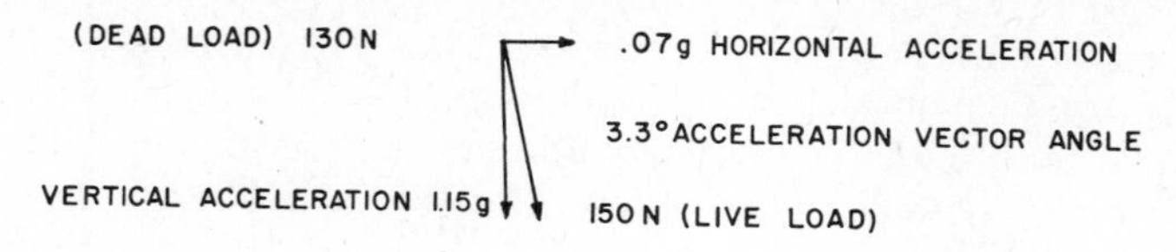

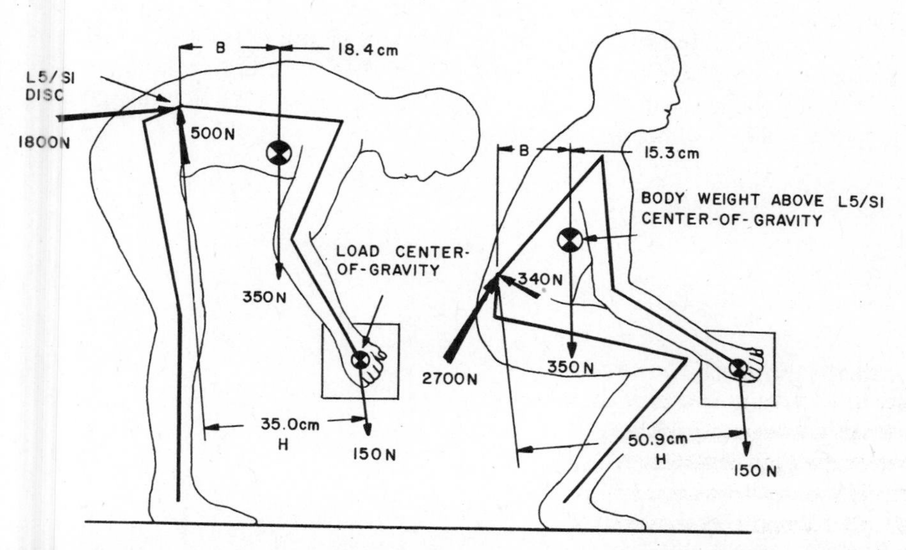

FIGURE 6.29 Comparison of predicted reactive L5/S1 forces during lifting of large object weighing 130 N (Park and Chaffin, 1974).

common dynamic load trajectory data when lifting a 150 N object that was too bulky to fit between the knees when attempting a squat lift posture. The F_C values were found to be much higher in the squat lift posture than in the stooped lift posture. This would suggest that any method of lifting that allows a person to minimize H is preferred over others with large H values. This conclusion was also reached by Andersson, Ortengren and Nachemson (1978) using intradiscal pressure measurements.

The biomechanical analysis also showed that the shear forces, however, were larger in the stooped posture compared with the squat posture. This suggests that maintaining the torso in a more erect (vertical) orientation is preferred. In this regard, the question of the best lordotic curvature and inclination angle for the lumbar spine while under such loads becomes important. Unfortunately, until more complex micromodels of the discs and facets are created, along with experimental validation, rules regarding the best or "safest" back posture to use in lifting must be regarded as speculative.

Three-Dimensional Low Back Modeling. The preceding has described a simple coplanar model of lifting that includes a single back-muscle force. Additional muscles have been considered by Schultz and Andersson (1981) in a three-dimensional analysis. An illustration of the forces considered in such a model is given in Figure 6.30. The internal forces are: posterior and lateral disc shear forces S_P and S_L; erector spinae muscle force F_M; disc compression force F_C; abdominal pressure resultant force F_A; rectus abdominus force A; and vertical and horizontal components of the lateral abdominal wall muscle forces V_R, H_R, V_L, and H_L on the right and left sides. Assuming the coordinate system shown in Figure 6.30, Schultz gives the following six equilibrium equations:

$$S_L = F_X \tag{6.48}$$

$$H_L + H_R + S_P = F_Y$$

$$F_C + F_A - F_M - A - V_L - V_R = F_Z$$

$$A\overline{ac} - F_A\overline{bc} - F_M\overline{dc} = M_X$$

$$V_L cf - V_R\overline{ce} = M_Y$$

$$H_L\overline{cf} - H_R\overline{ce} = M_Z$$

Since there are 10 unknown forces and six equations, the problem is statistically indeterminate. Schultz solved the problem by assuming that there are no antagonistic muscle actions, that the value of the abdominal pressure effect F_A is empirically predicted (equation 6.44), and that muscles act only in tension. These assumptions allow certain forces to be selectively zeroed, based on the following partitioning inequalities:

If $M_X > 0$,	then	$F_M = 0$	(no erector spinae)
$M_X < 0$,		$A = 0$	(no rectus abdominus)
$M_Y < 0$,		$V_R = 0$	(no right lateral flexors)
$M_Y < 0$,		$V_L = 0$	(no left lateral flexors)
$M_Z > 0$,		$H_R = 0$	(no right rotators)
$M_Z < 0$,		$H_L = 0$	(no left rotators)

This partitioning of the three-dimensional model allows six force estimates to be made, instead of two in a coplanar model, but the six that are included in the solution depend on the direction of the moments resulting from given input load and posture conditions.

An additional refinement of this spinal model was proposed by Schultz et al. (1982). This model further divided the torso muscles at the lumbar

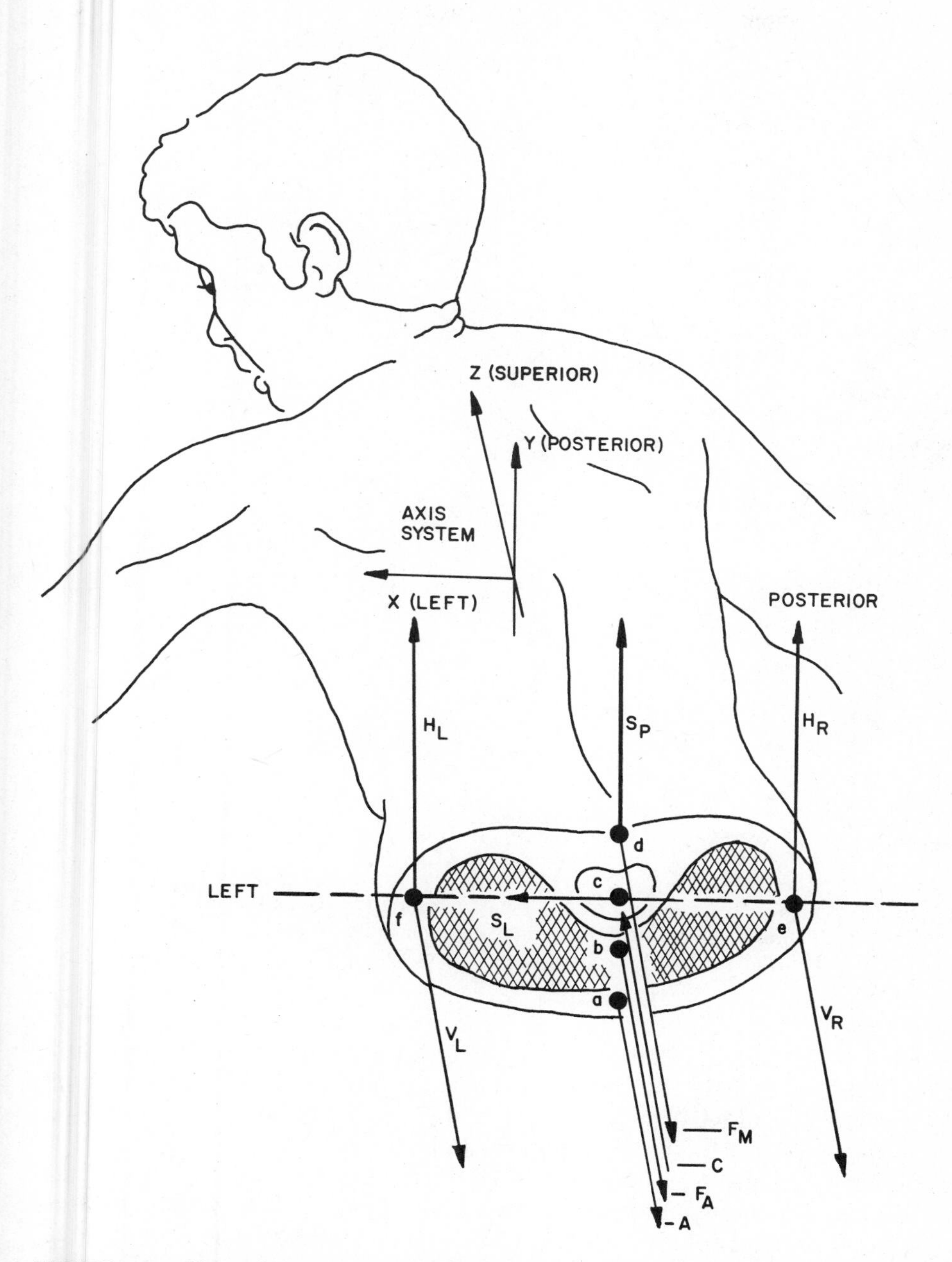

FIGURE 6.30 Three-dimensional low back model from Schultz (1981) (see text for variable names).

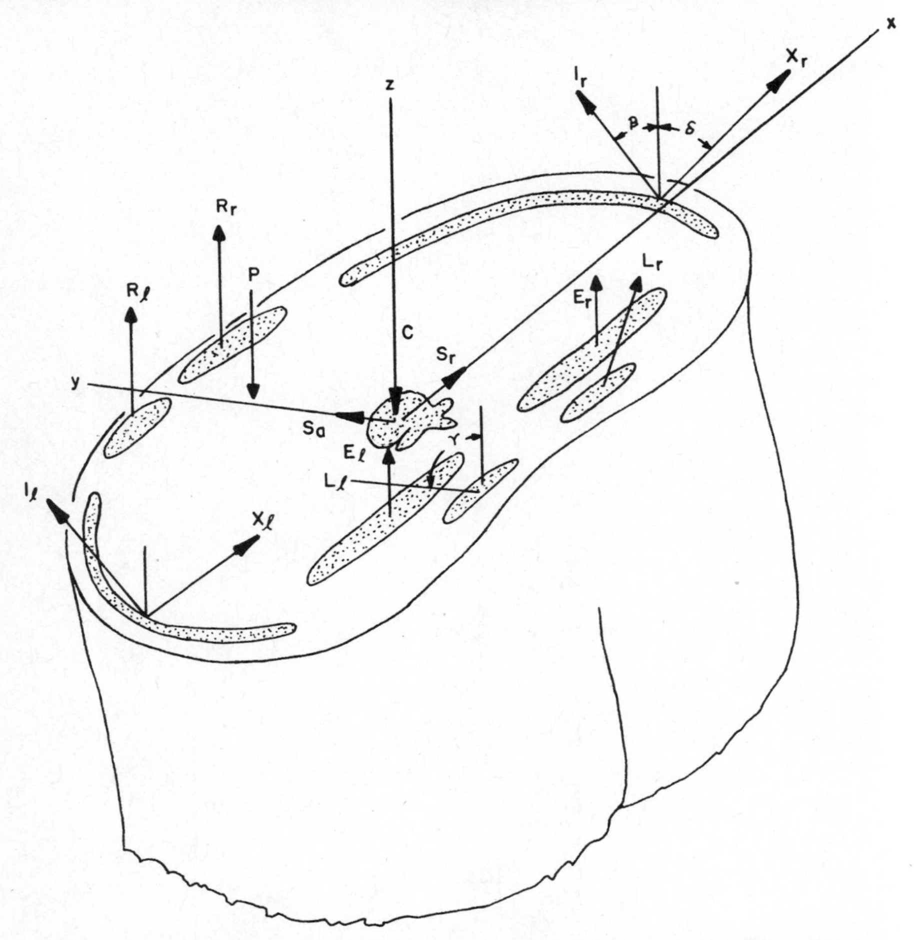

FIGURE 6.31 Schematic diagram of the 10-muscle model. The 10 unknown muscle forces and intra-abdominal force *P* are computed to predict the minimum compression force on the disc (adapted from Schultz, Andersson, Ortengren, Haderspeck, and Nachemson, 1982).

region into five pairs (left and right) as follows; rectus abdominus (R_1 and R_r); external oblique (X_1 and X_r); internal oblique (I_1 and I_r); erector spinal (E_1 and E_r); and latissimus dorsi (L_1 and L_r). The L3 motion segment was assumed to resist forces of compression, anterior–posterior shear, and lateral shear. Figure 6.31 depicts a force diagram of the system. This model includes 13 unknown internal forces, thus being statistically indeterminate.

The method of solution used was to assume that the muscles that created the smallest incremental compression force on the lumbar segment would contract first. In other words, the muscles with the largest appropriate moment arms contract first, increasing in intensity to create the reactive moment necessary to resist any external load moment. Maximum contraction strength limits for each muscle were assumed, based on their cross-sectional areas. If a single muscle was not capable of resisting an applied load moment, the muscle with the next largest moment arm would contribute. By adding this priority of muscle action and maximum values for each muscle's strength, unique solutions were possible for given external load and postural conditions.

A set of 25 standing and sitting posture validation experiments were performed on this 10-muscle model. Surface EMGs, intra-abdominal pressures, and intradiscal pressures were obtained for comparison with the model predictions. Schultz et al. (1982) reported simple linear correlations of about $r = 0.9$ between EMG amplitude levels and the predicted muscle forces for the erector spinae. The other more anterior trunk-muscle predicted forces were not as well correlated with the EMG levels, having r values of about 0.5 for the rectus abdominus and 0.2 for the oblique muscles. Intra-abdominal pressures were poorly correlated with that predicted $r = 0.24$. The intradiscal pressures, however, were in good agreement with the compression force predicted by the model, with $r = 0.94$. This latter result is depicted in Figure 6.32.

These models collectively demonstrate that the effect of external loads and postures on the lumbar spine can be successfully studied with various models of the torso. This is true even though the specific synergistic actions of muscles are more difficult to predict, as discussed in Section 6.2.4. Clearly, as more sophisticated models and bioinstrumentation methods become available, muscle actions and their effects on various motion segments will be known.

Dynamic Load Lifting Modeling of the Low Back. The preceding static models of torso muscle actions during lifting would not be complete without some discussion of dynamic effects. As presented earlier in the chapter, inclusion of inertial effects adds considerable complexity to the modeling. The dynamic models of Fisher (1967) and El-Bassoussi (1974) are noteworthy in this regard. A coplanar analysis of maximum load lifting was completed recently (Garg, et al, 1982), the results of which indicate that the inertial forces during the first accelerating phase of a lift can add considerably to the maximum L5/S1 compression force. The experiments required six male college students to first choose the maximum load they believed they could safely lift from the floor to carrying height. Once the maximum loads were chosen, they repeatedly lifted the loads while being filmed. The films were used to estimate the body segment and load acceleration components, which were then analyzed with a model similar to that described in equations 6.40 and 6.41. In addition to the moments and reactive forces at the extremity articulations, the analysis included use of the simple L5/S1 model (equations 6.45, 6.46, and 6.47). For each specific posture in the lifting sequence, the dynamic moment at the L5/S1 disc was estimated and then used as input in

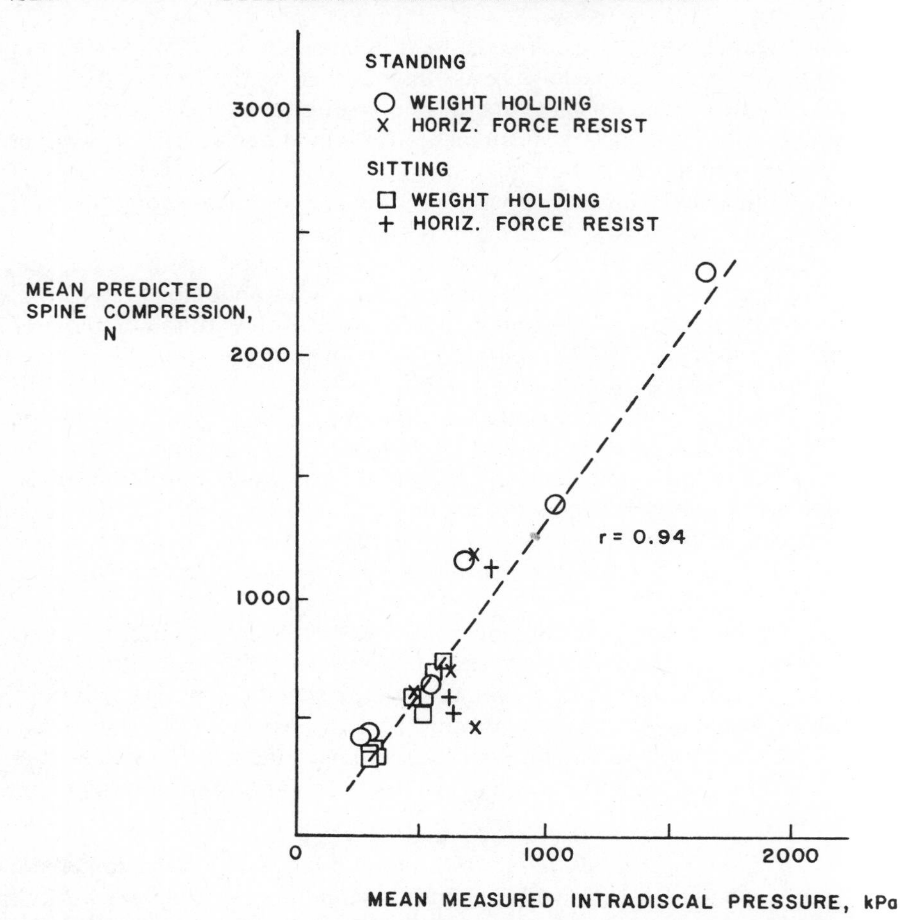

FIGURE 6.32 The predicted compression forces on the spine using the 10-muscle model compared to measured intra-discal pressures. Values are the means for four subjects performing various static exertions (from Schultz, Andersson, Ortengren, Haderspeck, and Nachemson, 1982).

the L5/S1 model to predict the compression force necessary to maintain equilibrium.

The lifting conditions and resulting average compression force predictions are illustrated in Figure 6.33. The results indicate that inertial forces increase the compression forces on the L5/S1 disc considerably. A static analysis showed that the compressive forces in these lifts were far below the NIOSH permissible limit. The dynamic analysis, however, indicated that the peak forces slightly exceeded the limit. This analysis also showed that the subjects chose loads that effectively compensated for the size of the box—the larger the box the lower the loads. Finally, peak forces were incurred early in such

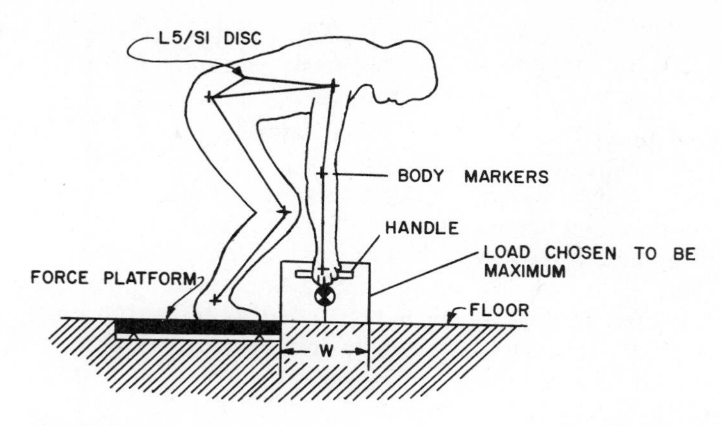

DESIGNATION	BOX SIZE	W(cm)	AVERAGE LOAD CHOSEN (N)
———— (solid)	SMALL	38	321 N
— — — — (dashed)	MEDIUM	51	289 N
----------- (dotted)	LARGE	64	271 N

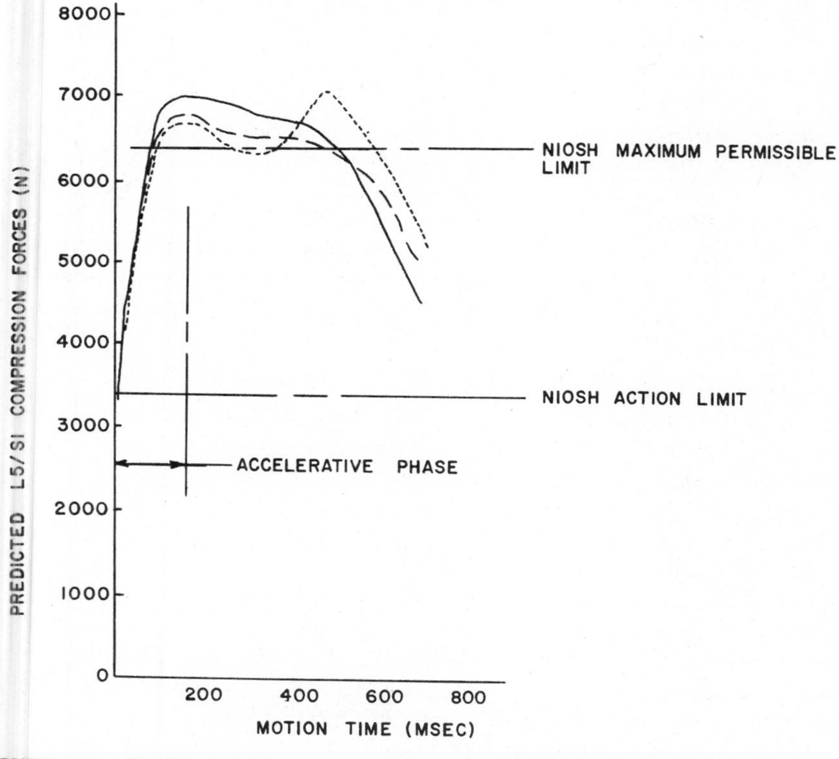

FIGURE 6.33 Predicted L5/S1 compression forces while lifting maximal acceptable loads (Freivalds et al, in press).

a motion, when the load was still close to the floor. Thus, instructions to move close to the load before lifting appear warranted to minimize the load moment arm H at the beginning of the lift.

Dynamic Push/Pull Modeling of the Low Back. A similar analysis of pushing and pulling carts was completed by Lee (1982). The multisegmental biodynamic model used for this analysis was described in equations 6.40 and 6.41, and in Section 6.4.3. As in the preceding dynamic lifting analysis, the joint moments were estimated for various postures chosen by subjects while pulling carts—that is, while walking backwards. These moments were treated as inputs to the simple L5/S1 model described in equations 6.45, 6.46, and 6.47. These yielded estimates of compressive forces on the L5/S1 disc at various periods in the pulling act. In pushing, the external load moments on the L5/S1 disc were often found to act in extension. Thus, the stabilizing muscle required was assumed to be the rectus abdominus muscle, as depicted in Figure 6.30, rather than the erector spinae. To determine whether antagonism existed between these muscles, EMG's of both muscles were obtained during the experiments. No interaction was found for the load range used, and thus a single muscle model with either the erector spinae or rectus abdominus appeared appropriate. Handle heights, speed of walking, and cart resistance were varied in the experiments.

Typical predicted peak compression forces on the L5/S1 disc are depicted in Figure 6.34. These show that cart pushing appears to result in significantly less compression force than cart pulling. This result is consistent with the abdominal pressures measured by Davis and Stubbs (1978), who showed that tensing of the anterior abdominal muscles creates higher gastric pressures when pushing than when pulling. Such increased pressures provide additional spinal stabilization at higher push forces and thus assist in maintaining the spinal compression forces at values lower than when pulling loads. Also, when pushing, the subjects would incline the torso more than when pulling, thus using the body weight more effectively to assist in counteracting the push force on the hands.

In this latter regard, it must be noted that the body posture assumed by a person performing such acts is directly dependent on the coefficient of friction of the shoe/floor interface discussed earlier in the chapter. Lee had his subjects in high-traction shoes walking on a high-traction floor; thus the use of body weight (torso inclination) was feasible. If a low coefficient of friction exists, then presumably the person would stand more erect to minimize the required coefficient of friction. In so doing, the back compression forces would probably be different than Lee has indicated.

It can be concluded from the initial dynamic cart pushing and pulling modeling efforts that much more research (both modeling and experimental) needs to be accomplished. Only after such research will it be possible to objectively predict that a specific pushing or pulling activity performed in a given work environment is hazardous.

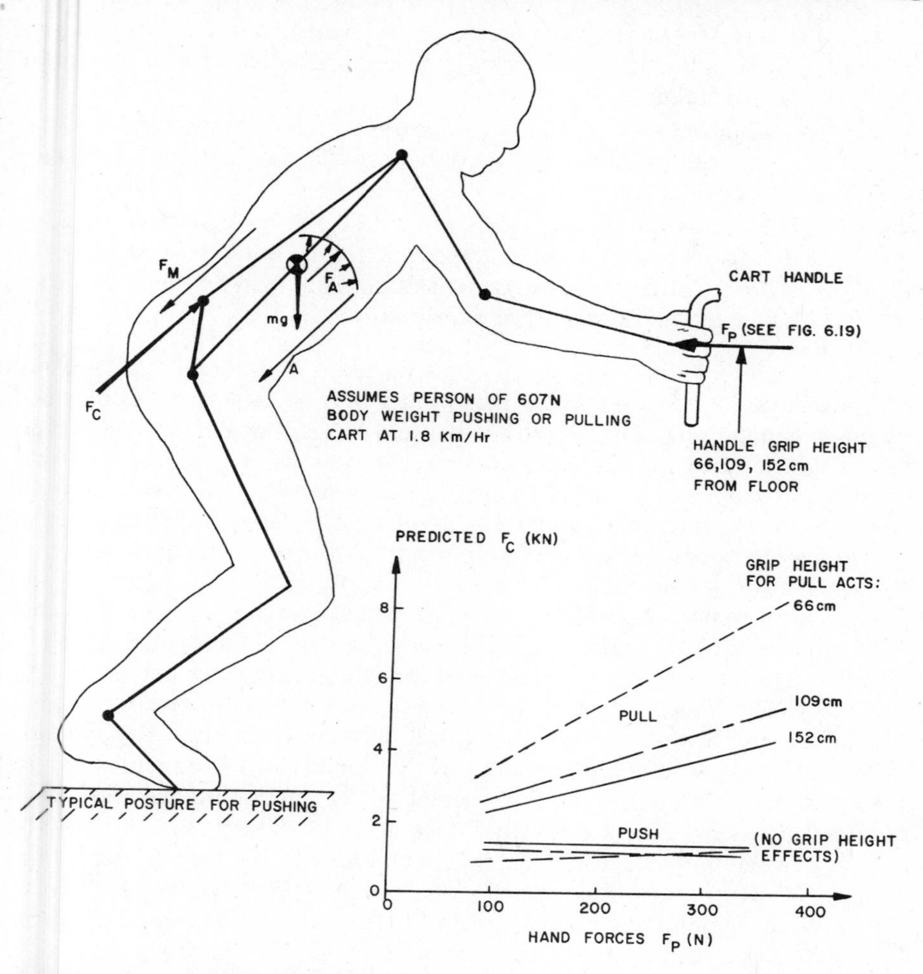

FIGURE 6.34 Predicted peak L5/S1 compression forces for pushing and pulling of cart (Lee, 1982).

6.5.2 Biomechanical Models of the Wrist and Hand

Though incidents of low back pain are prevalent and often result in severe disability for workers, recent studies indicate repetitive manual exertions of the hand cause cumulative trauma to the wrist of many workers—according to Tanzer (1959); Tichauer (1966); Wilson and Wilson (1957); Inglis, Straub, and Williams (1972); Hadler (1977); and Armstrong and Chaffin (1978). The disorders of concern are usually described as follows:

1. Wrist tenosynovitis or tendinitis, in which the finger flexor or extensor tendons and/or surrounding synovia become inflamed and are painful with movement.
2. Ganglionic cysts that develop as swollen and often painful nodules on finger tendons at either the wrist or adjacent to other bony articulations of the fingers.
3. Carpal tunnel syndrome (CTS), in which the median nerve is compressed as it passes through the carpal tunnel at the base of the palm and wrist, resulting in loss of sensation, pain, skin dryness, and atrophy of muscles in the nerve distribution area.

These types of disorders most often develop slowly and thus are not reported as an occupationally induced injury in many states. It is conceded by most medical specialist in orthopedics and occupational health, however, that abnormal usage of the hand precipitates these disorders, and some believe that specific patterns of manual activity are the basic cause of the disorders (Tichauer, 1966; Armstrong and Chaffin, 1978; Hadler, 1977). Epidemiology to support this belief is incomplete, but a biomechanical theory has evolved that associates certain hand exertions and postures with abnormal intra-wrist tissue forces. A brief and somewhat simplistic biomechanical model of the wrist is now presented as the basis for this theory.

The anatomy of concern is illustrated in Figure 6.35 for the act of gripping an object. The fingers are flexed by tendons inserted on the palmar side of the middle and distal phalanges (note that only one is shown). Each tendon is surrounded by a long thin-walled tube (tendon sheath) filled with synovial fluid to reduce sliding friction at the point where the tendon passes through confined areas or around bony trochleas at joints. The seven carpal wrist bones between the metacarpal and forearm bones are spanned on the palmar side by the flexor retinaculum ligament and on the dorsal side by the extensor retinaculum ligment. As shown in the cross-section in Figure 6.35, these ligaments band the wrist and contain most of the long tendons that cross the wrist, various arteries, veins, and—on the palmar side—the median nerve. The area enclosed on the palmar side by the flexor retinaculum ligament and the carpal bones is referred to as the *carpal tunnel*.

From a biomechanics perspective, as a load is applied to the palmar side of the fingers in grasping a tool, lever, or material, load moments result at each of the finger and wrist joints. Simple static equilibrium conditions have been used by several investigators to estimate the amount of tension required in the finger flexor tendons for various loads (Armstrong, 1976; Chao, et al, 1976; and Flatt, 1971). Armstrong (1976) proposed the tendon force values given in Figure 6.36. If a person was squeezing on an object with maximum effort, a typical finger load F_L would be about 150 N (Ohtsuki, 1981). Thus, the effector flexor tendon force would be between 420 and 645 N for the profundus and superficialis tendons combined from each finger. This magnitude of tensile force in the tendon, if repeated often, may be sufficient to

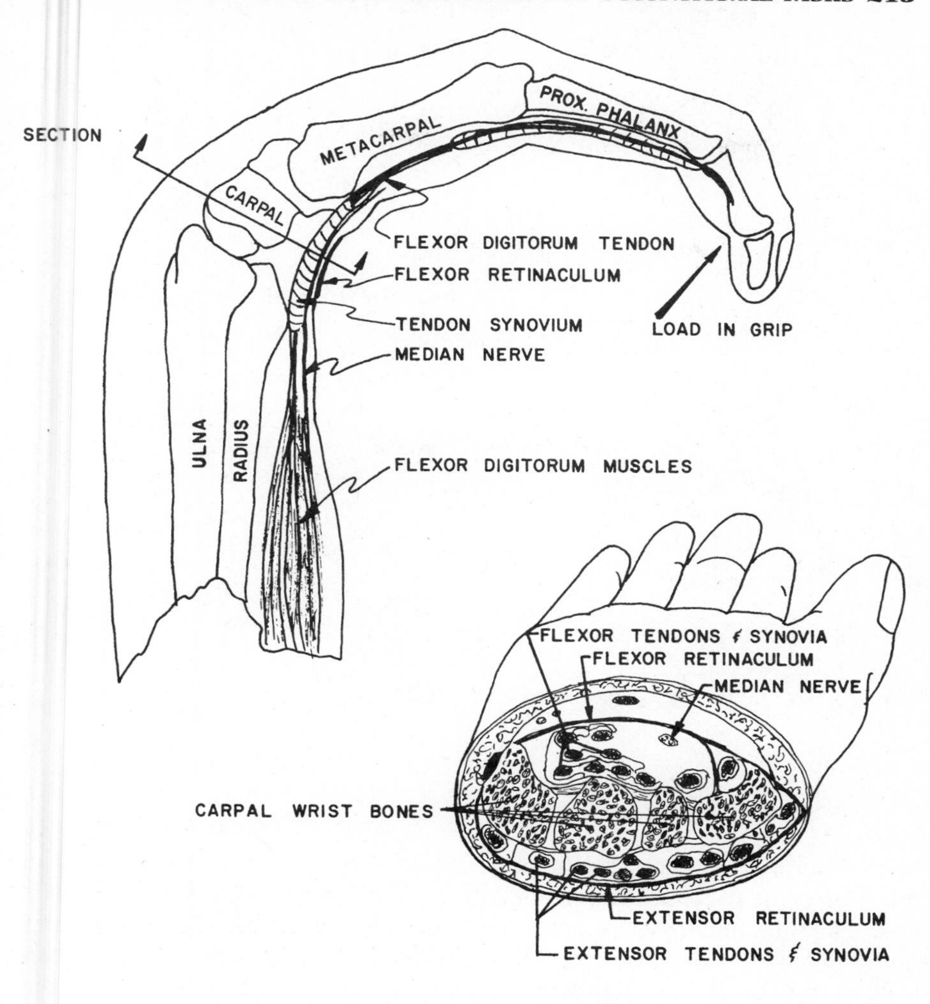

FIGURE 6.35 Anatomy of wrist is shown in an expanded cross-section along with single finger flexor tendon and synovium.

cause the tendon to gradually stretch, as shown in laboratory experiments (Goldstein, 1981). This phenomenon is referred to as *residual strain* by Abrahams (1967).

In essence, as a tensile load is applied to a tendon, it elongates (strains) about 1 to 2% of its no-load length. When the load is released, the tendon recovers. If a load is applied too often or too quickly, recovery is not complete, and residual strain results (adding as much as 1% to a tendon's length

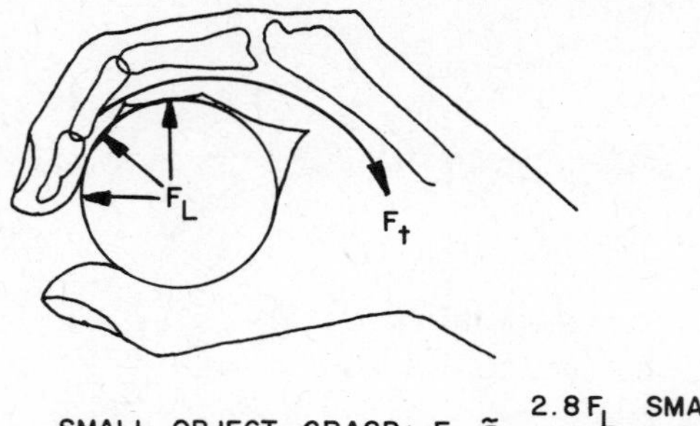

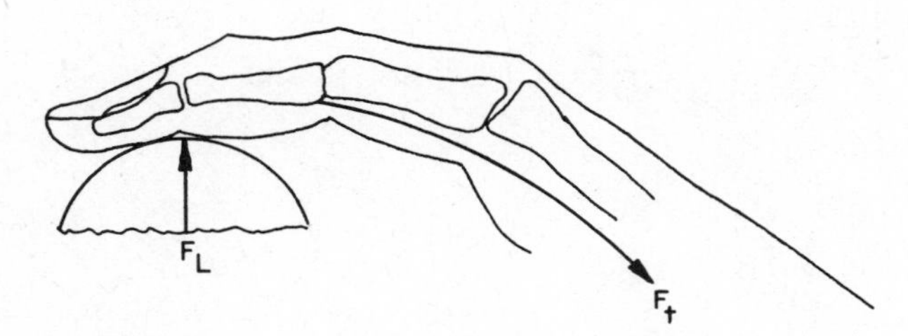

FIGURE 6.36 Ratios of finger flexor tendon forces to finger loads (Armstrong, 1976).

after a number of cycles). With rest, the tendon slowly recovers, but the question remains: Did the residual strain weaken the tendon, and/or did it create an inflammatory reaction and resulting proliferation of fibrous tissues in the tendon? Much more research is necessary on this subject before hazardous grip forces and frequency of exertion limits can be defined. Yet the preceding biomechanical etiology may explain tendinitis and ganglionic cysts.

If the wrist is in a neutral posture, the supporting tissues (synovia and flexor retinaculum ligament) and the adjacent median nerve will not be greatly stressed. Such is not the case, however, if the exertion takes place with a deviated wrist. In such a case, simple biomechanical concepts indicate the need for caution. These concepts are now presented.

When the wrist angle is deviated, particularly in flexion or extension, the long finger tendons are forced to curve around either the bones of the wrist or the flexor or extensor retinaculii. Figure 6.35 displays such a situation, where a finger flexor tendon is wrapped around the supporting flexor retinaculum ligament. This posture also discloses how the median nerve can be directly compressed when a load is applied to the fingers while gripping.

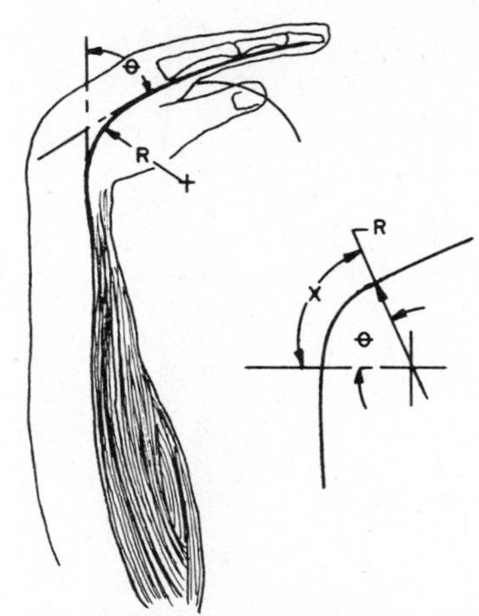

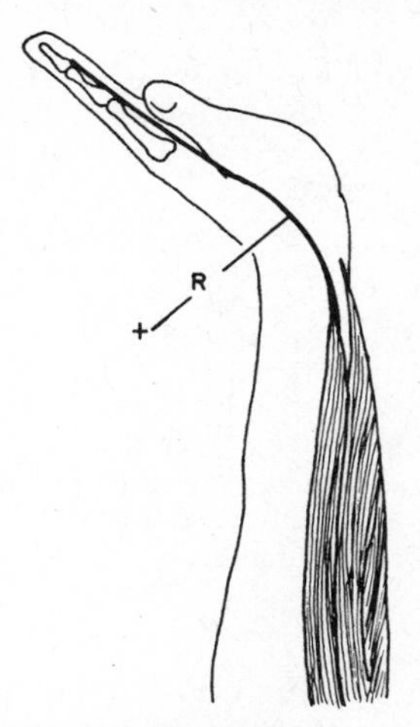

IN THE FLEXED WRIST, THE TENDONS ARE SUPPORTED ON THEIR PALMAR SIDE BY THE RECTINACULAR AND CARPAL LIGAMENTS.

IN THE EXTENDED WRIST, THE EXTRINSIC FINGER FLEXOR TENDONS ARE SUPPORTED ON THEIR DORSAL SIDE BY THE CARPAL BONES.

FIGURE 6.37 Finger flexor tendons depicted as wrapped around bone or ligaments of wrist (Armstrong and Chaffin, 1978).

Landsmeer (1962) proposed several kinematic models of the finger tendons when passing around a joint. Armstrong and Chaffin (1978) proposed the simple "pulley model" of the wrist depicted in Figure 6.37 as being a reasonable representation. The radii of curvature of the tendons in this model are depicted as constant throughout the normal range of wrist flexion and extension, with the arc length X of the curvature dependent upon both the radius R and the wrist deviation angle θ, giving

$$X = R\theta \tag{6.49}$$

where

X is the tendon arc distance around the pulley (sometimes referred to as the tendon displacement during wrist motion)
R is the radius of tendon curvature around the supporting tissues
θ is the angle of deviation of the wrist from a straight wrist (in rad)

By measuring the tendon displacement during motion of the wrist in flexion and extension on a select set of small and large cadaver hands, Armstrong and Chaffin (1978) reported values for the average radius of curvature of both the profundus and superficial flexor tendons (Table 6.1). As can be seen, the larger wrists were found to have larger radii of curvature. Bio-

TABLE 6.1
Estimated Values for the Radii of Curvature R of the Superficial and Profundus Flexor Tendons at the Wrist (from Armstrong and Chaffin, 1978)

Anthropometric Population Represented[a]	Wrist Thickness[a] (mm)	Radii of Curvature R (mm)			
		Profundus		Superficialis	
		Extension	Flexion	Extension	Flexion
5 %ile Female	31.8	8.9	15.0	10.7	16.8
95 %ile Male	44.8	12.0	18.1	14.4	20.5

[a] Population represented, based on wrist thickness measured by caliper at wrist crease compared with data and methods of Garrett (1970a, 1970b).

mechanically, this could partially explain why people with small wrists are thought to be at higher risk of wrist injury when placed on jobs requiring frequent manual exertions. It is also interesting that the radii of curvature during extension of the wrist, when the tendons wrap over the trochleas of the radius and carpal bones, are smaller than during wrist flexion, indicating that extension may be more hazardous than flexion. A biomechanical development of these kinematic results follows.

Continuing with the analogy of the tendons wrapped around a pulley, Figure 6.38 depicts the forces involved. Figure 6.36 shows that as the grip or pressing load F_L is increased on the fingers, the tendon forces F_T are approximately three to four times the load on each finger, depending on hand posture, external load orientation, and load application points. If F_T is known, the internal forces can be predicted, as follows.

The supporting forces of the tendon acting normal to the arc of the pulley (F_N in Figure 6.40) depend on the length X of the arc distance. In equation 6.49 this was the product $R\theta$. The value of F_N has been shown by Williams and Lissner (LeVeau, 1977) to be

$$F_N = \frac{F_T e^{\mu\theta}}{R} \tag{6.50}$$

where

F_N is the normal supporting force per unit of arc length (N/mm)
F_T is the average tendon force in tension (N)
μ is the coefficient of friction between the tendon and its supporting synovia (see Frankel and Burstein, 1970)
θ is the contact angle or wrist deviation angle (in rad)

In normal healthy synovial joints the value of μ can be considered to be small, on the order of 0.003 to 0.004 according to Fung (1981). With μ as-

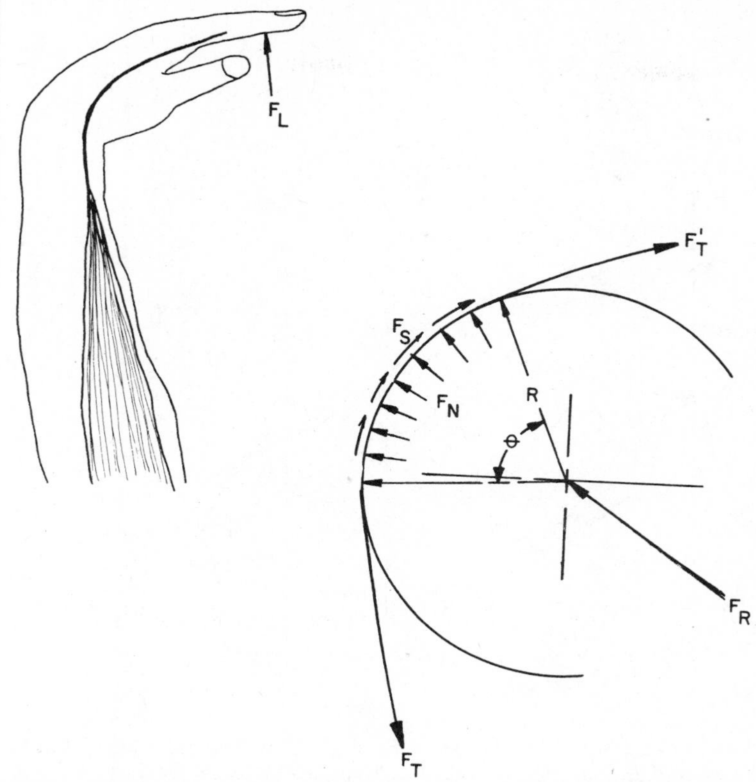

FIGURE 6.38 Forces acting at wrist due to finger flexor tendons being deviated from neutral position (from Armstrong and Chaffin, 1978).

sumed small, tendon shear forces are small and $F_N = F_T$. This also simplifies equation 6.50 to

$$F_N = \frac{F_T}{R} \tag{6.51}$$

Assuming values of F_T for the profundus tendon expected in various work situations, and using R values from Table 6.1, estimates of the supporting force F_N are depicted in Figure 6.39. These values indicate that the normal forces along the length of the synovium supporting the tendon curvature are higher for the person with a small wrist than for a person with a large wrist, especially while extending the wrist in a gripping action. We may conjecture that if the supporting synovia became inflamed, the coefficient of friction μ would increase. This would also increase the shearing forces F_S as the tendons attempt to slide through their synovial tunnels, since the shear forces are generally proportional to F_N:

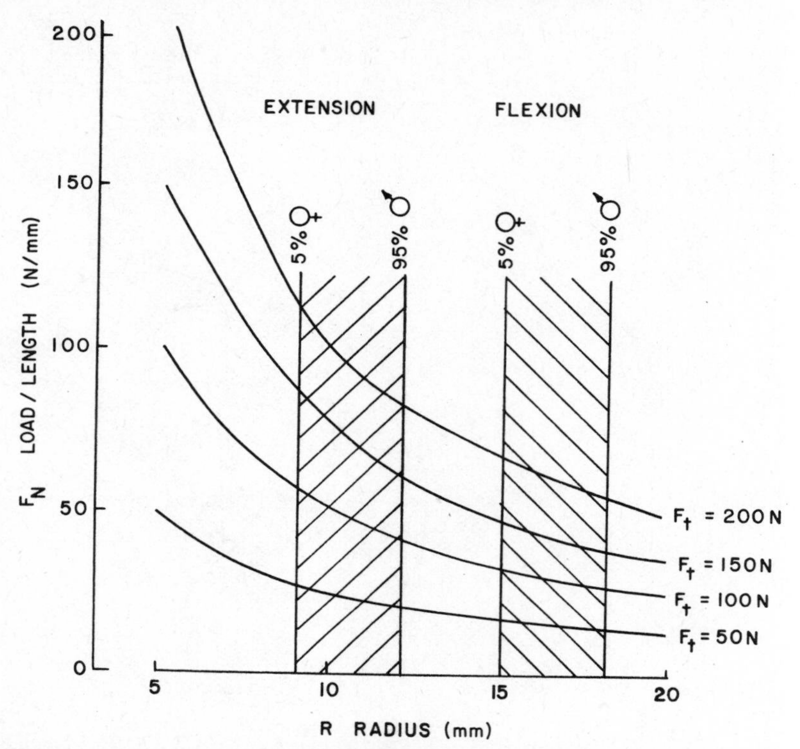

FIGURE 6.39 Force per unit length of the flexor digitorum profundus tendon on adjacent wrist structures is shown as a function of tendon curvature and load. Tendon curvatures for 5 and 95 percentile male and female wrists from values depicted in Table 6.1 (from Armstrong and Chaffin, 1979).

$$F_S = F_N \mu \tag{6.52}$$

Such an event could put the person with a small wrist at higher risk of more severe tendon and synovial trauma with repeated exertions than the person with a larger wrist though, again, epidemiology has not confirmed this hypothesis.

The total radial supporting force F_R in Figure 6.38 represents the force acting on adjacent structures—for example, the ligaments, bones, and median nerve enclosed in the carpal tunnel. The magnitude of this force depends on the amount that the tendon wraps around the pulley. The resulting relationship is

$$F_R = 2F_T \sin \frac{\theta}{2} \tag{6.53}$$

The effect of the wrist angle θ is therefore important in this regard. Figure

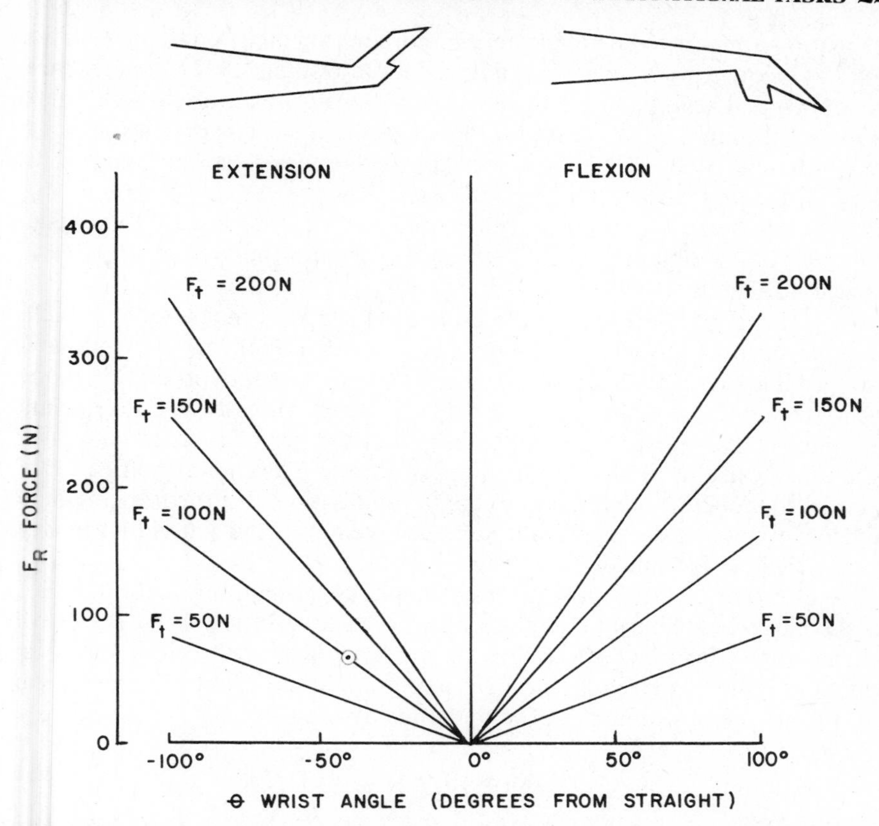

FIGURE 6.40 The resultant force exerted by a tendon on adjacent wrist structures as a function of wrist angle and tendon load. Resultant force is independent of tendon and wrist size (from Armstrong and Chaffin, 1979).

6.40 depicts typical F_R values. Inspection of these graphical values reveals large supporting forces during a grip exertion performed with a deviated wrist posture. Once again, the epidemiological and experimental data are not complete enough to relate specific F_R values to a risk of trauma. What currently exists is circumstantial clinical evidence indicating that repetitive grip-type exertions with a deviated wrist increase the risk of all types of cumulative trauma disorders of the wrist. The results from the biomechanical model are consistent with such clinical impressions, and provide a biomechanical rationale until further research is concluded.

6.5.3 Muscle Strength Prediction Modeling

As described earlier, skeletal muscles produce moments as well as reactive forces at various joints. The ability to use the muscles to produce moments at each joint appears to be the primary factor limiting many common exertion levels, as shown in a study by Andersson and Schultz (1979). In essence, this implies that when a person is attempting to lift, push, or pull with maximal effort, the moments created at each joint due to the application of the load on the hands as well as body weight must be less than or equal to the muscular moment strengths at each joint. Thus, ligament-produced moments, joint-friction levels, and other less well-defined means to produce additional joint moments appear to be of little importance in such maximal physical efforts.

Using the concept that at each joint their exists a measurable muscle-produced strength moment that cannot be exceeded by the moments created by external loads, a biomechanically based human strength prediction model can be developed. Such a model for this presentation will be coplanar and for isometric exertions. A noncoplanar model has been developed and used at the University of Michigan for the past decade (Garg and Chaffin, 1975). Dynamic models for whole-body strength prediction await further kinematic and physiological data regarding human movements and individual muscle strengths when dynamically contracted.

One of the first static coplanar (sagittal plane) strength prediction models was developed by Chaffin (1969) for analysis of load-lifting activities. It will be used as the basis for the following presentation. With reference to the concept of a muscle strength moment limitation at each joint, the model can be expressed as a simple set of inequalities at each joint:

$$-S_j < M_{j/L} < S_j \tag{6.54}$$

where

S_j is the muscle-produced moment strengths at each joint j (with negative limits in extension and positive values in flexion)

$M_{j/L}$ are the moments acting at each joint j due to external loads L on the hands and body segment weights.

The $M_{j/L}$ values can be computed from equations 6.25 and 6.26, which produce estimates of the moments at each joint for given postures, anthropometry, and external loads. The joint moment strengths S_j are obtained by measurements, using methods described in Chapter 4, or by referring to population values that have been developed for some major muscle functions (see Table 4.4, Chapter 4). In this latter regard, it must be recalled that muscle strength moments vary over the range of motion of a joint. Thus, joint angles must be known in order to predict S_j values, as shown in Figure 6.41. In fact, because muscles often span two joints, the angle at adjacent joints must be considered in predicting S_j values. Table 6.2 presents a synthesis of values for S_j from the moment strength studies of Clarke (1966), Schanne (1972) and Burggraaf (1972). The data were gathered on a limited

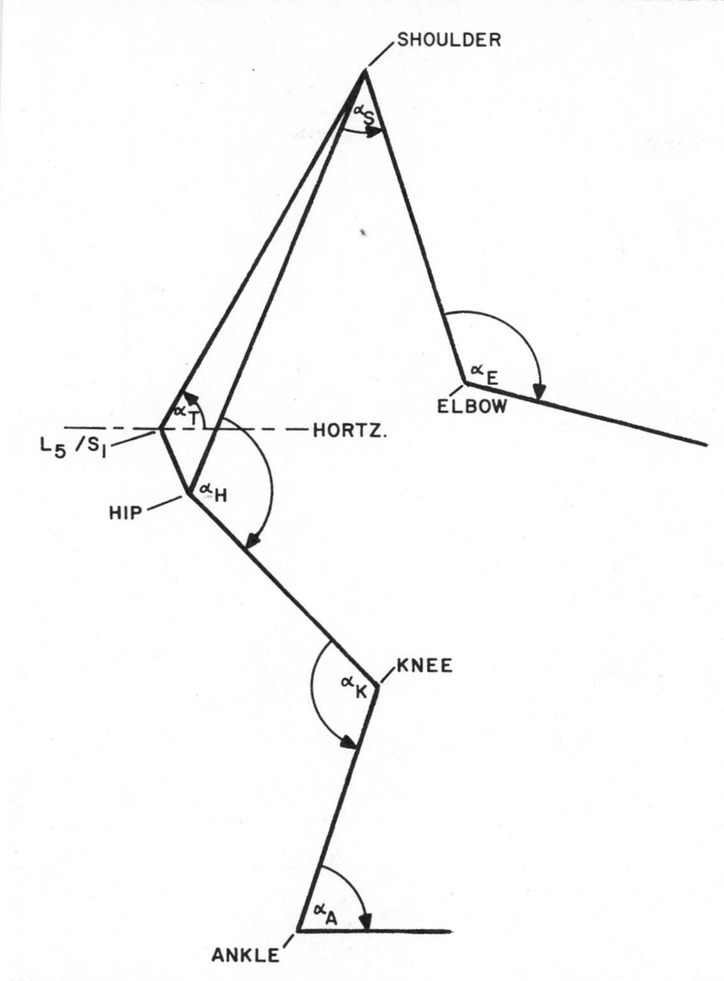

FIGURE 6.41 Angles used in strength prediction equations presented in Table 6.2.

number of college age men and women. Thus the form of predictions is probably reasonable that the S_j^* predictions vary over the postures used in an unbiased fashion—but the magnitudes are not necessarily representative of worker populations. For this reason, the strength data of Stobbe (1982), which is based on 30 men and 30 women working in industry, were used to adjust the mean prediction values and estimate a coefficient of variation about the mean predictions. The coefficient of variation, when multiplied by the mean predicted S_j^*, yields an estimate of the standard deviation SD of the population strengths in the chosen posture. (Recall that ± 1 SD = 68% of the population and ± 2 SD = 95% of the population, based on normally distributed strengths.)

TABLE 6.2

Joint Moment-Strength Mean Prediction Equations S_j^* for Exertions in Sagittal Plane from Data of Clarke (1966), Schanne (1972), and Burggraaf (1972), Corrected for Population Strengths of Stobbe (1982)

Strength	Primary/ Adjacent Joint	S_j^* Predicted Mean Strength (N m)[a]	G = Gender Adjustment		Coefficient of Variation ($SD/\bar{x}$)		Source
			Male	Female	Male	Female	
Elbow flexion	Elbow/ shoulder	$S_E = [336.29 + 1.544\alpha_E - 0.0085\alpha_E^2 - 0.5\alpha_S][G]$	.1924	.1011	.2458	.2629	Shanne
Elbow extension	Elbow/ shoulder	$-S_E = [264.153 - 0.575\alpha_E - 0.425\alpha_S][G]$	.2126	.1153	.2013	.3227	Shanne
Shoulder flexion	Shoulder/ elbow	$S_S = [227.338 + 0.525\alpha_E - 0.296\alpha_S][G]$	.3017	.1488	.2311	.2634	Shanne
Shoulder extension	Shoulder	$-S_S = [204.562 - 0.099\alpha_S][G]$	.4957	.2485	.3132	.3820	Shanne
Torso flexion	L_5/S_1	$S_T = [141.179 + 3.694\alpha_T][G]$	.3134	.1853	.2932	.3965	Shanne
Torso extension	L_5/S_1	$-S_T = [3365.123 - 23.947\alpha_T][G]$	.2467	.1380	.3152	.3455	Shanne
Hip flexion	Hip	$S_H = [-820.21 + 34.29\alpha_H - 0.11426\alpha_H^2][G]$	.1304	.0871	.2729	.3364	Clarke
Hip extension	Hip	$-S_H = [3338.1 - 15.711\alpha_H + 0.04626\alpha_H^2][G]$	.0977	.0516	.4016	.3779	Clarke
Knee flexion	Knee	$S_K = [-94.437 + 6.3672\alpha_K][G]$	.1429	.0851	.2934	.3212	Clarke
Knee extension	Knee	$-S_K = [1091.9 - 0.0996\alpha_K + 0.17308\alpha_K^2 - 0.00097\alpha_K^3][G]$	.0898	.0603	.3503	.3466	Clarke
Ankle extension	Ankle	$-S_A = [3356.8 - 18.4\alpha_A][G]$	.0816	.0489	.3307	.2745	Burggraaf

[a] See Figure 6.43 for reference angles (in degrees).

[b] The gender adjustment is calculated as the ratio of the mean male or female population strength over the predicted strength value at the standardized posture, adjusted by the conversion from in.-lb to Newton-meters (N m). The mean male or female population strength ($\bar{x}$) was obtained from Stobbe's tests of industrial populations. The predicted strength value was obtained by evaluating the full strength prediction equation at Stobbe's test posture. The original equations predicted strength in in.-lb. The gender adjustment incorporates a conversion factor to change the units of measurement from in.-lb to N m. The conversion is 1 in.-lb = 0.1130 N m.

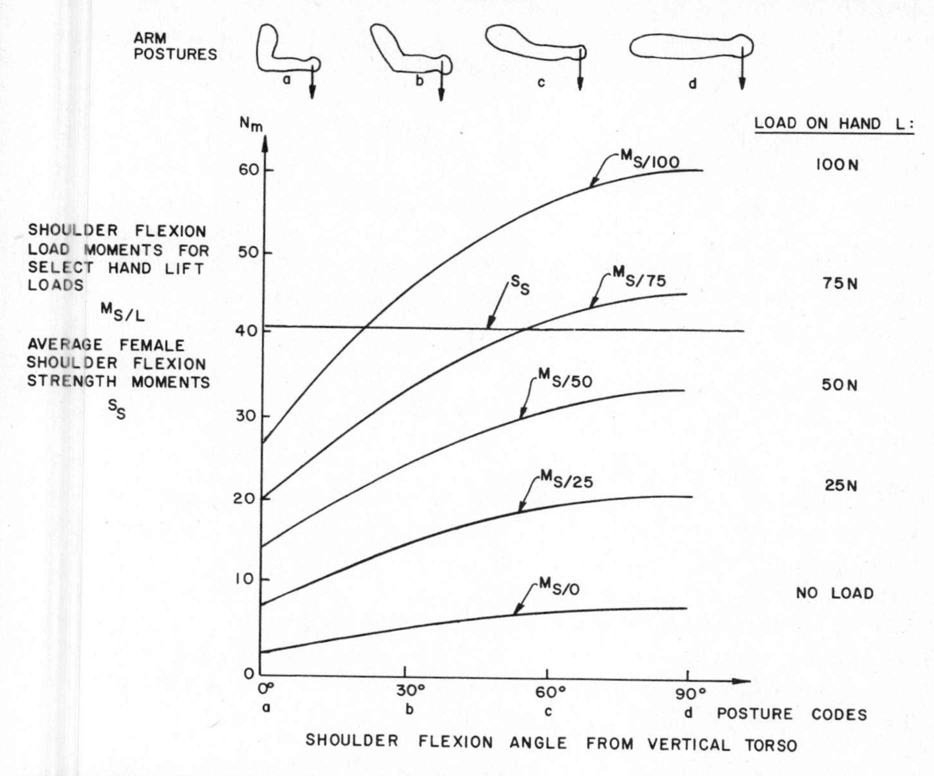

FIGURE 6.42 Shoulder flexion moment required of average anthropometric women to lift select loads versus predicted average shoulder flexion strength values in the same postures.

A graphical use of these strength values to predict how much load can be lifted in one hand by an average women is shown in Figure 6.42, assuming the strength limitation is in the shoulder flexion strength moment and not in the elbow flexion strength moment. As depicted, as the load is moved away from the body, the load moments' $M_{S/L}$ increase while the strength moments' S_S remains relatively constant. Where the $M_{S/L}$ and S_S values are equal, the average female lifting capability in that arm position is predicted. In other words, when the upper arm is at about 25°, 100 N can be lifted, but when raised to 90°, only 69 N can be held.

By setting the load moments $M_{j/L}$ equal to the strength S_j at each joint, and rearranging the M_j equations (equation 6.25) to solve for loads that can be held, it is possible to predict the maximum strength capability for the whole body. In essence, the minimum value of those loads predicted for each muscle strength-moment limit is the capability of the whole-body exertion. The logic of this can be expressed as follows. Recall from equation 6.26 that the load moments at joint j for given hand loads L are

$$M_{j/L} = M_{j/L=0} + \overline{jh}(L) \qquad \text{(repeated eq. 6.26)}$$

where $M_{j/L=0}$ are the load moments due to body weight only (equation 6.25) and $\overline{jh}(L)$ represents the additional moment at each joint due to an external load in the hands. If $M_{j/L}$ is set equal to the moment strengths S_j, and equation 6.26 is rearranged in order to calculate the maximum load on the hands L_j for each joint moment strength, then

$$L_j = \frac{[S_j - M_{j/L=0}]}{\overline{jh}} \tag{6.55}$$

This results in a maximum hand load for each muscle moment strength. Inspection is necessary to find the minimum value of these L_j values, which then represents the strength capability of the whole body.

Reference to the shoulder strength comparison in Figure 6.42 offers an example for the logic of this analysis. Consider the situation where the arm is completely extended at the elbow and horizontal (posture code *d* in Figure 6.42). The no-load moments and the average predicted strength values (assuming average female anthropometry) are

Elbow: $M_{E/L=0} = 2.2$ N m $\quad S_E = 30$ N m

Shoulder: $M_{S/L=0} = 8.0$ N m $\quad S_S = 44$ N m

The resulting load-lifting capability for the average female in such a posture for each limit is

Elbow: $L_E = (30 - 2.2)/.24 \text{ m} = 116 \text{ N}$

Shoulder: $L_S = (44 - 8.0)/.52 \text{ m} = 69 \text{ N}$

Thus, the load-lifting limitation in the extended arm posture is in fact due to shoulder flexion strength and not elbow flexion strength, and approximately 69 N would be predicted as the average female limit for such an act. Analysis with the arm held close to the body (posture code *a* in Figure 6.42) reveals that the limitation is about equally divided between the shoulder flexion and elbow flexion strengths. The values for this posture are

Elbow: $M_{E/L=0} = 2.2$ N m $\quad S_E = 40$ N m

Shoulder: $M_{S/L=0} = 2.2$ N m $\quad S_S = 41$ N m

The resulting load-lifting capabilities based on each limit are

$$\text{Elbow:} \quad L_E = (40 - 2.2)/.25 \text{ m} = 155 \text{ N}$$

$$\text{Shoulder:} \quad L_S = (41 - 2.2)/.25 \text{ m} = 155 \text{ N}$$

It should be clear from these examples that even in a static task it is not a simple matter to predict which muscle strength will be most stressed. Alterations in posture and/or the direction of load application will greatly affect both the resulting load moments and the moment strengths. Because of these complexities it is necessary to use the type of analysis just described if one is to understand the distribution of forces throughout the muscular system.

Because the computations associated with strength modeling are tedious and redundant, computer programs are desirable. The sagittal-plane strength model described was programmed and used by Chaffin (1972) to analyze astronaut lunar exploration activities for the United States Space Agency in the early 1970's (Martin and Chaffin, 1972) and is now available on several personal computers. A three-dimensional version also has been programmed for analysis of common industrial materials handling tasks (Garg and Chaffin, 1975). This model uses several postures and chooses the one that predicts the greatest strength. Figure 6.43 presents the logic involved in such a program. Comparison of strength predictions from these models with whole-body strength exertion data shows that the models account for about 70% of the variation in population strengths for a wide variety of tasks. The prediction values also tend to be unbiased—that is, they neither over- or under-predict population strengths, thus allowing the models to be used in the design and analysis of future manual tasks in industry.

6.6 FUTURE DEVELOPMENTS IN OCCUPATIONAL BIOMECHANICAL MODELS

This chapter has only begun to describe the breadth of biomechanical modeling activities now underway. Many of these efforts have direct application in an industrial setting. A few important questions for new students of biomechanical modeling are raised here:

1. When a person falls or an object strikes the body, what mechanical tolerance to such impacts can be expected? The answer will provide the basis for improving hard hats, face shields, foot guards, fall tethering systems, and so on.
2. What body organs are affected, and to what degree, by whole-body vibrations? This knowledge will be of value in improving vehicle seat-suspension systems of all types.

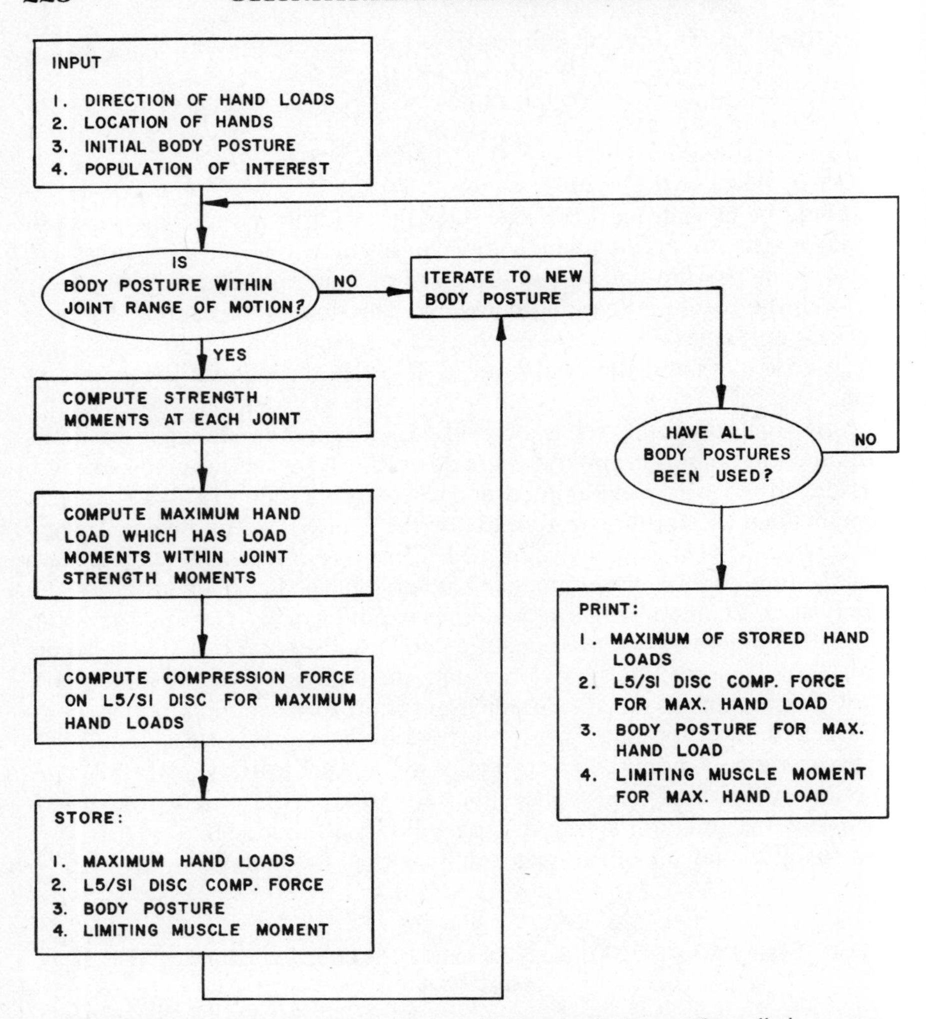

FIGURE 6.43 Macro-logic diagram of general biomechanical strength-prediction program.

3. What vibration energy levels can be tolerated by the musculoskeletal system when manipulating a power tool? Localized upper-extremity vibration can be reduced by careful design of hand tools, but improved human-tolerance data are needed to effectively assist in this endeavor.

Many other basic model developments are needed to extend the rather simple models presented in this chapter. In particular, models that describe internal muscle actions and joint/tissue reactions to movement and external force loadings are needed to develop musculoskeletal etiologies for the many disorders that are now idiopathic. In additon, such models would provide the means to predict the physical capabilities of people suffering acute or chronic musculoskeletal deficits, thus improving treatment and rehabilitation efforts for them.

It is clear that we know very little about the mechanical tolerance of various tissues when repeatedly stressed. Bone, muscle, ligament, tendon, and cartilage degeneration is common when frequent stress-loading occurs for periods of weeks, months, and even years. The limits to such frequent loadings are not well understood. Biomechanical models will continue to serve a vital role in interpreting and using the limited data that exist, as well as indicating the nature of new experiments that must be performed.

Our existing biomechanical modeling of occupational tasks has been largely static and confined to very simple planar motions. Newer, three-dimensional, high-speed motion sensing and recording systems make it feasible to expand the existing models to include more general manual tasks in industry. These new models will necessarily be complex, but with the aid of computer graphics they should be quite useful to engineers and various occupational health and safety professionals concerned with the prevention of mechanical trauma in the workplace.

REFERENCES

Abrahams, M., "Mechanical Properties of Tendon In-Vitro," *Med. Biol. Eng.*, **5**, 433–443 (1967).

Amis, A. A., D. Dowson, and V. Wright, "Analysis of Elbow Forces Due to High-Speed Forearm Movements," *J. Biomech.*, **13**, 825–831 (1980).

Andersson, G. B. and A. B. Schultz, "Transmission of Moments Across the Elbow Joint and the Lumbar Spine," *J. Biomech.*, **12**, 747–755 (1979).

Andersson, G. B. J., R. Ortengren, and A. Nachemson, "Studies of Back Loads in Fixed Spinal Postures and in Lifting," in C. G. Drury, Ed., *Safety in Manual Materials Handling*, USDHEW (NIOSH) 78-185, Cincinnati, 1978, pp. 26–33.

Armstrong, J. R., *Lumbar Disc Lesions*, Williams and Wilkins, Baltimore, 1965.

Armstrong, T. J., "Circulatory and Local Muscle Responses to Static Manual Work," unpublished Ph.D. thesis (Industrial Health), University of Michigan, Ann Arbor, 1976.

Armstrong, T. J. and D. B. Chaffin, "An Investigation of the Relationship between Displacements of the Finger and Wrist Joints and the Extrinsic Finger Flexor Tendons," *J. Biomech.*, **11**, 119–128 (1978).

Asmussen, E. and E. Poulson, "On the Role of the Intra-Abdominal Pressure in Relieving the Back," *Communications of Danish Nat'l Assoc. Infant. Paralysis*, **28**, 1–11 (1968).

Ayoub, M. M. and J. W. McDaniel, "Effects of Operator Stance on Pushing and Pulling," *AIIE Trans*, **6**, 185–195 (1974).

Badger, D. W., F. N. Dukes-Dobos, and D. B. Chaffin, "Prevention of Low Back Pain in Industry," *A NIOSH Symposium Report*, Physiology and Ergonomics Branch of National Institute for Occupational Safety and Health, Cincinnati, 1972.

Bartelink, D. L., "The Role of Abdominal Pressure in Relieving the Pressure on the Lumbar Intervertebral Discs," *J. Bone Jt. Surg.* **39B**, 718–725 (1957).

Burggraaf, J. D., *An Isometric Biomechanical Model for Sagittal Plane Leg Extension*, unpublished M.S. thesis (Industrial Engineering), University of Michigan, Ann Arbor, 1972.

Chaffin, D. B., "A Computerized Biomechanical Model: Development and Use in Studying Gross Body Actions," *J. Biomech.*, **2**, 429–441 (1969).

Chaffin, D. B., "On the Validity of Biomechanical Models of the Low Back for Weight Lifting Analysis," *ASME Proceedings*, 75-WA-Bio-1, *Amer. Soc. of Mech. Eng.*, New York, 1975.

Chaffin, D. B. and T. J. Armstrong, "Carpal Tunnel Syndrome and Selected Personal Attributes," *J. Occup. Med.*, **21**(7), 481–486 (1979).

Chaffin, D. B., R. K. Schultz, and R. G. Snyder, *A Prediction Model of Human Volitional Mobility*, SAE Paper 720002, Soc. of Auto. Eng,, Detroit, 1972.

Chao, E., J. Opgrande, and F. Axmear, "Three-Dimensional Force Analysis of Finger Joints in Selected Isometric Hand Functions," *J. Biomechanics*, **9**, 387–396 (1976).

Clarke, H. H., *Muscle Strength and Endurance in Man*, Prentice-Hall, Englewood Cliffs, 1966, pp. 39–51.

Crowninshield, R. D., "Use of Optimization Techniques to Predict Muscle Forces," *J. of Biomech. Eng.*, **100**, 88–92 (1978).

Crowninshield, R. D. and R. A. Brand, "A Physiologically Based Criterion of Muscle Force Prediction in Locomotion," *J. Biomech.*, **14**(11), 793–801, 1981.

Davis, P. R. and D. A. Stubbs, "Force Limits in Manual Work," *Applied Ergonomics*, **9**, 33–38 (1978).

Dempster, W. T., *Space Requirements of the Seated Operator*, WADC-TR-55-159, Aerospace Med. Res. Lab., Wright-Patterson AFB, Ohio, 1955.

El-Bassoussi, M. M., *A Biomechanical Dynamic Model for Lifting in the Sagittal Plane*, unpublished Ph.D. thesis (Industrial Engineering), Texas Technological University, Lubbock, 1974.

Evans, F. G. and H. R. Lissner, "Studies on the Energy Absorbing Capacity of Human Lumbar Intervertebral Discs," *Proceedings of the Seventh Stapp Car Crash Conference*, Springfield, Ill., 1965.

Farfan, H., *Mechanical Disorders of the Low Back*, Lea and Febiger, Philadelphia, 1973.

Fick, R., *Handbook der Anatomic und Mechanik der Gelenke*, Von Gustav Fisher, Jena, Germany, 1904.

Fiorini, G. T. and D. McCommond, "Forces on the Lumbo-Vertebral Facets," *Annals of Biomedical Eng.*, **4**(4), 354–363 (1976).

Fisher, B. O., *Analysis of Spinal Stresses During Lifting*, unpublished M.S. thesis (Industrial Engineering), University of Michigan, Ann Arbor, 1967.

Flatt, A. E., *The Pathomechanics of Ulnar Drift*, Final Report SRS Grant RD2226M, University of Iowa, Iowa City, 1971.

Frankel, V. H., and A. H. Burstein, *Orthopaedic Biomechanics*, Lea and Febiger, Philadelphia, 1970, pp. 145–155.

Fung, Y. C., *Biomechanics*, Springer-Verlag, New York, 1981, pp. 406–412.

Garg, A. and D. B. Chaffin, "A Biomechanic Computerized Simulation of Human Strength," *AIIE Trans*, March, 1–15 (1975).

Garg, A., D. B. Chaffin, and A. Freivalds, "Biomechanical Stresses from Manual Load Lifting," *IIE Trans*, **14**, 272–280 (1982).

Garrett, J. W., *Anthropometry of the Hands of Male Air Force Flight Personnel*, AMRL-TR-69-42, Wright-Patterson AFB, Ohio, 1970a.

Garrett, J. W., *Anthropometry of the Air Force Female's Hand*, AMRL-TR-69-26, Wright-Patterson AFB, Ohio, 1970b.

Goldstein, S., "Biomechanical Aspects of Cumulative Trauma to Tendons and Tendon Sheaths," unpublished Ph.D. thesis (Bioengineering), University of Michigan, Ann Arbor, 1981.

Gordon, E. E., "Natural History of the Intervertebral Disc," *Arch. Phys. Med.*, 750–763 (1961).

Hadler, N. M., "Industrial Rheumatology," *Arthritis and Rhematism,* **20**(4), 1019–1025 (1977).

Inglis, A. E., L. R. Straub, and C. S. Williams, "Median Neuropathy in the Wrist," *Clin. Orthop.*, **83,** 48–54 (1972).

Kraus, H., G. H. Robertson, and H. F. Farfan, *On the Mechanics of Weight Lifting,* Proceedings of New England Bioengineering Conference, University of Vermont, 1973.

Koremer, K. H. E. and D. E. Robinson, *Horizontal Static Forces Exerted by Men Standing in Common Working Postures on Surfaces of Various Tractions,* AMRL-TR-70-114, Aerospace Medical Research Laboratory, Wright-Patterson AFB, Ohio, 1971.

Krusen, F., C. M. Ellwood, and F. J. Kottle, *Handbook of Physical Medicine and Rehabilitation,* Saunders, Philadelphia, 1965.

Kulak, R. F., T. B. Belytschko, A. B. Schultz, and J. O. Galante, "Nonlinear Behavior of the Human Intervertebral Disc Under Axial Load," *J. Biomech.*, **9,** 377–386 (1976).

Landsmeer, J. M. F., "Power Grip and Precision Handling," Ann. Rheum. Dis., **21,** 164–170 (1962).

Lanier, R. R., "Presacral Vertebrae of White and Negro Males," *Am. J. Phys. Med.*, **25,** 343–420 (1939).

Lee, K. S., *Biomechanical Modeling of Cart Pushing and Pulling,* unpublished Ph.D. thesis (Industrial and Operational Engineering), University of Michigan, Ann Arbor, 1982.

LeVeau, B., *Biomechanics of Human Motion,* 2nd ed., Saunders, Philadelphia, 1977, pp. 150–156.

MacConaill, M. A. and J. V. Basmajian, *Muscles and Movements,* Williams and Wilkins, Baltimore, 1969, pp. 198–207.

Martin, J. B. and D. B. Chaffin, "Biomechanical Computerized Simulation of Human Strength in Sagittal Plane Activities," *AIIE Trans.*, **4**(1), 19–28 (1972).

Miller, D. I. and R. C. Nelson, *Biomechanics of Sport,* Lea and Febiger, Philadelphia, 1976, pp. 31–33 and 235–342.

Morris, J. M., D. B. Lucas, and B. Bressler, "Role of the Trunk in Stability of the Spine," *J. Bone Jt. Surg.*, **43A,** 327–351 (1961).

Nachemson, A., "Low-Back Pain, Its Etiology and Treatment," *Clin. Med.*, 18–24 (1971).

Nachemson, A. and G. Elfstrom, "Intravital Dynamic Pressure Measurements in Lumbar Disc," *Scand. J. Rehab. Med.* (Supp. 1), 1–39 (1970).

National Institute for Occupational Safety and Health, *A Work Practices Guide for Manual Lifting*, Tech. Report No. 81–122, U.S. Dept. of Health and Human Services (NIOSH), Cincinnati, OH, 1981.

Ohtsuki, T., "Inhibition of Individual Fingers During Grip Strength Exertion," *Ergonomics,* **24**(1), 21–36 (1981).

Park, K. S. and D. B. Chaffin, "A Biomechanical Evaluation of Two Methods of Manual Load Lifting," *AIIE Transaction,* **6**(2), 105–113 (1974).

Perey, O., "Fractures of the Vertebral Endplate in the Lumbar Spine," *Acta Ortho. Scand.*, Supp 25, 1957.

Plagenhoef, S., *Patterns of Human Motion,* Prentice-Hall, Englewood Cliffs, 1971, pp. 48–55.

Rowe, L. M., "Low Back Disabilities in Industry: Updated Position," *J. Occup. Med.*, **13,** 476–483 (1971).

Schanne, F. A., *A Three-Dimensional Hand Force Capability Model for the Seated Operator,* unpublished Ph.D. thesis (Industrial Engineering) University of Michigan, Ann Arbor, 1972.

Schultz, A. B. and B. J. G. Andersson, "Analysis of Loads on the Lumbar Spine," *Spine,* **6**(1), 76–82, 1981.

Schultz, A. B., B. J. G. Andersson, R. Ortengren, K. Haderspeek, and A. Nachemson, "Loads on the Lumbar Spine," *J. Bone Jt. Surg.*, **64-A,** 713–720 (1982).

Schultz, A. B., D. N. Warwick, M. H. Berkson, and A. L. Nachemson, "Mechanical Properties of Human Lumbar Spine Motion Segments," *J. Biomech. Eng. Trans.*, ASME **101**, 46–52 (1979).

Seireg, A. and R. J. Arvikar, "The Prediction of Muscular Load Bearing and Joint Forces in the Lower Extremities," *J. of Biomech.*, **8,** 89–102 (1975).

Shames, I. H., *Engineering Mechanics: Statics and Dynamics,* 2nd Ed., Prentice-Hall, Englewood Cliffs, 1967.

Smith, A. DeF., E. M. Deery, and G. L. Hagman, "Herniations of the Nucleus Pulposus: A Study of 100 Cases Treated by Operation," *J. Bone Jt Surg.*, **26,** 821–833, 1944.

Sonoda, T., "Studies of the Strength for Compression, Tension, and Torsion of the Human Vertebral Column," *J. Kyoto Prefect. Med. Univ.*, **71,** 659–702 (1962).

Stobbe, T., unpublished Ph.D. thesis (Industrial Engineering) University of Michigan, Ann Arbor, 1982.

Tanzer, R. C., "The Carpal Tunnel Syndrome," *J. Bone Jt Surg.*, **41A,** 626–634 (1959).

Thieme, F. P., *Lumbar Breakdown Caused by Erect Posture in Man,* unpublished Anthropometric Paper, No. 4, University of Michigan, Ann Arbor, 1950.

Tichauer, E. R., *Biomechanics Monograph,* E. F. Byars, R. Contini and V. L. Roberts, Eds., Amer. Soc. Mech. Eng., N.Y., 1967, p. 155.

Tichauer, E. R., "A Pilot Study of the Biomechanics of Lifting in Simulated Industrial Work Situations," *J. Safety Res,* **3**(3), 98–115 (1971).

Tichauer, E. R., "Some Aspects of Stress on Forearm and Hand in Industry," *J. Occup. Med.*, **8,** 63–71 (1966).

Wilson, R. N. and S. Wilson, "Tenosynovitis in Industry," *Practitioner,* **178,** 612–625 (1957).

7

METHODS OF CLASSIFYING AND EVALUATING MANUAL WORK

7.1 TRADITIONAL METHODS

7.1.1 Historical Perspective

Describing manual activities with reference to standard categories of effort has been recognized for many years as the first step in organizing the labor of groups of workers. This usually has involved the following sequence of events:

1. Development of a preferred method of performing a job. This assumes that the task or job can be performed using different methods. To develop a preferred method an analyst must (1) state the objective of the operation, (2) identify available methods to accomplish that objective, (3) use feasible methods on a trial basis, and (4) select the best method to meet the objective.
2. Preparation of a standard practice. This requires that the preferred method be stated *formally*. This includes (1) sketching the workplace, with the machines or tools used and with relevant dimensions of objects handled, (2) listing abnormal working conditions that could affect work performance (e.g., dust, heat, illumination), and (3) tabulating the sequence of motions required of a worker to complete the operation.
3. Determination of a time standard. This is accomplished either by having a skilled worker perform the operation while being timed—that is, a *time study* is performed—or by consulting normative time

data available in the literature for each motion required and adding these values to predict the operation's standard time—that is, a *predetermined time system* is used.

4. Training of the worker. Instruction sheets, verbal directions, and various training aids are developed and used to assure that workers understand the job requirements and that preferred work methods are followed.

These four steps evolved from a concern by management that labor was not well utilized—that time was being wasted. As the cost of labor has increased, more emphasis has been given to assuring that each step is fully implemented by the industrial or production engineering functions in a well-managed plant. Within the last decade an additional concern has been shown—assurance that the operation is accomplished safely. It is this special concern for worker safety and health that makes occupational biomechanics important when performing a job analysis today. To better understand how traditional job analysis and biomechanics blend together, a brief description of traditional approaches follows.

Early professional development in labor planning focused on procedures necessary to perform motion analysis in order to predict the time necessary to produce specific goods or services. Many professionals have contributed to these early motion and time analysis methods, as discussed by Barnes (1968) and Niebel (1972). One of the early major contributors to the field was Frederick W. Taylor, sometimes referred to as the father of modern time analysis. Taylor's study in 1898 of shoveling rice, coal, and iron ore was a rigorous, comprehensive investigation to determine the optimal shovel size, shoveling frequency, and rest allocation necessary to maximize the total output of a group of workers (Copley, 1923). Taylor's evaluation showed, for instance, that a shovel that would allow about 100 N of iron to be lifted each time resulted in maximizing the total daily output for a group of physically strong workers. By a combination of (1) careful selection and training of workers for such work, (2) provision of special shovels for different weight materials, and (3) paying bonuses for above average output, Taylor was able to demonstrate that 140 men could perform the same amount of work previously done by 400 to 600 workers (Copley, 1923).

These results, along with others from similar studies, allowed Taylor to state certain basic *principles* on the management of manual labor. He referred to them as the *Principles of Scientific Management* and published them in a book with the same name in 1929 (Taylor, 1929). These principles are:

> *First*. The development of a science is necessary for each element of a man's work, thereby replacing the old rule-of-thumb methods.
>
> *Second*. The selection of the best worker for each particular task is required, and then training, teaching, and developing the workman; in place of the former

practice of allowing the worker to select his own task and train himself as best he could.

Third. The development of a spirit of hearty cooperation between the management and the men in the carrying on of the activities in accordance with the principles of the developed science is required.

Fourth. The division of the work into almost equal shares between the management and the workers should be accomplished, each department taking over the work for which it is the better fitted; instead of the former condition, in which almost all of the work and the greater part of the responsibility were thrown on the men.

It is clear that both the first and second principles advocate the development of specific knowledge of job demands and worker capabilities, respectively. Taylor's method of acquiring this knowledge was based on measuring the time required of selected workers to perform a sequence of specifically defined manual tasks comprising a job. This method of studying manual labor persists today as one of his greatest contributions to work analysis.

It is interesting to note that his third and fourth principles advocated a labor-management cooperative program for developing and implementing productivity improvements, much as now advocated by those interested in improving the quality of work life. To improve the biomechanical aspects of work requires this same spirit of cooperation between labor and management. The reasons for this are obvious—hazardous biomechanical stresses are not easily detected except by knowledgeable workers on the job. Further, the correction of such stresses often requires intimate knowledge of work methods and economically feasible, alternative methods available to accomplish the job with reduced biomechanical stress. Cooperative labor-management studies in plants are often the best means by which biomechanical problems can be accurately described and comprehensive solutions developed.

About the same time that Taylor's studies were being published, Frank and Lillian Gilbreth were advocating that manual activities in industry be carefully categorized to minimize fatigue and monotony so as to best maximize productivity and employment of the handicapped (Barnes, 1968). One of the earliest and most quoted studies performed by Frank Gilbreth was of bricklaying (Gilbreth, 1911). By using photographs of the bricklayers in his construction business he was able to demonstrate that fatigue and wasted motions were not necessary for high output if (1) the bricks were first sorted for good and bad bricks, (2) they were oriented with the best side facing out, and (3) they were presented to the bricklayer at a comfortable working height by means of an adjustable scaffold. Gilbreth's work methods resulted in well over double the output per man-hour than had previously been reported in the industry.

Two specific methodological contributions to the study of manual labor must be attributed to the Gilbreths. One is the use of *micromotion study,* a term coined by them in a paper in 1912 to the American Society of Mechanical Engineers (Gilbreth, 1912). In this procedure, great attention is given to categorizing individual motions, referred to as *elemental motions.* These were carefully timed and became the basis later for one of the first predetermined time systems. Once the necessary elements of a job were known, the Gilbreths advocated trying out various sequences in simulations of the job. To accomplish this they set up training schools and advocated that large companies establish motion analysis laboratories in which mock-ups of new jobs could be "tried-out" and alternative method sequences evaluated.

The Gilbreths are also credited with the first use of a *cyclegraph or chronocyclegraph* to compare alternative motion sequences. In the latter, a blinking light (flashing at a given rate) is attached to an arm or hand and a photograph is made with the camera shutter open for a prolonged period. The path of motion is transcribed by a sequence of dots on the film. By measuring the number and displacements of the dots, the Gilbreths were able to measure and compare different job-motion requirements. Today this technique has been enhanced by modern video and computer techniques to allow whole-body motion studies (see Chapter 5).

It should be clear that the general approach and investigative methods of work analysis developed by Taylor and the Gilbreths early in this century are directly applicable to occupational biomechanics. Without the means to perform systematic work classification, time study, and motion analysis in industry it would be difficult to apply various biomechanical concepts to reduce injurious musculoskeletal stress.

What follows is a brief description of one contemporary work analysis system that has been derived from these earlier efforts. It is followed by a presentation of several new work analysis systems that have been used to evaluate biomechanical stress levels in various jobs.

7.2 TRADITIONAL WORK ANALYSIS SYSTEM

In 1924, Segur developed time-prediction equations for various classes of motions from film analysis of industrial operations during World War I (Niebel, 1972). Similar data were published by Holmes, Engstrom, Barnes, Quick, and others during the 1930's. What these authors showed is that the time required of people to perform certain basic or *elemental motions* is about the same for different people. Thus, the time to perform a complex manual job can be predicted by describing the job as a sequence of elemental motions that have known time requirements, and then summing the normative times required for each elemental motion. Most of the normative elemental motion times have been derived by performing frame-by-frame analysis of films to determine the time taken in various industrial operations. The resulting elemental time values have become known as *predetermined elemental motion times*. Because different groups of experts performed the analyses, slightly different motion classifications were used, and hence different motion-time prediction systems developed, many of which exist today. Table 7.1, developed by Barnes (1968), describes the background of several popular predetermined motion-time systems.

Because it is often necessary today to be able to plan staffing requirements for a new task before it becomes operational, predetermined moion-time systems are more often used rather than time-study techniques. The latter obviously requires direct observation of a skilled operator. A survey of the use of such techniques by Karger and Bayha (1965) disclosed that about two-thirds of firms sampled in the United States use some form of work measurement system, and that about 56% of these used a predetermined motion-time system. They also disclosed that of the many predetermined motion-time systems available today, the Methods-Time Measurement (MTM) system is used about twice as often as other systems, and has an international organization that assists its members in utilizing the system.

What follows is a brief description of the MTM predetermined motion-time system. It is not meant to be a thorough discourse on the use of this particular system, but rather should be viewed as an example of one popular system used by thousands of practitioners in industries throughout the world. For a thorough understanding of the use of such systems the reader should consult textbooks by Barnes (1968), Niebel (1972), Karger and Bayha (1965), and Konz (1979).

7.2.1 MTM-1: An Example of a Predetermined Motion–Time System

The Methods–Time Measurement system developed by Maynard and others in the late 1940's is based on film and time studies of various industrial jobs. It relies on a description of manual activities with reference to a well-defined set of basic elemental motions. It should be noted that to use this particular motion classification system with accuracy and consistency requires training in a special course of 24 to 80 classroom hours offered by the MTM Association.

The resulting time values for each elemental motion in the MTM procedure are given in units of one hundred-thousandths of an hour (0.00001 hr), and are referred to as one time-measurement unit (TMU). This fraction of an hour was chosen as the basic unit of measure because production rates in industry are often quoted in units produced per hour—one TMU equals 0.0006 min or 0.036 sec. It should also be noted that the resulting time values published in tables are proprietary; written permission to duplicate them is required from the MTM Association, Fairlawn, New Jersey (USA).

As an example of the time required to move various objects about in the workplace, MTM-predicted time values are presented in Table 7.2 for dif-

TABLE 7.1
Some Traditional Motion-Time Data Systems (Barnes, 1968, pp. 472–473)

Name of System	Date First Applied	First Publication Describing System	Publication Containing Information about System	How Data Were Originally Obtained	System Developed by
Motion–time analysis (MTA)	1924	Data not published, but information concerning MTA published in *Motion-Time Analysis Bulletin*, a publication of A. B. Segur & Co.	"Motion-Time-Analysis" by A. B. Segur, in *Industrial Engineering Handbook,* H. B. Maynard, editor, McGraw-Hill Book Co., New York, pp. 4–101 to 4–118, 1956.	Motion pictures micromotion analysis, kymograph	A. B. Segur
Body member movements	1938	*Applied Time and Motion Study* by W. G. Holmes, Ronald Press Co., New York, 1938	*Applied Time and Motion Study* by W. G. Holmes, Ronald Press Co., New York, 1938	Not known	W. G. Holmes
Motion–time data for assembly work (get and place)	1938	*Motion and Time Study,* 2nd ed., by Ralph M. Barnes, John Wiley & Sons, New York, 1940, Chapters 22 and 23	*Motion and Time Study: Design and Measurement of Work*, 6th ed., by Ralph M. Barnes, John Wiley & Sons, New York, 1968, Chapter 30	Time study, motion pictures of factory operations, laboratory studies	Harold Engstrom and H. C. Geppinger, Bridgeport Plant of General Electric Co.

The work-factor system	1938	"Motion-Time Standards" by J. H. Quick, W. J. Shea, and R. E. Koehler, *Factory Management and Maintenance*, **103**(5), 97–108 (May 1945)	*Work-Factor Time Standards*, by Joseph H. Quick, James H. Duncan, and James A. Malcolm, Jr., McGraw-Hill Book Co., New York, 1962 *Ready Work-Factor Time Standards,* by J. A. Malcolm, Jr. et al., Haddonfield, N. J., 1966	Time study, motion pictures of factory operations, study of motions with stroboscopic light unit	J. H. Quick W. J. Shea R. E. Koehler
Elemental time standard for basic manual work	1942	"Establishing Time Values by Elementary Motion Analysis," by M. G. Schaefer, *Proceedings Tenth Time and Motion Study Clinic,* IMS, Chicago, pp. 21–27, November 1946	"Establishing Time Values by Elementary Motions" by M. G. Schaefer, *Proceedings Tenth Time and Motion Study Clinic,* IMS, Chicago, November, 1946. Also "Development and Use of Time Values for Elemental Motions" by M. G. Schaefer, *Proceedings Second Time Study and Methods Conference,* SAM-ASME, New York, April, 1947	Kymograph studies, motion pictures of industrial operations, electric time-recorder studies (time measured to 0.0001 min)	Western Electric Co.
Methods–time measurement (MTM)	1948	*Methods–Time Measurement* by H. B. Maynard, G. J. Stegemerten, and J. L. Schwab, McGraw-Hill Book Co., New York, 1948	*Methods-Time Measurement* by H. B. Maynard, G. J. Stegemerten, and J. L. Schwab, McGraw-Hill Book Co., New York, 1948	Time study, motion pictures of factory operations	H. B. Maynard G. J. Stegemerten J. L. Schwab

TABLE 7.1 ***(Continued)***

Name of System	Date First Applied	First Publication Describing System	Publication Containing Information about System	How Data Were Originally Obtained	System Developed by
Basic motion time study (BMT)	1950	Manuals by J. D. Woods & Gorton, Ltd., Toronto, Canada, 1950	*Basic Motion Timestudy* by G. B. Bailey and Ralph Presgrave, McGraw-Hill Book Co., New York, 1958	Laboratory studies	Ralph Presgrave G. B. Bailey J. A. Lowden
Dimensional motion times (DMT)	1952	"New Motion Time Method Defined" by H. C. Geppinger, *Iron Age*, **171**(2), 106–108 (January 8, 1953)	*Dimensional Motion Times* by H. C. Geppinger, John Wiley & Sons, New York, 1955	Time study, motion pictures, laboratory studies	H. C. Geppinger
Predetermined human work times	1952	"A System of Predetermined Human Work Times" by Irwin P. Lazarus, Ph.D. thesis, Purdue University, 1952	"Synthesized Standards from Basic Motion Times," *Handbook of Industrial Engineering and Management*, W. G. Ireson and E. L. Grant, editors, Prentice-Hall, Englewood Cliffs, N.J., pp. 373–378, 1955	Motion pictures of factory operations	Irwin P. Lazarus

TABLE 7.2

Predicted Move-Time Data in Which a *Move* is Defined as a Motion of the Hand Required to Transport an Object (from MTM Association for Standards and Research, Fairlawn, New Jersey 07410)

Distance Moved (cm)	Time TMU A	Time TMU B	Time TMU C	Time TMU Hand in Motion B
0 to 2	2.0	2.0	2.0	1.7
4	3.1	4.0	4.5	2.8
6	4.1	5.0	5.8	3.1
8	5.1	5.9	6.9	3.7
10	6.0	6.8	7.9	4.3
12	6.9	7.7	8.8	4.9
14	7.7	8.5	9.8	5.4
16	8.3	9.2	10.5	6.0
18	9.0	9.8	11.1	6.5
20	9.6	10.5	11.7	7.1
22	10.2	11.2	12.4	7.6
24	10.8	11.8	13.0	8.2
26	11.5	12.3	13.7	8.7
28	12.1	12.8	14.4	9.3
30	12.7	13.3	15.1	9.8
35	14.3	14.5	16.8	11.2
40	15.8	15.6	18.5	12.6
45	17.4	16.8	20.1	14.0
50	19.0	18.0	21.8	15.4
55	20.5	19.2	23.5	16.8
60	22.1	20.4	25.2	18.2
65	23.6	21.6	26.9	19.5
70	25.2	22.8	28.6	20.9
75	26.7	24.0	30.3	22.3
80	28.3	25.2	32.0	23.7

Allowance: Weight (kg.) up to	Constant (TMU)	Factor	Case and Description
1	0	1.00	A Move object to other hand or against stop
2	1.6	1.04	
4	2.8	1.07	
6	4.3	1.12	
8	5.8	1.17	B Move object to approximate or indefinite location
10	7.3	1.22	
12	8.8	1.27	
14	10.4	1.32	
16	11.9	1.36	C Move object to exact location
18	13.4	1.41	
20	14.9	1.46	
22	16.4	1.51	

ferent types of human *moves* common in industry. The time values presented in Table 7.2 vary, depending on (1) three types of terminal conditions, (2) the load being transported, and (3) the length of motion. The effect of the terminal conditions and length of motion are self-explanatory. The load effect is more complicated in that it assumes that the weight held in one hand adds an initial time delay (referred to as a "Constant" in Table 7.2) plus a proportional delay (referred to as a "Factor"). Thus, if one moves a load of

4-kg mass with one hand a distance of 10 cm to an indefinite location, the move time would be 6.8(1.07) + 2.8 = 10.1 TMU, or 0.36 sec. If two hands are used, the load is considered to be halved when choosing specific table values. Also, if an object is pushed or pulled along a shelf or table, the load is considered to be the average horizontal force required during the move, that is, the object's weight multiplied by the expected coefficient of friction between the object and its supporting surface.

The procedure requires looking-up in tables similar time values for the following standardized elemental motions:

1. Reach—a motion of the unloaded hands or fingers.
2. Position—small motion necessary when aligning object to be released at end of motion.
3. Release—either a distinct motion of fingers or the release of an object at the end of a motion without such an overt motion.
4. Disengage—an involuntary (rebound) motion often required when two objects suddenly come apart under exertion.
5. Grasp—an overt motion necessary to gain control of an object.
6. Eye focus travel—the time required for the eyes to move and accommodate to provide visualization of an object.
7. Turn apply pressure—the manipulation of controls, tools, and objects necessary to turn an object by rotation of the hand about the long axis of the forearm.
8. Body, leg/foot motion—the motion of transporting the body with values given per step for varied conditions.
9. Simultaneous motions—rules are given so that some motions can be performed together—for example, both the right and left hand reach to an object—therefore only the greatest of the two time values are used in the standard time prediction.

Each of the above motions has several specific task conditions that must be evaluated before choosing a time value from the MTM tables, much like those described earlier for the *move* times given in Table 7.2. These job conditions are coded on the job analysis form. Some example codes and their interpretations are given in Table 7.3.

An example MTM job analysis is given in Table 7.4 using the standard coding notations to specify the elemental motion conditions expected of a worker performing the job. Note that the brackets indicate simultaneous motions, and thus only the greatest time is entered in the center column to be used in predicting the total time of the task. In the example, a tote box weighing 200 N (22 kg) is lifted to a work bench and one part weighing 10 N (1 kg) is removed to the bench from those other parts jumbled in the box.

TABLE 7.3
Coding Conventions for MTM Motion Analysis (Barnes, 1968, p. 503)

Code Example	Interpretation of Codes
R8C	Reach, 8 in., Case C
R12Am	Reach, 12 in., Case A, hand in motion at end
M6A	Move, 6 in., Case A, object weighs less than 2.5 lb
mM10C	Move, 10 in., Case C, hand in motion at the beginning, object weighs less than 2.5 lb
M16B15	Move, 16 in., Case B, object weighs 15 lb
T30	Turn hand 30°
T90L	Turn object weighing more than 10 lb 90°
AP1	Apply pressure, includes regrasp
G1A	Grasp, Case G1A
P1NSD	Position, Class 1 fit, nonsymmetrical part, difficult to handle
RL1	Release, Case 1
D2E	Disengage, Class 2 fit, easy to handle
EF	Eye focus
ET14/10	Eye travel between points 14 in. apart where line of travel is 10 in. from eyes
FM	Foot motion
SS16C1	Sidestep, 16 in., Case 1
TBC1	Turn body, Case 1
W4P	Walk four paces

7.2.2 Benefits and Limitations in Contemporary Work Analysis Systems

In reviewing the Methods Analysis Chart in Table 7.4 it should be noted that it documents several important aspects of manual activities in industry. First is the sketch of the workplace, with major hand motions and distances noted. Such a sketch can serve as the basis for planning changes in the workplace layout necessary to reduce musculoskeletal stresses. Second, the methods analysis indicates the relative balance of work time between the left and right arms. If an imbalance is great it may indicate the basis for certain asymmetric symptoms in workers; for example, carpal tunnel syndrome is often reported in the wrist of the dominant hand used in jobs. Third, the analysis describes the extent of static holding activities required of one hand while the other hand performs useful activities. Such static postural requirements are often the cause of muscle fatigue and should be avoided through improved workpiece fixturing, tool design, and workplace layout, as discussed in Chapter 10. Last, the motion times predicted by such a system provide the means of comparison with motion times measured in laboratory and field experiments. This comparison assures that such motions

TABLE 7.4
Example MTM Analysis of Lifting a Tote Box from a Pallet to a Workbench and Removing a Part from the Box (courtesy J. Foulke)

Activity Description	Left Hand Elements	Motion Time (TMU)	Right-Hand Elements or Body Motions
Sidestep to pallet.		17.0	SS12C1
Stoop to tote box, and			S
Reach to tote box during stoop.	~~R10B~~	29.0	~~R10B~~
Grasp handles on tote box,	G1A	2.0	G1A
Arise, and		31.9	AS
Lift tote box during arise.	~~M10B$\frac{22}{2}$~~		~~M10B$\frac{22}{2}$~~
Sidestep towards bench with tote box, and		23.7	SS12C
Move tote box during sidestep.	M12B$\frac{22}{2}$		M12B$\frac{22}{2}$
Release tote box on bench.	RL1	2.0	RL1
Reach into tote box.		12.9	R10C
Grasp part in box.		7.3	G4A
Move part to bench.		10.6	M18B
Release part on bench.		2.0	RL1
Total time required:		138.4 TMU or 4.98 s.	

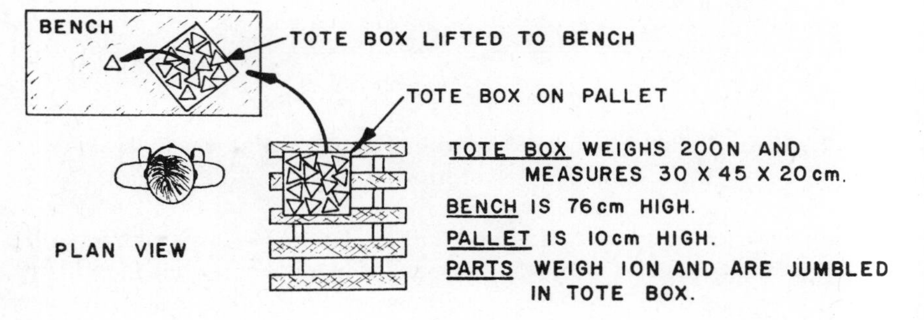

observed in the laboratory are consistent with those occurring in industry. In this regard, it is one thing to study the kinesiological aspects of well-trained and selected athletes for sports biomechanics applications. It is another thing to infer that the results of such studies are applicable to conditions and workers in industry. Only through detailed methods analysis and comparison of resulting elemental motion-time values and motion patterns can one begin to make comparisons and apply laboratory biomechnical data to solve real occupational biomechanical problems in industry.

Despite the preceding benefits of the traditional work analysis methods, several major limitations must be recognized. First, traditional work analysis systems rarely record the exact loads expected to be handled. Normally, an estimate of the "average" load is used as a basis for the time predictions.

This is because the analysis system often requires only approximate load values for time prediction purposes. It is not unusual for load estimates to be in error by over 50 N. In certain postures, such an error may make the difference between an acceptable and unacceptable biomechanical job stress. Thus the actual loads handled should be noted for biomechanical purposes, with particular attention to extremely heavy loads that may occasionally be handled.

A second limitation of existing job analysis systems is that they do not require postural data per se. The normal sketch of the workplace accompanied by a hand-motion diagram (the latter being optional in most systems) provides the necessary initial data, but this needs to be augmented by data describing extreme or awkward postures that could be potentially stressful to to the musculoskeletal system. It is for this reason that the Gilbreths in the early part of the century used photographs of their workers. This allowed them to systematically analyze a job, identify awkward postures and inefficient motions, and recommend corrections. From a biomechanics point of view, a picture is truly worth a thousand words. A good methods analysis augmented with selected photographs or sketches of potentially overstressing postures is the preferred approach.

This leads to perhaps the greatest biomechanical shortcoming of traditional job analysis procedures. Because the primary intent of the existing methods analysis schemes is to predict the time necessary to perform a job, activities that seldom occur in a job may be ignored entirely in the analysis. Yet these occasional exertions may be the most biomechanically stressful to perform. An example of this occurs when a worker must load stock into a machine, remove scrap, or service (adjust) a machine he or she is operating. The traditional methods analysis may ignore these tasks if they account for less than 5 to 10% of the operator's total workday. Yet in many follow-up accident reports these undocumented but required job tasks account for a large number of injuries on the job. Unfortunately, because these tasks are undocumented work methods, engineers, methods analysts, and safety and health personnel may not be aware of the hazards inherent in them, and thus alternative and safer work methods are not introduced until a catastrophe calls attention to the unsafe act or condition.

Despite these limitations, however, a biomechanical analysis of a job should begin with a careful examination at the job to determine the existing work methods used by a worker. If the job is highly repetitive, a detailed methods analysis similar to the MTM system or its equivalent should be used. If the job is more unstructured, with a variety of different manual tasks, a work sampling procedure or time study may be used to document the job requirements.

Once the general work methods are documented, however, special initiatives will be necessary to (1) identify work postures that are potentially hazardous to the musculoskeletal system, either by filming or sketching the postures, (2) measure the actual loads that are handled, with special attention given to both average and peak loads, and (3) assure that even occasional tasks are included in the analysis by conducting detailed discussions with both supervisors and workers involved in the execution of the job.

7.3 CONTEMPORARY BIOMECHANICAL JOB ANALYSIS SYSTEMS

Because of the need for more specific job data than provided by traditional methods analysis systems, several contemporary approaches have been advocated to improve the documentation of biomechanical requirements in jobs. A description of a few of these approaches follows.

7.3.1 Physical Stress Checklists and Surveys

When first evaluating a variety of jobs composed of differing manual requirements, a checklist of physical tasks can often be utilized to document the general job physical requirements. Figure 7.1 illustrates one such checklist developed by Harman in 1951. It has been used by the United States Civil Service Commission to improve the job placement process for individuals with physical impairments (Smith, Armstrong, and Lizza, 1982).

A more informative survey approach has been advocated by the United States Department of Labor. In this case a job analyst directly observes the job and fills in a form noting the average frequency of occurrence and weight handled during the performance of different manual tasks (U.S. Dept. of Labor, 1972). This procedure was modified by Koyl and Hanson (1973) to include the recording of the hours during the day that each activity was performed, as displayed in Figure 7.2, as well as whether the left or right hand performed the activity. By comparison of these data with a clinical assessment of a person's mobility and strength, Koyl and Hanson proposed that placement of individuals on specific jobs could be enhanced. The United States Department of Labor has used such data to subjectively classify the physical demands of jobs with regard to the general criteria given in Table 7.5.

It should be clear that such physical stress survey data are important when identifying jobs that could be potentially hazardous to a worker's musculoskeletal system. They are particularly helpful when jobs are composed of tasks that are *not* highly repetitive, for which a traditional methods analysis is not appropriate. In this sense, the survey results indicate where special in-depth studies, some of which are described in the following, are warranted. Such surveys are especially useful when combined with musculoskeletal injury data analysis to motivate more intensive evaluations.

7.3.2 Manual Lifting Analysis

Because manual lifting is so prevalent, and because it is often the cause of serious musculoskeletal disability, a special evaluation of lifting tasks is often

U.S. CIVIL SERVICE COMMISSION
WASHINGTON, D.C.

JOB ANALYSIS FOR PHYSICAL FITNESS REQUIREMENTS

(TITLE OF POSITION)	(DEPARTMENT OR AGENCY)
(NAME AND LOCATION OF ESTABLISHMENT)	(OCCUPATIONAL CODE NO.)

I. FUNCTIONAL FACTORS (Check all functional factors involved in performance.)

A. FINGERS	B. ARMS	C. LEGS AND FEET	D. BODY TRUNK
1. Pressing	1. Pressing	1. Balancing	1. Bending
2. Holding	2. Holding	2. Turning	a. Flexion
3. Pulling	3. Pulling	3. Pushing	b. Extension
4. Picking	a. straight	4. Squatting	2. Lifting
5. Grasping	b. hand over hand	5. Running	3. Torsion
6. Reaching		6. Pressing	4. Rotation
7. Touching	4. Flexion	7. Climbing	5. Flexion and extension of cervical spine
8. Cutting with shears	5. Extension	a. stairs	
	6. Rotation	b. ladders	
	7. Hammering	8. Walking	
	8. Throwing	9. Lifting	
	9. Lifting	10. Holding	
		11. Prolonged standing	

FIGURE 7.1 Checklist of job physical requirements (proposed by Harman, 1951, p. 84).

warranted. One comprehensive approach to lifting analysis has been advocated by the National Institute for Occupational Safety and Health in the United States (NIOSH, 1981). This government research agency has proposed that manual lifting tasks be evaluated by documenting the following job variables:

1. *Weight of the object lifted*—determined by direct weighing. If this varies from time to time, the average and maximum weights are recorded.
2. *Position of load* with respect to the body—measured at *both the starting and ending points* of a lift in terms of horizontal and vertical coordinates. The horizontal location from the body (H) is measured from the midpoint of a line joining the ankles to the midpoint at which the hands grasp the object while in the lifting position. A rule of thumb is $H = \left(\frac{W}{2} + 15\right)$ cm, where W is the width of the object measured along a horizontal axis (Figure

JOB TITLE: ______________________________

____

JOB LOCATION: ______________________________

PHYSICAL FACTORS:

	No.	Item	Group
	1	1- 5	LIFTING (POUNDS)—INCLUDES PUSHING AND PULLING EFFORT WHILE STATIONARY
	2	6- 10	
	3	11 - 25	
	4	26- 50	
	5	51 -100	
	6	100 +	
	7	1- 5	CARRYING (POUNDS) — INCLUDES PUSHING AND PULLING EFFORT WHILE WALKING
	8	6- 10	
	9	11- 25	
	10	26- 50	
	11	51- 100	
	12	100+	
	13	R	FINGERING
	14	L	
	15	R	HANDLING
	16	L	
	17	R	BELOW SHOULDERS — REACHING
	18	L	
	19	R	ABOVE SHOULDERS — REACHING
	20	L	
	21	R	THROWING
	22	L	

	No.	Item	Group
	23	SITTING	
	24	TOTAL TIME ON FEET	
	25	STANDING	
	26	WALKING	
	27	RUNNING	
	28	JUMPING	
	29	LEGS ONLY	CLIMBING
	30	LEGS AND ARMS	
	31	R	WHILE SITTING — TREADING
	32	L	
	33	R	WHILE STANDING — TREADING
	34	L	
	35	STOOPING	
	36	CROUCHING	
	37	KNEELING	
	38	CRAWLING	
	39	RECLINING	
	40	TWISTING	
	41	WAITING TIME	

____
JOB ANALYST'S NAME

FIGURE 7.2 Physical demands analysis worksheet for noting hours involved in various physical tasks (Koyl and Hanson, 1973).

7.3), assuming the object is lifted close to the front of the body. The vertical component is determined by measuring the distance from the floor to the point at which the hands grasp the object. The coordinate system is also illustrated in Figure 7.3. If the four values vary from task to task (e.g., stacking cartons on top of each other), the job is separated into individual lifting tasks and each is evaluated separately.

3. *Frequency of lift*—recorded on the job analysis sheet in average lifts/minute for high-frequency lifting. A separate frequency should be entered for each distinguishable lifting task if performed at a frequency greater than once every five minutes.

4. *Period* (*or duration*)—the total time engaged in lifting. This is defined as either being more or less than one hour for the purposes of this procedure.

The resulting object weight and hand coordinate data obtained by such an analysis are compared with biomechanical and psychophysical strength

TABLE 7.5
Strength Requirement Classification Criteria adopted by U.S. Department of Labor (from Smith, Armstrong, and Lizza, 1982)

Degree of Strength	Amount of Lifting/ Carrying	Posture; Other Activities
Sedentary Work	Occasional: 10 lb maximum	Primarily sitting; walking and standing occasionally at most
Light work	20 lb maximum; 10 lb (or less) frequently	Significant amount of walking or standing or Primarily sitting, but requiring pushing and pulling of arm and/or leg controls
Medium work	50 lb maximum; 25 lb (or less) frequently	Unspecified
Heavy work	100 lb maximum; 50 lb (or less) frequently	Unspecified
Very heavy work	Over 100 lb allowed; 50 lb (or more) frequently	Unspecified

limits. Data on lifting frequency and duration data are used to predict the metabolic energy requirements of a job and compared to population work capacity norms. The lifting limits predicted by the NIOSH *Work Practices Guide* are presented in Chapter 8 as part of the general discussion of manual materials handling.

To record job lifting data, a *Physical Stress Job Analysis Sheet* has been developed by NIOSH. It is shown in Figure 7.4, filled-in for a lifting task requiring a 200-N stock reel to be lifted once every shift to the top of a punch press, as illustrated in Figure 7.5. It is assumed that the lift is performed by grasping the object near the center and lifting it while standing in front of the press. Both the dimensions of the stock reel and the press make this task difficult. Using the limits described in Chapter 8, this lift would exceed the NIOSH *Action Limit* (predicted to be 70 N), thus requiring careful selection and training of workers for the task. In fact, the lifting requirement of 200 N almost exceeds the *Maximum Permissible Limit* of 212 N predicted by the NIOSH method. If it exceeded it, either the load would have to be reduced or the lifting postures modified. In this latter regard, an analysis of this task is included in the NIOSH *Work Practices Guide* and shows that the lifting stress can be greatly reduced by having the worker grasp the reel by its perimeter and lift it along the side of the press instead of standing in front of the press. This allows the worker to stand much closer to the center of mass (reducing H to approximately 20 cm).

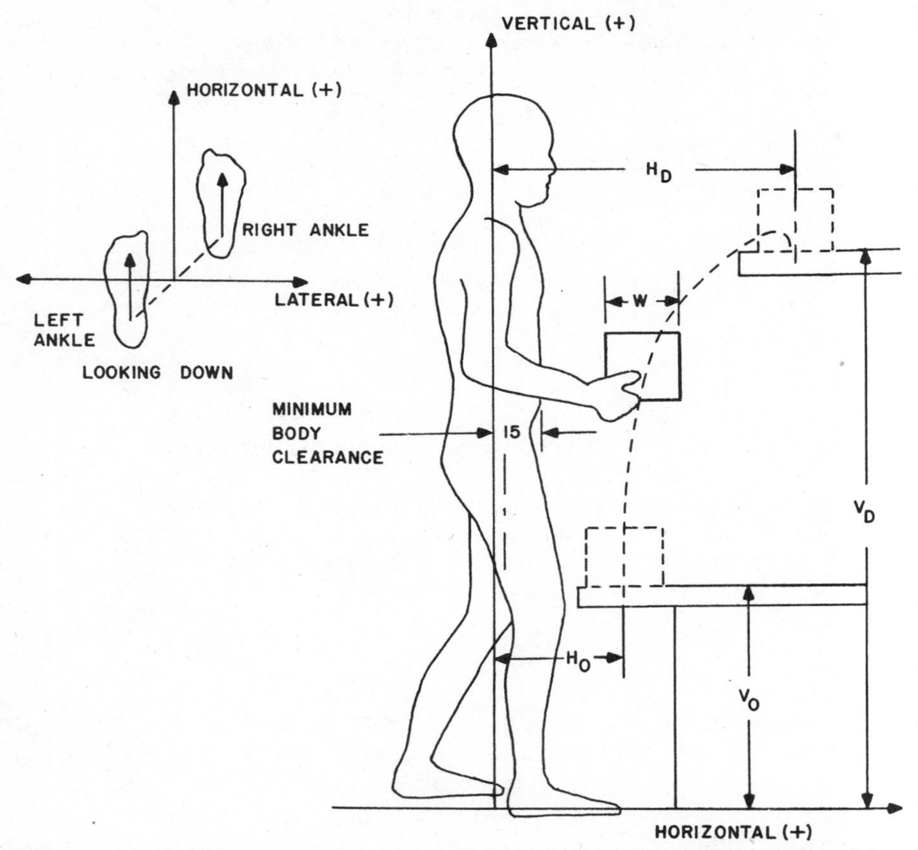

FIGURE 7.3 Graphic representation of vertical and horizontal coordinates (adapted from NIOSH *Work practices Guide to Manual Lifting,* 1981).

When a lifting evaluation is completed for each lifting task in a job, the resulting predicted *Action Limits* (AL) and *Maximum Permissible Limits* (MPL) are added to the coding forms (Figure 7.4) in the columns provided. These limits then can be compared line by line to determine the relative stress of each task analyzed. Because the *H* distance is so critical in determining the resulting biomechanical stress, particularly to the low back, it is important that the job analyst measure *H* at a point in the lift at which the load effect is expected to be great. This is usually at or near the origin of the lift, where the inertial effect of accelerating the mass upward is great. If a great deal of control over the destination of the load is required, however, such as when achieving a precise position or when handling a heavy object that is fragile, it may be necessary to measure *H* at the destination. This decision requires a biomechanical judgment by the analyst as to what seg-

PHYSICAL STRESS JOB ANALYSIS SHEET

DEPARTMENT Fabrication DATE 2-18-80

JOB TITLE Punch Press ANALYST'S NAME EJB

TASK DESCRIPTION	OBJECT WEIGHT (NEWTONS)		HAND LOCATION				TASK FREQ	AL	MPL	REMARKS
			ORIGIN		DESTINATION					
	AVE	MAX	H cm	V cm	H cm	V cm				
Load Stock	200	200	53	38	53	160	0			

FIGURE 7.4 NIOSH Job Lifting Analysis Form filled in for lifting task depicted in Figure 7.5. Note: The zero entry for task frequency denotes that the stock reel is loaded at a frequency of less than once every five minutes.

ment of the lift should be considered the point of maximum loading and unloading of the musculoskeletal system for analysis purposes.

7.3.3 Job Static Strength Analysis

Because the NIOSH *Work Practices Guide for Manual Lifting* applies only to symmetric (both hand) lifting of loads in the sagittal plane, a more comprehensive job physical stress analysis scheme may be necessary. One such scheme relies on the static strength prediction model described in Chapter 6. This model compares the load moments produced at various body joints during the execution of a large variety of manual exertions with the static strength moments obtained from tests of over 3000 workers in the United States. Thus it predicts the proportion of the population capable of performing the exertion with reference to various static strength norms.

This strength prediction methodology has provided the means to evaluate a variety of manual exertion data obtained by direct observations of workers. The job analysis procedure is generally the same as that adopted by NIOSH. A job is described as a series of physical tasks. These tasks denote the load vector direction operating on the hands (i.e., "lift" is a vertical downward load vector, "push" is a sagittal plane, horizontal vector toward the body, "pull" is a sagittal plane, horizontal vector away from the body, etc.). For

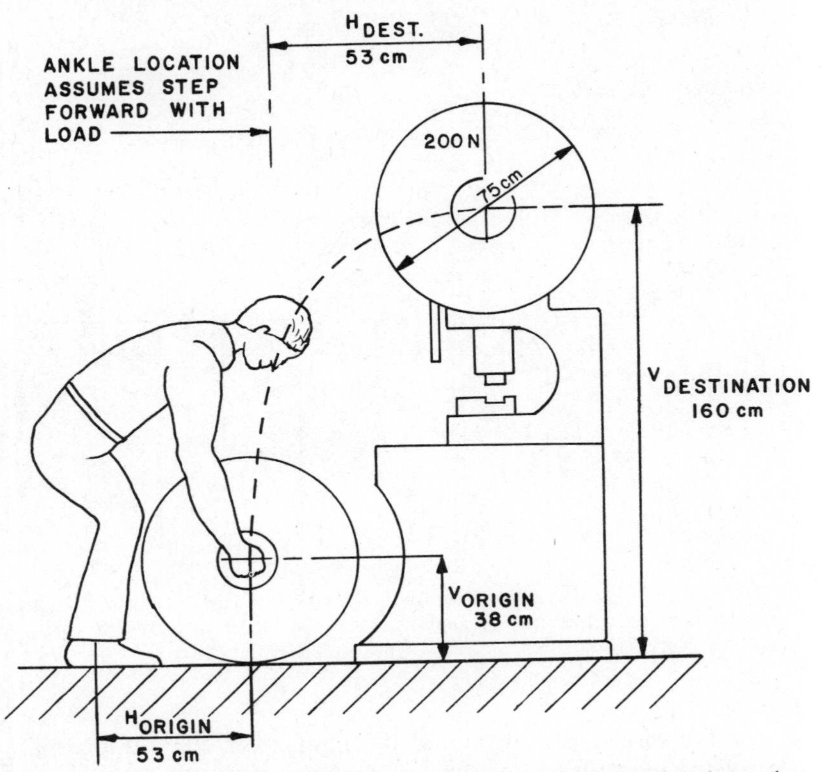

FIGURE 7.5 Example of lifting stock into punch press. It is assumed that the worker steps forward with the load to place it atop press—i.e., H remains constant, while V changes.

each task, a general posture of the legs and torso is designated (i.e., standing, sitting, squatting, stooping, etc.), and the three-dimensional coordinates of the hands relative to the midpoint between the ankles are recorded. Though specific postural data are preferred for such an analysis, these approximate postures can be used as initial input in a strength analysis. If a strength model is computerized, it then iterates through alternative postures to determine what maximizes the number of people capable of performing the task (as described in the algorithm presented in Figure 6.44).

A coding form has been derived by Chaffin et al. (1977) to tabulate these data for input into a computerized version of the strength prediction methodology described in Chapter 6. The coding form for this analysis is presented in Figure 7.6.

One example of the use of this more robust strength evaluation methodology was reported by Chaffin et al. (1977). This case involved rolling stock reels to and from machines and lifting them into the machines. The pushing action required to roll the reels is coded as task number 01 on the

Col (1-3) Date	(9-18) Plant	(19-25) Dept.	(26-32) Analyst	(33) Rep.	(34) Non Rep.	(35-42) Job Classification	(43-61) Job Location	(62-81) Job Title	Subdescription and Comments (82-101)
7/76	H	WIRE	DBC.		✓			WIRE	REEL HAND

(1-3) TASK NUMBER	(4-5) TASK CODE	(6) BODY POSTURE CODE	(7-16) OBJECT	FORCE (N)		Hand Location (cm)											
						Origin						Destination					
						Right Hand			Left Hand Displacement*			Right Hand			Left Hand Displacement*		
				(17-19)	(20-22)	23	26	29	32	35	38	41	44	47	50	53	56
				AVG.	MAX.	25	28	31	34	37	40	43	46	49	52	55	58
						V	L	H	V	L	H	V	L	H	V	L	H
01	03	03	REEL	15	22	35	15	50	0	-30	0	40	15	60	0	-30	0
02	01	01	REEL	200	290	40	15	40	0	-30	0	100	15	40	0	-30	0

*Note: Left hand coordinates are measured relative to right hand location.

TASK LOAD TYPES OF STRENGTH ANALYSES

CODE	TASK	DIRECTION OF LOAD	MOTION
01	Lift	↓	↑
02	Lower	↓	↓
03	Push	←	→
04	Pull In	→	←
05	Pull Right	→Left	←Right
06	Pull Left	→Right	←Left
07	Pull Down	↑	↓
08	Hold	↓	O
09	Torque (R)	CCW	CW
10	Torque (L)	CW	CCW

GENERAL POSTURE DATA DESCRIPTORS

CODE	POSTURE	DESCRIPTION
01	Stand	Body is in upright position with no significant deviation from the vertical. Included angles at knee, hip and trunk are near 180°.
02	Sit	Body is seated
03	Squat	Body is in a crouched position with significant bending of the knees (included angle < 150°). Slight to moderate trunk flexion (bending forward) will occur.
04	Deep Squat	Similar to squat, however included angle at knee is less than 100°
05	Stoop	Trunk is flexed forward with slight bending of the knees.
06	Lean	Joint angles at knee, hip and trunk remain at or near 180°. Lower leg angle with respect to the floor is allowed to deviate forward or backward from 90°.
07	Split	One foot is significantly forward of the other foot. Included angle at knee of the forward leg deviates from 180°. Rear leg remains straight.

FIGURE 7.6 Job strength evaluation form (Chaffin et al., 1977).

example coding form in Figure 7.6. The task of lifting the reel to place it in the machine is coded as task number 02. The tasks are not unlike that described before in loading the punch press (Figure 7.5). In this case the reel weighs 290 N and is raised to only about 100 cm instead of 160 cm in the punch press analysis. The strength analysis reported by Chaffin et al. (1977) showed that the peak push force of 22 N required to roll the reel to the machine was so low that over 97% of men and women could be expected to be able to perform the task. The analysis also showed that the stooping and squatting required to push the reels stressed the hip extensor muscles more than other muscle groups. The reel lifting task (01) was more stressful, with one-third of the men and only a few women having the necessary strength to perform the lift. The limiting muscle action in this task was found to be shoulder abduction, particularly for female workers. An analysis of health data for musculoskeletal problems indicated that workers performing these jobs had injuries that were over ten times more serious (in terms of days lost for medical reasons) than the average injuries reported for the plant. It was also shown that these serious injuries most frequently involved the lower extremity, shoulders, and back. The authors report that the combination of biomechanical and epidemological evaluation prompted mechanization to assist the operators in performing these jobs.

7.3.4 Job Postural Evaluation Method

Another approach to recording potentially stressful postures, referred to as posture targeting, was proposed by Corlett, Madeley, and Manenica (1979). This procedure requires the job analyst to observe a worker at random times during the workday (i.e., a work sampling study is performed). At the time of each observation the analyst records the angular configuration of various body segments with the aid of the "body diagram" displayed in Figure 7.7. The angular data are recorded by simply placing a small X on each of the postural targets when a body segment deviates from the erect anatomical position shown in Figure 7.7. The concentric circles on each target represent 45°, 90°, and 135° angular deviations of a joint from that shown, with the arrow at the center of the target indicating the front of the body. The radial lines indicate the amount of deviation from the sagittal plane, viewed from overhead. For example, if a person were holding the forearm horizontally and at 45° off the mid-sagittal plane, the forearm diagram would have an X marked as indicated in Figure 7.8a. If a person were observed many times during the workday in a seated position, but with the upper arms abducted and flexed and the torso leaning toward the left side, the resulting composite postural diagram would look like that shown in Figure 7.8b. Counting off the number of X marks in a certain zone of a diagram, or simply observing how the marks cluster together, provides insight as to possible stressful postures. Though Corlett et al. did not combine these postural data with external load data for biomechanical analysis, such would be possible by

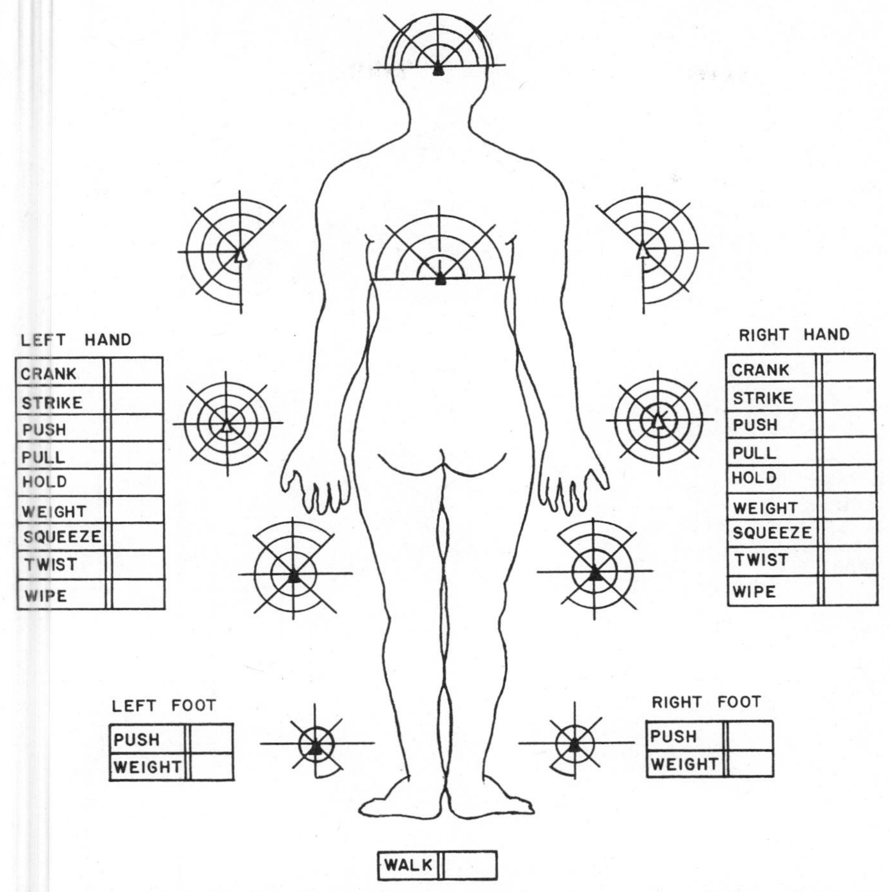

FIGURE 7.7 Body diagram, showing each target adjacent to its associated body part. Deviations from the standard position shown are marked on the neighboring diagram, otherwise no mark is made with each observation of the worker (Corlett, Madeley, and Manenica, 1979).

noting the load magnitudes on the activity lists provided with the body diagram. In its simplest form the procedure documents job postures in such a way that the analyst can easily identify the most frequent and potentially stressful ones for more detailed biomechnical analysis. Corlett and Manenica (1980) have also demonstrated that this procedure is useful in evaluating workplace layouts when combined with worker reports of localized musculoskeletal pain obtained at several intervals during a workday.

It should be mentioned that in 1974, Priel proposed a system to allow postures to be numerically defined and recorded. From repeated observation of workers, the basic posture of the body within a three-dimensional coordinate system, the levels at which joints and limbs are located, and the

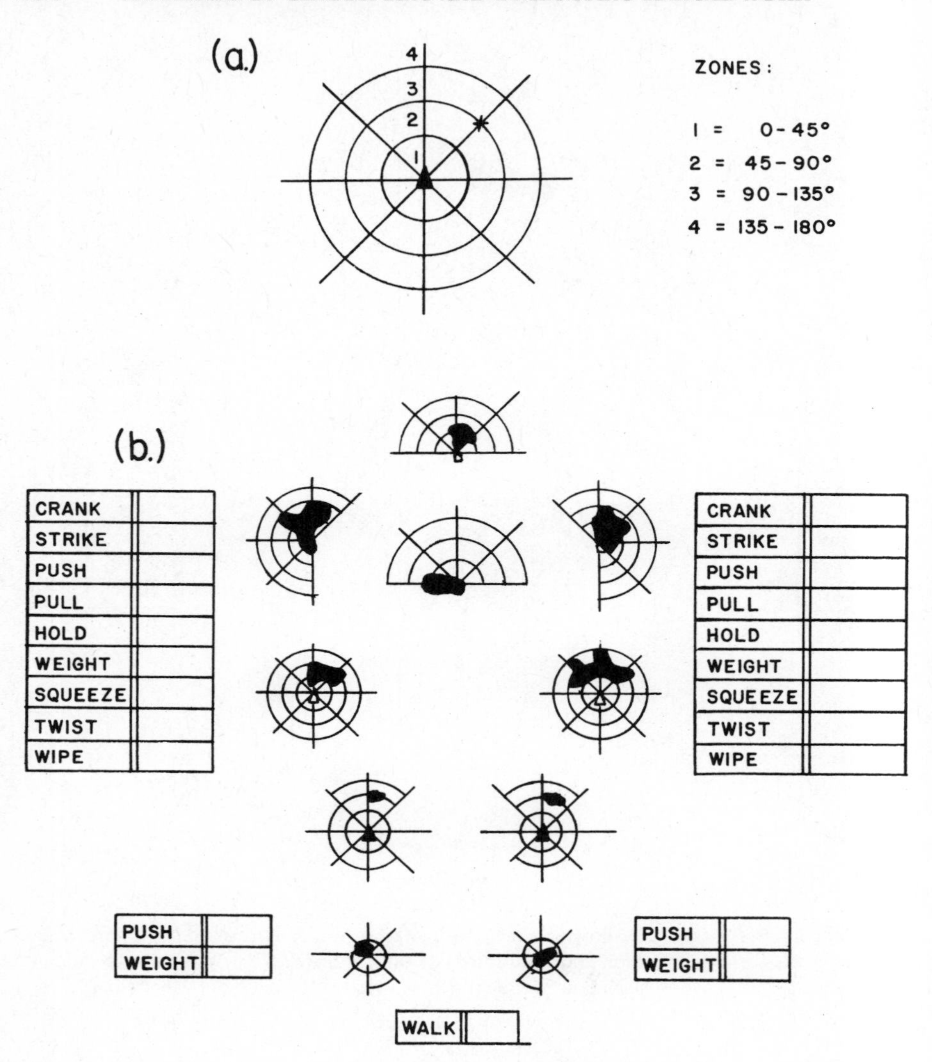

FIGURE 7.8a A body target for the forearm marked to indicate that the forearm is held horizontal (45°–90°) and at 45° from mid-saggital plane. Figure 7.8b is a composite body diagram indicating a seated operator, who frequently leans the torso toward the left, abducts both arms and flexes the forearm, particularly the right (Corlett, Madeley, and Manenica, 1979).

direction and amount of movements are determined and recorded on a "posturegram." Another postural observation system was developed in Finland—the so-called OWAS system. The Ovaco Working Posture Analysis System (OWAS) is a practical method for identification and evaluation of unsuitable working postures (Karhu et al., 1977; Karhu et al., 1981). The method consists of two parts. Part one is an observation technique for the evaluation of working postures. It can be used in time-and-motion studies in the daily routine and gives reliable results after a short training period of those performing the studies. The second part of the method is a set of criteria for the redesigning of working methods and working places. The criteria are based on evaluations made by experienced workers and ergonomic experts. The criteria take into consideration factors such as health and safety, but the main emphasis is placed on the discomfort caused by working postures. This method has been extensively used by a steel company that participated in its development. OWAS has now been further developed and can be used in two types of investigations: basic OWAS investigations and applied OWAS investigations. The basic OWAS is used when a task involves the whole body. The applied OWAS is used when a work task, or the work, is performed in a sitting or standing position and most of the work is done with the hands. A method similar to the OWAS is a technique developed by Berns and Milner (1980) for analyzing moving work postures (TRAM). Both OWAS and TRAM record the postures at regular intervals on a recording sheet, which contains a number of basic postures (back, arms, and legs). By using various combinations of the basic posture, a wide range of postures can be described. Included in the recording is an approximation of a weight or force that acts externally on the body, as well as longer static positions.

A new Swedish system called ARBAN was developed by Holzmann (1982). ARBAN is a method for ergonomic analysis of work, including work situations involving greatly differing body postures and loads. The method consists of four different steps: (1) recording of the work place situation on a videotape or film, (2) coding of the posture and load situation in a number of closely spaced "frozen" situations, (3) computerization, and (4) evaluation of the results. The computer calculates the total ergonomic stress for the whole body based on heuristic rules regarding the relative stress of specific acts, as well as for different parts of the body. The results are presented as ergonomic stress/time curves, with the heavy load situations occurring at the peak of the curve.

The observation techniques can be used together with discomfort/comfort scales (Corlett and Bishop, 1976) and different biomechanical analyses techniques. Grieve (1979) has suggested a method to determine potential mechanical constraints in a static exertion based on equilibrium (body balance) considerations, resulting in the so-called *Postural Stability Diagram*.

7.3.5 Upper Extremity Postural Analysis

Because of the prevalence of upper extremity and shoulder pain in industry, particularly in bench work, a specific postural analysis procedure has been advocated by Armstrong, Foulke, Joseph and Goldstein (1982). This procedure requires filming an operator then coding the posture of the upper extremity from each film frame. The coding procedure for each posture involves noting the angular deviations of the shoulder (in reference to three axes), the elbow (in reference to two axes) and the wrist (in reference to two axes). The hand posture is designated from a choice of six distinct postural categories. If forceful hand exertions are required in the job, surface electrode EMG recordings are suggested to estimate the exertion level, using the procedure described in Chapter 5. The hand grip forces are then recorded for each posture photographed. The point of force loading on the hand is noted by a code—0 for palm contact and 1 to 5 for digits 1 to 5 when applying force. The form used to record these data is displayed in Figure 7.9.

Armstrong et al. (1982) described the use of this system to evaluate work methods and tools used to process poultry. The analysis indicated that changes in workplace layout, meat-cutting knife design, and worker training would reduce the stresses on the wrist and shoulder necessary to hold the turkey carcass and perform specific cuts.

7.3.6 Trunk Flexion Analysis

A measurement device for recording movements and postures in the sagittal plane has been developed by Ortengren and Andersson (Nordin et al. 1983). The instrument consists of a pendulum potentiometer as a transducer, a five-level analog digital converter, control circuits, and nine digital registers. Together the units form a portable battery-powered system of 1 kg that can be worn on the back in a small harness. Although the analyzer has the potential for measuring movement of any body segment, it has so far been adapted for measuring trunk movements in the sagittal plane (forward flexion). The portable unit is placed on the back of a subject and calibrated to measure trunk flexion from upright to 90°.

The range of forward flexion is divided into five intervals. Forward flexion of more than 90° is recorded in the interval (73–90°). The analyzer records the total amount of time that the subject spent in each interval of forward flexion as well as the number of times that the amount of flexion changed from one interval to another (borderline passages) in one direction of flexion. The flexion analysis has been tested and found to be accurate and simple to use (Nordin et al. 1983; Hultman et al. 1983).

Changes in work techniques and work design can easily be measured and quantified. Continuous recordings can be made without an observer. This is not, however, to be recommended, as valuable information about the work cycle will be lost. The advantage of a flexion analyzer unit is that it is non-invasive, reliable, easily transported, and the data are obtained immediately after testing. The limitation of the instrument in its present form is obvious—it records a trunk flexion but does not indicate a possible simultaneous external weight. Observation techniques and a trained observer are always needed for biomechanical analysis of manual materials handling.

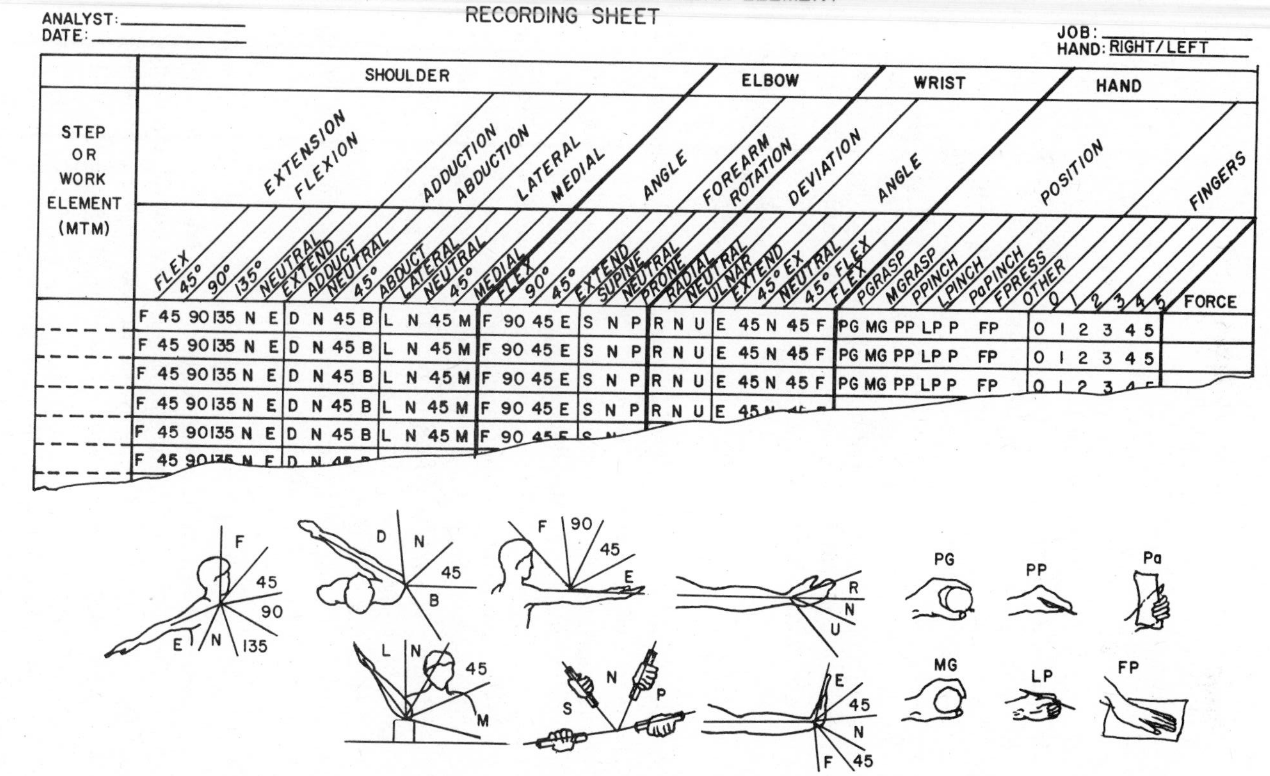

FIGURE 7.9 Form for recording upper extremity postures where analyst circles appropriate posture codes for each observed work element. (Armstrong, Foulke, Joseph, and Goldstein, 1982).

7.4 FUTURE IMPACT OF OCCUPATIONAL BIOMECHANICS ON WORK MEASUREMENT SYSTEMS

It should be clear from the preceding that traditional work measurement systems are limited in their ability to provide the data necessary to evaluate and improve the biomechanical aspects of various jobs. Even the most detailed predetermined motion-time system, such as the MTM-1 system, does not include data regarding postural and manual load-handling factors. Despite these limitations, however, it must be realized that these traditional motion-time analysis systems provide the fundamental data structures and procedures necessary to evaluate potential job-related biomechanical problems. The traditional systems emphasize a rigorous analysis and tabulation of human motions. By including postural and load data in such analyses, excellent biomechanical evaluations can be performed.

Unfortunately it is not possible at this time to specify an "optimal" analysis method for future occupational biomechanical studies. In fact, from the limited attempts to develop such a system it would appear that several different job biomechanical analysis schemes will be desirable in the future. Some considerations that are meant to assist in the choice of future analysis procedures are

1. The degree of standardization of manual tasks comprising a job.
2. The length of repetitive work cycles in a job.
3. The proportion of a workday involved in manual effort that is potentially overstressful to the musculoskeletal system.
4. The available time to observe and record manual activities—since an MTM-1 analysis may require an analyst to spend 350 times the length of a job's cycle time to perform, according to Magnusson (1972).
5. The degree of sophistication to be utilized in evaluating the biomechanical job data.
6. The expected end use of the analysis—for example, to redesign a machine, tool, or workplace; select and place workers on stressful jobs; modify work methods, and so on.

At present, several different biomechanical job analysis procedures are being used, some of which have been described. None would appear to be well validated in the field (i.e., they have not been widely used and publicly critiqued by various practitioners in the field), though all have been successfully used to solve well-defined occupational biomechanical problems in specific industry.

With the advent of more portable and robust motion and human force measuring systems, as described in the chapter on *Bioinstrumentation*

(Chapter 5), it can be expected that new job evaluation procedures will be feasible. Combining these measurement systems with various existing motion description systems will provide an exciting and beneficial development in the field of occupational biomechanics in the future.

REFERENCES

Armstrong, T. J., J. A. Foulke, B. S. Joseph, and S. A. Goldstein, "Investigation of Cumulative Trauma Disorders in a Poultry Processing Plant. *JAIHA*, **43**(2), 103–115 (1982).

Barnes, R. M., *Motion and Time Study,* John Wiley & Sons, New York, 1968, pp. 10–20 and 487–510.

Berns, T. A. R. and N. P. Milner, "TRAM—A Technique for the Recording and Analysis of Moving Work Posture," in Methods to Study Work Posture, N. P. Milner, Ed., ERGO-LAB, Stockholm, Sweden, Report 80:23, 1980, pp. 22–26.

Chaffin, D. B., G. D. Herrin, W. M. Keyserling, and A. Garg, "A Method for Evaluating the Biomechanical Stresses Resulting from Manual Materials Handling jobs," *AIHA,* **38**(Dec.), 662–675 (1977).

Copley, F. B., *Frederick W. Taylor,* Vol. I, Harper and Bros., New York, 1923, pp. 10 and 56.

Corlett, E. N. and R. P. Bishop, "A Technique for Assessing Postural Discomfort," *Ergonomics,* **19,** 175–182 (1976).

Corlett, E. N., S. J. Madeley, and I. Manenica, "Postural Targeting: A Technique for Recording Working Postures," *Ergonomics,* **22**(3), 357–366 (1979).

Corlett, E. N. and I. Manenica, "The Effects and Measurement of Working Postures," *Applied Ergonomics,* **11**(1), 7–16 (1980).

Gilbreth, F. B., *Motion Study,* Van Nostrand, Princeton, 1911, p. 88.

Gilbreth, F. B., "The Present State of the Art of Industrial Management," *Trans. ASME,* **34,** 1224–1226 (1912).

Grieve, D. W., "The Posture Stability Diagram (PSO): Personal Constraints on the Static Exertion of Force," *Ergonomics,* **22,** 1155–1164 (1979).

Harman, B., *Physical Capabilities and Job Placement*, Nordisk Rotogravyr, Stokholm, 1951, p. 84.

Holzmann, P., "ARBAN—A New Method for Analysis of Ergonomic Effort," *Appl. Ergon,* **13,** 82–86 (1982).

Hultman, G., M. Nordin, and R. Ortengren, "The Influence of a Preventive Educational Program on Trunk Flexion in Janitors," *Appl. Ergon.* (1983), in press.

Karger, D. W. and F. H. Bayha, *Engineered Work Measurement,* Industrial Press, New York, 1966, pp. 89–402.

Karhu, O., R. Hārkōnen, P. Sorvali, and P. Vepsäläinen, "Observing Working Postures in Industry," *Appl. Ergon,* **12,** 13–17 (1981).

Karhu, O., P. Kansi, and I. Kuorinka, "Correcting Working Postures in Industry," *Appl. Ergon,* **18,** 199–201, 1977.

Konz, S., *Work Design,* Grid Publishing, Columbus, 1979, pp. 103–144.

Koyl, F. F. and P. Marsters-Hanson, Age, Physical Ability and Work Potential (unpublished contract report), Manpower Administration, U.S. Dept. of Labor, Washington, D.C., 1973.

Magnusson, K., "The Development of MTM-2, MTM-V, and MTM-3," *J. of Methods—Time Measurement,* **17**(Feb.), 11–23 (1972).

National Institute for Occupational Safety and Health, "*Work Practices Guide for Practices Guide for Manual Lifting,*" Technical Report 81-122, NIOSH, Cincinnati, 1981, pp. 129–144, (reprinted: Amer. Industrial Hygiene Assoc., Akron, OH., 1983).

Niebel, B. W., *Motion and Time Study,* Irwin, Homewood, Ill. 1972, pp. 417–462.

Nordin, M., R. Örtengren, and G. B. J. Andersson, "Measurement of Trunk Movement during Work," *Spine* (1983), in press.

Priel, V. C., "A Numerical Definition of Posture," *Human Factors,* **16,** 576–584 (1974).

Smith, P., T. J. Armstrong, and G. D. Lizza, "IE's Can Play Crucial Role in Enabling Handicapped Employees to Work Safely, Productively," *Industrial Engineering* (April 1982).

Taylor, F. W., *The Principles of Scientific Management,* Harper and Bros., New York, 1929, p. 36.

U.S. Department of Labor, *Handbook for Analyzing Jobs,* Superintendent of Documents, U.S. Government Printing Office, Washington, D.C., No. 2900-0131, 1972.

8

MANUAL MATERIALS HANDLING LIMITS

8.1 INTRODUCTION

The act of manually lifting, pushing, or pulling an object has been of continual concern to those planning efficient use of a workforce, and to those attempting to prevent unnecessary injury and illness in industry. The recent proliferation of industrial robots undoubtedly will decrease the number of workers performing manual materials handling jobs, especially if the jobs are highly structured and repetitive. Though estimates are not available as to how fast this displacement of manual labor will occur, it is evident that many manual acts will *not* be readily automated. Automation will be difficult in jobs that are unstructured, particularly in the service industries—for example, building construction, mechanical repair of equipment, baggage and package handling, police protection, and fire fighting, to name a few. A recent report from the National Institute for Occupational Safety and Health (NIOSH) stated that approximately one-third of the U.S. workforce is presently required to exert significant strength as part of their jobs (NIOSH, 1981).

This report also presented the following statistics:

1. Overexertion was claimed as the cause of lower back pain by over 60% of people suffering from it.
2. Overexertion injuries of all types in the U.S. occur to about 500,000 workers per year (about 1 in 20 workers each year).

3. If the overexertion injuries involve low-back pain with significant lost time, less than one-third of the patients eventually returned to their previous work.
4. Overexertion injuries account for about one-fourth of all reported occupational injuries in the United States, with some industries reporting that over half of the total reported injuries are due to overexertion.
5. Approximately two-thirds of overexertion injury claims involved lifting loads, and about 20% involved pushing or pulling loads.

Collectively these observations indicate that manual materials handling activities are now, and will continue to be, prevalent in many industries, and that such acts are associated with either causing or aggravating preexisting musculoskeletal disorders in a large number of workers.

It also became clear from a review of pertinent literature that a comprehensive program of control would be necessary (Troup, 1978). Herrin et al. (1974), proposed that several distinct groups of factors needed to be considered simultaneously in the prevention of musculoskeletal disorders related to manual materials handling. These factors, which define a manual material handling system, were grouped as follows (each is defined in Table 8.1):

Worker characteristics
Material/container characteristics
Task characteristics
Work practices

Herrin et al. concluded that of all the different types of manual materials handling acts performed in industry, research findings were most conclusive regarding the act of manually lifting of loads that are symmetrically balanced in front of the body. Based on this, a multidisciplinary team of specialists in epidemiology, bimechanics, work physiology, and ergonomics were assembled by NIOSH to develop a *Work Practices Guide to Manual Lifting* (NIOSH, 1981). It represents the first comprehensive approach to the control of the adverse effects of manual materials handling in industry. The basis and use of the *Guide* will be presented in Section 8.2.4 of this chapter.

The NIOSH sponsored review of the research literature in the area also revealed large "gaps in knowledge" regarding asymmetric (one-handed or side) lifts and pushing and pulling acts in industry (Herrin et al. 1974). Specific human performance limits for these acts are discussed in this chapter, but must be viewed with caution because of the limited research base.

8.2 LIFTING LIMITS IN MANUAL MATERIALS HANDLING

Many human performance limits are produced through a consensus of experts. Such was the case when the International Labor Organization (ILO)

TABLE 8.1
Characteristics of Major Components Affecting Manual Materials Handling System (Herrin et al., 1974)

Worker Characteristics

Physical: include general worker measures, such as: age; sex; anthropometry; postures.

Sensory: measures of worker sensory processing capabilities, such as: visual; auditory; tactual; kinesthetic; vestibular; proprioceptive.

Motor: measures of worker motor capabilities, such as: strength; endurance; range of movement; kinematic characteristics; muscle training state.

Psychomotor: measures of worker capabilities interfacing mental and motor processes, such as: information processing; reaction/response time; coordination.

Personality: measures of worker values and job satisfaction by attitude profiles; attribution; risk acceptance; perceived economic need.

Training/experience: measures of the worker education level in terms of formal training or instruction in manual material handling skills; informal training; work experience.

Health status: measures from worker general health appraisal, such as: previous medical complaints; diagnosed medical status; emotional status; regular drug usage; pregnancy; diurnal variations; deconditioning.

Leisure time activities: measures of the person choosing to be involved in physical activities during leisure hours, such as: holding a second job or regular participation in sports.

Material/Container Characteristics

Load: measure of force; weight, pushing/pulling force requirements, mass moment of inertia.

Dimensions: measures of size of unit workload, such as: height; width; breadth when indicating the form as rectangular, cylindrical, spherical, etc.

Distribution of load: measure of the location of the unit load CG with respect to the worker for one hand or two handed carrying.

Couplings: measures of simple devices used to aid in grasping and manually manipulating the unit load, such as: texture; handle size, shape, and location.

Stability of load: measures of load CM location consistency, as a concern in handling liquids and bulk materials.

Task Characteristics

Workplace geometry: measures of the spatial properties of the task, such as: movement distance; direction and extent of path; obstacles, nature of destination.

Frequency/duration/pace: measures of the time dimensions of the handling task including frequency, duration and required dynamics of activity over the short term and long term.

Complexity: measures of combined or compounding demands of the load, such as: manipulation requirements of movement; objective of activity; precision of motion tolerance; number of kinetic components.

Environment: measures of added deteriorative environmental factors, such as: temperature, humidity; lighting; noise; vibration; foot traction; seasonal toxic agents.

TABLE 8.1 (*Continued*)

Work Practices Characteristics

Individual: measures of operating practices under the control of the individual worker, such as: speed and accuracy in moving objects; postures (i.e., lifting techniques) used in moving objects.

Organizational: measures of work organization, such as: physical plant size; staffing of medical/hygiene/engineering/and safety functions; and utilization of teamwork.

Administrative: measures of administration of operating practices, such as: work and safety incentive system; compensation scheme; safety training and control; hygiene and safety surveys; and medical aid and rescue; long work shifts; rotation; personal protective devices.

issued an *Information Sheet on Manual Lifting* in 1962. Using a consensus of medical experts, the limits summarized in Table 8.2 were issued. The motivation for such limits at the time was that the ILO experts believed manual lifting contributed to (1) a threefold incidence in spinal, knee, and shoulder injuries, (2) a tenfold incidence in elbow injuries, and (3) about a fivefold increase in hip injuries (ILO, 1962). The ILO limits recognize that people are different (gender and age groups are assigned different values). Unfortunately, the limits have no regard to the many other factors in the workplace that affect a person's lifting capability.

Possibly because the ILO limits are not comprehensive, or perhaps because the scientific rationale for the limits was not well stated, they appear to have had little effect on musculoskeletal injury and illness in industry (NIOSH, 1981). With cognizance of this failure, NIOSH assembled research literature and experts representing four different approaches to the development of the new *Work Practices Guide to Manual Lifting*. This new *Guide* was to be based on consideration of:

TABLE 8.2
ILO Suggested Limits for Occasional Weight Lifting in Newtons[a] (ILO, 1962)

Age (years)	Men	Women
14–16	143	96
16–18	181	115
18–20	222	134
20–35	240	143
35–50	202	125
Over 50	153	96

[a] 4.45 N = 1 lbf or 9.81 N = 1 kgf = 1 kp.

1. Epidemiology of musculoskeletal injury.
2. Biomechanical concepts.
3. Physiological considerations.
4. Psychophysical (muscular strength) lifting limits.

Although other attempts have been made to provide guidelines on limits in manual materials handling, the NIOSH guide is, in our view, the most comprehensive and will, therefore, be discussed here. It is beyond the scope of this chapter, however, to review the entire NIOSH development effort (four years were required and 400 pieces of literature were reviewed). What follows are the recommendations resulting from this important activity.

8.2.1 Scope of NIOSH Work Practices Guide for Manual Lifting

As stated earlier, adequate and consistent research findings were available to support a recommendation on symmetric (two-handed) lifting of loads in the sagittal plane. The resulting recommendations, therefore, are limited to:

1. Smooth lifting (no sudden acceleration effects).
2. Moderate width objects (hand separation of less than 75 cm).
3. Unrestricted lifting postures (no bracing of the torso).
4. Good couplings (hand holds are secure, and shoe/floor slip potential is low).
5. Favorable temperature conditions.

The *Guide* that was issued focuses on those task and material container characteristics that best define a hazardous lifting act. These factors were defined and given a variable designation, as follows:

1. Weight of object lifted (L).
2. Location of object center of mass (or hand grip center) measured horizontally from a point on the floor midway between the ankles (H).
3. Location of object center of mass (or hand grip center) measured at beginning (origin) of lift (V).
4. Vertical travel distance of hands from origin to destination (release) of object (D).
5. Frequency of lifting (in lifts per minute) averaged over period of lifting (F).
6. Duration of the period during which lifting takes place (less than one hour or on an eight-hour basis).

Once an evaluation of a job has been made with reference to these variables, if a lifting task is found to be hazardous, control procedures are recommended. In other words, the determination of the existence of a lifting hazard depends on the collective effect of all of these variables.

Depending on the nature of the lifting activities, limitations may be based on the biomechanical and muscle strength demands of the task. This is especially true in *infrequent* lifting of large or heavy objects requiring awkward postures. If more moderate size or lighter weight objects are lifted *frequently, but for less than a one-hour period,* then limits based on an acceptable psychophysical work load appear to be most appropriate with the studies of Snook (1978) and Ayoub et al. (1980) being cited most often. If *frequent lifting for an entire work day* (defined as eight hours) is necessary, then cardiovascular/metabolic based muscle fatigue studies provide the limitations. In this latter case, the work physiology studies cited in Astrand and Rodahl (1970), Bonjer (1971), Kamon and Ayoub (1976), Garg et al. (1978), and Petrofsky and Lind (1978), and many others provide the basis for limiting such sustained, frequent lifting.

From an epidemiological perspective, the NIOSH *Guide* cites studies revealing that musculoskeletal injury rates (i.e., number of injuries per man-hours on the job) and severity rates (i.e., number of hours lost due to injury per man-hours on the job) increased significantly when

1. Heavy objects are lifted (i.e., L is large).
2. The object is bulky (i.e., H is large).
3. The object is lifted from the floor (i.e., V is small).
4. Objects are frequently lifted (F is high).

The four different criteria used in developing the *Guide*—epidemiological, psychophysical (strength), biomechanical, and physiological (muscle fatigue)—indicate that no single task characteristic acts independently to influence the hazard level. All are interactive, and often are multiplicative. This means that the recommendations that define a potentially hazardous lifting task needed to consider the collective effects due to each of the four criteria.

8.2.2 Definition of Lifting Hazard Levels

From population studies of strength, anthropometry, and aerobic work capacity, it is obvious that a large variation in lifting capability exists in any normal group of workers. Because of this, the NIOSH recommendations are based on two levels of hazard. The first level establishes an *Action Limit* (AL), wherein an increased risk of injury and fatigue for *some* individuals exists if not carefully selected and trained for the lifting task found to exceed the limit. Specifically, the *Action Limit* is based on:

1. Epidemiological data indicating that *some* workers would be at increased risk of injury on jobs exceeding the AL.

2. Biomechanical studies indicating that L5/S1 disc compression forces can be tolerated by most (but not all) people at about the 3400 N level, which would be created by conditions at the AL.
3. Physiological studies disclosing that the average metabolic energy requirement would be 3.5 kcal/min for jobs performed at the AL.
4. Psychophysical studies showing that over 75% of women and 99% of men could lift loads at the AL.

The second level of hazard in the *Guide* establishes a *Maximal Permissible Limit* (MPL). This limit is based on:

1. Epidemiological data indicating that musculoskeletal injury rates and severity rates are significantly higher for *most* workers placed on jobs exceeding the MPL.
2. Biomechanical studies indicating that L5/S1 disc compression forces cannot be tolerated over the 6400 N level in most workers, which would be created at the MPL.
3. Physiological studies disclosing that the metabolic energy expenditure rate would exceed 5.0 kcal/min for most workers frequently lifting loads at the MPL.
4. Psychophysical studies showing that only about 25% of men and less than 1% of women workers have the muscle strength to be able to perform lifting above the MPL.

Thus, the AL and MPL permit lifting tasks to be classified into three hazard categories for control planning:

1. Those *above the MPL* should be considered as unacceptable, and engineering controls should be sought to redesign the lifting conditions.
2. Those *between the AL and MPL* are unacceptable without administrative or engineering controls, thus requiring careful employee selection, placement and training, and/or job redesign.
3. Those conditions *below the AL* are believed to represent nominal risk to most workers.

This categorization is illustrated in the *Guide* by considering the act of occasional (F less than .2/min) lifting of objects from a pallet height (V = 15 cm) to carrying height (D = 60 cm). Figure 8.1 depicts the AL and MPL defined maximum weight values that can be lifted for varying object sizes (H is varied between 15 and 80 cm). For such occasional lifts, fatigue avoidance is not of concern. Rather, as described in Chapters 4 and 6, the muscle strength and predicted L5/S1 back compression forces define the limits. Both limits are sensitive to the size of the object. Thus, if the person can maintain the load center of mass (or its effective load on the hands) close to the body, then a weight of greater magnitude can be lifted than if located away from the body. Similarly, varying V, D, or F will influence the weight maximum values for the task.

To allow consideration of the collective effect of the task variables, a prediction equation was defined by the authors of the NIOSH *Guide*. To determine the maximum weight lifting value for a job at the AL, the equation is:

$$\text{AL} = 392\left(\frac{15}{H}\right)(1 - (0.004\,|\,V - 75\,|\,))\left(0.7 + \frac{7.5}{D}\right)\left(\frac{1 - F}{F_{\max}}\right) \quad (8.1)$$

and for the MPL:

$$\text{MPL} = 3\text{AL} \quad (8.2)$$

where

AL and MPL are the maximum weight lifting values (in Newtons) for the given job conditions.

H is the horizontal distance (cm) from the load center of mass at the origin of the vertical lift to the midpoint between the ankles, with a minimum value of 15 cm (body interference) and a maximum value of 80 cm (reach distance for most people).

V is the vertical distance (cm) from the load center of mass at the origin of the vertical lift measured from the floor, with no minimum value and a maximum of 175 cm (upward reach for most people).

D is the vertical travel distance (cm) of the object assuming a minimum value of 25 cm and a maximum of 200 cm minus the vertical origin height V. Note: if the distance moved is small (D less than 25 cm), the effect is nominal, so D is set equal to 25 cm.

F is the average frequency of lifting (lifts/minute) with a minimum value for occasional lifts of 0.2 (once every five minutes) and a maximum value defined by both the period of lifting (less than 1 hour or for 8 hours) and whether the lifting involves only arm work or significant body stabilization or movement. The maximal values F_{max} are given in Table 8.3.

Inspection of the prediction equation 8.1 reveals that under optimum conditions, 392 N can be lifted. This would occur (1) for occasional lifts (F less than 0.2), (2) when the load is held close to the body ($H = 15$ cm), (3) at carrying height ($V = 75$ cm), and (4) when the load is not lifted far (D less than 25 cm). Any deviation from these optimal conditions results in a predicted decrease in lifting capability. In essence, each lifting task parameter

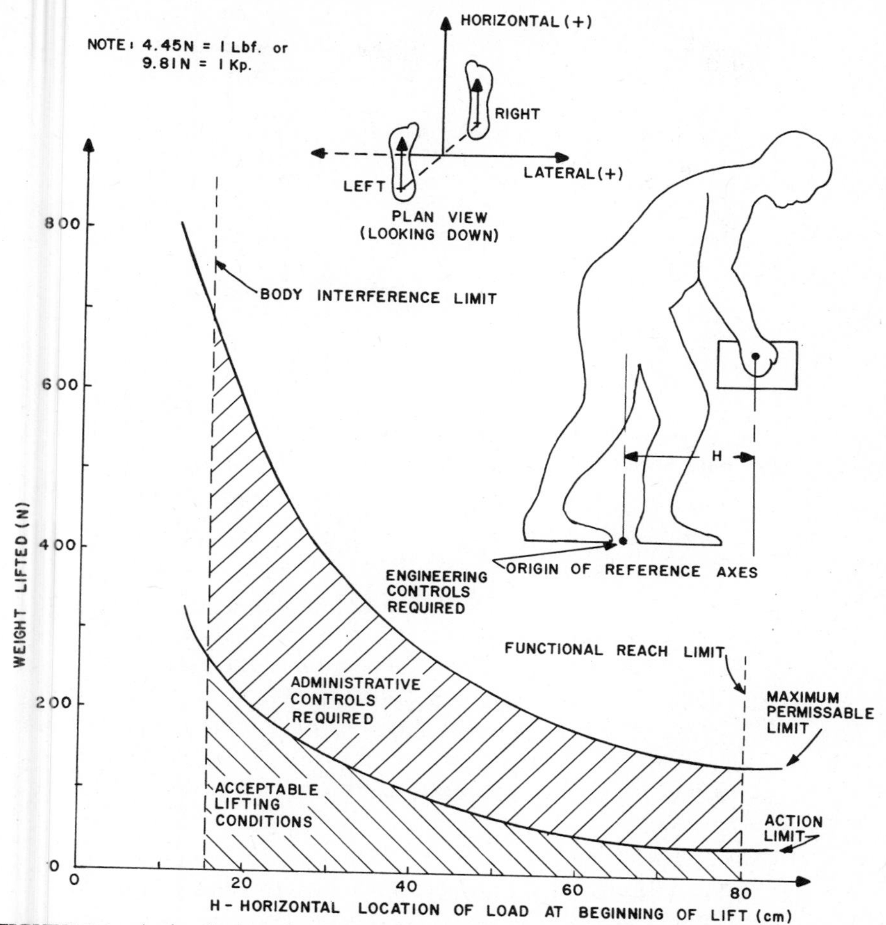

FIGURE 8.1 Action limit and maximum permissible limit for different horizontal location of loads lifted from floor (V = 15 cm) to knuckle height (D = 60 cm) on an infrequent basis (F < .2).

in the equation has a multiplicative discounting effect. Thus equation 8.1 can be expressed as

$$AL = 392(HF)(VF)(DF)(FF) \tag{8.3}$$

where

HF is the discounting factor due to the *horizontal* location of the load at the beginning of the lift.

TABLE 8.3

Maximum Frequency of Lifts/Minute Allowed F_{max} for Different Postures (Hand Vertical Locations) and Lifting Periods (from NIOSH: *Work Practices Guide for Manual Lifting*, 1981)

Duration of Lifting Period	$V > 75$ cm (Standing)	$V \leq 57$ cm (Stooped)
1 hr	18	15
8 hr	15	12

VF is the discounting factor due to the *vertical* location of the load at the beginning of the lift.

DF is the discounting factor due to the *distance* the load is lifted.

FF is the discounting factor due to the *frequency* of the lifts.

All of the discounting variables have maximum values of 1.0, which are achieved at the optimum conditions (equation 8.1). The values of the discounting factors are given in Figure 8.2. Inspection of the graphs reveal that the horizontal location H and frequency of lift F factors can exhibit the greatest discounting effect. Thus job evaluations must give these two factors careful consideration. The next most important factor is the vertical location of the load V at the initiation of the vertical lift, followed by the distance D that the load is moved.

8.2.3 NIOSH Recommendations for Control Lifting Hazards

The NIOSH *Guide* contains several examples of how jobs can be evaluated with reference to the MPL and AL values. Such an evaluation is the first phase of an effective control program (as described in Chapter 7). The *Guide* continues by describing engineering and administrative controls that are often appropriate in reducing the hazard levels. The *engineering controls* involve either reducing the weight of the load lifted or changing workplace/container dimensions to optimize the H, V, and D effects. Concomitant concern is required to minimize the frequency or period of lifting.

Administrative controls advocated in the *Guide* are of two types—improved worker selection and placement strategies, and improved worker training. Objective assessment of a person's specific capability to perform heavy lifting is recommended (as described in Chapter 13). Worker training is also recommended to reduce lifting stresses. Such training requires that an individual have personal knowledge of:

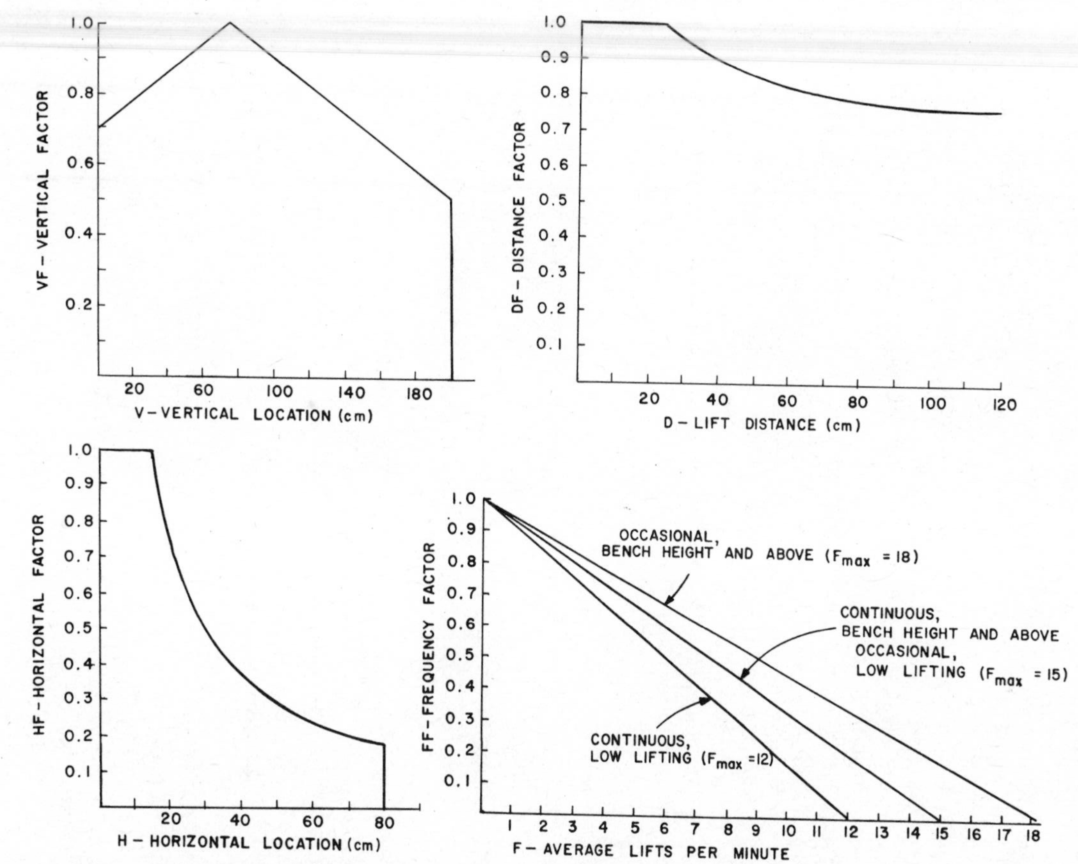

FIGURE 8.2 Graphical depiction of discounting factors in equation 8.1 (from NIOSH *Work Practices Guide for Manual Lifting*, 1981).

TABLE 8.4

Considerations in Developing Manual Material Handling Worker Training Program
(adapted from NIOSH: *Work Practices Guide for Manual Lifting*, 1981)

The individual involved in lifting loads above the *Action Limit* on a job should be made aware of:

1. Past injury experience in job or similar jobs within organization.
2. Basic principles of biomechanics (e.g., use of levers, gravity, friction, momentum etc.), which would help in reducing stresses.
3. Kinesiological effects on the body (e.g., how muscles stabilize spine, move extremities, create pressure within torso).
4. Awareness of individual body strengths and weaknesses (e.g., his or her own lifting capacity).
5. Avoidance of an unexpected situation (e.g., foot slipping, tripping, sudden loss of grip, snagging clothing, use of foot guards, ect.)
6. Development of lifting skills (e.g., keeping body close to load, not twisting with load, using smooth lifts, keeping the load within individual capacity, planning the motion trajectory, etc.)
7. Use of handling aids (e.g., how hoists, lifting platforms, conveyors, hooks, etc. can be used to reduce stresses).

---

1. The risk of injury in the job due to lifting in a careless or unskilled fashion.
2. Lifting methods by which one can reduce unnecessary stress.
3. His or her physical capacities to perform required lifts.

Such training involves several specific considerations, a few of which are listed in Table 8.4. Further discussion of personal work methods used in lifting can be found in papers by Davies (1976) and Garg and Herrin (1979).

8.2.4 Comments on the Status of the NIOSH Lifting Guide

The NIOSH *Guide* is a recent effort to control one aspect of manual materials handling problems—namely, that associated with the simple act of lifting a load in the sagittal plane. It is an attempt to be more comprehensive than previous efforts relative to (1) job evaluation methods, (2) criteria used for the limits, and (3) control strategies.

It is too short a period since publication of the *Guide* to discern its effect on controlling musculoskeletal disorders in industry. Recent prevention programs advocated by Snook (1978), Chaffin and Park (1973), Tichauer (1978), Davis and Stubbs (1978), and Garg and Ayoub (1980) concentrate on setting upper limits to the load that can be lifted by most workers in various postures and with different frequencies. In this sense they advocate an MPL. These proponents suggest, through reference to biomechanical, psychophysical, and restrospective epidemiological studies, that re-engineering of jobs to reduce heavy load lifting could significantly decrease both the incidence and severity of musculoskeletal disorders. The magnitude of such improvement in the injury and illness statistics is not estimated, however, nor have controlled, longitudinal, demonstration experiments been performed.

The use of improved employee selection and training procedures, (i.e., employment of an Al with administrative controls) also has advocates—Davies (1976), Garg and Herrin (1979), and Keyserling et al. (1980) to name a few. Once again, however, longitudinal controlled experiments have not been performed to allow precise estimates of the impact of various types of administrative control programs. (See also Chapter 13.)

In summary, one must conclude that the NIOSH *Guide* is another attempt to deal with a very complicated and serious occupational health and safety problem. It differs from past attempts in that it synthesizes data from many different studies, and advocates both engineering and administrative controls. Students of occupational biomechanics should be thoroughly knowledgeable about this new attempt, and should assist in evaluation of its application and its improvement in the future.

8.3 LOAD PUSHING AND PULLING CAPABILITIES

Unlike the act of sagittal-plane lifting, pushing and pulling capabilities have been studied only within a very limited scope. Furthermore, estimates of the number of injuries that occur during pushing or pulling of loads are not complete, though approximately 20% of overexertion injuries have been associated with pushing and pulling acts (NIOSH, 1981). This underestimates the seriousness of these problems, however, in that foot-slip potential also is very high while performing a pushing or pulling act, as discussed in Section 6.4.3 on biomechanical models. If the foot does slip, the probability of a person falling and suffering an impact injury, possibly to the head, is high. It should be remembered that one of the leading causes of non-vehicle related deaths in industry is slipping and/or falling (National Safety Council, 1978).

It would appear from the work of Fox (1967) and Kroemer and Robinson (1971) that the effect of friction (μ) on push and pull static strength capability is of primary importance. They collectively showed that healthy young males can only exert a mean force of approximately 200 N if μ is aboout 0.3. With μ greater than 0.6, the mean push or pull strength capability increases to 300 N for the same group, according to Kroemer and Robinson (1971). Bracing one foot with the use of the back to apply force (rather than using the hands) further increased the static push force capabilities, according to Kroemer (1969) and Fox (1967).

Martin and Chaffin (1972), Ayoub and McDaniel (1974), Lee (1982), and Davis and Stubbs (1978) reported that the vertical height of the handle against which one pushes and pulls is of critical importance. In the experimental strength study of Ayoub and McDaniel (1974), the elbows and rearward knee were kept straight for the exertions. This resulted in the recommendation that the optimal height for a handle to be pushed or pulled should be approximately 91 to 114 cm (i.e., about hip height for males), above the floor. Davis and Stubbs (1978) developed recommendations for pushing and pulling limits based on abdominal pressure measurements. These experiments also disclosed a larger force capability when the hands were at hip height than when raised to shoulder or above.

By using a biomechanical strength model, such as described in Chapter 6, Martin and Chaffin (1972) predicted that maximum push or pull forces could be obtained on a high-traction surface with the hands in a slightly lower position (about 50 to 90 cm) from the floor. This lower posture allows a person to lean further forward when pushing, or backward when pulling, effectively utilizing body weight to assist in applying the hand forces. Unfortunately, with such extreme postures, the risk of falling forward or backward is greatly increased if the foot suddenly slipped. In this regard, Lee (1982) performed a set of dynamic push and pull experiments (Chapter 6) and found that the predicted compression forces were less when the hands were approximately 109 cm above the floor, compared with either 152 cm or 66 cm.

It was also reported by Kroemer and Robinson (1971) that a large, 95 percentile male, pushing with less than maximum force, would require a rearward foot–ankle location of approximately 165 cm from the fixture being pushed. This result was for subjects utilizing a high-traction surface and lean-forward posture. A recent experimental study of push and pull postures by Chaffin et al. (1983) showed that an average-sized man would probably require approximately 120 cm of horizontal clearance from the rearward ankle to the hands or fixture being pushed. For pulling activities, Ayoub and McDaniel (1974) reported that it is best to place the ball of the forward foot close to or under the handle being pulled, thus allowing the person to lean back significantly, pivoting on the forward foot. If a slip does occur, the rearward foot would presumably catch the person before falling. Figure 8.3 depicts postures that would allow high push and pull capabilities.

An attempt at combining the various data available to form the basis for a design limit is presented in Table 8.5. The values selected assumed that a person (1) can achieve postures similar to those in Figure 8.3, (2) is required to exert the force levels occasionally for a short period (less than 6 sec), and (3) has a coefficient of friction of at least 0.5 at the feet. The strength values represent what at least 90% of the designated population achieved. A typical ratio of the standard deviation divided by the mean for such data is 0.3 (coefficient of variance). Thus, mean strength values are about double the 90 percentile limits quoted in Table 8.5.

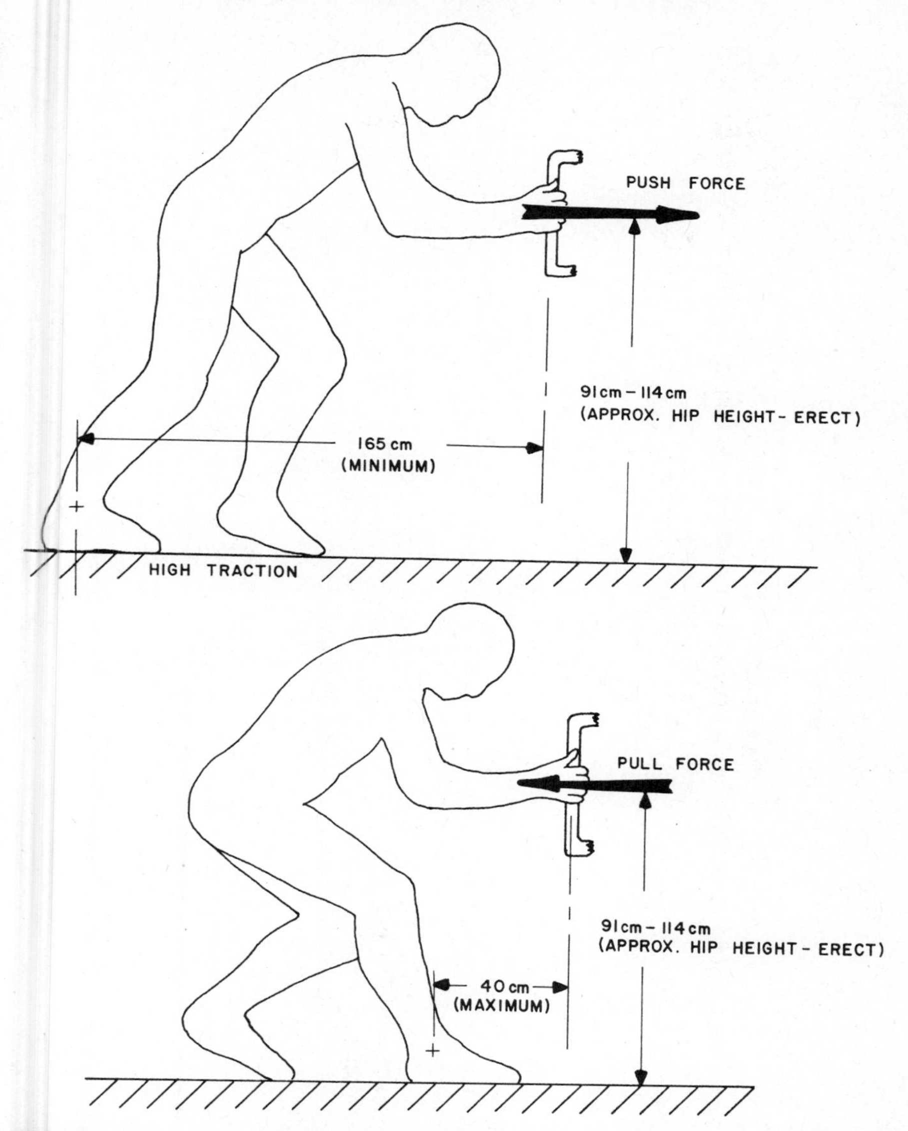

FIGURE 8.3 Typical postures and dimensions required for maximum push and pull forces (adapted from Chaffin, et al, 1983).

As Kroemer and Robinson (1971) reported, if the back or foot is based against an object, average push or pull strengths for young adult males can be expected to reach 500 N. As discussed, handle heights higher than those displayed in Figure 8.3 would decrease the strength values, as would reduced traction.

TABLE 8.5
Approximate Limits (N) to Accomodate Workers Performing Occasional Pushing and Pulling Activities in Good Postures and on Varied Traction Surfaces

Sources	Criteria Applied	Approximate Population Age (Years)	No. of Subjects	Pushing		Pulling	
				Male	Female	Male	Female
Davis & Stubbs (1978)	Abdominal pressure limit of 12kN/m^2	20–60	No Data	235	ND	392	ND
Lee (1982)	Required $\mu = 0.5$ and L5/S1 comp. force <3400 N	NA	Model value	200	ND	200	ND
Kroemer & Robinson (1971)	Static strengths capable of being exceeded by 95% of male subjects, $\mu = 0.6$	18–25	28	200	ND	ND	ND
Snook (1978)	Psychological peak forces capable of being exceeded by 90% of male and female workers on high-traction surface	30 (avg)	78	230	190	270	220
Ayoub & McDaniel (1974)	Static strengths of 50% of subjects on high-traction surface	19–23	46	360	230	400	290

In summary, the need to understand pushing and pulling activities in industry is recognized since many overexertion and fall injuries appear to be related to such activities. It is also clear that many biomechanical factors interact to alter push and pull capabilities. At present, there is only a limited amount of data and modeling of these common activities; thus, any design limits must be carefully interpreted.

8.3.1 Foot-Slip Prevention

Foot-slip related injuries are common in push and pull type of manual exertions as well as in other work activities, particularly when a sudden change in body movement is involved. As discussed briefly in Section 6.4.3, a "safe" coefficient of friction is dependent on the type of task being performed. If heavy carts are being pushed or pulled, the coefficient of friction required between the shoe sole and floor may be greater than 0.8. If people are walking on a level surface with a normal pace and no external loads, a reasonable, safe coefficient of friction would be about 0.5, measured under static conditions.

It should be noted that considerable controversy has developed recently in the scientific community regarding the basic interpretation and use of such *static* coefficient of friction values for design purposes. Some experts in the field now believe that *dynamic* coefficient of friction values are more appropriate because a person's foot is moving as it strikes the ground. These experts point out that with certain combinations of floor material, shoe material, and surface conditions, the dynamic coefficient of friction will vary considerably from that measured by static means alone. Further, the means to measure the dynamic coefficient of friction is far from standardized (Andres, 1983).

Because of these complexities, the designation of "safe" floor and shoe conditions is not possible now with any degree of precision. After reviewing a great deal of data obtained by various researchers, the French National Institute of Safety proposed the qualitative guidelines shown in Table 8.6. They are presented here to illustrate some of the options (and trade-offs) that must be considered in this matter.

8.4 ASYMMETRIC LOAD HANDLING

Based on both biomechanical models and experimental strength studies, symmetric load handling (i.e., where the load is moved in the mid-sagittal plane with both hands) is recommended, as opposed to asymmetric handling (i.e., one hand and/or with load at side of body), discussed by Tichauer (1978). Unfortunately, the studies of asymmetric load handling are few in number, because of the experimental and biomechanical modeling complexities associated with three-dimensional force analysis. It is clear, how-

TABLE 8.6

General Consensus Guidelines for Floor and Shoe Materials as Proposed by French National Institute of Safety (from *National Safety News*, 1974)

Floor		Surface Condition of the Floor		
		Dry	Wet	Greasy
Ceramic tile	Recommended	Plastic (vinyl) and neoprene. Supple rubber (white non-hardened) preferably with shallow tread.	Plastic (dense vinyl) and neoprene. Non-crepe and non-hardened rubber, preferably with pronounced tread. Leathers.	Plastic (except dense vinyl). Vegetable tanned leather.
	Not recommended	Leathers (especially vegetable tanned). Hardened rubber. Crepe rubber.	Hardened rubber Crepe rubber.	Plastic (dense vinyl only) and neoprene. Leathers (except that advised above). All types of rubbers.
Synthetic silcrete slab and corundum concrete slab	Recommended	Plastic (vinyl) and neoprene. Supple rubber (white, non-hardened).	Plastic (dense vinyl) and neoprene Non-crepe and non-hardened rubber, preferably with pronounced tread. Leathers.	Plastic (dense vinyl). Hardened rubber. Leathers.
	Not recommended	Leathers (especially vegetable tanned). Hardened rubber. Crepe rubber.	Hardened rubber. Crepe rubber.	Non-hardened rubber. Neoprene.

Concrete	Recommended	Plastic (vinyl) and neoprene. Supple rubber (white, non-hardened).	Plastic (dense vinyl) and neoprene. Non-crepe rubber and non-hardened rubber, preferably with pronounced tread. Leathers.	Plastic (dense vinyl). Hardened rubber. Leathers.
	Not recommended	Leathers (especially vegetable tanned). Hardened rubber. Crepe rubber.	Hardened rubber. Crepe rubber. Supple rubber.	Supple or non-hardened rubber. Neoprene.
Roughened concrete	Recommended	Plastic (vinyl). Supple rubber (white, non-hardened). Crepe rubber. Chrome retanned leather.	Plastic (dense vinyl) and neoprene. Non-crepe and non-hardened rubber, preferably with pronounced tread. Crepe rubber. Leathers.	Plastic (dense vinyl). Hardened rubber. Leathers.
	Not recommended	Vegetable tanned leather. Hardened rubber.	Hardened rubber.	Non-hardened rubber. Neoprene.
Wood	Recommended	Plastic (vinyl) and neoprene. Supple rubber (white, non-hardened), preferably with shallow tread.	Plastic (dense vinyl) and neoprene. Non-crepe and non-hardened rubber, preferably with pronounced tread. Leathers.	Hardened rubber. Leathers.
	Not recommended	Leathers (especially vegetable tanned). Hardened rubber. Crepe rubber.	Hardened rubber. Crepe rubber.	Plastic (dense vinyl) and neoprene. Non-hardened rubber.

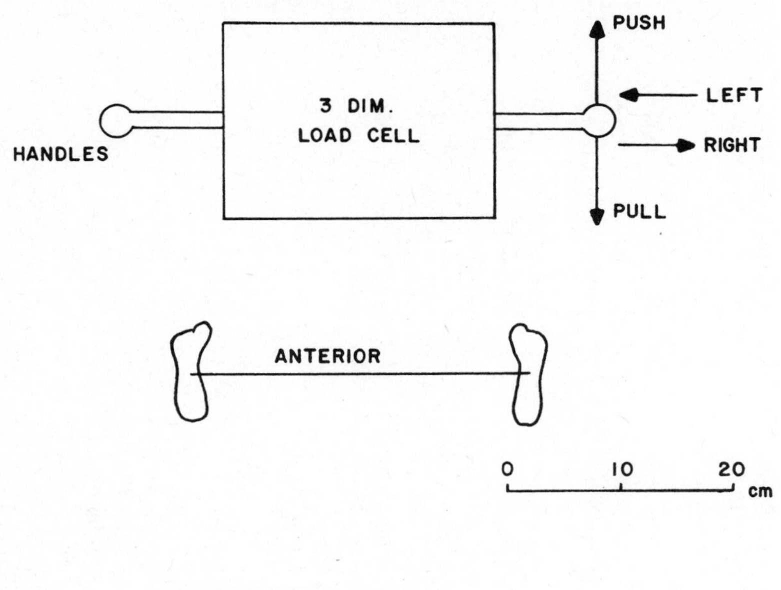

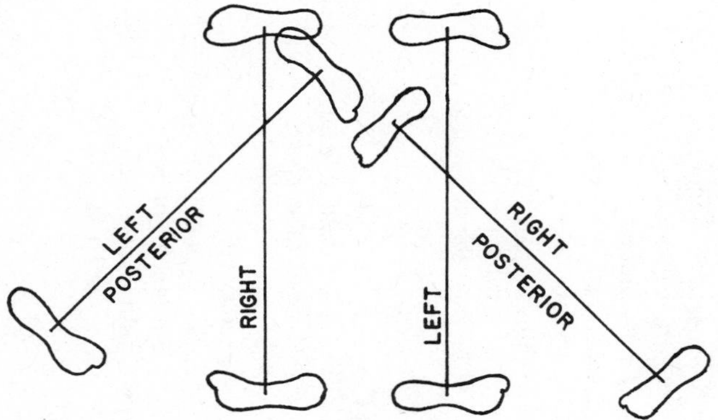

FIGURE 8.4 Plan view of foot positions relative to load cell used by Warwick et al. (1980) to measure static strengths during symmetric and asymmetric exertions.

ever, that asymmetric load handling results in asymmetric muscle activities and loads on the spine. These aspects have been discussed in Chapter 6. Asymmetry in load also influences performance strength and can result in a poorer postural stability.

One recent study compared asymmetric pushing and pulling static strengths of 29 male subjects age 21 to 73 years (Warwick, Novak, Schultz, and Berkson, 1980). The subjects performed isometric exertions against a load cell mounted at shoulder (142 cm) and knee (60 cm) heights, and with the cell either (1) directly in front of the body (''anterior'' position), (2)

TABLE 8.7
Mean Static Forces of Male Subjects in Various Symmetric Asymmetric Exertions with Foot Locations Depicted in Figure 8.4
(Warwick et al., 1980)

Hands Used	Foot Positions	Lift Up	Press Down	Push Forward	Pull Backward	Push Right	Push Left
(a). Mean Magnitudes of the Forces Exerted with the Measurement Handles at Shoulder Height (N)							
Both	Anterior	386	340	222	170	156	167
	Right	239	246	257	193	117	141
	Right posterior	239	240	236	176	112	117
Left	Right	103	142	147	135	88	87
	Right posterior	77	119	143	105	74	101
Right	Right	136	160	190	147	112	113
	Right posterior	135	161	184	138	85	95
(b) Mean Magnitudes of the Forces Exerted with the Measurement Handle Piece at Knee Height (N)							
Both	Anterior	275	372	212	225	194	179
	Left	127	321	278	218	172	139
	Right	134	333	278	179	136	154
Left	Left	158	237	188	175	137	113
	Right	113	217	169	160	114	113
Right	Left	107	107	170	163	141	108
	Right	101	272	200	147	135	162

directly to the left or right side of the body (''left'' or ''right'' position), or (3) behind and to the left or right (''left posterior'' or ''right posterior'' position). A plan view of the foot placements relative to the load cell placement is depicted by the investigators in Figure 8.4. In essense, the ''anterior'' position of the feet required a rather symmetric mid-sagittal plane exertion, while the ''right'' or ''left'' foot placements required the torso to be twisted towards the load cell. The ''posterior'' foot placements required even greater torso twisting to reach the handles on the load cell. Maximum isometric exertions were measured in six orthogonal directions as indicated in the figure.

A summary of the mean strength values is given in Table 8.7. Inspection of this table reveals the extreme (fivefold) effect that posture and force direction can have on strength performance. In general, the activities required in an asymmetric posture resulted in decreased strengths (about 20%) when the load cell was laterally to the left or right of the person. A decrease of approximately 26% was found when the load cell was behind and to the right of the person, compared with when it is directly in front. Clearly, the precise effect of posture cannot be discussed independently of the force direction

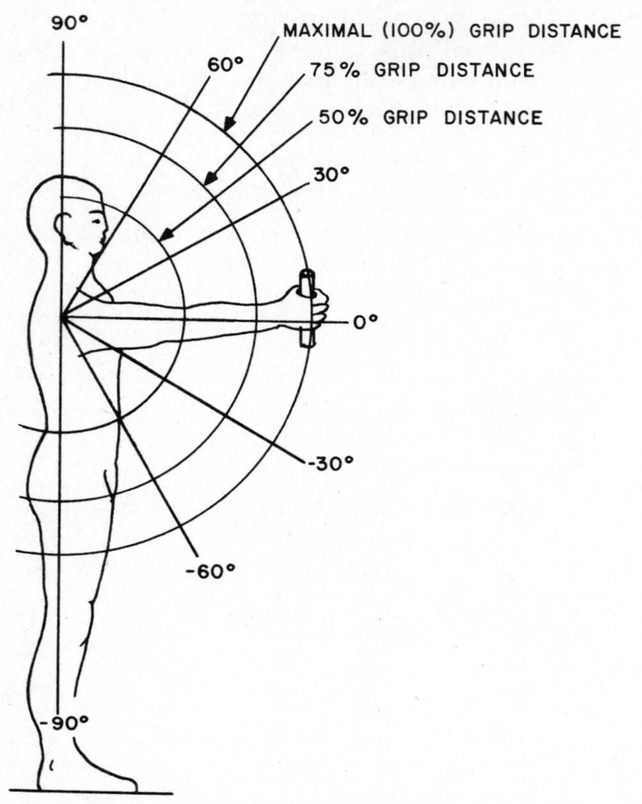

FIGURE 8.5 Arm postures used in one arm strength study. The results are given in Table 8.7 (Rohmert, 1966).

(i.e., lift, press down, push, etc.) as a large interaction is evident in the mean values.

A study of one-handed strengths in the standing position was made by Rohmert (1966). Five healthy young males of average anthropometry were used in the experiment. Right-hand strengths were measured in various positions, as indicated in Figure 8.5. Six strengths were measured while the subjects stood erect with feet parallel and 30 cm apart. The means of the resulting strength data are shown in Table 8.8.

The results once again demonstrate the complexity of asymmetric exertions, with postures and force directions interacting to create a large (almost fivefold) variation in the mean values. Clearly, if the left hand and the feet were allowed to assume different "bracing" configurations, the values generally would be expected to increase and be more varied than in Table 8.8.

The consideration of bracing a seated person to allow one-arm maximum push and pull exertions is experimentally addressed by Laubach (1978). In

TABLE 8.8
Maximal Right-Handed Static Forces Exerted on a Vertical Hand-Grip by Standing Young Male Subjects (Rohmert, 1966)

Type of Exertion	Arm Angle (deg)	Location of the Handgrip (see Figure 8.5): At Percentages of Maximal Grip Distance 50%	75%	100%
		Force (N)		
Push outward, horizontal	30	71	108	142
	0	133	156	178
	−30	125	135	142
	−60	125	142	160
Pull inward, horizontal	30	85	98	116
	0	102	116	129
	−30	125	134	138
	−60	102	125	151
Push to the left, horizontal	30	156	136	107
	0	187	147	107
	−30	187	151	116
	−60	147	136	116
Push to the right, horizontal	30	107	98	93
	0	136	111	89
	−30	147	120	98
	−60	111	102	93
Lift, vertical	30	125	107	85
	0	151	116	80
	−30	222	178	125
	−60	280	227	182
Press down, vertical	30	338	258	182
	0	249	178	147
	−30	156	147	136
	−60	173	160	142

his experiment, aircraft seats were used with full shoulder and lap belts to brace the torso. The seat back was tilted at various angles and the seat moved fore and aft to alter leg postures. The resulting static strengths of Air Force recruits were found to be greatly affected in a complex fashion by each of these changes. It should also be remembered that in some situations the support to the body is reduced—for example, at large back-rest angles—and, therefore, a push-off effect becomes less effective.

In summary, asymmetric exertions appear to be more hazardous to the musculoskeletal system than symmetric exertions, as discussed by Kumar (1980). It is conceded, however, that biomechanical and epidemiological data do not exist to qualify this belief and to define specific limits for asymmetric activities. Likewise, strength performance is decreased in asymmetric materials handling tasks, but norms for the population are not yet available, due to limited experimental and biomechanical modeling results. Perhaps even more important is the lack of knowledge regarding dynamic loading of the musculoskeletal system, which often is concomitant with an asymmetric lift of an object (e.g., when performing a lift with one hand to stoop down and pick up a small, light object from the floor). Many times, acute low-back pain is associated with such dynamic, asymmetric activities.

8.5 SUMMARY OF MANUAL MATERIALS HANDLING LIMITS

Though there is a new, comprehensive program from NIOSH to control problems associated with symmetric load lifting, it is too early to predict its effectiveness. "Safe lifting" programs have been advocated in the past by various groups with apparent mixed long-term success, despite the acknowledged urgent need to control injuries to the musculoskeletal system today. Perhaps both the biomechanics knowledge and the expertise necessary to apply it did not exist in the past. Perhaps it still doesn't. Clearly, when discussing any materials handling task other than symmetric load lifting in the sagittal plane, large "gaps in knowledge" become evident.

Hopefully, this text will not only assist those concerned with control of injuries associated with manual materials handling, but will motivate the necessary research to better define the basis for reasonable human limits for such activities in the future.

REFERENCES

Andres, R. O., "European Laboratory Techniques and Devices for Dynamic Coefficient of Friction Measurements," Center for Ergonomics Technical Report, The University of Michigan, Ann Arbor, MI, 1983.

Astrand, P. O. and K. Rodahl, *Textbook of Work Physiology,* McGraw-Hill, New York, 1970.

Ayoub, M. M. and McDaniel, J. W. "Effect of Operator Stance on Pushing and Pulling Tasks," *AIIE Tr.,* **6,** 185–95 (1974).

Ayoub, M. M., A. Mital, G. M. Bakken, S. S. Asfour, and N. J. Bethea, "Development of Strength and Capacity Norms for Manual Materials Handling Activities: The State of the Art," *Human Factors,* **22**(3), 271–283 (1980).

Bonjer, F. H., "Temporal Factors and Physiological Load," in W. T. Singleton, J. G. Fox and D. Whitfield, Eds., *Measurement of Man at Work,* Taylor Francis, London, 1971.

Chaffin, D. B., R. O Andres, and A. Garg, "Volitional Postures during Maximal Push/Pull Exertions in the Sagittal Plane," *Human Factors,* **25**(5), 541–550 (1983).

Chaffin, D. B. and K. S. Park, "A Longitudinal Study of Low-Back Pain as Associated with Occupational Lifting Factors," *AIHA J,* **34,** 513–525 (1973).

Davies, B. T., "Training in Manual Handling and Lifting," in C. G. Drury, Ed., *Safety in Manual Materials Handling,* U.S.D.H.H.S. (NIOSH), No. 78-185, Cincinnati, OH, 1976.

Davis, P. R. and D. A. Stubbs, "Performance Capacity Limits," *Applied Ergonomics,* **9**, 33–38 (1978).

Fox, W. F., "Body Weight and Coefficient of Friction Determinants of Pushing Capability," Human Engineering Special Studies Series, No 17, Lockheed Co., Marrietta, GA, 1967.

Garg, A. and M. M. Ayoub, "What Criteria Exist for Determining How Much Load Can Be Lifted?," *Human Factors,* **22**(4), 475–486 (1980).

Garg, A. and D. B. Chaffin, "A Biomechanical Computerized Simulation of Human Strengths," *AIIE Tr.,* **7,** (March) 1–15 (1975).

Garg, A., D. B. Chaffin, and G. D. Herrin, "Prediction of Metabolic Rates for Manual Materials Handling," *AIHA J.,* **39**(8), 661–674 (1978).

Garg, A. and G. D. Herrin, "Stoop or Squat: A Biomechanical and Metabolic Evaluation," *AIIE Tr.,* **11**(4), 293–302 (1979).

Herrin, G. D., D. B. Chaffin, and R. S. Mach, *Criteria for Research on the Hazards of Manual Materials Handling,* Workshop Proceedings on Contract CDC-99-74-118, U.S. Dept. of Health and Human Services (NIOSH), Cincinnati, OH, 1974.

International Labour Organization, *Maximum Permissible Weight to be Carried by One Worker,* Information Sheet No. 3, Geneva, Switzerland, 1962.

Kamon, E. and M. M. Ayoub, *Ergonomics Guide to Assessment of Physical Work Capacity,* Am. Indust. Hygiene Assoc., Akron, OH, 1976.

Keyserling, W. M., G. D. Herrin, and D. B. Chaffin, "Isometric Strength Testing as a Means of Controlling Medical Incidents on Strenuous Jobs," *JOM,* **22**(5), 332–336 (1980).

Kroemer, K. H. E., *Push Forces Exerted in 65 Common Work Positions,* AMRL-TR-68-143, Aerospace Medical Research Laboratory, Wright-Patterson Air Force Base, Ohio, 1969.

Kroemer, K. H. E. and D. E. Robinson, *Horizontal Static Forces Exerted by Men Standing in Common Working Postures on Surfaces of Various Tractions,* AMARL-TR-70-114, Aerospace Medical Research Laboratory, Wright-Patterson Air Force Base, Ohio, 1971.

Kumar, S., "Physiological Responses to Weight Lifting in Different Planes," *Ergonomics,* **23**(10), 987–993 (1980).

Laubach, L. L., "Human Muscular Strength," in Webb Associates, Ed., *Anthropometric Source Book,* NASA No. 1024, U.S. National Aeronautics and Space Administration, Washington, D.C., 1978.

Lee, K., *Biomechanical Modeling of Cart Pushing and Pulling,* unpublished doctoral dissertation, University of Michigan, Ann Arbor, MI, 1982.

Martin, J. B. and D. B. Chaffin, "Biomechanical Computerized Simulation of Human Strength in Sagittal Plane Activities," *AIIE Tr.,* **4,** 19–28 (1972).

National Institute for Occupational Safety and Health, *A Work Practices Guide for Manual Lifting,* Tech. Report No. 81-122, U.S. Dept. of Health and Human Services (NIOSH), Cincinnati, OH, 1981.

National Safety Council, *Accident Facts,* National Safety Council, Chicago, IL, 1978.

National Safety News, *Shoe Sole Slipperiness Standard Status,* National Safety Council, Chicago, IL, August, 1974.

Petrofsky, J. S. and A. R. Lind, "Comparison of Metabolic and Ventilatory Responses of Men to Various Lifting Tasks and to Bicycle Ergometry," *J. of Applied Physiol.,* **45**(1), 60–63 (1978).

Rohmert, W., *Maximalkräfte von Männern im Bewegungsraum der Arme und Beine,* Köln, Germany: Westerdeutscher Verlag, 1966.

Schanne, F. A., *Three-Dimensional Hand Force Capability Model for the Seated Operator,* unpublished doctoral dissertation, University of Michigan, Ann Arbor, MI, 1972.

Snook, S. H., "*The Design of Manual Handling Tasks,*" *Ergonomics,* **21**(12), 963–986 (1978).

Tichauer, E. R., *The Biomechanical Basis of Ergonomics,* Wiley-Interscience, New York, 1978.

Troup, J. D. B., "Manual Materials Handling—The Medical Problem," in C. G. Drury, Ed., *Safety in Manual Materials Handling,* U.S.D.H.H.S. (NIOSH) No. 78-185, Cincinnati, OH, 1978.

Warwick, D., G. Novack, and A. Schultz, "Maximum Voluntary Strengths of Male Adults in Some Lifting, Pushing and Pulling Activities," *Ergonomics,* **23**(1), 49–54, 1980.

9

GUIDELINES FOR SEATED WORK

9.1 GENERAL CONSIDERATIONS IN SITTING POSTURES

Sitting has been defined as a body position in which the weight of the body is transferred to a supporting area mainly by the ischial tuberosities of the pelvis and their surrounding soft tissues (Schoberth 1962). Depending on the chair and posture, some proportion of the total body weight will also be transferred to the floor, as well as to the backrest and armrests of the chair (Figure 9.1).

When considering biomechanical aspects of sitting, the spine is particularly important, but the lower and upper extremities must also be considered. Although there are large individual variations among people in the shape of the spine when assuming different standing and sitting postures and when using different chairs, there are common distinguishing features (Åkerblom, 1948; Keegan, 1953; Schoberth, 1962; Carlsöö, 1963, 1972; Andersson et al., 1979). To understand some of these features, a short review of the gross anatomy of the spine is now presented.

Functionally, the vertebral column consists of four parts (Figure 9.2). Two mobile segments, the cervical and lumbar spine, are below and above the relatively immobile thoracic spine. The lumbar spine is attached to the sacrum, which is almost completely fixed to the pelvis. When a person stands erect, the vertebral column is normally straight in the anteroposterior aspect and curved in the later aspect, producing a compound curvature referred to as *cervical lordosis, thoracic kyphosis,* and *lumbar lordosis*. The lumbar curve is lordotic partly because the vertebrae and discs are thicker anteriorly than posteriorly and partly because the upper surface of the sacrum is at an angle to the horizontal plane. As the sacrum is fixed to the pelvis, it follows that a rotational movement of the pelvis influences the shape of the lumbar

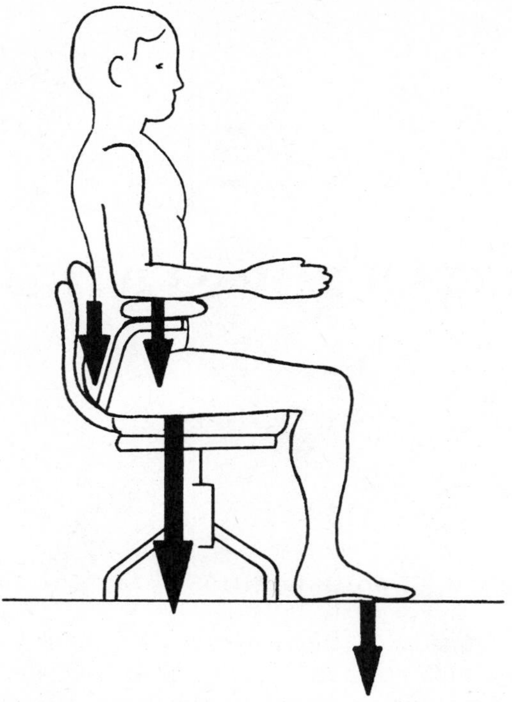

FIGURE 9.1 When sitting, the main part of the body weight is transferred to the seat. Some weight is also transferred to the floor, the backrest, and the armrests.

spine. A forward rotation of the pelvis causes the lumbar spine to move toward increased lordosis in order to maintain an upright trunk posture. When the pelvis is tilted backward, on the other hand, the lumbar spine tends to flatten, and sometimes a kyphosis can develop. When the knees and hips are flexed in sitting, the pelvis is rotated backward and the lumbar lordosis flattens (Figure 9.3).

Three different general types of sitting postures can be distinguished—anterior, middle, or posterior sitting posture (Figure 9.3). They have been defined by Schoberth (1962) according to the location of the center of mass of the body, which affects the proportion of body weight transmitted to the floor by the feet. The postures also are different with respect to the shape of the lumbar spine.

In the middle posture the center of mass is directly above the ischial tuberosities and the floor supports about 25% of the body weight. When relaxed in a middle posture, the lumbar spine is either straight or in slight kyphosis. The anterior (forward leaning) posture can be reached from the middle posture either by a forward rotation of the pelvis with the spine straight, by assuming a slight kyphosis, or by use of little or no rotation of

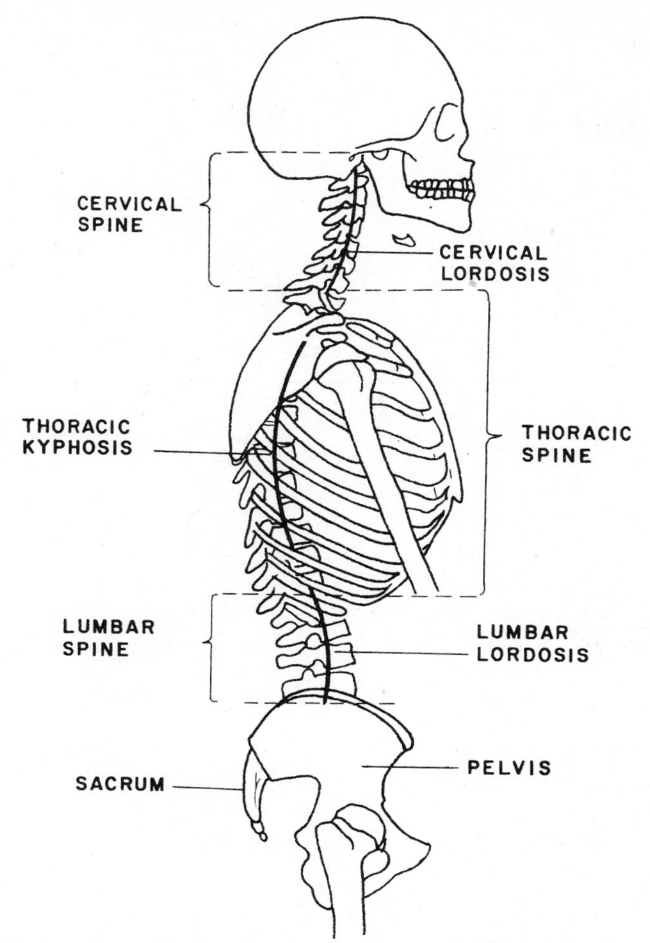

FIGURE 9.2 The spine is divided into four segments—the cervical, thoracic, and lumbar segments, and the sacrum. In the lateral view there is a lumbar lordosis, a thoracic kyphosis, and a cervical lordosis.

the pelvis but with large induced kyphosis of the spine. In this posture the center of mass is in front of the ischial tuberosities and the floor supports more than 25% of the body weight. In the posterior (backward leaning) posture, less than 25% of the body weight is supported by the floor and the center of mass is behind the ischial tuberosities. The posterior posture is often obtained by a backward rotation of the pelvis and simultaneous kyphosis of the spine.

A group of muscles at the back of the thighs—the hamstring muscles—influence the configuration of the lumbar spine and pelvis because they run

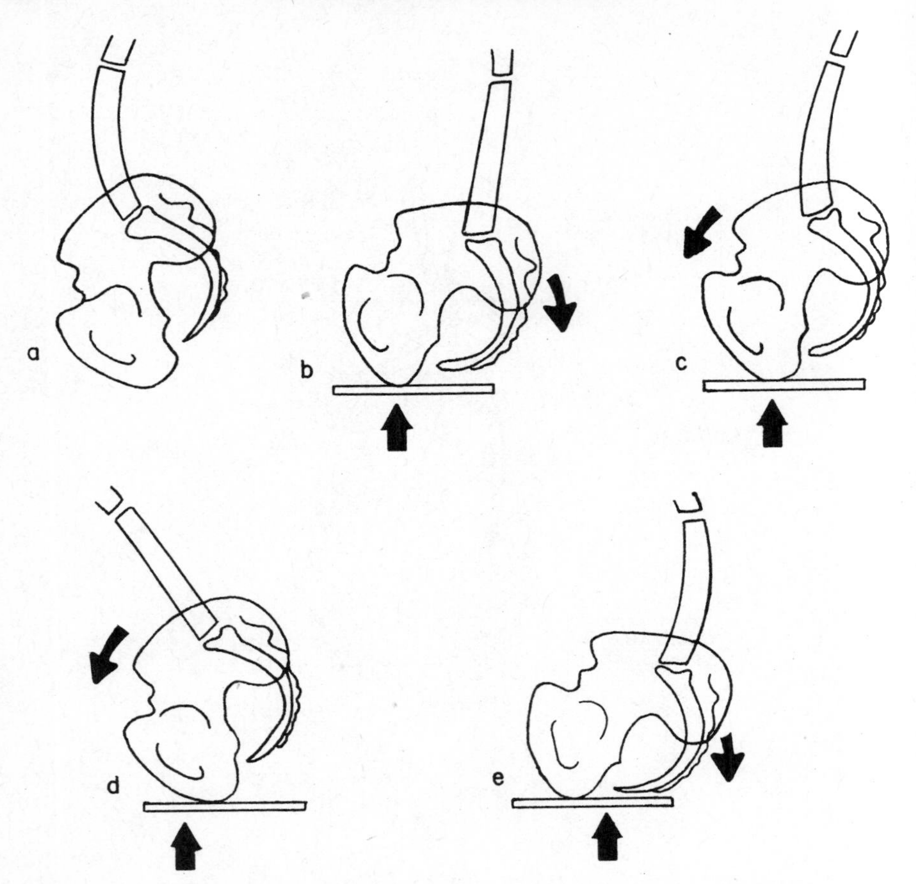

FIGURE 9.3 The pelvis and lumbar part of the spine when (a) standing; (b) sitting relaxed, unsupported in the middle position; (c) sitting erect, unsupported in the middle position; (d) sitting in the anterior posture, (e) sitting in the posterior posture.

from the lower limb to the pelvis, crossing both the hip and knee joints (Figure 9.4). This muscular aspect will be discussed later in the chapter.

When work is performed in sitting postures, the location and slope of the work area have a major influence on the postures of the neck, shoulders, and upper extremities. Thus, the seat should not be considered without taking the work to be performed into account. This will be discussed in depth in Chapter 10.

In general, the posture of a seated person depends not only on the design of the chair, but also on individual sitting habits and on the task to be performed. Anterior sitting postures are adopted most often when desk work is performed, while posterior positions are assumed in chairs with large backrests that incline and are often preferred for resting. The height and

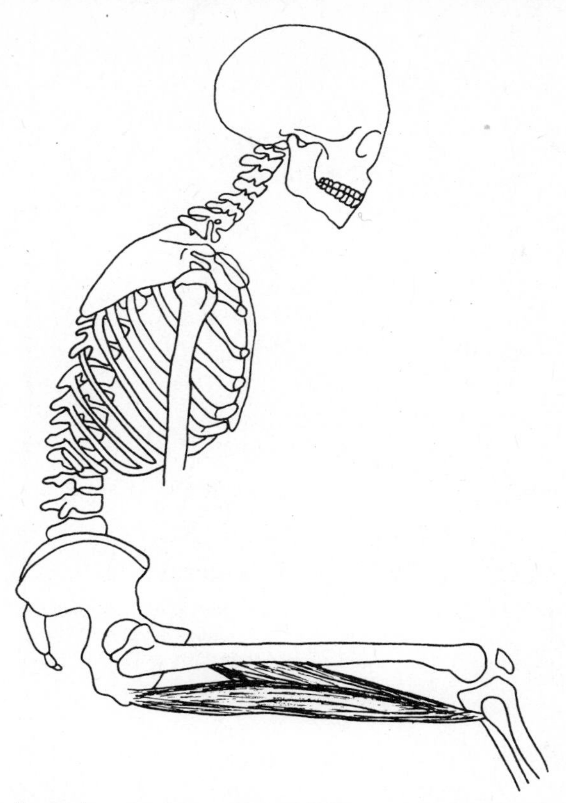

FIGURE 9.4 The hamstring muscles cross two joints (the knee and the hip). Movements of the lower limb can therefore affect pelvic rotation and influence the posture of the spine. (Adapted from Jonsson and Andersson, 1978.)

inclination of the seat of the chair, combined with the position, shape, and inclination of the backrest and the presence of other types of support, combine to influence the resulting posture. Obviously it is important to provide not only a "good" chair, but a chair that is functionally adapted to the task of the occupant. This is particularly important when seated work is considered, because even minor changes in the dimensions of the work space can change the required seated posture considerably.

As it is unlikely that there is a single ideal posture, and furthermore no body posture can be maintained indefinitely, it becomes important that alterations in one's posture also be permitted by the chair. This basic criterion of a good chair was stated by Vernon (1924), and has later been stressed by several investigators (Åkerblom, 1948; Keegan, 1953; Floyd and Roberts, 1958; Carlsöö, 1963; Kroemer and Robinette, 1969; Engdahl, 1971; Kroemer, 1971). To facilitate sitting and standing, an intermediate posture, *semisitting,* is desirable. In *semisitting*, a higher than normal chair is used, usually with

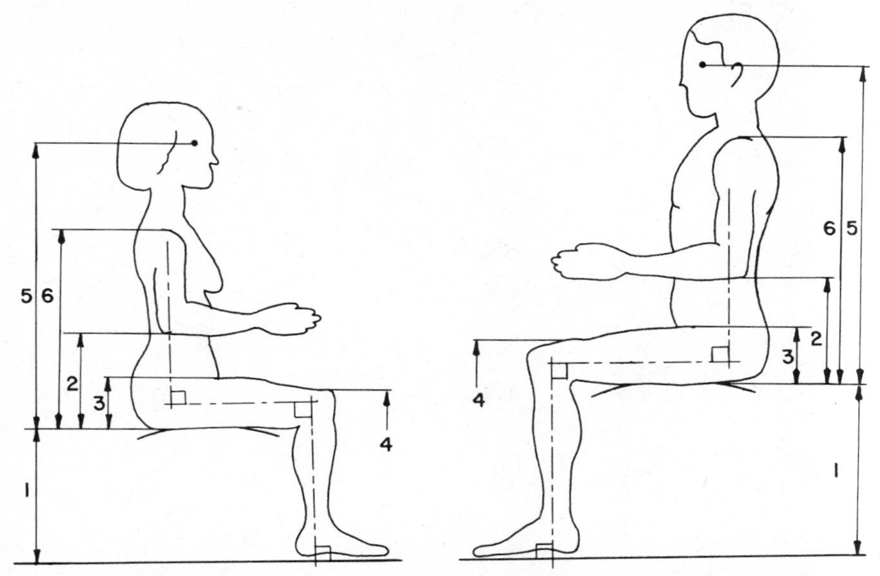

FIGURE 9.5 Illustration of postures and vertical antropometric measurements relevant to sitting. Definitions in Table 9.1: (1) sitting height, (2) elbow height, (3) thigh height, (4) patellar height, (5) orbital (eye) height, (6) shoulder height. (Adapted from Engdahl, 1977.)

a forward sloping seat that a person leans on, dividing the weight bearing between the buttocks and feet.

The advantages of providing a regular sitting posture are (1) that it provides stability required on tasks with high visual and motor control, (2) sitting is less energy consuming than standing, (3) it places less stress on the lower extremity joints, and, (4) it lowers the hydrostatic pressure on the lower extremity circulation. As will be apparent from this chapter, biomechanical considerations must be kept in mind to achieve these advantages, and not cause undue stresses to the back and shoulder as well as to the upper extremities when designing work in the seated posture.

9.2 ANTHROPOMETRIC ASPECTS OF SEATED WORK

General anthropometry has been discussed in Chapter 3. Some additional aspects are now presented as they relate to sitting postures, chairs, and the design of a seated work place.

Anthropometric measurements for the height of seated workers are defined in Table 9.1 and illustrated in Figure 9.5. Additional sagittal depth measurements are also defined in Table 9.1 and illustrated in Figure 9.6. The transverse breadth measurements are defined in Table 9.1 and are shown

TABLE 9.1
Definitions of Sitting Anthropometric Dimensions Displayed in Figures 9.5, 9.6, and 9.7

Illustrated in Figure 9.5

1. *Sitting height:* The vertical distance from the floor to the horizontal midsection of the back of the thigh of a subject sitting with the thigh in contact with the seat, with the popliteal fold 2–3 cm above the seat surface, with a knee flexion angle of 90°, and the side of the bare foot flat on the floor.
2. *Elbow height:* The vertical distance from the floor to the posterior tip of the olecranon when the arm is flexed to 90° at the elbow and the shoulder is in the 0 position. Can be measured from sitting height.
3. *Thigh height:* The vertical height from the floor to the highest part of the thigh. Can be measured from sitting height.
4. *Patellar height:* Vertical distance from the floor to the superior tip of the patella.
5. *Orbital height (eye height):* The vertical distance from the floor to the orbit when sitting with the spine straight. Can be measured from sitting height.
6. *Shoulder height:* The vertical distance from the floor to the superior aspect of the acromion. Can be measured from sitting height.
7. *Hand-grip height:* The vertical distance from the floor to the midpoint of the hanging fist.

Illustrated in Figure 9.6

1. *Internal sitting depth (buttock–popliteal):* The sagittal distance from the posterior aspect of the popliteal fold to the posterior aspect of the buttock.
2. *External sitting depth (buttock–patella):* The sagittal distance from the anterior aspect of the knee to the posterior part of the buttock.
3. *Abdominal depth (buttock–abdomen):* The sagittal distance from the anterior abdominal wall to the posterior part of the buttock.

Illustrated in Figure 9.7

1. *Buttocks width (sitting breadth):* The maximum transverse distance at the buttocks.
2. *Shoulder width (biacromial breadth):* The maximum transverse distance at the shoulders.
3. *External elbow width:* The maximum transverse distance between the tips of the olecrani when the arms are abducted to 90°.

in Figure 9.7. Relevant data from the literature are given in Tables 9.2 through 9.4 for all of these dimensions.

9.3 COMFORT ASPECTS OF SEATED WORK

Measurements of seated comfort have been based on (1) observations of body postures and movements (Grandjean et al., 1960; Branton and Grayson, 1967; Rieck, 1969; Wotzka et al., 1969), (2) observations of task perform-

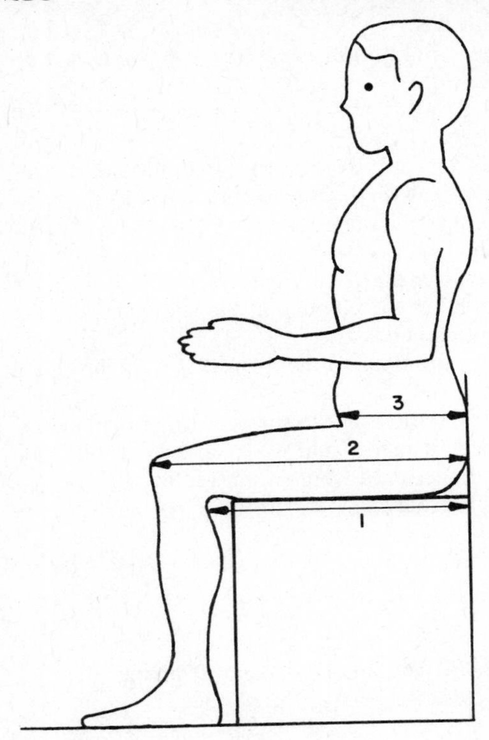

FIGURE 9.6 Sagittal anthropometric measurements: (1) internal sitting depth (buttock–popliteal); (2) external sitting depth (buttock–patella); (3) abdominal depth (buttock–abdomen). Definitions in Table 9.1.

ances (Jones, 1969), or (3) direct subjective ratings of general comfort, of using chair feature checklists, body area comfort rankings, and general comfort rankings (Wachsler and Learner, 1960; Barkla, 1964; Grandjean et al., 1969, 1973; Jones, 1969; Le Carpentier, 1969; Schackel et al., 1969; Wotzka et al., 1969; Hall, 1972). Often these different methods have been combined

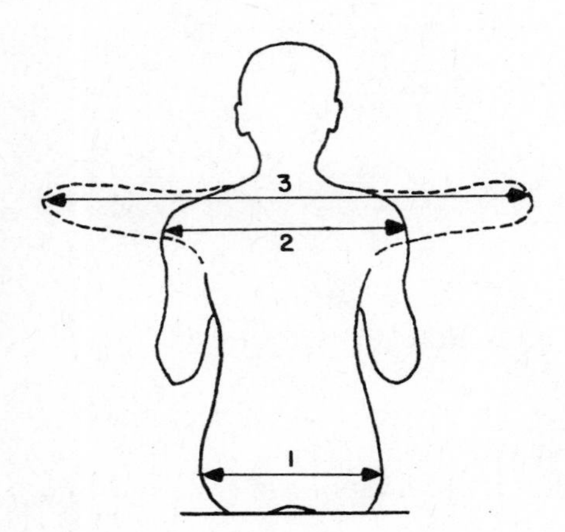

FIGURE 9.7 Transverse anthropometric measurements: (1) buttocks width (sitting breadth); (2) shoulder width (biacromial breadth); (3) external elbow width. Definitions in Table 9.1.

TABLE 9.2
Vertical Anthropometric Measurements in Sitting Postures (For Definitions, See Table 9.1 and Figure 9.5) (after Aldman and Lewin, 1977)

Measurement	Source	Sex	Age	Mean	SD	Percentiles 10	50	90	Subjects (n)
Sitting height	1	MF	18–70	42.4	3.2	38.5	42.1	46.5	947
Elbow height	1	MF	18–70	59.6	3.3	55.4	59.6	63.9	575
Thigh height	1	MF	18–69	56.9	3.5	52.4	56.9	61.6	834
	3	F	—	51.0	3.3	—	—	—	1166
Patellar height	2	M	17–26	55.0	3.2	—	—	—	561
	2	M	25–49	52.8	3.0	—	—	—	86
	2	F	25–49	49.7	3.1	—	—	—	77
Eye height	2	M	25–49	78.5	4.2				87
	3	F	—	73.1	3.3				1166
Shoulder height	1	MF	18–70	99.5	6.0	91.6	99.99	107.5	921

(Drury and Conry, 1982). To improve on the subjective methods, experts in seat evaluations (Jones, 1969) or test subjects with low-back pain (Hall, 1972) have been used. In spite of the numerous studies on seated comfort, no general agreement has been reached on which method is best in terms of precision or reliability (Schackel et al., 1969; Drury and Conry, 1982). Comfort is still an unexplored concept and often defined as the absence of discomfort (Floyd and Roberts, 1958; Wachsler and Lerner, 1960).

The time needed for each subject to sit in a given seat to make an evaluation has also been debated. Wachsler and Learner (1960) found comfort

TABLE 9.3
Horizontal (sagittal) Anthropometric Measurements in Sitting Postures (For Definitions, See Table 9.1 and Figure 9.6) (after Aldman and Lewin, 1977)

Measurement	Source	Sex	Age	Mean	SD	Percentiles 5	95	Subjects (n)
Buttock–Poplitea	2	M	25–49	48.2	3.0	44.2	52.8	87
	2	F	25–49	47.2	2.3	43.7	51.8	112
Buttock–Patella	2	M	70	58.3	3.0	53.6	63.7	163
	2	M	17–26	61.9	3.3	56.2	66.3	273
	2	M	25–49	59.3	3.0	54.4	64.2	87
	2	F	70	56.6	3.0	51.3	61.6	183
	2	F	25–49	56.9	3.0	52.1	61.5	282
	2	F	50–64	57.3	3.2	52.7	63.1	143
	1	MF	18–70	58.7	3.5	—	—	921
Buttock–Abdomen	2	M	25–49	24.1	3.1	19.5	29.0	87
	2	F	25–49	24.6	2.4	20.8	28.7	112

TABLE 9.4
Horizontal Breath Anthropometric Measurements in Sitting Postures (For definitions, see Table 9.1 and Figure 9.7) (after Aldman and Lewin, 1977).

Measurement	Source	Sex	Age	Mean	SD	Percentiles 5	Percentiles 95	Subjects (*n*)
Sitting breadth	2	M	17–26	34.3	2.2	33.4	42.0	586
	2	F	20–24	38.0	2.6	34.3	42.8	170
	2	F	25–49	39.1	2.9	34.6	43.7	279
	2	F	50–64	40.3	3.1	35.4	45.7	143
	1	MF	18–70	37.9	3.3			922
Biacromial Breadth	2	M	17–26	38.5	2.1	34.8	41.8	853
	2	M	25–49	35.7	1.8	32.2	39.0	85
	2	F	20–24	35.8	1.8			88
	2	F	25–49	36.2	1.6			279
	2	F	50–64	35.9	1.8			143
Elbow breadth	2	M	17–26	42.2	3.5	37.5	48.7	586

ratings after five minutes to be as reliable as those obtained after four hours, while Barkla (1964) found at least 30 minutes of sitting necessary for reliable ratings. In summary, comfort is a difficult concept to apply in evaluating a chair. Perhaps more reasonable is an approach that first evaluates a chair against anthropometric and physiologic data, followed by the use of comfort trials and direct workplace observations to evaluate final designs. Since comfort is task-dependent, laboratory comparisons of chairs are often not meaningful. For further reading, please refer to Oborne (1978) and Drury and Conry (1982).

9.4 THE SPINE AND SITTING

As stated in the introduction to this chapter, the configuration of the lumbar spine is altered when sitting down from standing. This creates spinal forces within the spine motion segments that need to be considered. The cervical spine is also influenced by sitting, as the field of vision needed to perform a task may require the head to be in a certain position. We will first concentrate on the lumbar spine and review some of the clinical and biomechanical data on the possible negative effects of various sitting postures to that area of the spine.

9.4.1 Clinical Aspects of Sitting Postures

There are several studies indicating an increased risk of low-back pain in subjects who perform work in a predominantly sitting posture (Hult, 1954;

Lawrence, 1955; Kroemer and Robinette, 1969; Partridge and Anderson, 1969; Magora, 1972). These studies also show an increase in back symptomology in subjects with back pain when required to sit for prolonged periods. Other studies, however, have not found indications of an increased risk of low-back pain in subjects with predominantly seated work tasks (Braun, 1969; Westrin, 1973; Bergquist-Ullman and Larsson, 1977; Svensson and Andersson, 1983). To further confound this matter, Kelsey (1975a,b) and Kelsey and Hardy (1975) found that men who spend more than half their workday in a car have a threefold increased risk of disc herniation. Whether this is due to the sitting postures or vibration was difficult to establish in these studies.

Bergquist-Ullman and Larsson (1977) found that those who did not sit except for short periods of time had longer sickness absence periods following acute low-back pain than others. The study is ambiguous on this point, however, as several other work-related factors associated with poor prognosis were more common in workers with mainly standing and walking work postures. Further, sitting work is often physically lighter. Lastly, several studies indicate the importance of changes in ones work posture. Postural fatigue and sickness absence decreases when such changes are required (Griffing, 1960; Kroemer and Robinette, 1969; Magora, 1972).

9.4.2 Radiographic Data

Radiographic studies have verified thoroughly that the pelvis rotates backward and the lumbar spine flattens when sitting (Åkerblom, 1948; Burandt, 1969; Carlsöö, 1972; Keegan, 1953; Schoberth, 1962; Umezawa, 1971; Rosemeyer, 1972; Andersson et al., 1979).

Åkerblom, Keegan, Schoberth and others found that the flattening of the lumbar lordosis in sitting can be prevented by the use of a well designed low-back support. Andersson et al. (1979) studied the influence of different types of lumbar supports placed at different levels of the lumbar spine on the lumbar lordosis angle, as well as the influence of changes in backrest inclination on that angle. Several different angles depicted in Figure 9.8 were measured from radiographs of 34 healthy males and females in different age groups.

When moving from a standing to an unsupported sitting position, the lumbar lordosis decreased by an average of 38°. This mainly occurred by backward rotation of the pelvis (average 28°). The remaining 10° were mostly changes in the vertebral body angles of the two lower lumbar segments. There were also small angular changes between L1 and L2 and between L2 and L3. The changes in the sacroiliac joint angle were about 4°.

When a back support was used, there was an increase in the total lumbar angle as well as in the individual lumbar vertebral body angles. An increase in backrest inclination from 90 to 110° had a slight decreasing effect on the total lumbar angle. The main postural change induced by an increase in the

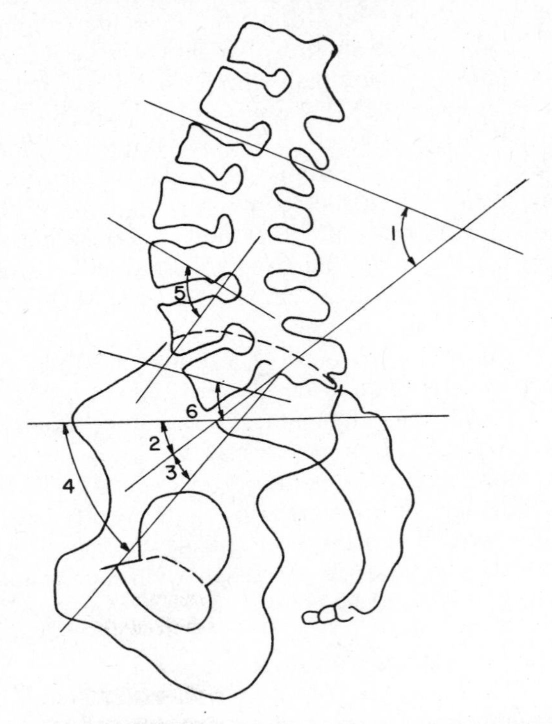

FIGURE 9.8 Angles measured from the radiographs: (1) total lumbar angle; (2) sacral horizontal angle; (3) sacral pelvic angle; (4) pelvic horizontal angle; (5) vertebral body angles L1–2, L2–3, L3–4, L4–5; (6) the L5–S1 angle. (Adapted from Andersson et al., 1979.)

backrest-seat angle was a rotation of the pelvis. The sacral-horizontal angle changed from a mean value of 34° to a mean value of 3°; the pelvic-horizontal angle, from a mean of 53° to about 22°. There was minimal motion in the sacroiliac joint and in individual lumbar vertebral body angles.

The amount of lumbar support had a marked influence on the total lumbar angle, which changed from a mean of 9.7° to 46.8° when the lumbar support was increased. The curvature of the lumbar spine when the lumbar support was 4 cm in front of the plane of the backrest (Fig. 9.13) closely resembled the lumbar curve of the standing position. The location of the lumbar support with respect to the level of the spine did not significantly influence any of the angles measured as long as the support was in the lumbar region.

Based on this study, the following conclusions can be made. The basis for the shape of the lumbar spine during sitting seems to be a result of rotation of the pelvis. To balance the trunk in standing, a lordosis is usually required as the sacral end-plate is almost always inclined forward. When sitting, the sacral end-plate is in a more horizontal position and the normal lumbar lordosis becomes flattened. Schoberth (1962) found that a sacral-horizontal

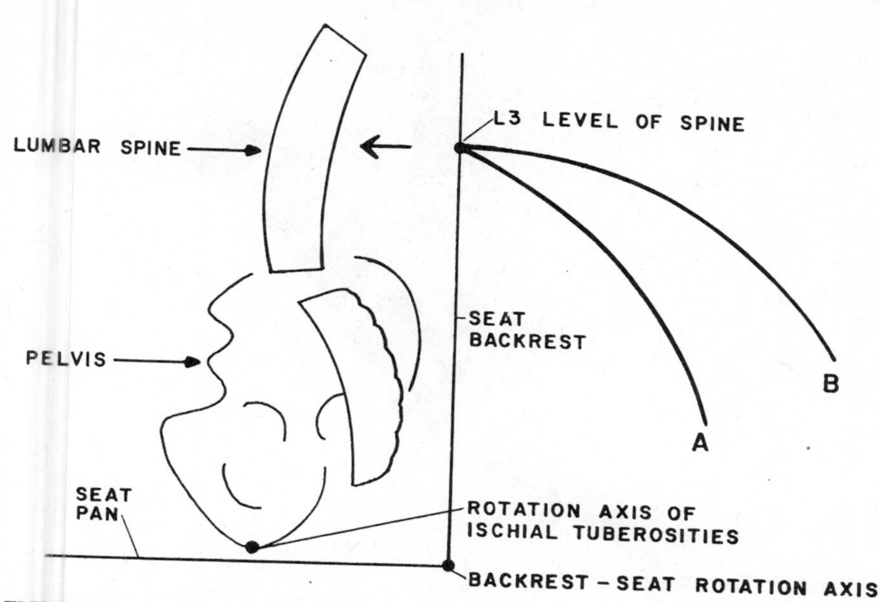

FIGURE 9.9 When a lumbar support is fixed and the backrest seat angle increased through an axis posterior to the ischial tuberosities, the lumbar pad moves along B, which rotates around the backrest seat axis, while, for example, L3 moves along A, which rotates around the ischial tuberosity axis (see text). (Adapted from Andersson et al., 1979.)

angle of 16° or more was a requirement for a normal lumbar lordosis in relaxed unsupported sitting.

In order to prevent flattening of the lumbar spine in sitting, suggestions have been made for changes in the seat or in the backrest. A forward tilted seat was proposed by Burandt (1969), Carlsöö (1963), and Mandal (1975; 1976), while Rosemeyer (1972) suggested fixation of the pelvis on the seat by a pillow support. More common, however, are suggestions of supports for either the whole back or the lumbar spine only. The influence of a lumbar support seems to be of greater importance in this respect than a particular backrest inclination, or the height of the support.

The location of the lumbar support with respect to the level of the lumbar spine was as mentioned of little importance to the shape of the lumbar curve in the study by Andersson et al. (1979). This finding is of practical importance, particularly when a chair permits changes in backrest–seat inclination. The backrest inclination is usually changed by rotation about a fixed axis which, according to Snorrason (1968), is located about 17 cm posterior to the rotational axis of the body at the ischial tuberosities (Figure 9.9). A simple geometric calculation shows that the lumbar support moves about 4.5 cm upward with respect to the lumbar spine when the inclination of the backrest is increased from 90 to 105°, almost a whole lumbar segment. Ad-

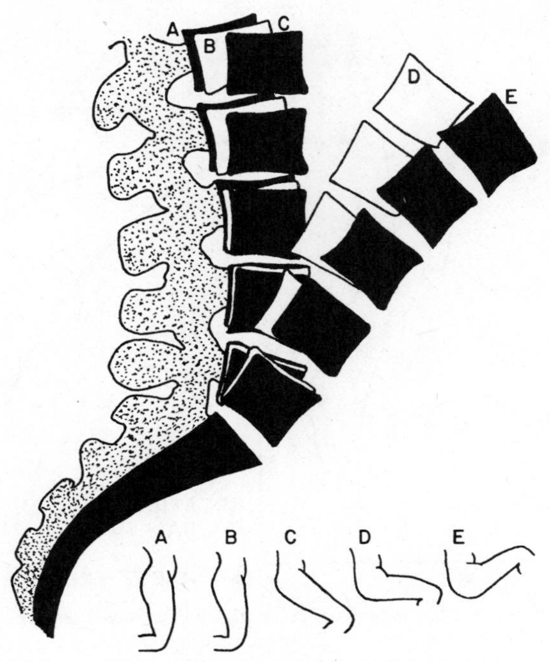

FIGURE 9.10 Tracings of roentgenograms of the lumbosacral spine of a subject in a lateral recumbent position with the sacrum superimposed in each tracing. The only variant in these positions is the decrease of the trunk thigh angle, the thorax and knees being maintained at constant positions. Particularly noteworthy is the great difference in the lumbar curve between position C at 135° and position D at 90°. (Adapted from Keegan, 1953.)

ditional backward inclination increases the upward movement of the lumbar pad even more, so that the suport is no longer placed where needed in the lower lumbar area.

The knee flexion angle also is important in seat design, as mentioned. The same is true for the hip flexion angle. Keegan (1953) obtained radiographs of a subject in a lateral recumbent position and found that the lumbar curve flattened when a 90° trunk–thigh angle was obtained, compared with a 135° angle (Figure 9.10). He attributed this to the combined actions of the various muscles that rotate the pelvis and thus influence the lumbar curve (Figure 9.11).

9.4.3 Disc Pressure Data During Sitting

In 1964, Nachemson and Morris published data on *in vivo* disc pressure measurements of subjects standing and sitting without support. The pressures measured when standing were found to be about 35% lower than those

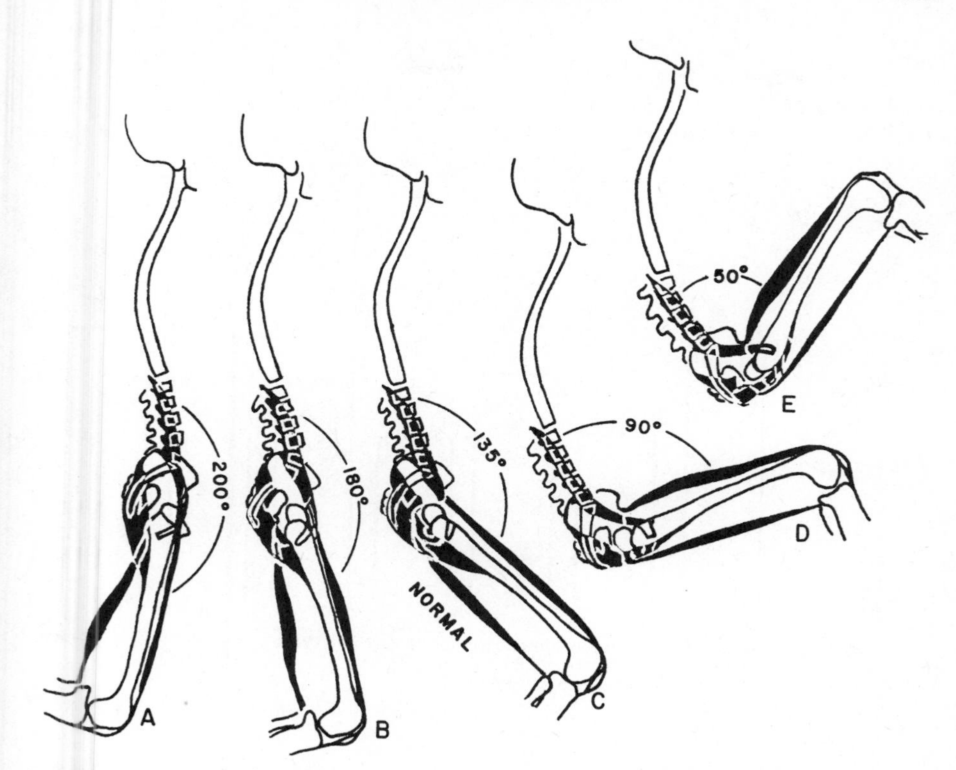

FIGURE 9.11 A series of tracings from roentgenograms of the lumbosacral spine, pelvis, and femur of one individual in the lateral recumbent position, only the angle between trunk and thighs being varied. A diagrammatic outline of the anterior and posterior thigh muscles is superimposed to show that the limited length of these muscles rotates the pelvis and alters the lumbar curve. Note the normal position of balanced muscle relaxation at 135°, with increase of the lumbar curve as the thighs are brought backward and decrease of this curve as the angle between the thighs and the trunk is reduced. (Adapted from Keegan, 1953.)

when sitting. These findings were later confirmed by Okushima (1970), Nachemson and Elfström (1970) and Tzivian et al. (1971).

In the early seventies, a series of studies were completed in which disc pressures were measured in the third lumbar discs of subjects standing and sitting in different chairs, and with different back supports (Andersson, 1974; Andersson et al., 1974a,b,c,d,e,f). These studies confirmed that the disc pressure is considerably lower in standing than in unsupported sitting (Figure 9.12). In different unsupported sitting postures, the lowest pressure was found when sitting with the back straight. The reasons for the increased pressure in certain postures are (1) an increase in the trunk load moment when the pelvis is rotated backward and the lumbar spine and torso are rotated forward and (2) the deformation of the disc itself caused by lumbar spine flattening.

FIGURE 9.12 Disc pressure measurements in standing and unsupported sitting postures. (Adapted from Andersson et al., 1974.)

When supports were added to the chair, disc pressure was found to be influenced by several of the support parameters studied. Inclination of the backrest (defined in Figure 9.13), resulted in a decrease in disc pressure (Figure 9.14) especially when tilting the backrest from vertical to 110 degrees. An increase in lumbar support (defined in Figure 9.13) resulted also in a decrease in disc pressure. The decrease was generally larger when the backrest–seat angle was small (Figure 9.14). Studies performed in an office chair placing the back support at different lumbar levels showed a slightly lower pressure when the support was at the level of the fourth and fifth lumbar vertebrae compared with the first and second. The use of arm rests always resulted in a decrease in disc pressure, less pronounced, however, when the backrest–seat angle was large (Figure 9.15).

The preceding results can be interpreted as follows: (1) When lumbar supports are used, part of the body weight is transferred to them when a person leans back, reducing the load on the lumbar spine caused by the upper body weight; (2) an increase in backrest inclination means an increase in load transfer to the backrest and a reduced disc pressure; (3) the use of arm rests supports the weight of the arms, reducing the disc pressure, and

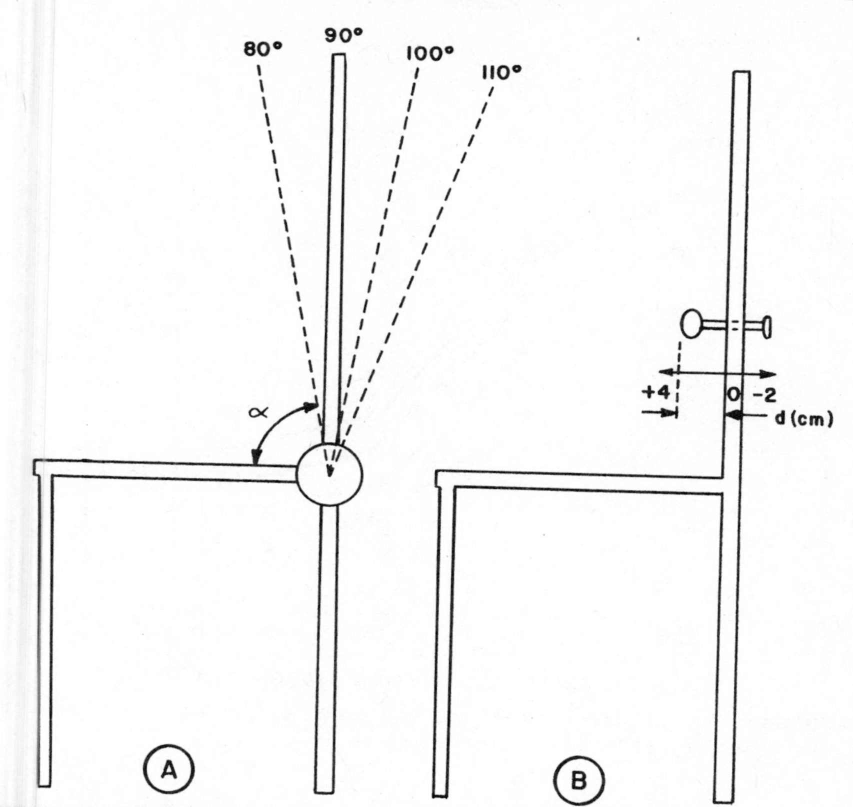

FIGURE 9.13 Definitions of the two support parameters used in the study. A. The backrest inclination is defined as the angle between the seat and the backrest. B. The lumbar support is defined as the distance between the front of the lumbar pad and the plane of the backrest. (Adapted from Andersson et al., 1974.)

(4) the use of a lumbar support changes the posture of the lumbar spine toward lordosis, and hence reduces the deformation of the lumbar spine and corresponding disc pressure.

Studies were also made of typical seated office work. When writing at a desk, a decrease in disc pressure was noted (Figure 9.16) compared with other tasks. This was expected since the arms can be well supported by the desk. Other office activities, such as typing and lifting a phone at arm's length, increased the pressure due to larger external load moments imparted to the spine during such tasks.

9.4.4 Muscle Activity

Electromyography has been used to study the activity of back muscles when sitting. Generally, similar activity levels have been recorded when standing

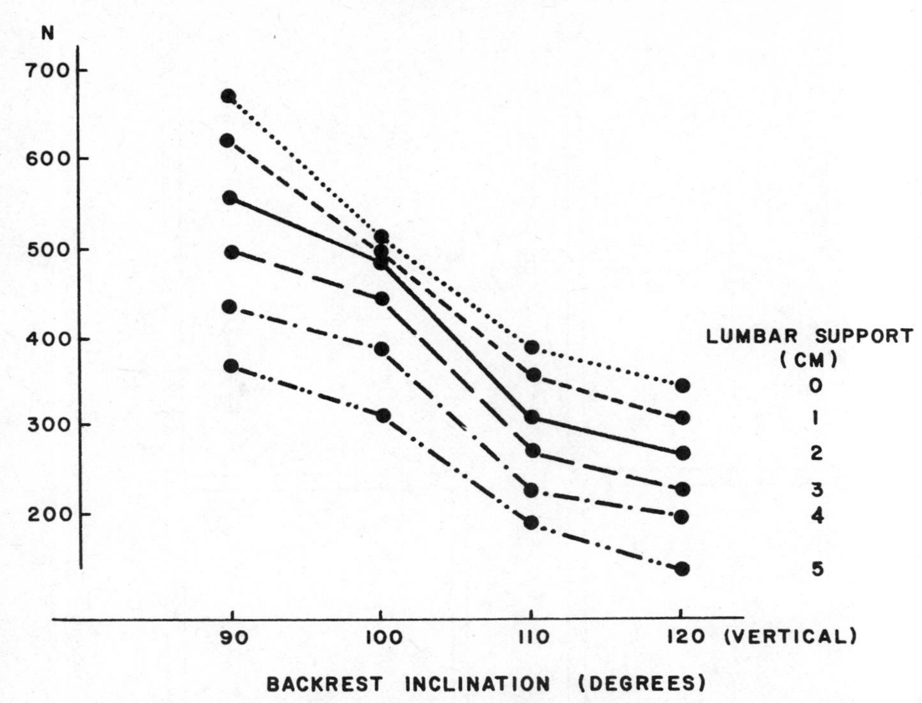

FIGURE 9.14 Disc pressures measured with different backrest inclinations and differently large lumbar supports. (Adapted from Andersson et al., 1974.)

and sitting (Floyd and Silver, 1955; Carlsöö, 1963; Rosemeyer, 1971; Andersson and Örtengren, 1974a; Andersson et al., 1974b). There is general agreement that in sitting the myoelectric activity decreases when (1) the back is slumped forward in full flexion (Åkerblom, 1948; Lundervold, 1951; Floyd and Silver, 1955; Schoberth, 1962; Jonsson, 1970); (2) the arms are supported (Carlsöö, 1963; Floyd and Ward, 1969; Rosemeyer, 1971; and Andersson et al., 1974a,b); or (3) a backrest is used (Åkerblom, 1948; Carlsöö, 1963; Knutsson, Lindh, and Telhag, 1966; Floyd and Ward, 1969; Rosemeyer, 1971, and Andersson et al., 1974a,b).

Of these different support parameters, backrest inclination has been found to be very important, with the EMG levels decreasing as the backrest–seat angle is inclined (Knutsson et al., 1966; Rosemeyer, 1971; Yamaguchi et al., 1972)—depicted in Figure 9.17 from Andersson et al. (1974). Andersson et al. (1974) also found that the myoelectric activity not only decreased in the lumbar region, but also in the thoracic and cervical areas of the spine when backrest inclination was increased. When the angle was greater than 110°, however, there was little further effect on the EMG levels. Knutsson et al. (1966) studied the effect of an additional lumbar support on EMG levels and

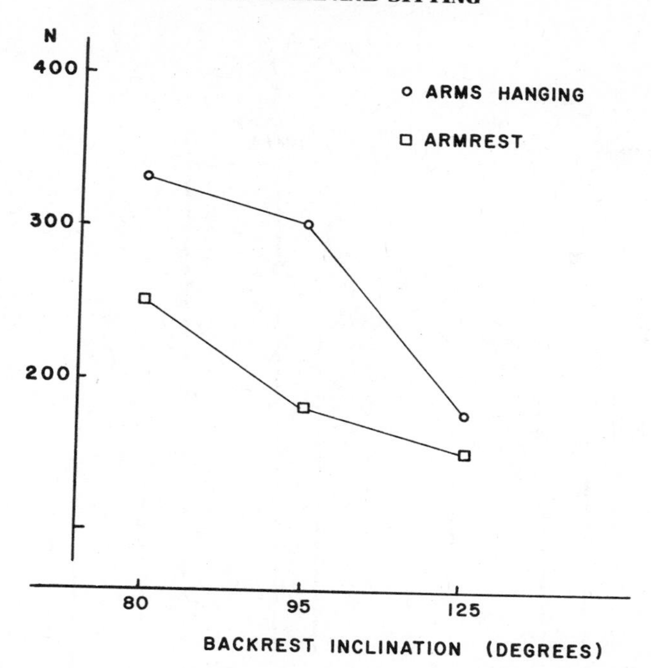

FIGURE 9.15 Influence of armrest on the disc pressure. (Adapted from Andersson et al., 1974.)

found that muscle activity was reduced, both when the support was in front of and behind the plane of the backrest. Andersson et al. (1974), on the other hand, found the influence of a lumbar support to be small (Figure 9.18).

Yamaguchi et al. (1972) found that muscle activity decreased when the seat inclination was increased backward. The effect of different locations of the backrest was studied by Lundervold (1951a,b; 1958) and Floyd and Roberts (1958). They found that the myoelectric activity was less when the back support was located in the lumbar region than in the thoracic, and thus confirmed a finding by Åkerblom (1948) that a support in the lumbar region was as "effective" as a full back support.

Carlsöö (1963) studied the importance of the angle of the knees on EMG activity. Flexion beyond 90° was found to increase the myoelectric activity, while there was a decrease with extension of the knees. The effect of both the height of the seat and of the table has also been investigated. Too high or too low a seat has been found to increase muscle activity (Lundervold, 1951a,b; 1958; Laurig, 1969), and the vertical distance between the seat and the table appears also to be an important factor (Lundervold, 1951a,b; 1958; Schoberth, 1962; Carlsöö, 1963; Floyd and Ward, 1969; Laurig, 1969; Chaffin, 1973; Andersson et al., 1974). The vertical distance is particularly important for the shoulder muscles, as discussed in Chapter 10.

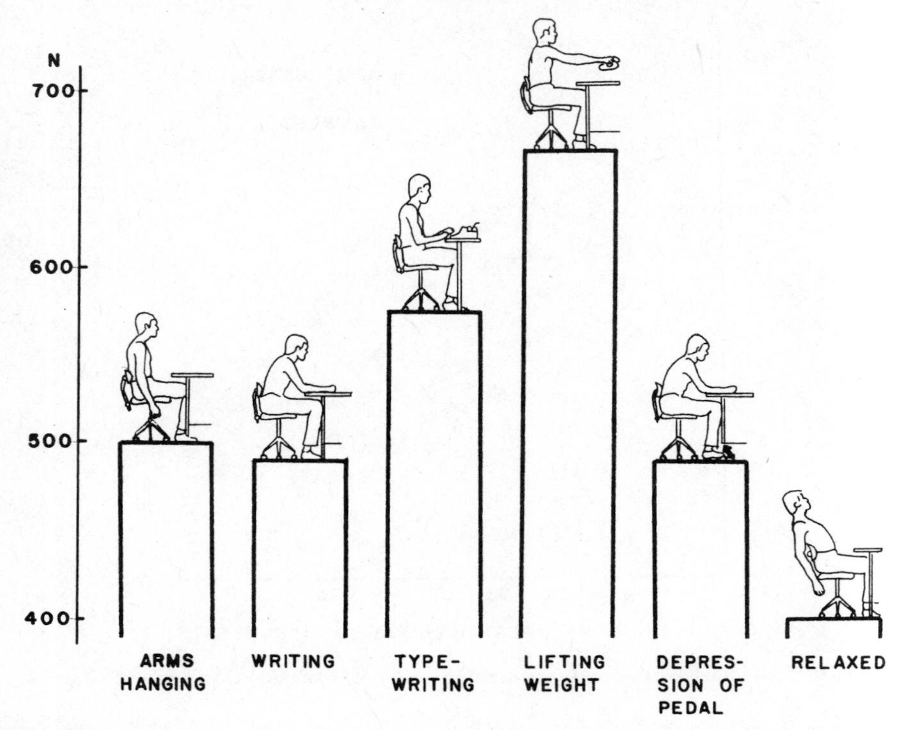

FIGURE 9.16 Disc pressure measured in an office chair during simulated work activities. (Adapted from Andersson et al., 1974.)

9.4.5 Discussion of the Spine in Sitting Postures

It appears that in spite of the fact that hard data is still not available, there are several indicators that sitting can cause back pain. It appears from radiographic, disc pressure, and myoelectric data that the load on the spine increases when sitting without a lumbar support, compared with standing. Reduced low-back stress levels can be expected, however, by the use of proper back supports. The most important factor in reducing low-back stress is the backrest, and the most important parameter in backrest design is its inclination angle. By addition of a separate lumbar support, the stress on the back can be further reduced, particularly when sitting upright. Such support should be placed in the lumbar region to achieve a more normal lordotic curvature when sitting. In order to provide as much comfort as possible, the support should be adjustable in both height and size. The evidence presented also discloses that support of the arms as well as assuring that the seat is adjusted to the proper height can reduce low-back stress further.

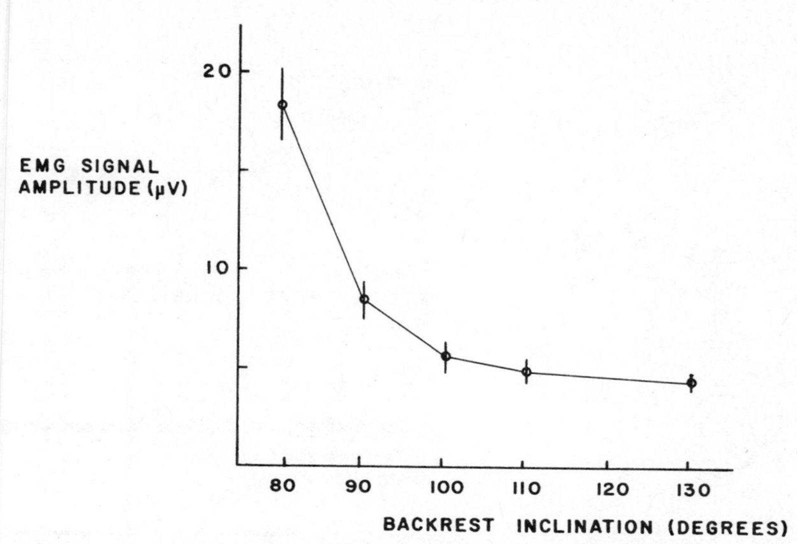

FIGURE 9.17 The myoelectric activity (EMG) decreases when the backrest inclination is increased. (Adapted from Andersson and Örtengren, 1974.)

9.5 THE SHOULDER AND SITTING

The placement of a work surface or the workpiece relative to the person doing sitting work is important not only because it influences the spine, but perhaps even more so because it influences the posture of the shoulders and load moments acting on the shoulders and upper torso.

A work surface placed above the height of the elbow requires the angle of the elbow to decrease or the shoulders to be abducted or lifted. Such postures increase the stress on the shoulder joints as well as on the muscles of the shoulder–neck area (Figure 9.18). When performing desk work, an abduction angle of 15 to 20° or less and an elevation angle of 25° or less should always be attempted (Figure 9.19). Further discussion of work surface heights will be given in Chapter 10.

9.6 THE LEGS AND SITTING

Leg support is critical to better distribute and reduce the load on the buttocks and the back of thighs. To do so, the feet should rest firmly on the floor or foot support so that the weight of the lower legs is not supported by the front part of the thighs resting on the seat. If pressure is applied to the thighs close to the knees it can create clinical problems in terms of swelling of the legs and pressure to the sciatic nerve (Figures 9.20). In general, it has been found that over a work day the legs increase in volume by 4% (Winkel, 1978;

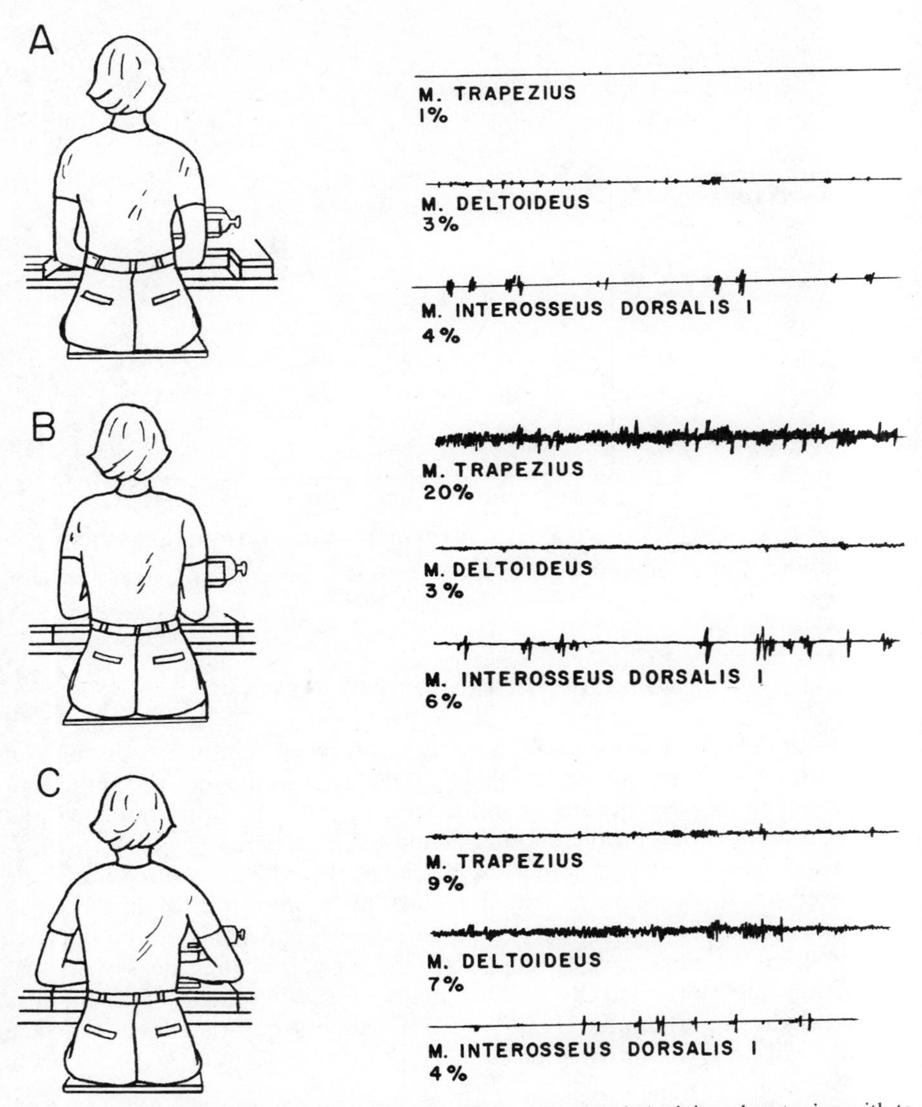

FIGURE 9.18 Electromyographic recording of shoulder muscle activity when typing with tables of different heights. The percent EMG refers to percentage of maximum voluntary contraction position: (A) optimal height; (B) too high, resulting in elevation of shoulders; (C) too high, compensated for by abduction of the arms. (Adapted from Hagberg, 1982.)

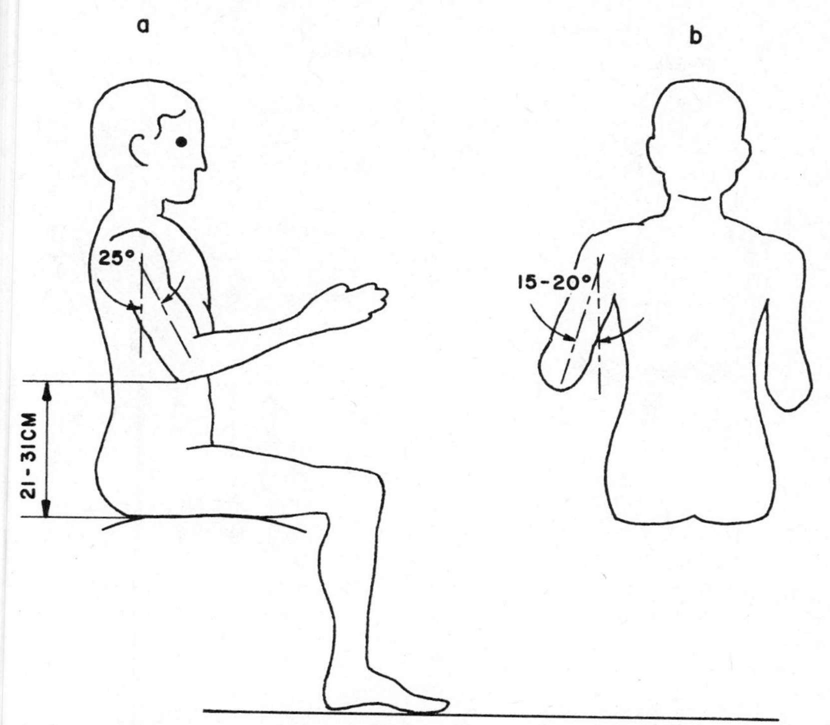

FIGURE 9.19 Abduction angles of 15–20° (b) and flexion angles of 25° (a) are acceptable. Further discussion of this in Chapter 10. (Adapted from Engdahl, 1978.)

1981). By requiring movements every 15 minutes, swelling was reduced to 2.3%. When a chair is too low, the knee flexion angle becomes large and the weight of the trunk is transferred to the seat-pan surface over a small area at the ischial tuberosities (Figures 9.21). The small knee and hip angles soon become uncomfortable, and the spine is flexed as the pelvis rotates backward (Keegan, 1953; Floyd and Roberts, 1958; Kroemer, 1971; Kroemer and Robinette, 1969). Further, the abdominal organs are compressed in this posture when one leans forward.

When a chair is too high, so that the feet do not reach the floor, the pressure on the back of the thighs becomes uncomfortably large (Åkerblom, 1948, 1969; Schobert, 1962; Bush, 1969). Therefore, the subject tends to slide forward to the front part of the chair. This allows the feet to be supported, but the backrest will not be used properly (Burandt and Grandjean, 1963; Kroemer, 1963, 1971), resulting in low-back pain if the posture is prolonged.

A very high chair can be used for semi-sitting postures. This provides some support of the torso in work postures where extreme mobility and

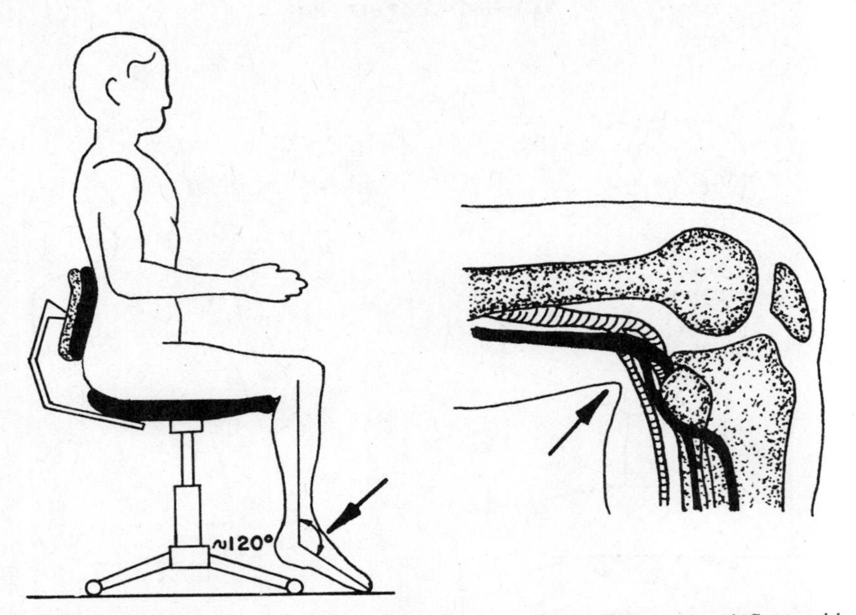

FIGURE 9.20 Too high a chair causes pressure at the popliteal fold, and can influence blood circulation and cause pressure on the nerve. It also requires that the foot be in an equine position, with the heel off the floor.

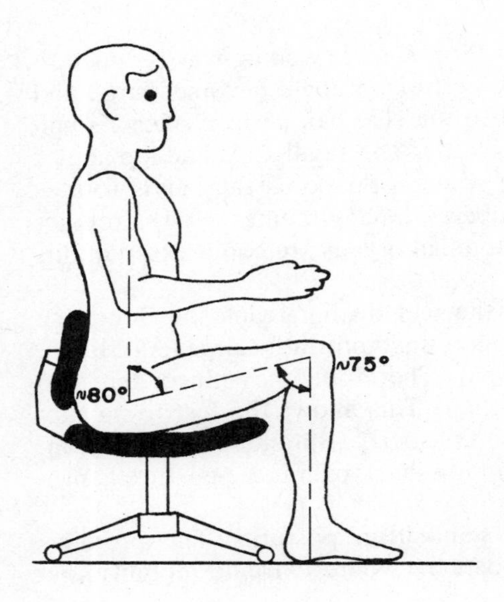

FIGURE 9.21 When the chair is too low, the knee flexion angle becomes small and the weight is transferred to a small area at the ischial tuberosities. (Adapted from Jonsson and Andersson, 1978.)

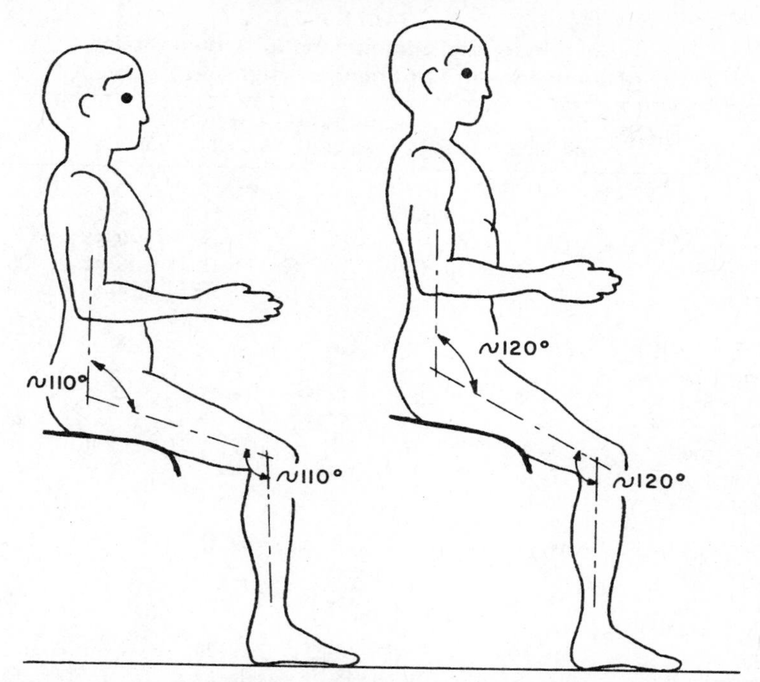

FIGURE 9.22 A semi-sitting posture allows for rapid changes between sitting and standing and preserves the lumbar lordosis. (Adapted from Engdahl, 1978.)

reach is required (Laurig, 1969) (Figure 9.22). Transitions between sitting and standing are obviously rapid and simple when using the semi-sitting posture, and lumbar lordosis is somewhat preserved. Unfortunately, the legs are more stressed in this posture than when completely seated.

9.7 THE SITTING WORKPLACE

For a more complete review on this aspect, the reader is referred to textbooks on ergonomics. A few areas will be covered here as they relate to occupational biomechanics. It goes without saying that the seated workplace as a whole should be adjusted to the worker. This requires not only a good chair, but also a good work surface height and work space design. Other factors of importance not to be discussed here are lighting and esthetics. The purpose of work space design is to faciliate work performance while ensuring good posture and preventing physical disability.

9.7.1 The Chair

Because of many different opinions and user requirements, available chairs vary widely today. Compare, for example, the driver's seat in an automobile

TABLE 9.5
Recommended Dimensions for Office Chairs
(Dimensions in Centimeters, Angles in Degrees)

Feature	BS 3079 and 3893[a]	CEN[b]	Diffrient et al.	Danero & Zelnik	Grandjean	DIN[c]	SS[d]
Seat							
Height	43–51	39–54	35–52	36–51	38–53	42–54	39–51
Width (breadth)	41	40–	41	43–48	40–45	40–45	42–
Length (depth)	36–47	38–47	33–41	39–41	38–42	38–42	38–43
Slope angle	0–5	0–5	0–5	0–5	4–6	0–4	0–4
Backrest							
Top height	33–		—	—	48–50	32–	—
Bottom height	20–		—	—	—		—
Center height	—	17–26	23–25	19–25	30	17–23	17–22
Height	—	10–	15–23	10–20		22–	22–
Width (breadth)	30–36	36–40	33	25	32–36	36–40	36–40
Horizontal radius	31–46	min. 40	31–46	—	40–50	40–70	40–60
Vertical radius	convex	—	—	—	—	70–140	convex
Backrest–seat angle	95–105		35–100	95–105	—		—
Armrest							
Length	22	20–	15–21	—	—	20–28	20–
Width (breadth)	4	4–	6–9	—	—	—	4–
Height	16–23	21–25	18–25	20–25	—	21–25	21–25
Interarmrest width	47–56	46–50	48–56	46–51	—	48–50	46–
Feet number	—		—	—	5	—	
Base number	—		—	—	40–45	—	

[a] British standards.
[b] European standards.
[c] German standards.
[d] Swedish standards.

with an office chair or a work chair in a factory. Irrespective of use, it is important to be able to adjust any chair to meet basic anthropometric dimensions of the worker, as discussed in Section 9.2, as well as of the workplace. A number of recommendations have been published in this regard, and are different for various countries. This is not surprising, since anthropometric dimensions vary greatly between countries. In the review here, a more explanatory approach of why different dimensions are critical will be used, with the main emphasis being on the specification of office and industrial chairs.

Table 9.5 summarizes some published chair dimension recommendations. It should be emphasized that the dimensions are different from the anthropometric measurements. The following seat dimensions are important: (1) seat height, (2) seat width (breadth), (3) seat length (depth), and (4) seat slope, defined in Figure 9.23. In addition, such data as (5) the shape of the seat, its (6) frictional properties, (7) softness, (8) adjustability, and (9) cli-

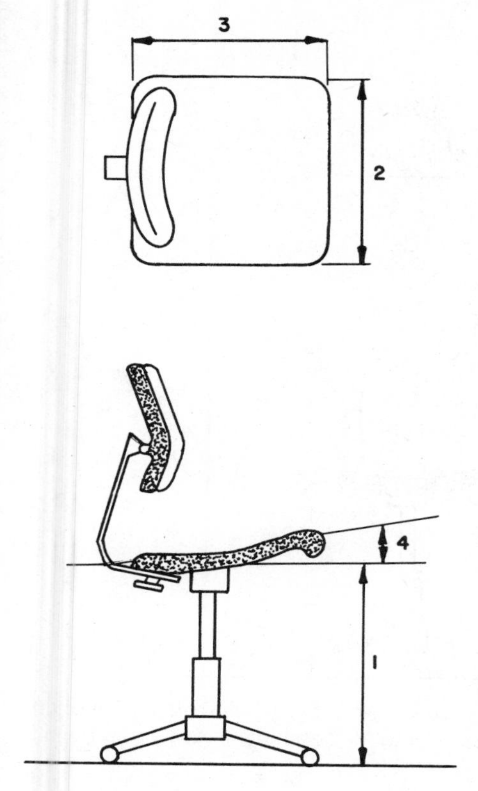

FIGURE 9.23 Seat dimensions: (1) Seat height; (2) seat width; (breadth); (3) seat length (depth); (4) seat slope.

matic comfort are also specified in some of the standards. The importance of correct sitting height has already been outlined. This requires a proper seat height, adjustable to the individual user. Suggestions are usually that the seat surface should be 3 to 5 cm below the knee fold when the lower limb is vertical. When adjustability is not available, foot supports can be used with higher than normal chairs. The width of the seat should be sufficient to accomodate the user population. The edges of the seat should not be detectable during ordinary sitting work.

The length (depth) of the seat is also important, as discussed. It must be possible to use the backrest by assuring that the seat pan is not too deep. Pressure should be avoided on the back of the thigh near the knees. A free area between the back of the lower limb and the seat pan is also useful to facilitate arising and leg movements. About 10 cm is suggested as the minimum clearance. The front part of the seat should be contoured. Horizontal seats should be avoided. In some cases a forward slope can be advantageous, reducing the load on the spine for semi-sitting postures with raised seats. More often a backward slope of about 5° is suggested for normal sitting. The slope facilitates the use of the backrest and prevents sliding on the seat

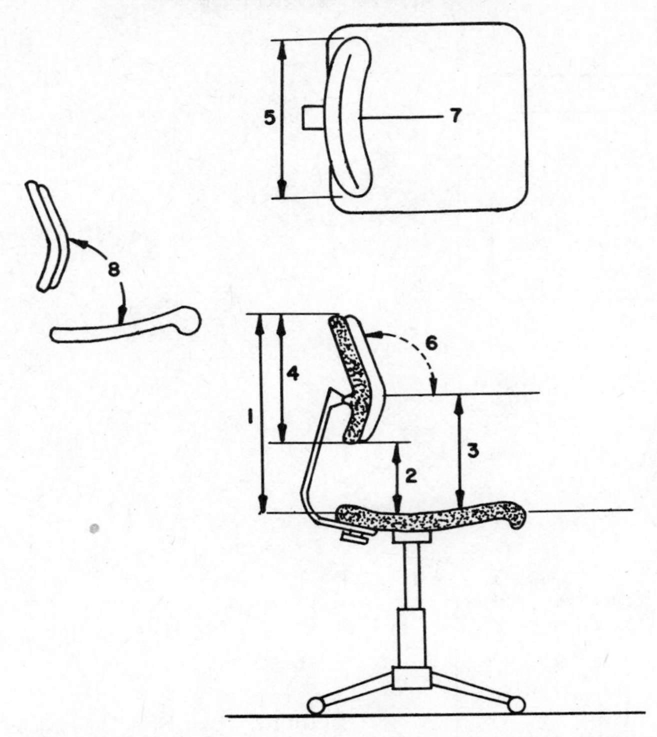

FIGURE 9.24 Backrest dimensions: (1) backrest top height; (2) backrest bottom height; (3) backrest center height; (4) backrest height; (5) backrest width (breadth); (6) backrest horizontal radius; (7) backrest vertical radius; (8) backrest-seat angle.

surface when one moves around in the chair. Sliding can also be prevented by seat contour and high surface friction.

For the backrest, the following dimensions are important: (1) backrest top height, (2) backrest bottom height, (3) backrest center height, (4) backrest height, (5) backrest width, (6) backrest horizontal radius, (7) backrest vertical radius, and (8) backrest–seat angles. Other factors are (9) pivoting and recline possibility, (10) softness, (11) adjustability, and (12) climatic comfort (for definitions see Figure 9.24). Most of these features have already been discussed. The backrest should be adjustable both in horizontal and vertical places. It should support the lumbar spine and not restrict movement of the spine or arms. The width should allow all users to be supported without arm interference. The shape should be convex from top to bottom to conform to the normal lumbar lordosis, and concave from side to side to conform to human anatomy and support the occupant in the chair. A spring-loaded pivoting action should allow the backrest to follow natural body movements but maintain body support. When reclining is permitted, a control system

TABLE 9.6
Recommended Dimensions for Office Tables
(Dimensions in Centimeters, Angles in Degrees)

Feature	DIN[a] Adjustable	Final	SS[b] Adjustable	SS Fixed	CEN[c] Adjustable	Fixed
Office Table						
Top height	65–75	72	67–78	71–73	67–77	72
Bottom height	—	65–	60–	63–	—	65–
Table slope	—	—	—	—	—	—
Typewriter Table						
Top height	60–68	65	59–67	63–67	60–68	65
Bottom height	62–	62–	54–65	61–	62–	62–
Table slope	—	—	—	—	—	—

[a] German standards.
[b] Swedish standards.
[c] European standards.

must be used so that a firm support is given when the backrest is in the vertical position.

Another chair feature is the armrests, and the following dimensions are important: (1) length, (2) width, (3) height, (4) width between armrests, and (5) distance from armrest front to seat front. When the armrests are too high, the occupant must raise the shoulders and abduct the arms. This is also the case when the armrest-to-armrest width is too large. If the armrest is too low, on the other hand, it can be used only by either sliding forward or leaning to one side. Proper placement of the armrests is also important to reduce pressure on the seat surface and load on the spine, and facilitate rising from the chair. On the other hand, armrests can be obstacles in some work situations. When they are too wide or too high, they can prevent sliding of the chair under a table, and when too long (too short distance from front armrest to front seat) they make sliding under the table and rising more difficult.

Factors such as the number of feet on the chair, the base diameter, and the use of casters or wheels are necessary safety considerations. Five feet or castors are suggested with a minimum radius of 300 mm to prevent tipping, and a maximum horizontal radius of 350 mm to prevent tripping over the base of the chair. Castors should be used when chair movements are often required at the workplace.

9.7.2 The Table (Work Surface)

Because of different factors such as the size of the workpiece, motions demanded by the task performer, and overall work layout, the height of the work surface cannot be the same for all types of work. And for each type

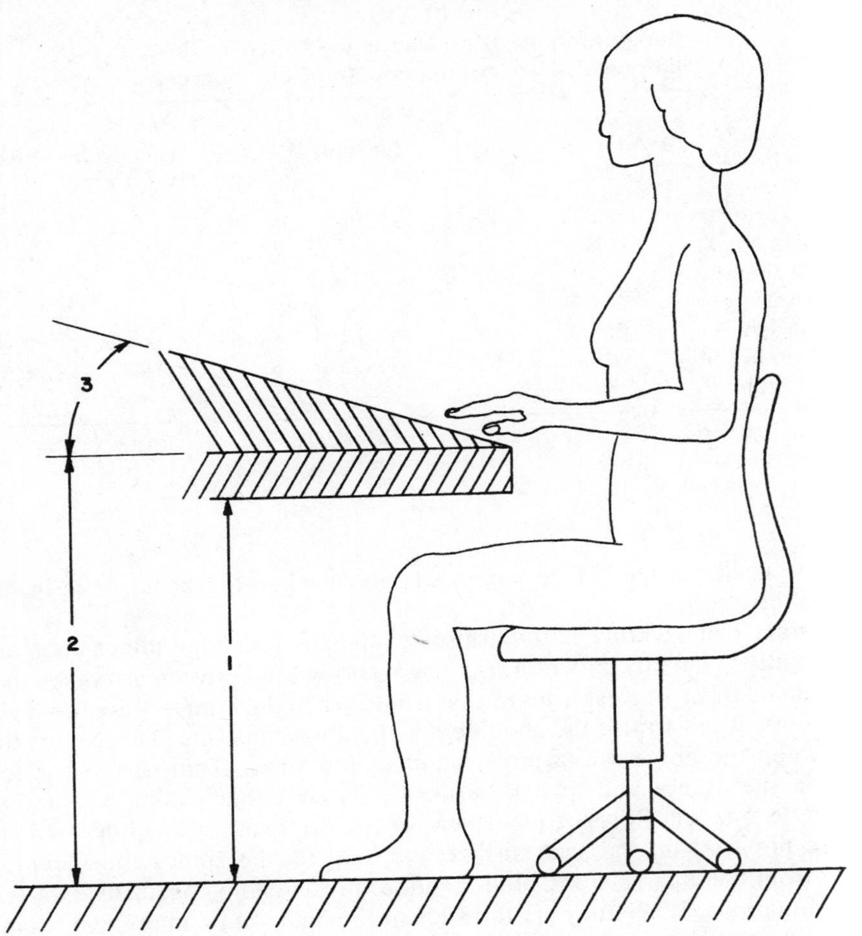

FIGURE 9.25 Table (work surface) dimensions: (1) bottom height; (2) top height; (3) surface slope.

of work, adjustability is advantageous to ensure proper fit to the worker. The table (work surface) dimensions outlined in this respect are (1) work surface bottom height, (2) work surface top height, and (3) work surface slope. In addition, the work surface should be large enough to accomodate work objects and the friction high enough to prevent sliding of these objects. When controls are used, they must be placed within an optimum work area in all planes.

The work surface bottom height is critical to ensure sufficient leg room. Based on anthropometric dimensions, standards have been agreed upon, again different in different countries (Table 9.6, Figure 9.25). The table top height is also included in these standards. The field of vision is of utmost importance here to prevent forward flexion of the neck and of the trunk. A focal distance ranging from 20 to 40 cm is common. The height of the table should always be functionally related to the position of the elbow. This requires either an adjustable table or a high table and an adjustable footrest. Too low a table causes kyphosis of the lumbar spine, increasing the load. Too high a table, on the other hand, causes abduction of the arms and elevation of the shoulder as well as kyphosis of the neck, causing fatigue in the shoulder and neck muscles. Chaffin (1973) found a 15° angle at the neck to be acceptable. Shoulders and elbow considerations have already been discussed. It is important to remember that the work surface height is not always the table height—in using a typewriter or word processor, for example, it is the keyboard height. A slope to the work surface is advantageous from the neck and back point of view, but must of course be related to the work task performed.

9.8 SUMMARY

The sitting work place is at present the most common one in the industrial world. Although sitting offers many excellent advantages, it should be obvious that the sitting work place needs to be thought out carefully so as not to introduce musculoskeletal problems to the workers. This chapter offers some guidance for that purpose, but needs to be supplemented by the information on work place design and function available in many ergonomic texts to be fully effective. The era when the work postured problem was solved by simply providing a chair should be over. With attention to detail in design, layout, and work method, seated work can be made much safer than is presently the case in many office and factories.

REFERENCES

Åkerblom, B., *Standing and Sitting Posture. With Special Reference to the Construction of Chairs,* Nordiska Bokhandeln, Stockholm, 1948, Doctoral dissertation.

Åkerblom, B., "Anatomische und Physiologische Grundlagen zur Gestaltung von Sitzen," in E. Grandjean, Ed., *Sitting Posture,* Taylor, London, 1969, pp. 6–17.

Aldman, B. and T. Lewin, *Anthropometric Measurements of Sitting Adults,* Report, Chalmers University of Technology and the University of Gothenburg, Gothenburg, Sweden, 1977.

Andersson, G. B. J., "On Myoelectric Back Muscle Activity and Lumbar Disc Pressure in Sitting Postures," doctoral dissertation, Gotab, Univ. of Gothenburg, Gothenburg, Sweden, 1974.

Andersson, G. B. J., and R. Örtengren, "Myoelectric Back Muscle Activity During Sitting," *Scand. J. Rehab. Med.,* Suppl. 3, 73–90 (1974a).

Andersson, G. B. J., B. Jonsson, and R. Örtengren, "Myoelectric Activity in Individual Lumbar Erector Spinae Muscles in Sitting. A Study with Surface and Wire Electrodes," *Scand. J. Rehab. Med.,* Suppl. 3, 91–108 (1974b).

Andersson, G. B. J., R. Örtengren, A. Nachemson, and G. Elfström, "Lumbar Disc Pressure and Myoelectric Back Muscle Activity During Sitting. I. Studies on an Experimental Chair," *Scand. J. Rehab. Med.,* **3,** 104–114 (1974c).

Andersson, G. B. J., and R. Örtengren, "Lumbar Disc Pressure and Myoelectric Back Muscle Activity During Sitting. II. Studies on an Office Chair," *Scand. J. Rehab. Med.,* **3,** 115–121 (1974d).

Andersson, G. B. J., and R. Örtengren, "Lumbar Disc Pressure and Myoelectric Back Muscle Activity During Sitting. III. Studies on a Wheelchair," *Scand. J. Rehab. Med.,* **3,** 122–127 (1974e).

Andersson, G. B. J., R. Örtengren, A. Nachemson, and G. Elfström, "Lumbar Disc Pressure and Myoelectric Back Muscle Activity During Sitting. IV. Studies on a Driver's Seat," *Scand. J. Rehab. Med.,* **3,** 128–133 (1974f).

Andersson, G. B. J., R. W. Murphy, R. Örtengren, and A. L. Nachemson, "The Influence of Backrest Inclination and Lumbar Support on the Lumbar Lordosis in Sitting," *Spine,* **4,** 52–58 (1979).

Barkla, D. M., "Chair Angles, Duration of Sitting, and Comfort Ratings." *Ergonomics,* **7,** 297 (1964).

Bergquist-Ullman, M. and U. Larsson, "Acute Low Back Pain in Industry," *Acta Orthop. Scand,* Suppl. 170 (1977).

Branton, P. and G. Grayson, "An Evaluation of Train Seats by Observation of Sitting Behavior," *Ergonomics,* **10,** 35 (1967).

Braun, W., "Ursachen des lumbalen Banscheiberverfalls," *Die Wirbelsäule in Forschung und Praxis,* **43,** 1969.

British Standards Institution. *British Standard 30–79: Anthropometric Recommendations for Dimensions of Non-adjustable Office Chairs, Desks and Tables,* London, 1959.

British Standards Institution. *British Standard 3893. Specifications for Office Desks, Tables and Seating,* London, 1965.

Burandt, U., Röntgenuntersuchung über die Stellung von Becken und Wirbelsäule beim Sitzen auf vorgeneigten Flächen," In E. Grandjean, Ed., *Sitting Posture,* Taylor, London, 1969. pp. 242–250.

Burandt, U. and E. Grandjean, "Sitting Habits of Office Employees," *Ergonomics,* **6,** 217 (1963).

Bush, C. A., "Study of Pressures on Skin Under Ischial Tuberosities and Thighs During Sitting." *Arch. Phys. Med.,* **50,** 207, (1969).

Carlsöö, S., "Writing Desk, Chair and Posture of Work," Dept. of Anatomy, Stockholm, 1963. (in Swedish).

Carlsöö, S., *How Man Moves,* Heinemann, London, 1972.

Chaffin, D. B., "Localized Muscle Fatigue–Definition and Measurement," *J. Occup. Med.,* **15,** 346 (1973).

Diffrient, N., A. R. Tilley, and J. C. Bardagjy, *Human Scale 1/2/3,* MIT Press, Cambridge, 1974.

Drury, C. G. and B. G. Conry, A Methodology for Chair Evaluation," *Applied Ergonomics,* **13,** 195–202 (1982).

Engdahl, S., *School-Chairs,* Swedish Furniture Research Institute, Report No. 24, Stockholm, 1971 (in Swedish).

Engdahl, S., *Anthropometric Measurements of Adults,* Swedish Furniture Research Institute, Report No. 29, Stockholm, 1974.

Engdahl, S., "Specification for Office Furniture," in B. Jonsson, Ed., *Sitting Work Postures,* National Board of Occupational Safety and Health (Sweden), No. 12, 1978, pp. 97–135 (in Swedish).

Engdahl, S., *Workchairs for Dentists*. Stockholm, SPRI—Report No. 10, 1977 (in Swedish).

Floyd, W. F. and D. F. Roberts, "Anatomical and Physiological Principles in Chair and Table Design," *Ergonomics* **2,** 2,1 (1958).

Floyd, W. F. and P. H. S. Silver, "The Function of the Erectores Spinae Muscles in Certain Movements and Postures in Man," *J. Physiol.* (*London*), **129,** 184 (1955).

Floyd, W. F. and J. S. Ward, "Anthropometric and Physiological Considerations in School, Office and Factory Seating, in E. Grandjean, Ed., *Sitting Posture,* Taylor, London, 1969, pp. 18–25.

Grandjean, E., *Fitting the Task to the Man,* International Publications Service, New York, 1980.

Grandjean, E., A. Boni, and H. Kretzschmar, "The Development of a Rest Chair Profile for Healthy and Notalgic People," in E. Grandjean, Ed., *Sitting Posture,* Taylor, London, 1969, pp. 193–201.

Grandjean, E., W. Hünting, W. Wotzka, and R. Schärer, "An Ergonomic Investigation of Multipurpose Chairs," *Human Factors,* **15,** 247–255 (1973).

Grandjean, E., M. Jenni, and R. Rhiner: "Eine Indirekte Methode of Erfassung des Komfortgefühls beim Sitzen," *Int. Z. Angew. Physiol.,* **18,** 101 (1960).

Griffing, J. P., "The Occupational Back," in *Modern Occupational Medicine,* No. 2, Chapter 2, Lea & Febiger, Philadelphia, 1960, pp. 219–227.

Hall, M. A. W., "Back Pain and Car-Seat Comfort," *Appl. Ergonomics,* **3,** 82 (1972).

Hult, L., "The Munkfors Investigation, *Acta Orthop. Scand.,* Suppl. 16 (1954a).

Hult, L., "Cervical, Dorsal and Lumbar Spinal Syndromes," *Acta Orthop. Scand.,* Suppl. 17 (1954b).

Jones, J. C., "Methods and Results of Seating Research," in E. Grandjean, Ed., *Sitting Posture,* Taylor, London, 1969, pp. 57–67.

Jonsson, B., "The Functions of Individual Muscles in the Lumbar Part of the Erector Spinae Muscle," *Electromyography,* **10,** 5 (1970).

Jonsson, B. and G. B. J. Andersson, "Functional Anatomical and Biomechanical Aspects on Sitting," in B. Jonsson, Ed., *Sitting Work Postures,* National Board of Occupational Safety and Health (Sweden) No. 12, 1978, pp. 6–17 (in Swedish).

Keegan, J. J., "Alterations of the Lumbar Curve Related to Posture and Seating." *J. Bone Joint Surg.,* **35-A,** 589 (1953).

Kelsey, J. L., "An Epidemiological Study of Acute Herniated Lumbar Intervertebral Discs," *Rheumatol. Rehabil.,* **14,** 144–145 (1975a).

Kelsey, J. L., "An Epidemiological Study of the Relationship Between Occupations and Acute Herniated Lumbar Intervertebral Discs," *Int. J. Epidemiol.,* **4,** 197–204 (1975b).

Kelsey, J. L. and R. J. Hardy, "Driving of Motor Vehicles as a Risk Factor for Acute Herniated Lumbar Intervertebral Discs." *Am. J. Epidemiol.,* **102,** 63–73 (1975).

Knutsson, B., K. Lindh, and H. Telhag, "Sitting—An Electromyographic and Mechanical Study," *Acta Orthop. Scand.,* **37,** 415 (1966).

Kroemer, K. H. E., "Über die Höhe von Schreibtischen, *Arbeitswissenschaft,* **2,** 132 (1963).

Kroemer, K. H. E., "Seating in Plant and Office," *Amer. Industr. Hyg. Ass. J.,* **2,** 2, 132 (1971).

Kroemer, K. H. E. and J. C. Robinette, "Ergonomics in the Design of Office Furniture," *Industr. Med. Surg.,* **38,** 115 (1969).

Laurig, W., "Der Stehsitz als Physiologisch Günstige Alternative zum Reinen Steharbeitsplatz," *Arbeitsmed Sozialmed Arbeitshyg,* **4,** 219 (1969).

Lawrence, J. L., "Rheumatism in Coal Miners. Part III. Occupational Factors," *Br. J. Ind. Med.,* **12,** 249–261 (1955).

Le Carpentier, E. F., "Easy Chair Dimensions for Comfort—A Subjective Approach," in E. Grandjean, Ed., *Sitting Posture.* Taylor, London, 1969, pp. 214–223.

Lundervold, A. J. S., "Electromyographic Investigations of Position and Manner of Working in Typewriting," *Acta Orthop. Scand.*, Suppl. 84 (1951a).

Lundervold, A. J. S., "Electromyographic Investigations During Sedentary Work, Especially Typewriting," *Brit. J. Phys. Med.*, **14,** 31 (1951b).

Lundervold, A. J. S., "Electromyographic Investigations During Typewriting," *Ergonomics*, **1,** 226 (1958).

Magora, A., "Investigation of the Relation Between Low Back Pain and Occupation. 3. Physical Requirements: Sitting, Standing and Weight Lifting," *Industr. Med. Surg.*, **41,** 5 (1972).

Mandal, Å. C., "Work-Chair With Tilting Seat," *Lancet,* 642–643, 1975.

Mandal, Å. C., "Work Chair With Tilting Seat", *Ergonomics*, **19,** 157–164 (1976).

Nachemson, A. and G. Elfström, "Intravital Dynamic Pressure Measurements in Lumbar Discs," *Scand. J. Rehab. Med.*, Suppl. 1, (1970).

Nachemson, A., and J. M. Morris, "In vivo measurements of intradiscal pressure", *J. Bone Joint Surg.*, **46A,** 1077 (1964).

Oborne, D. J., "Passenger comfort—an overview", *Applied Ergonomics,* **9**(3), 131–136 (1978).

Okushima, H., "Study on Hydrodynamic Pressure of Lumbar Intervertebral Disc," *Arch. Jap. Chir.*, **39,** 45 (1970).

Partridge, R. E. and J. A. Anderson, "Back Pain in Industrial Workers," *Proceedings of the International Rheumatology Congress,* Prague, Czechoslovakia, 1969, Abstract 284.

Rieck, A., "Über die Messung des Sitzkomforts von Atuositzen," in E. Grandjean, Ed., *Sitting Posture*. Taylor, London, 1969, pp. 92–97.

Rosemeyer, B., "Electromygraphiche Untersuchungen der Rucken-und Schultermuskulatur im Stehen und Sitzen unter Berucksichtigung der Haltung des Autofahrers," *Arch. Orthop. Unfallchir.*, **71,** 59 (1971).

Rosemeyer, B., "Eine Methode zur Beckenfixierung im Arbeitssitz," *Z. Orthop.*, **110,** 514 (1972).

Schackel, B., K. D. Chidsey, and P. Shipley, "The Assessment of Chair Comfort," in E. Grandjean, Ed., *Sitting Posture,* Taylor, London, 1969, pp. 155–192.

Schoberth, H., *Sitzhaltung, Sitzschäden, Sitzmöbel,* Springer, Berlin, 1962.

Schoberth, H., "Die Wirbelsäule von Schulkindern—orthopädische Forderungen an Schulsitze," in E. Grandjean, Ed., *Sitting Posture,* Taylor, London, 1969, pp. 98–111.

Snorrason, E., "Easy Chair Problems," *Tidskr for Danske Sygehuse,* **44,** No. 22, 1968, pp. 3–15 (in Danish).

Svensson, H. O. and G. B. J. Andersson, "Low Back Pain in 40–47 Year Old Men: Work History and Work Environment Factors," *Spine*, **8,** 272–276 (1983).

Tzivian, I. L., V. H. Rayhinstein, V. F. Motov, and F. F. Ovseychik, "Results of Clinical Study of Pressure within the Intervertebral Lumbar Discs," *Ortop. Traumatol Protez.*, **6,** 31 (1971).

Umezawa, F., "The Study of Comfortable Sitting Postures," *J. Jap. Orthop. Assoc.*, **45,** 1015 (1971).

Vernon, H. M., as cited by B. Åkerblom, 1948, pp. 46–47.

Wachsler, R. A. and D. B. Learner, "An Analysis of Some Factors Influencing Seat Comfort," *Ergonomics*, **3,** 315 (1960).

Westrin, C. G., "Low Back Sick-Listing. A Nosological and Medical Insurance Investigation," *Scand. J. Soc. Med.*, Suppl. 7 (1973).

Winkel, J., "Leg Problems from Longlasting Sitting," in B. Jonsson, Ed., *Sitting Work Postures,* National Board of Occupational Safety and Health (Sweden), No. 12, 1978, pp. 72–78 (in Swedish).

Winkel, J., "Swelling of lower leg in sedentary work—a pilot study", *J. Human Ergol.*, **10,** 139–149 (1981).

Wotzka, G., E. Grandjean, U. Burandt, H. Kretzschmar, and T. Leonard, "Investigations for the development of an auditorium seat", in E Grandjean, Ed., *Sitting Posture*, Taylor, London, 1969, pp. 68–83.

Yamaguchi, Y., F. Umezawa, and Y. Ishinada, "Sitting Posture: an Electromyographic Study on Healthy and Notalgic People," *J. Jap. Orthop. Assoc.*, **46,** 277 (1972).

10

BIOMECHANICAL CONSIDERATIONS IN MACHINE CONTROL AND WORKPLACE DESIGN

10.1 INTRODUCTION

Though occupational biomechanics is often associated with manual materials handling activities requiring near maximal exertions, such is not always the case. Sophisticated injury analysis methods have shown that frequent or sustained exertions, particularly when performed in awkward postures, are associated with a variety of musculoskeletal disorders. A contemporary example is the increasing number of musculoskeletal complaints from workers operating video display units in constrained and often awkward postures (Hunting, Laubli and Grandjean, 1982). This chapter briefly discusses some biomechanical factors that should be considered when designing a machine or laying-out a workplace for operators of varied size and strength capabilities so that postures associated with musculoskeletal trauma are avoided.

10.1.1 Localized Musculoskeletal Injury in Industry

As shown in earlier chapters, external loads on the musculoskeletal system induce high muscle, tendon, and joint forces. Because these activities are volitional during work, peak tissue stresses are usually well within the physical capacity or strength of the tissues, provided the forces are of short duration and rest periods are adequate. Prolongation of tissue loads, or performance of very frequent exertions can be harmful, however, resulting in diminished functional capacity. In muscle, a fatigue process develops along with acute pain, and mechanical damage due to tissue fatigue and interference with nutrition can occur in tendons and joints. Thus, if frequently repeated or continuous activities are performed for several weeks or longer, chronic pain and tissue degeneration can develop.

The concept of physiological fatigue is best exemplified in skeletal muscle tissue. A number of studies have measured the length of time a volitional exertion can be sustained (i.e., the muscle endurance time) for exertions set at levels proportional to the strength of the muscle (or group of muscles) involved. Based on these experiments, muscle endurance–force relationship curves can be made (Figure 10.1). If frequent exertions are required for given contraction periods with controlled rest periods between each exertion, then the endurance–force relationship displayed in Figure 10.1 is shifted toward the upper right quandrant, with the amount of change dependent upon the length of each exertion period and interspersed rest period. Figure 10.2 displays a family of curves developed by Lee (1979) for the forearm-flexion muscles, using a constant exertion of 2.5 s, with interspersed rest periods of varied lengths.

The variation in muscle endurance times between (1) individuals, (2) muscles tested, and (3) specific experimental conditions is large in these experiments. Thus, the prediction of the endurance time for a muscle group depends upon a number of conditions, only some of which can be easily measured and controlled. Despite this, it is clear from both human and in vivo animal studies that muscle endurance can only be maintained in sustained or frequent exertions by either assuring that the contractile force is low, or by interjecting adequate rest periods.

The discussion of the exact mechanism for the muscle endurance–force relationship is beyond the scope of this book. Needless to say, it is dependent upon alterations in muscle cell metabolism, muscle cell membrane functions, depletion of certain metabolic substates, and accumulation of intermediate metabolites and heat. During the first stages of fatigue, such alterations can be considered to produce tissue irritation, implying that the effect is minor and reversible. As a fatigue process continues, however, the irritant stimulus is detected at a conscious level, and is referred to as ''muscle'' or ''ischemic pain.''

Caldwell and Smith (1966) have described the pain perception process of prolonged static exertions. They asked a group of subjects to grade the pain they perceived during sustained contractions using the following scale: 1 (just noticeable), 2 (moderate), 3 (severe), 4 (very severe), and 5 (intolerable). Hand grip exertions were set at 25%, 40%, and 55% of each subject's demonstrated isometric grip strength. The mean values of the resulting pain ratings are displayed in Figure 10.3, and disclose a consistent relationship with contraction time and relative load. The test–retest correlation coeffi-

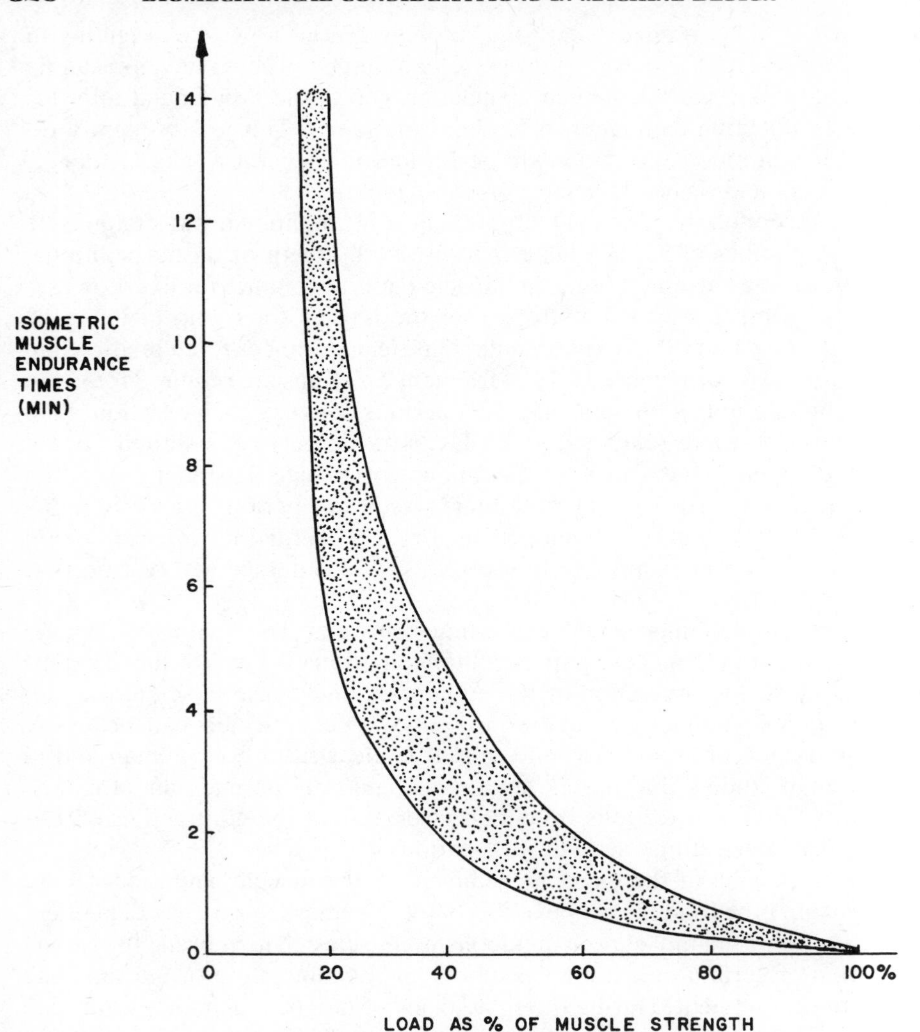

FIGURE 10.1 Composite of muscle endurance–load relationships for different muscles and experimental conditions (adapted from Schutz, 1972).

cient of the pain ratings ranged from 0.48 to 0.80, with middle intensity pain levels 2, 3, and 4 being most consistently repeated.

The acute pain of a muscle exertion can be thought of as a precursor or warning sign. If the pain causing exertion is continued on a daily basis, adaptation may occur—that is, the muscle becomes stronger and gains in endurance with a concomitant decrease in pain levels. If the exertion is so heavy that pain does not decrease or even increases, however, not only is the muscle tissue unable to adapt, but an inflammatory process can be

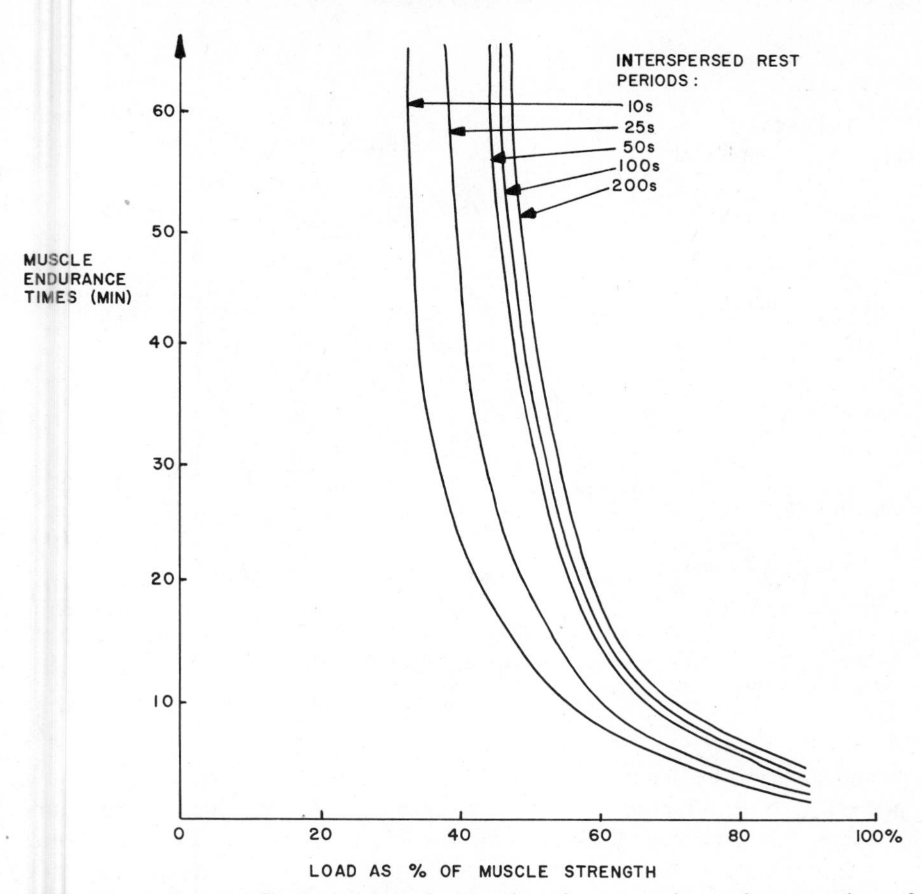

FIGURE 10.2 Forearm flexor muscle endurance times in consecutive static contractions of 2.5 sec duration with varied rest periods. The shorter the rest period, the shorter is the endurance time at each percentage level of muscle strength (adapted from Park, 1979).

initiated in tendon and joint tissues involved in the exercise. Such inflammation can develop as a reaction to mechanical strain of the tissues when there is not adequate rest to allow physiological adaptation. In particular, inflammation of tendons, or *tendinitis,* can occur when a tendon has been repeatedly tensed. Swelling from inflammation of the tendon may itself contribute to pain. With further exertion, the tendon collagen fibers become separated. At the point of greatest stress, often where the tendon passes around an adjacent bone or ligament structure, the collagen fibers can be pulverized, leaving debris containing calcium salts. These calcium salts and circulatory fluids within the injured tendons produce further swelling and pain (Cailliet, 1981). If the harmful work activity is continued, the degeneration will involve the surrounding tendon synovia and possibly the bursa

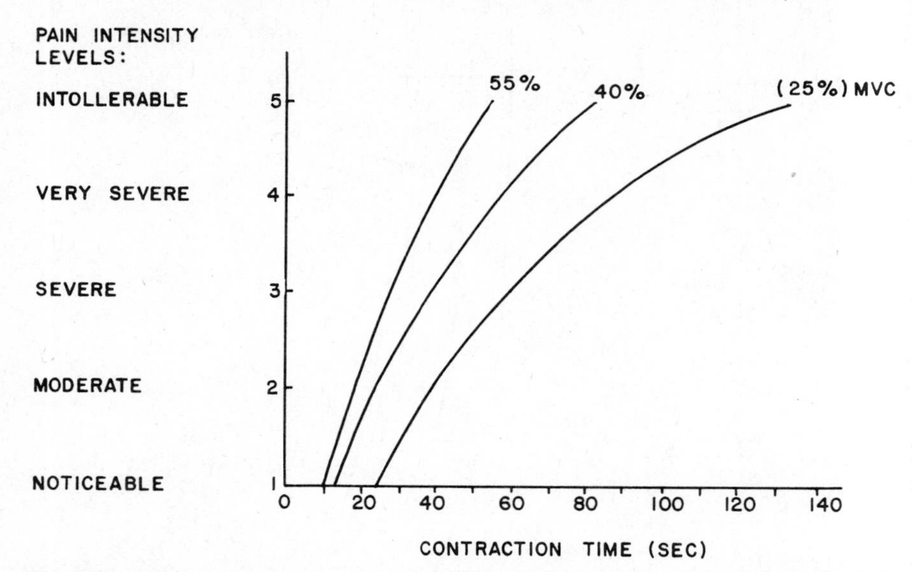

FIGURE 10.3 Mean time of appearance of various pain intensities for three static loads relative to subject's MVC (Maximum Voluntary Contraction). Note that the pain intensity level is higher at short contraction periods when the exertion is at a larger percentage of MVC (adapted from Caldwell and Smith, 1966).

membrane of the adjacent joint. The latter structure at the shoulder is shown in Figure 10.4, adapted from Cailliet (1981). Since the synovia and bursa provide a low-friction sliding mechanism for tendons, muscles, and bones, any inflammation and resulting degeneration of these tissues can produce limitation of motion, and pain. If restricted mobility remains, muscles and fascia shorten, producing a muscle contracture. A permanent functional disability is often the final result.

The sequence of events leading to a permanent dysfunction is shown in Figure 10.5. The figure also illustrates that natural somatic reactions occur because of motion related pain. Thus, the joint can be immobilized by antagonistic muscle actions. These muscle contractions "splint" the joint and often lead to spastic and painful muscles. They also promote fibrous reactions or adhesions between adjoining sliding tissues in the synovia, bursa, and joint capsular articulating surfaces (Cailliet, 1981).

Avoidance of the conditions described requires the recognition that the musculoskeletal system is well suited to produce repeated motion at low force loads. When the loads become high, are prolonged or frequent in nature, or involve awkward postures, however, the muscles, tendons, synovia, bursa, and bone of the articulating structures become painful, inflammed, and begin to degenerate. What follows is a discussion of various workplace and machine control layout considerations meant to be helpful in reducing

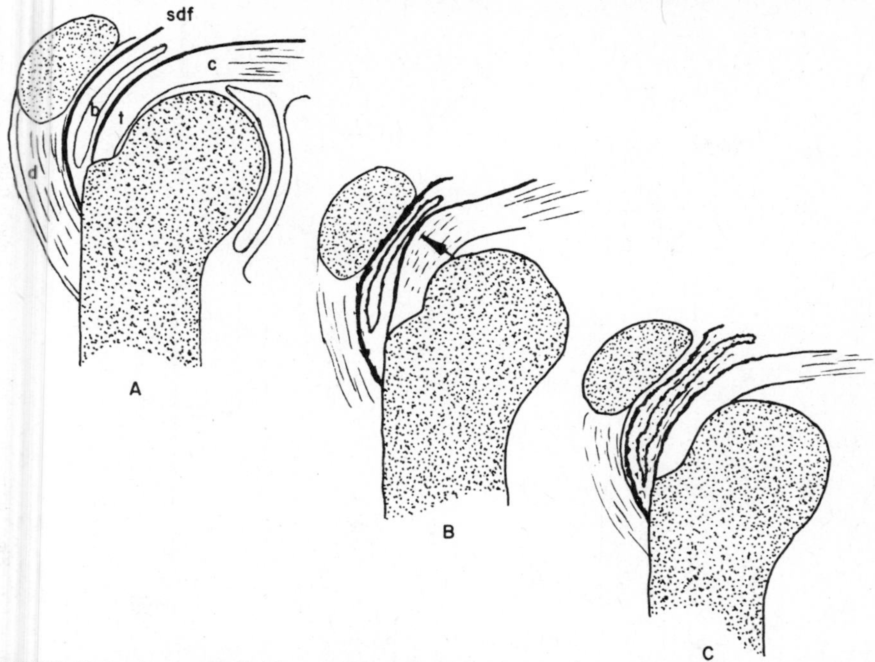

FIGURE 10.4 Acute tendinitis and inflammation of contiguous tissues. A. The tissues contained between the rotator cuff (c) and its tendons (t), and the subdeltoid fascia (sdf) lining the undersurface of the deltoid muscle (d). The subdeltoid fasica is rich in blood vessels and sympathetic nerves. Between the subdeltoid fascia and the fascia covering the cuff is loose connective tissue, within which is located the subdeltoid bursa (b). B. Acute bulging of the tendon compresses and causes inflammation and swelling of the fascial tissues and the bursa. This is the acute mechanical phase during which there is severe pain and limitation of motion. C. The bulging of the tendon has subsided, but the resultant fascia and bursal inflammation persists, causing stiffness of the shoulder. A can go to B then to C and reverse through the entire phase back to A. The tendon remains frayed, and degenerative changes remain if stage C is reached, (Cailliet, 1981).

the incidence and severity of these types of problems. Similar considerations are given for hand tool design in Chapter 11 and for seat design in Chapter 12.

10.2 GUIDELINES FOR WORKPLACE AND MACHINE CONTROL LAYOUT

It should be apparent from the preceding discussion that if frequent or sustained muscle exertions are required, care must be taken to minimize the external load required and avoid unnatural postures. This section attempts to develop some practical guidelines regarding the placement of machine controls and general workplace layout. Because seated operator activities

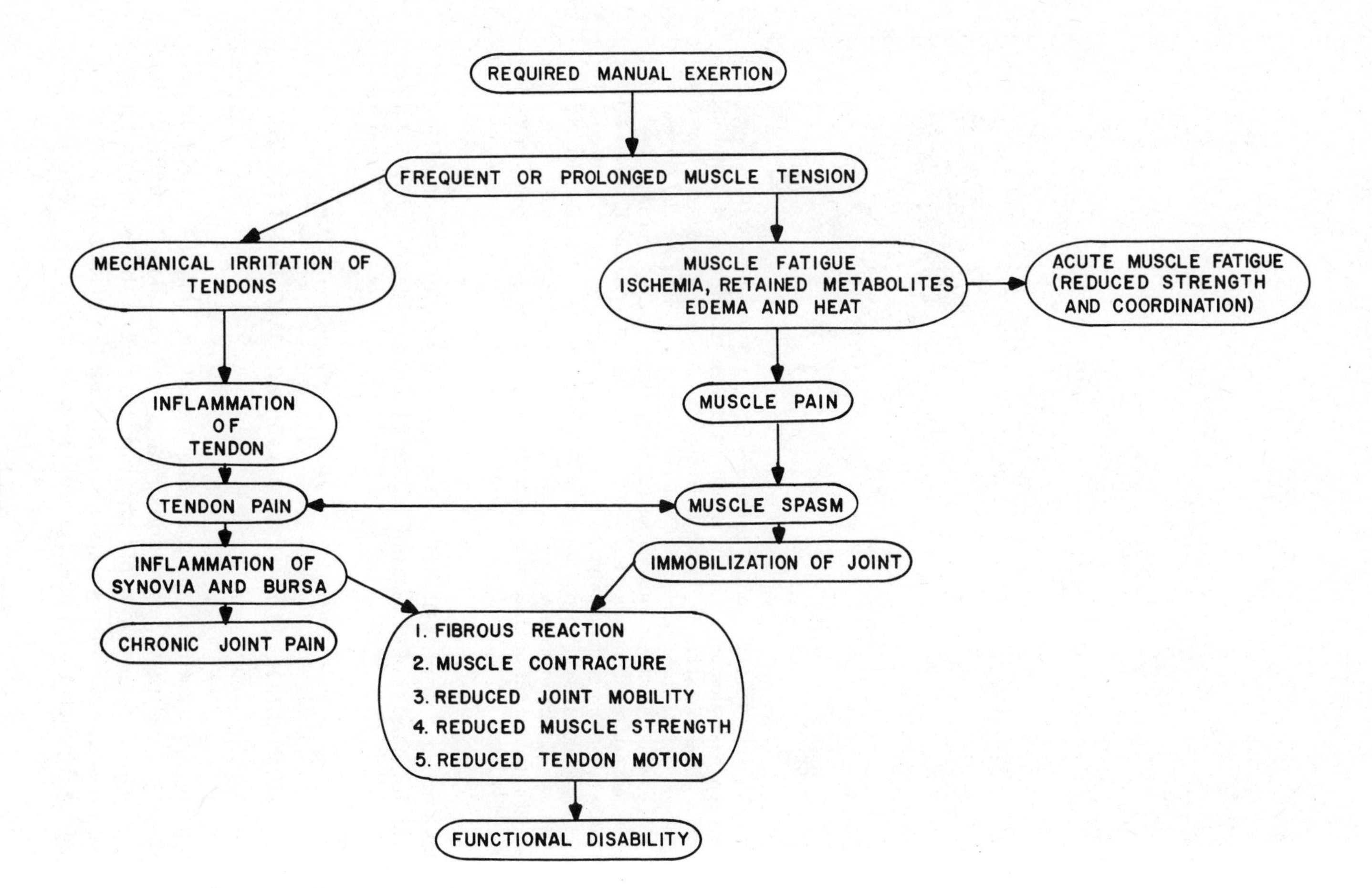

FIGURE 10.5 Sequence of events producing muscoloskeletal pain and functional disability because of frequent or prolonged muscle tension.

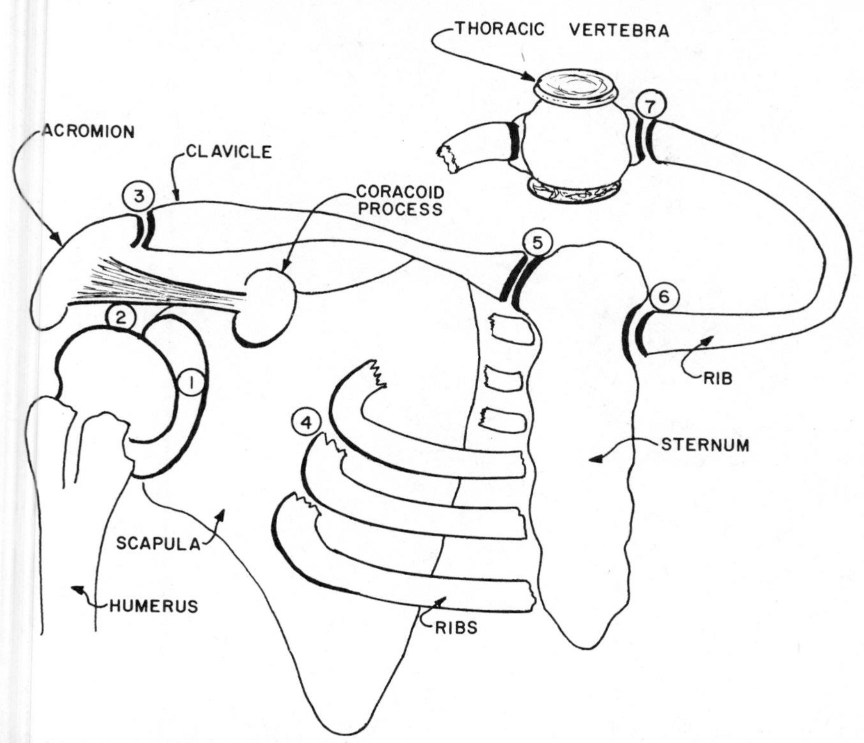

FIGURE 10.6 The joints of the shoulder girdle: (1) glenohumeral; (2) suprahumeral; (3) acromioclavicular; (4) scapulocostal; (5) sternoclavicular; (6) costosternal; (7) costovertebral (adapted from Cailliet, 1981).

are so prevalent in industry today, shoulder, arm, and neck problems are first discussed in that context. Following that is a general discussion of machine control locations and work surface heights.

10.2.1 Shoulder-Dependent Overhead Reach Limitations

The shoulder joint represents one of the most complex biomechanical structures one can imagine. The ball-in-socket arrangement of the glenohumeral joint (which is really a ball in a comparative loose socket) provides a large amount of mobility for the arm, and is the true *shoulder joint* per se. Mobility for the arm is further enhanced by six additional joints that comprise the trunk–arm complex sometimes referred to as the *shoulder girdle*. Figure 10.6 illustrates the joints involved in arm motion. These allow the glenohumeral joint to translate and rotate, providing additional reach capability.

In providing such extreme mobility, intrinsic stability has been sacrificed. In fact, the stability of the shoulder joint during normal motions is largely

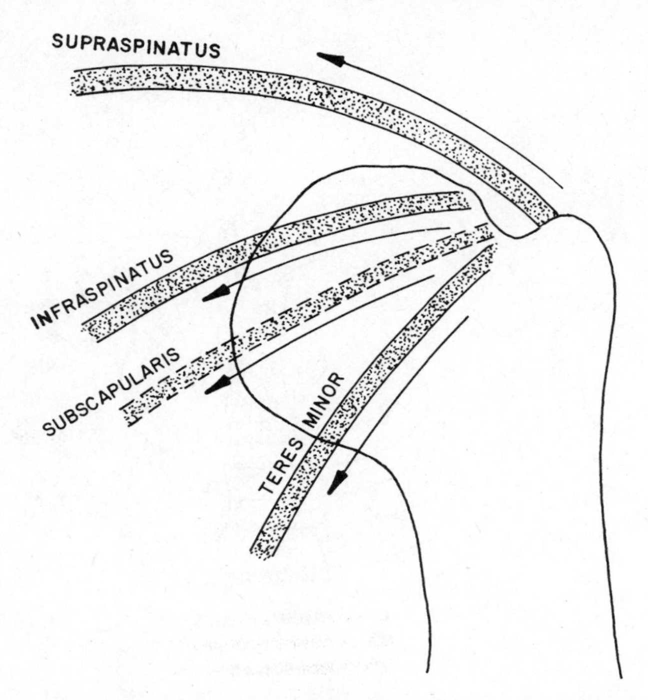

FIGURE 10.7 Rotator cuff mechanism. The supraspinous (supraspinatus) muscle pulls the head of the humerus into the glenoid and slightly rotates the humerus into abduction. The infraspinous (infraspinatus) muscle also rotates the head and slightly pulls it down. The teres minor muscle pulls in a more downward direction. The subscapular (subscapularis) muscle pulls the head into the glenoid, but its main rotatory action is to internally rotate the humerus about its longitudinal axis (adapted from Cailliet, 1981).

dependent upon the ligaments and joint capsules, and on the extensive musculature of the upper torso. In particular, the head of the humerus is held firmly in the glenoid socket of the scapula by the combined actions of the rotator cuff muscles depicted in Figure 10.7 and the pectoral (breast) muscles. Motion and forceful arm exertions are mainly accomplished by the larger shoulder and arm muscles depicted in Figure 10.8. In simplified form, from Quiring and Warfel (1967) and Perry (1978), arm abduction in the coronal plane is accomplished by the deltoid muscle, with assistance of the rotator cuff muscles. Arm flexion forward in the sagittal plane is accomplished mainly by the anterior deltoid, coracobrachialis, and biceps brachii muscles. Adduction and medial rotation inward relies on the pectoralis major. Extension and rearward adduction depends on the posterior deltoid, long head of the triceps brachii, and teres major. Of course, if the arm begins its motion from an extreme position—for example, it is hyperextended and a forward flexion or abduction is attempted—different combinations of mus-

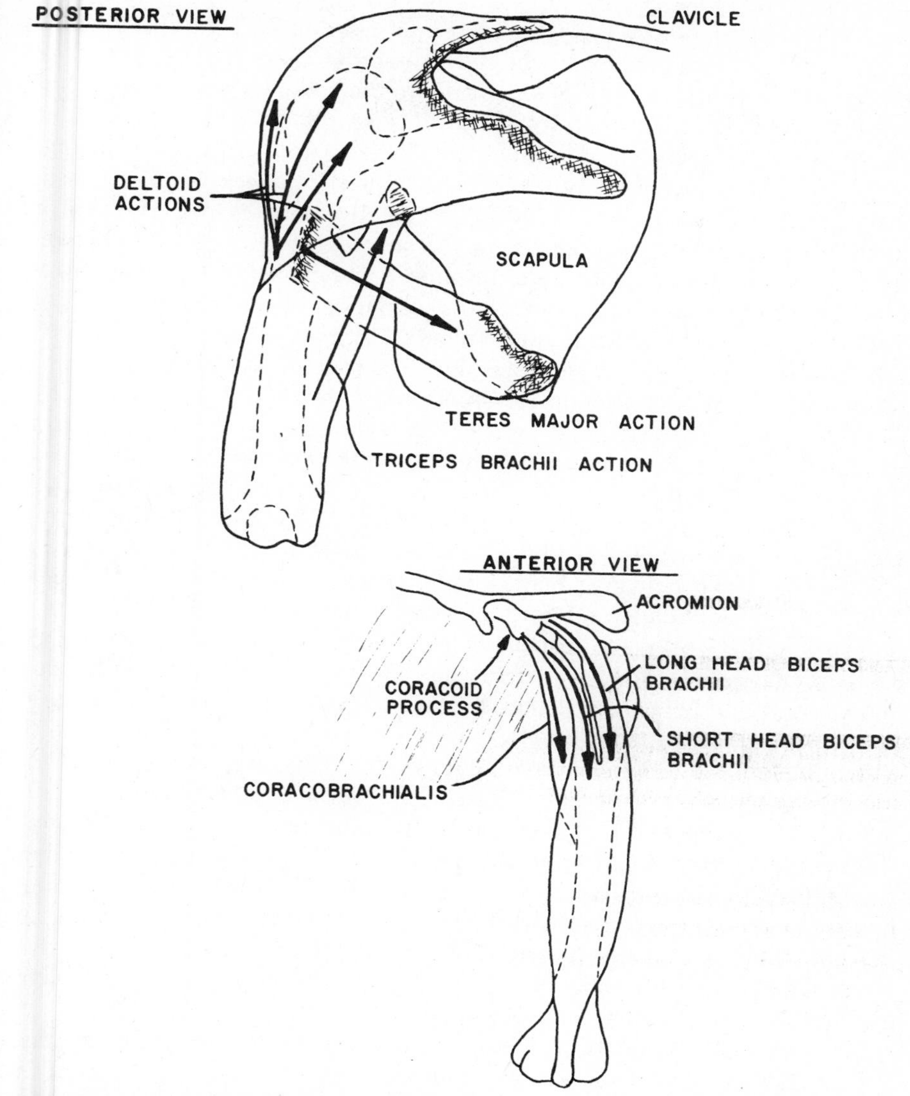

FIGURE 10.8 Major muscular actions (ignoring rotator cuff muscles) responsible for arm motions at shoulder as seen from posterior (top) and anterior (bottom) views.

cles are used as the arm moves through the full arch-of-motion. Also complicating the kinesiological description is the fact that when the scapula is raised, lowered, or rotated, it alters the line-of-action of the muscles and their respective mechanical advantages. This in turn changes the relative muscle actions necessary to complete a given arm motion or exertion.

Because of these complexities the shoulder girdle has not been biomechanically well modeled, as discussed by Engin (1980). One final complication must also be acknowledged in this regard. Shoulder pain per se may be caused by various conditions extrinsic to the shoulder joint. Owen (1969) has listed these extrinsic pain sources as (1) visceral lesions (e.g., gallbladder disease, ruptured viscus, diaphragmatic irritation, coronary insufficiency, etc.), (2) cervical spine disorders (e.g., whiplash, cervical muscle spasms, disc-related radiculitis, etc.), and (3) brachial plexus lesions (e.g., entrapment of brachial nerve plexus and/or the subclavian artery or vein between the first rib and the scalenus anterior muscle and clavicle). Because of these concerns, for the present it appears that empirical observations as to shoulder postures and arm loads that are found to be highly fatiguing or otherwise injurious to various shoulder tissues must be relied upon to guide design of the workspace until more complex models are developed and validated.

Of particular concern in the specification of arm work requirements is that the hands not have to reach frequently or be held for sustained periods above shoulder height. Jobs that require such elevated arm activities have been shown to create "degenerative tendinitis" in the biceps and supraspinatus muscles (Bjelle, Hagberg, and Michaelson, 1979; Herberts et al. 1981). If the arm is held in an elevated posture (e.g., when welding overhead), shoulder muscle fatigue and biceps tendinitis has been identified as a major concern in the workplace, especially for older workers who have reduced joint mobility (Herberts, Kadefors, and Broman, 1980). It has also been shown by Hagberg (1981a), using an EMG analysis technique, that the upper part of the trapezius muscle rapidly fatigues and becomes painful when the arm is flexed and held at 90° (shoulder height). These studies showed that the rate of muscle fatigue development in several shoulder muscles was 10 to 20 times slower when the shoulder load moment was below 20% of the shoulder flexion strength moment, compared with when the load moment was approximately 40% of the strength moment. This latter result should indicate a major concern to workspace designers when coupled with the fact that the strength moment for the shoulder in flexion and abduction is much lower when the arm is elevated than when in lower positions, as depicted in Figure 10.9 and discussed in Chapter 4.

The effect of holding the arm in various elevated positions has been measured by Chaffin (1973) using EMG frequency spectra shifts and a subjective pain rating to measure endurance times of young healthy males. Figures 10.10 and 10.11 depict these findings. From these studies it is concluded that sustained, elevated arm work, especially if supporting a load, must be minimized to avoid shoulder muscle fatigue and associated tendinitis problems (Hagberg, 1981b). It should also be clear that with repeated exertions of short duration (perhaps only a few seconds), similar muscle fatigue will develop if the relative load on the muscles is over approximately 40% of the expected strength and rest periods between contractions are smaller than about 10 times the contractile period (see Figure 10.2 and Lee, 1979).

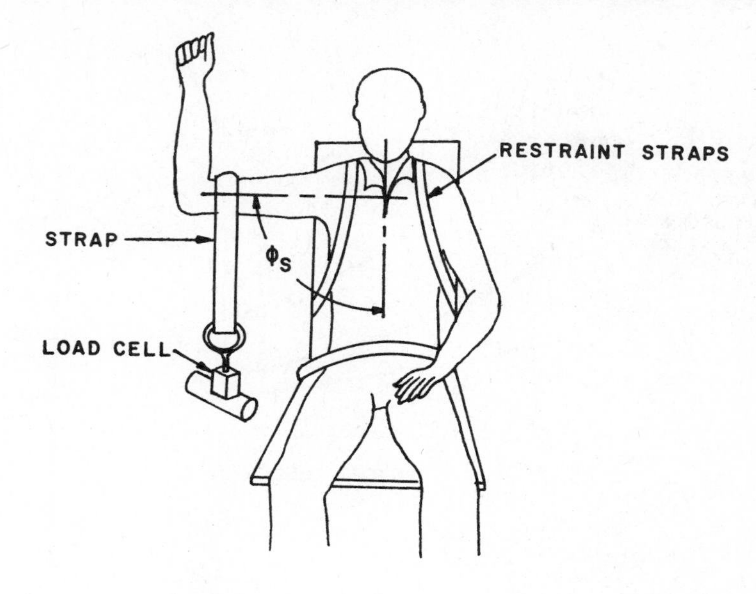

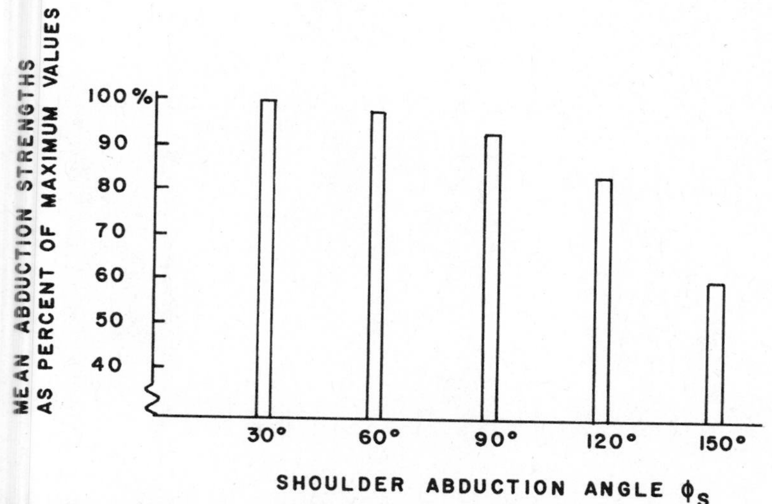

FIGURE 10.9 Shoulder abduction strengths decrease as the arm is abducted from the torso (adapted from Stobbe, 1982).

Further, even without a hand load, any elevation of the arm in abduction or forward flexion above about 90° greatly increases the stress on various tendon–ligament–capsular tissues, as indicated by greatly increased passive resistive moments towards the end of the volitional range-of-motion of the shoulder (Engin, 1980). In this regard, it is known that acute tendinitis of the shoulder muscles can be induced by high-velocity arm motions, as in

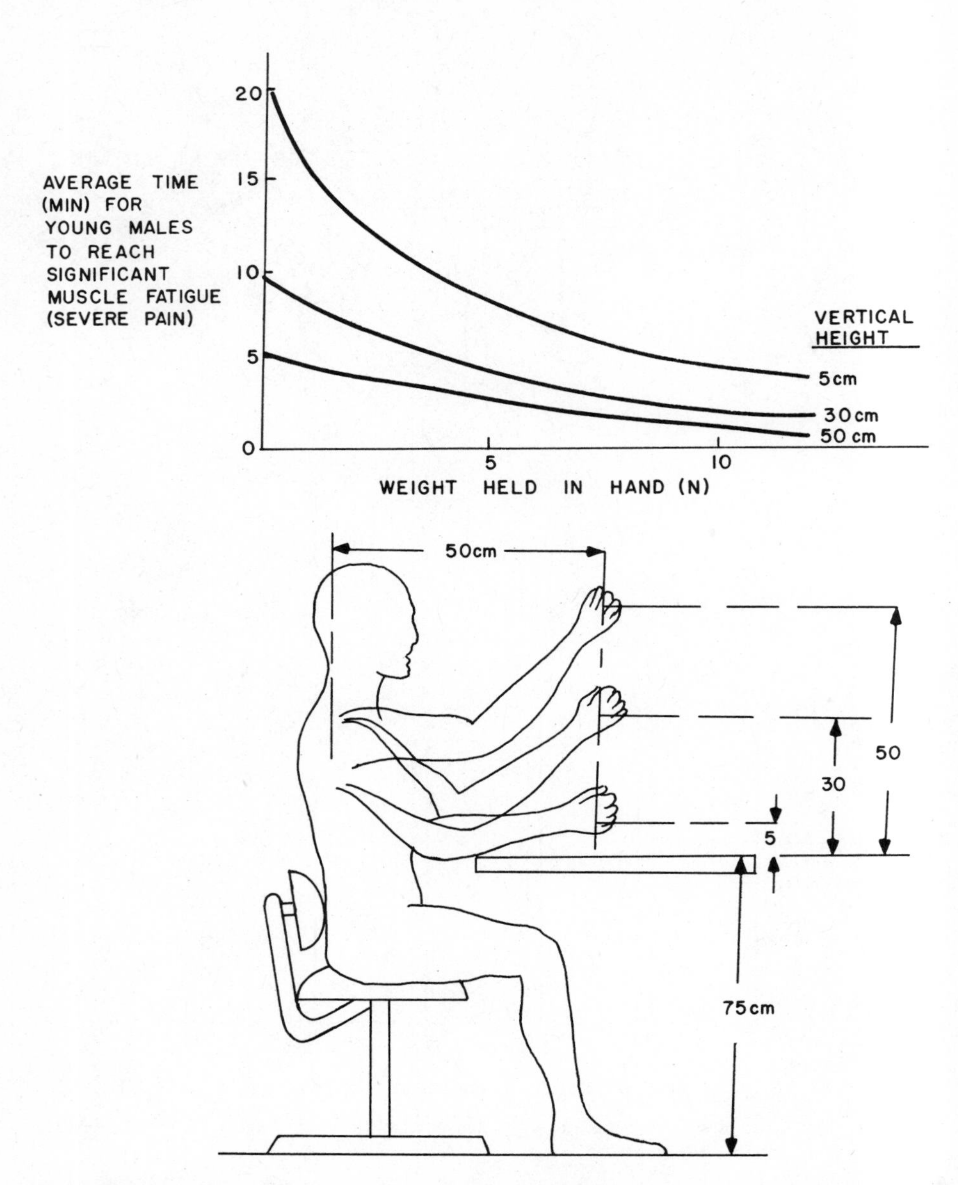

FIGURE 10.10 Expected time to reach significant shoulder muscle fatigue for varied arm flexion postures. The larger the flexion angle, the earlier fatigue will develop (Chaffin, 1973).

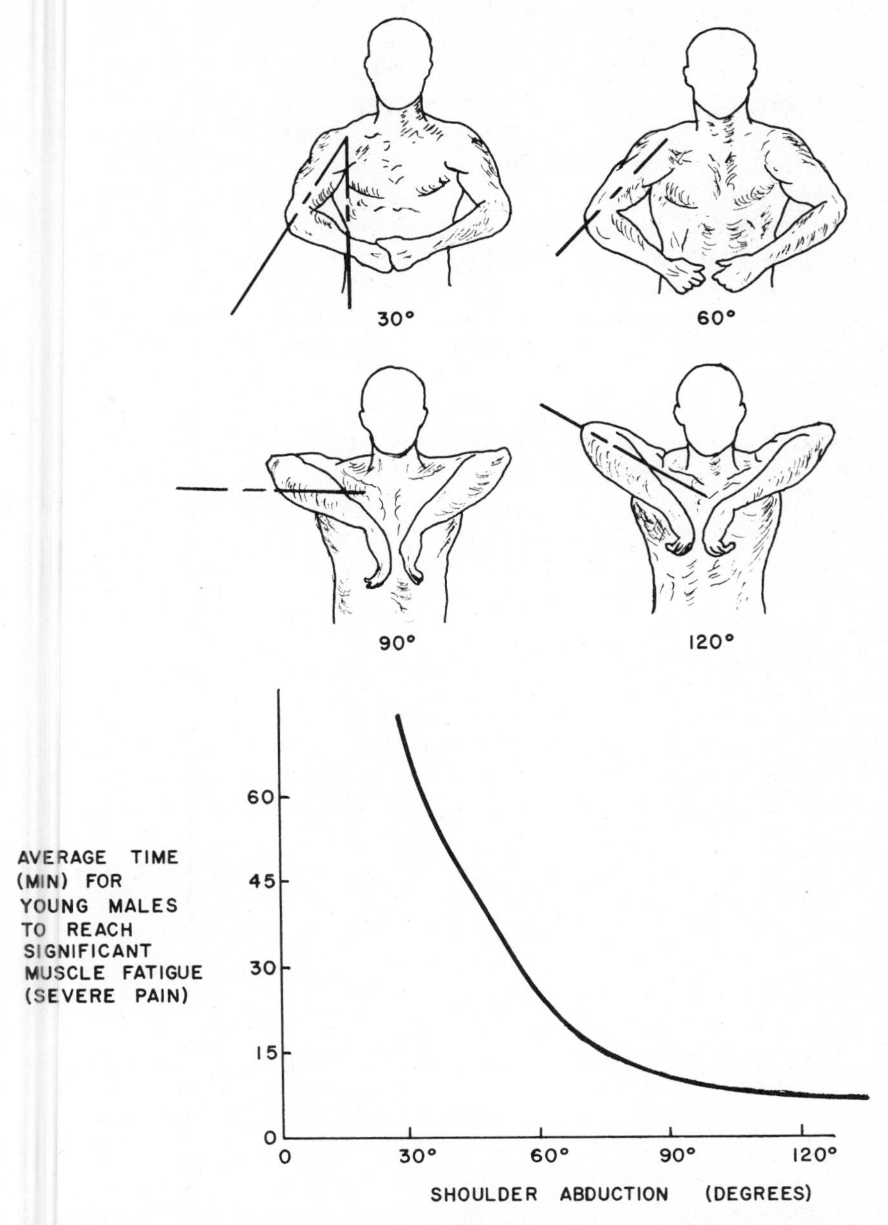

FIGURE 10.11 Expected time to reach significant shoulder muscle fatigue for varied arm abduction angles (Chaffin, 1973).

tossing materials (Nolan, 1979). It is believed that such motions result in sudden and excessive strain on specific tendons as particular muscles contract to provide the acceleration and deceleration necessary to execute the gross motion while maintaining joint integrity. High-velocity motions (i.e., those that are executed with motion times less than normal predetermined motion times, as discussed in Chapter 7) should be carefully scrutinized to determine if they are harmful. This is especially true if rapidly moving a load held in the hands. Though epidemiological evidence is not sufficient to support a strong conclusion, it is generally known that the use of piece-part pay incentive plans in industry (in which the worker receives a monetary bonus for each part produced over a standard number) encourages high-speed arm motions that can be harmful to the musculoskeletal system if the motions approach the range-of-motion of the shoulder joint.

10.2.2 Shoulder- and Arm-Dependent Forward Reach Limits

When reaching forward, the shoulder joint is flexed and the elbow becomes extended. If a load is held in the hands, the load moments at the elbow and shoulder can become large relative to the flexor strength moments required at both joints (see biomechanical discussion in Section 6.5.3 and arm reach models described by Anderson, Ortengren, and Schultz, 1980). If the arm and forearm are elevated to nearly horizontal when reaching forward, a load of only 56 N held in the hands will create a load moment at the shoulder equivalent to the flexor strength moment predicted for the average female, while a 115 N load will equal the shoulder lifting strength of an average male. Clearly, from the preceding discussions, such loads can only be held in these positions for a few seconds before shoulder fatigue develops. With weaker or older subjects these endurance limits are even lower.

In this regard, varied forward reach arm postures and hand load conditions were used in a study of shoulder muscle fatigue (Chaffin, 1973). The results are shown in Figure 10.12 as the mean endurance times of five healthy young males. The endurance times were based on EMG frequency shifts and subjective pain levels.

In the same study by Chaffin (1973) an adjustable elbow support was also used. This external body support reduced the load moment on the shoulders. The resulting muscle fatigue then developed in the elbow flexor muscles instead of the shoulder muscles, with the mean endurance times as shown in Figure 10.13.

It should be clear from these data that loads of even nominal amounts cannot be supported for sustained periods, especially if the arm or forearm is elevated in a forward reach posture. This means that the use of a padded forearm support is important to relieve the load moments at the shoulder and elbow, as advocated by Chaffin (1973), Grandjean (1980), and Konz (1979). Power handtools used frequently during a workday (often used when driving screws repeatedly in an assembly) or for sustained periods (when

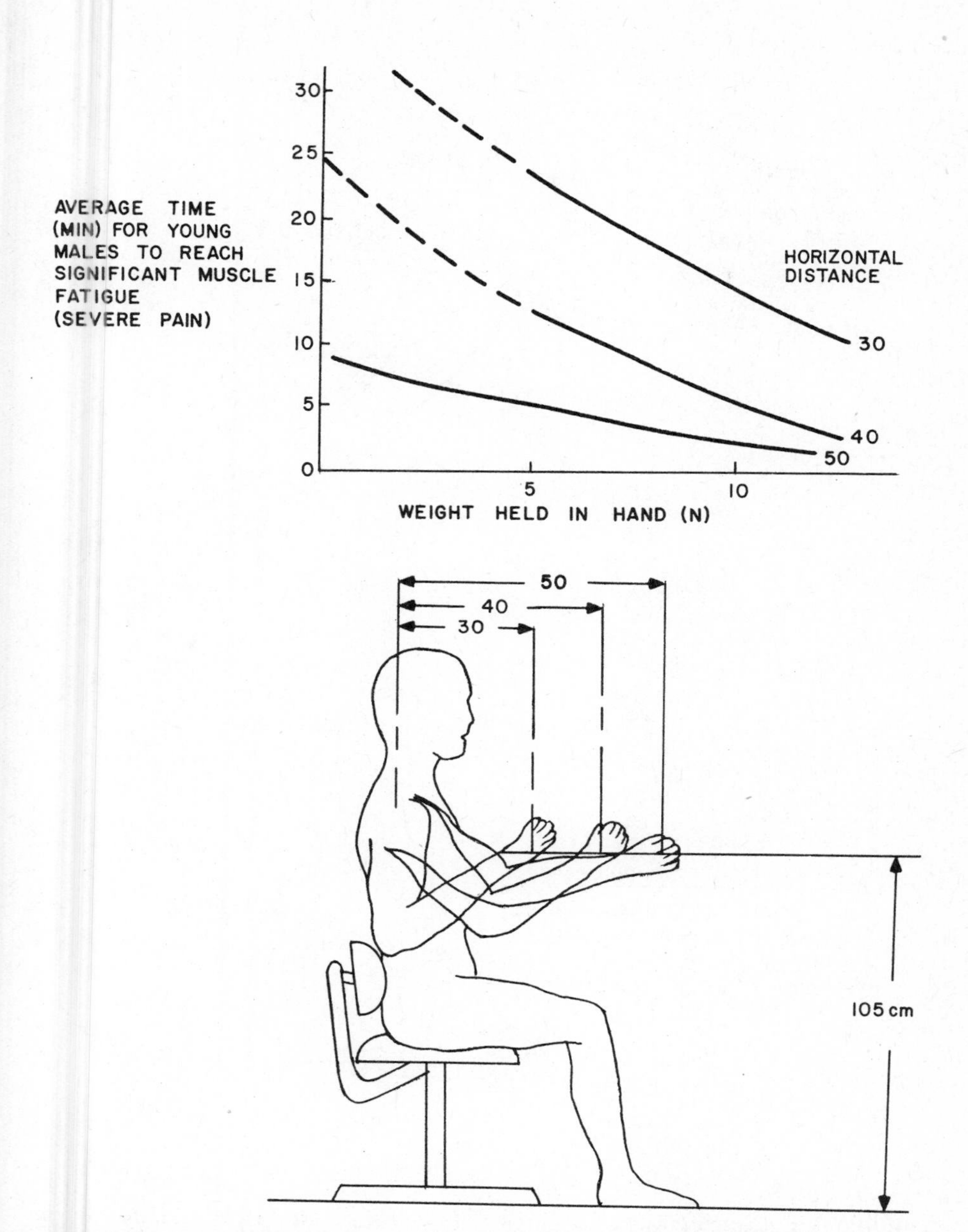

FIGURE 10.12 Expected time to reach significant shoulder muscle fatigue for different forward arm reach postures. The elbow is unsupported. The greater the reach, the shorter the endurance time (Chaffin, 1973).

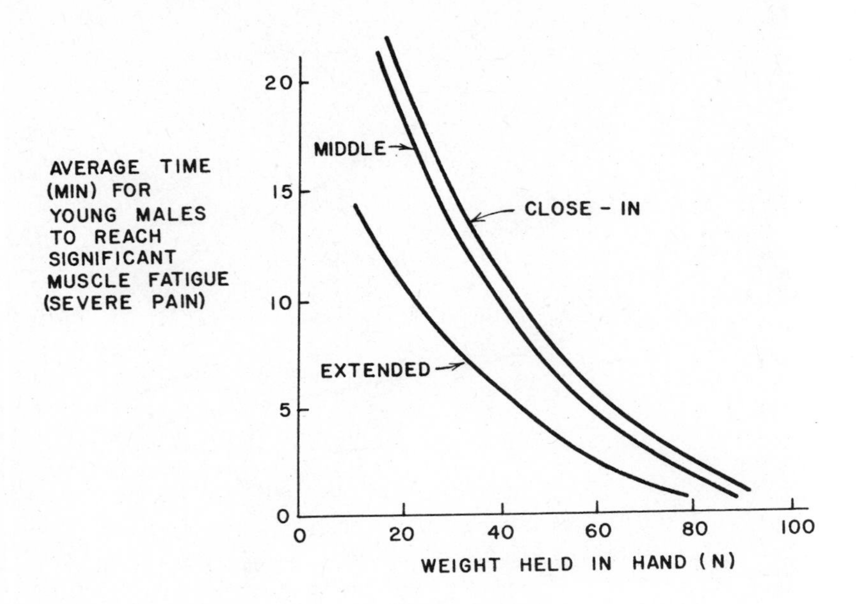

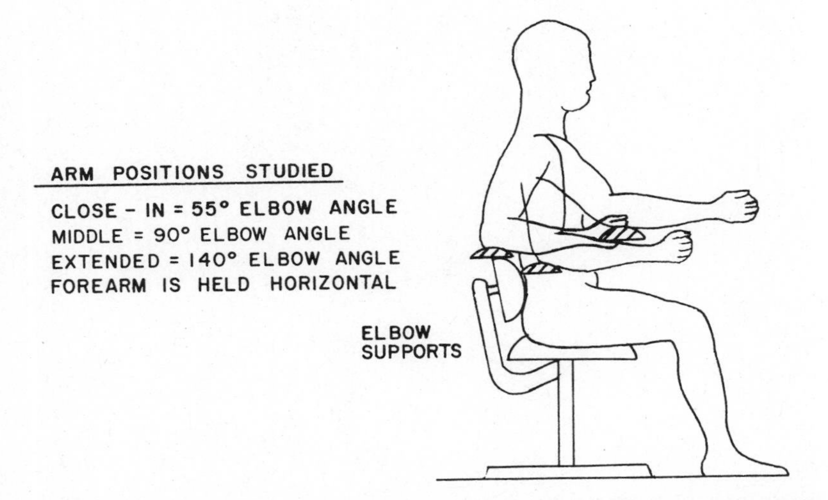

FIGURE 10.13 Expected time to reach significant shoulder and arm muscle fatigue for different arm postures and hand loads with the elbow supported. The greater the reach, the shorter the endurance time (Chaffin, 1973).

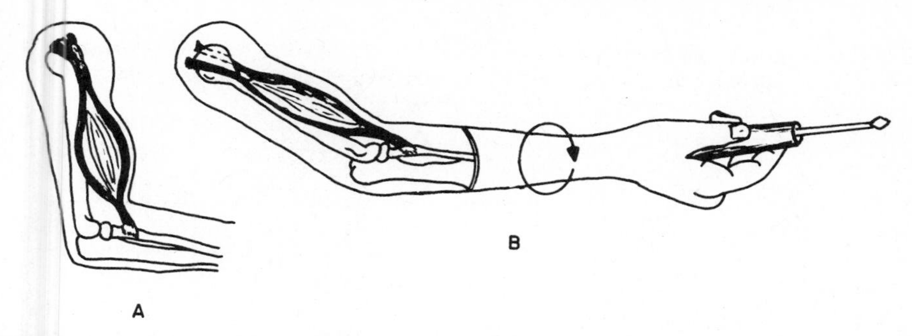

FIGURE 10.14 Forceful outward rotation (supination) of the forearm and hand is best achieved with the elbow flexed (A). When extended (B), the biceps brachii is not positioned to assist in rotation of ulna, and produces high joint compression forces (Tichauer, 1978).

welding, spraying, or grinding) should be suspended from a tool balancer device to minimize the weight effect (Konz, 1979). Also, workpieces or assemblies should not have to be supported by one hand while the other performs a required operation. Good workplace design provides adjustable fixtures that support the workpiece in proper orientations for the operator, taking account of both visual and manual task requirements (Chaffin, 1973).

In this latter regard, it should be recognized that arm/forearm postures are often dictated by hand orientation around the long axis of the forearm. If the hand is supinated (palm-up), then the arm will normally be adducted and close to the torso. If the task requires the hand to be prone (palm-down), then the arm normally will be more abducted and elevated. If the hand is located in a position that already requires the arm to be elevated, then using a prone hand posture may further the arm elevation. Realization of this interdependence between hand orientation and arm postures is important, particularly in designing handles on tote boxes, hand-tool configurations, and controls on machines.

In reference to this latter point, if a control or a screw must be turned in a clockwise direction, requiring outward rotation or supination of the hand about the long axis of the forearm, it is also important that the elbow be flexed to 90° or less (Tichauer, 1978). This provides good mechanical advantage for the biceps brachii (see Figure 6.7), which is a major outward rotator of the forearm as well as an elbow flexor muscle. According to Tichauer (1978), by having the elbow flexed, biceps contraction rotates the proximal head at the ulna about the radius without producing a large compression stress within the elbow joint, as is the case when the elbow is extended. This is illustrated in Figure 10.14. Clearly, the flexed elbow reduces the tendon force requirement on the biceps tendon as well as assisting the supinator and brachio-radialis muscles that are also involved in forearm supination. When studying the effects of elbow angle in operations requiring frequent supination exertions, Tichauer (1978) compared injuries of two

TABLE 10.1
Mean Values of Forearm Rotational Torque Strengths (N m) Obtained with Varied Hand Locations for Five Male Subjects (Rohmert, 1966)

Direction of Exertion	Shoulder Angle	Location of Handgrip as Percentage of Maxmimal Grip Distance		
		50%	75%	100%
Rotate clockwise (supination)	30°	17.5	16.6	7.9
	0	18.4	15.9	5.8
	−30°	15.6	13.3	4.9
	−60°	11.6	9.1	6.7
Rotate counterclockwise (pronation)	30°	16.9	15.0	10.9
	0	17.9	15.5	10.9
	−30°	20.8	18.2	11.8
	−60°	22.8	20.1	13.7

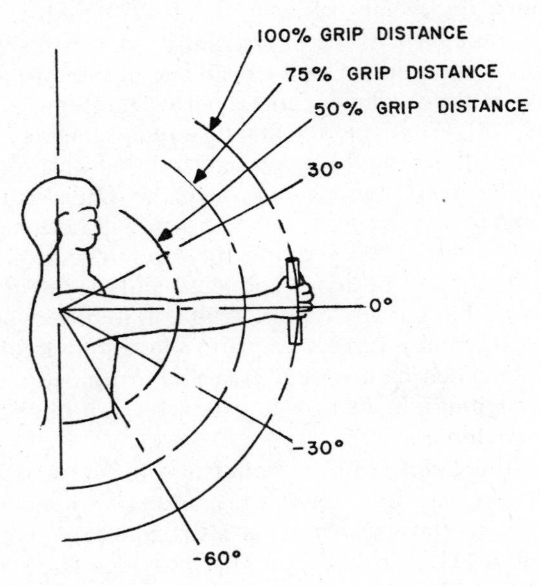

groups of workers. Twenty-three subjects with elbow pain were found among workers who performed repeated forearm supination activities with the elbow extended to 130° or more, while no cases were reported for workers using an elbow angle of 85° or less.

Rotational strength performance of the forearm is also enhanced by having the elbow flexed, as reported by Rohmert (1966). In his studies, five healthy males attempted to rotate a vertical handle placed at various locations relative to their reach and shoulder locations. Table 10.1 presents the means of the static strength values obtained. What is clearly evident in these values is a trend towards lower strengths when the elbow is extended to provide maximum reach. This result was consistent at all four shoulder angles. Thus from biomechanical, strength, and epidemiological points of view, it appears that work activities involving forearm rotation should be performed with the elbow flexed.

When attempting to perform a maximal push or pull exertion with one hand in a seated posture, the resulting strength performance is very dependent upon shoulder and elbow angles. Because of the interaction of strength performance and posture, biomechanical strength models are important in understanding the complexities of such three-dimensional exertions. In this regard, the three-dimensional strength model described in Section 6.5.3 was used to predict one-handed exertions of the human operator performing various seated static exertions (Garg and Chaffin, 1975). The validation of this model was performed using one-handed strength data obtained from Schanne (1972) and Thordsen et al. (1972). This disclosed that strength values could be predicted with an average correlation coefficient of between 0.7 and 0.9, depending on the specific activity being simulated. Some resulting one-handed strength predictions for an average male in a seated posture are presented in Figure 10.15. Inspection of these predictions once again discloses how complex asymmetric strengths are to rationalize. In general, exertions in a lateral direction (to the left or right) with the hand close to the body require arm humeral rotation at the shoulder, and result in much lower values than exertions in the sagittal plane (in and out). These latter type of exertions use arm flexion and extension, involving the stronger biceps, brachialis, and triceps muscles of the arm.

It is also clear in Figure 10.15 that when the arm is close to its reach limit (elbow is straighter or locked in extension), push and pull forces can become quite high, limited only by torso strengths and balance. In this regard, if the seat and foot support do not brace the person, the reactive load moments and forces created at the torso and legs will become limiting—that is, the person will fall-off the seat or the seat will tip over, as discussed in Chapter 9.

The consideration of bracing a seated person to allow one-arm maximum push and pull exertions is experimentally addressed by Laubach (1978). In his experiment, aircraft seats were used with full shoulder and lap belts to brace the torso. The seat back was tilted at various angles and the seat was moved fore and aft to alter leg postures. The resulting static strengths of Air Force recruits were found to be greatly affected by each of these changes in a complex fashion.

It must be concluded from these results that specification of an optimal machine control location or the best position to locate stock is not simply determined. The direction of the exertion must first be determined (i.e., lift, push-inward, pull to left, rotate, etc.). Then the normative muscle endurance and strength curves presented in Figures 10.9 through 10.15 should be consulted for general guidance. A biomechanical strength model determination,

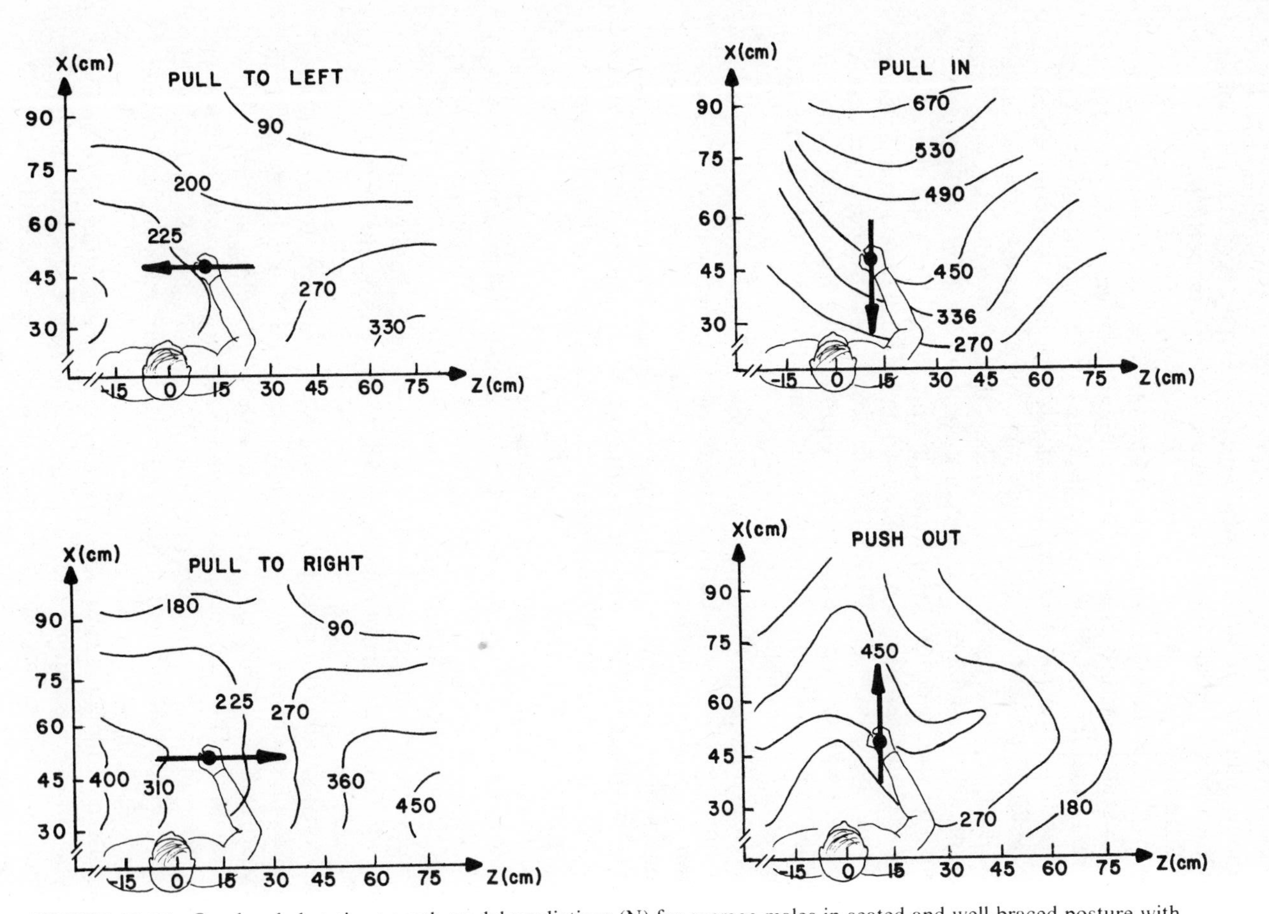

FIGURE 10.15 One-handed static-strength model predictions (N) for average males in seated and well-braced posture with hand located in horizontal plane 25 cm above hip joints (Chaffin, 1972).

as described in Chapter 6, may also be warranted. Finally, because of the various complexities described, if the manual function is strenuous (i.e., it involves a frequent or forceful exertion), a mock-up should be constructed and representative workers should perform the required exertions to determine if further corrections are necessary to control potential musculoskeletal problems.

10.2.3 Neck/Head Posture Work Limitations

Often, when performing precise motions, a worker is required to lean forward to obtain good vision of the work area. Some common examples of such postural requirements occur when inspecting parts for small visual defects, assembling equipment with small or intricate parts, or reading text material written on paper or on a VDU.

The requirement of work postures with the neck flexed forward can occur because of a combination of (1) the seat height being too high, (2) the seat placement being too far back from the work area, (3) a workbench or table being too low, or (4) the visual demands of the task requiring a specific eye location (e.g., to look into a near-vertical microscope or into a part assembly lying flat on a workbench, or to view a meter or VDU poorly located relative to the relaxed seated eye location of an individual).

From a biomechanics viewpoint, the problem has been described in general by Kumar and Scaife (1979). In essence, there are two moment equilibrium problems to solve in which the cervical extension muscle force F_C and the lumbar extensor muscle force F_L are resolved separately. The relevant postures and variables are defined in Figure 10.16. The values for F_C and F_L are found from the moment equations:

$$F_C = F_H \frac{a_H}{a_C} \tag{10.1}$$

and

$$F_L = \frac{F_T b_T + F_H B_H}{b_L} \tag{10.2}$$

where

F_H is the weight of the head and neck
F_T is the weight of the torso
F_C is the force due to the cervical extensor muscles
F_L is the force due to the lumbar extensor muscles
a_C is the moment arm of the cervical extensor muscles

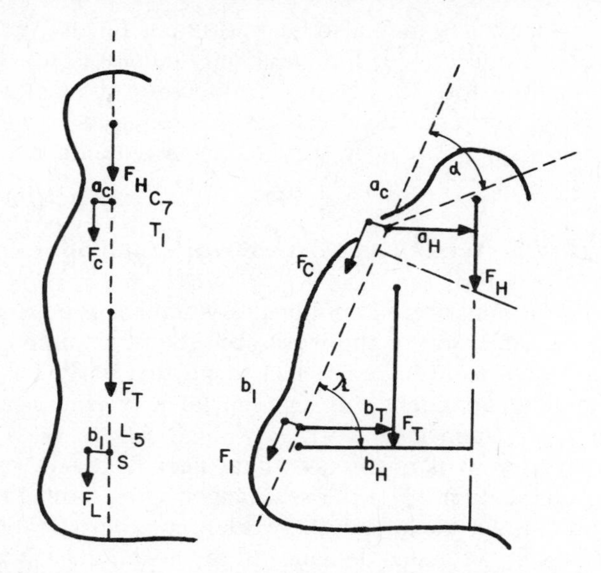

FIGURE 10.16 Forces acting on the spine in resting and forward inclined postures (Kumar and Scaife, 1979).

b_L is the moment arm of the lumbar extensor muscles
a_H and b_H are the moment arms of the head weight
b_T is the moment arm for the weight of the torso

Since the values of the muscle moment arms a_C and b_L are relatively constant for a given posture, the values of F_C and F_L become proportional to the load moments $F_H a_H$ and $F_T b_T + F_H b_H$, respectively. The magnitude of these moments primarily depends on the size of the angles α and λ depicted in Figure 10.16. Thus, for even a 30° inclination angle from the vertical, the moment and corresponding muscle force values are at 50% of the values achieved at 90° (horizontal). In other words, even slight forward lean postures can produce significant muscle contractions. This is illustrated in Figure 10.17, with reference to an average woman's head anthropometry and neck strength values determined by Snyder et al. (1975).

It appears that the cervical extensor muscles may produce muscle fatigue symptoms and pain quite quickly when head inclination angles become significant. Kumar and Scaife (1979) noted subjective reports of neck discomfort during the workday, and found these to be related to neck inclination angle. Using EMG frequency shifts in the cervical muscles and a subjective pain rating, Chaffin (1973) found that the endurance time of five young healthy females was greatly decreased when the neck inclination angle exceeded 30°, as shown in Figure 10.18.

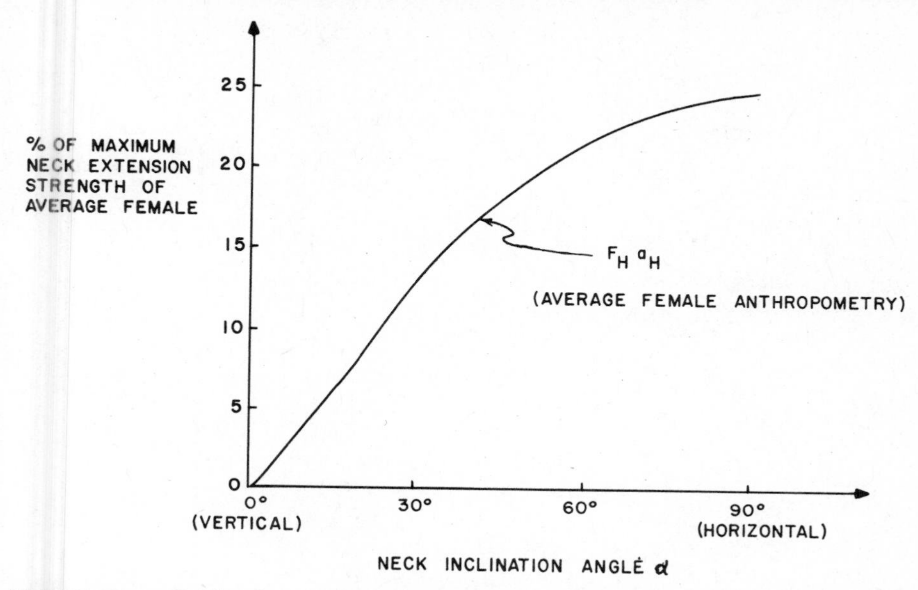

FIGURE 10.17 Predicted proportion of neck muscle strength required to support the head at varied inclination angles (for average female).

As discussed earlier in this chapter, acute muscle fatigue due to prolonged static exertions, if recurrent for months or years, can be a precursor of other more serious and chronic musculoskeletal disorders. The cervical spine is no exception. For this reason, it is imperative that neck pain symptoms be considered as having a potential occupational biomechanics etiology. The worker complaining of such should be evaluated to determine if the head is inclined more than a normal 20 to 30° for any prolonged period of time. If so, adjustments should be made in the chair, workbench, and workpiece location to provide vision with a more vertical neck and head orientation, or frequent mini-rest periods should be scheduled to allow the worker to move the head into a relaxed, upright posture, The latter is particularly necessary if the work being done is highly scheduled or in other ways mentally demanding. Such mental concentration has a tendency to result in postural muscles being "over-tensed," further contributing to the muscle fatigue process.

10.2.4 Torso Postural Considerations in Workbench Height Limitations

As indicated in the Section 10.2.3 (equation 10.2), the lumbar extensor muscles are required to stabilize the torso when standing in a stooped posture. As the torso is inclined from vertical with angle λ (see Figure 10.16), the

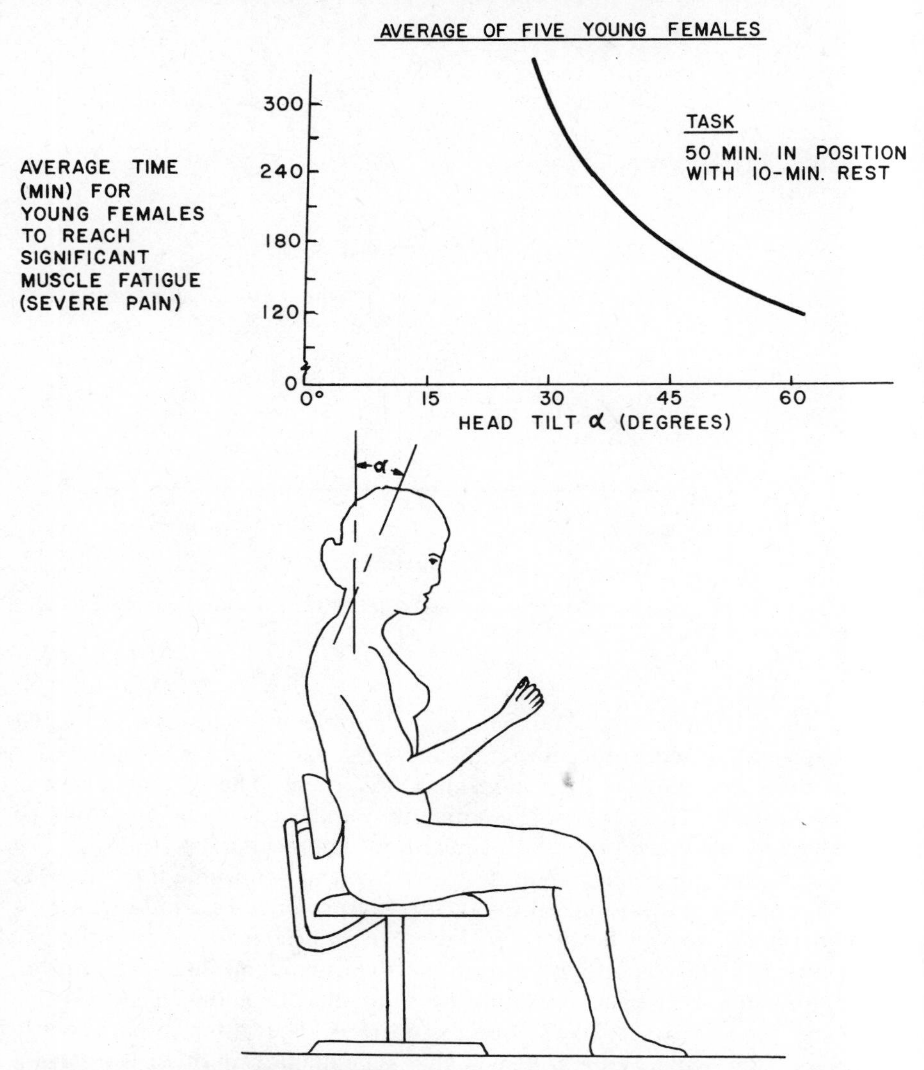

FIGURE 10.18 Neck extensor fatigue vs. head tilt angle. The more the head is tilted, the earlier fatigue will develop (Chaffin, 1973).

load moment at the lumbar discs increases as the sine of λ. Thus, at approximately 30° of torso inclination, the load moment is 50% of its maximum value achieved at 90° (horizontal).

Often, prolonged, forward stooped postures are necessary for workers performing work in a standing or seated posture when the workbench is too low. EMG's of the lumbar erector spinae muscles have been obtained on workers, and have corroborated muscle fatigue of the low back when the

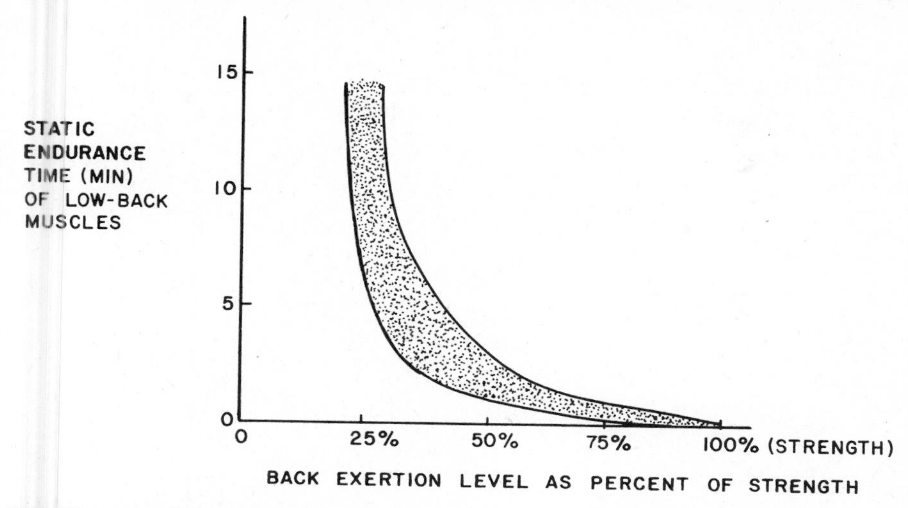

FIGURE 10.19 Range of back muscle endurance values obtained using various torso forward inclination postures and loads to vary exertion levels (Jorgensen, 1970).

workers were required to be in stooped postures for prolonged periods of the day (Habes, 1980). Subjective discomfort levels in the lower back have also been directly correlated with the load moment at the hip joint by Boussenna, Corlett, and Pheasant (1982). A study by Jorgensen (1970) revealed that back muscle endurance times in various stooped postures decreased appreciably when the postures required more than about 30% of isometric back strength (Figure 10.19). From these latter results, the author speculates that most men and 85 to 90% of women are capable of maintaining a 20° stooped posture during the day. Because the muscle endurance–tension curve in Figure 10.19 is so asymptotic at about 30% of strength, and because the load-moment increases rapidly for each degree of torso inclination above 20°, any additional flexion would become increasingly difficult for people to tolerate.

It must also be recognized that when a workbench is too low, a person not only leans forward, but also may lower and rotate the shoulders forward, causing fatigue, inflammation, and pain in the levator scapulae muscles (Figure 10.20).

Conversely, concern must be shown for a workbench that is too high. Tichauer (1968) found that the metabolic cost of packing groceries increased while performance decreased if the work height was such that the arms had to be abducted to an angle of 20° or more. Figure 10.21 displays these results, which correspond to the shoulder muscle fatigue studies discussed earlier in this chapter (see Figure 10.11). Once again, prolonged arm abduction and shoulder elevation cannot be tolerated by even the most fit workers.

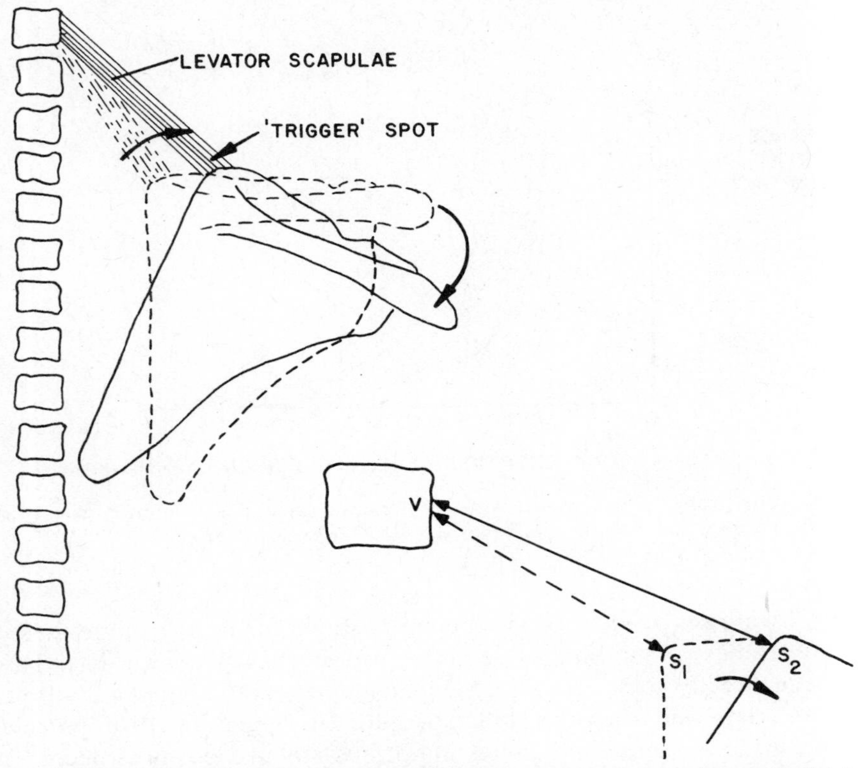

FIGURE 10.20 Levator scapulae "trigger" zone in postural fatigue syndrome. With lateral downward rotation of the scapula (curved arrows), the superior medial angle of the scapula moves the insertion of the levator scapulae muscle (V-S_1 to V-S_2). The muscle under this traction becomes ischemic, inflamed, and thus tender and painful at the "trigger" site (Cailliet, 1981).

In short, adjustment of the height of a workbench if standing and the seat-workbench combination when seated is necessary to accommodate a large proportion of the population. Konz (1979) reviewed various recommendations in this regard, and concluded that having the work area about 5 cm below the elbow when standing or seated in an erect posture was optimal. This provides adequate clearance for necessary elbow and forearm motions and requires little stooping or shoulder rotation if a person needs to rest or stabilize the forearm on the workbench.

10.2.5 Biomechanical Considerations in Design of VDT Workstation

Because of the considerable increase in the use of VDT's (Video Display Terminals), a few words will be specifically addressed to the design of such workplaces. The basis for these recommendations is given earlier in this

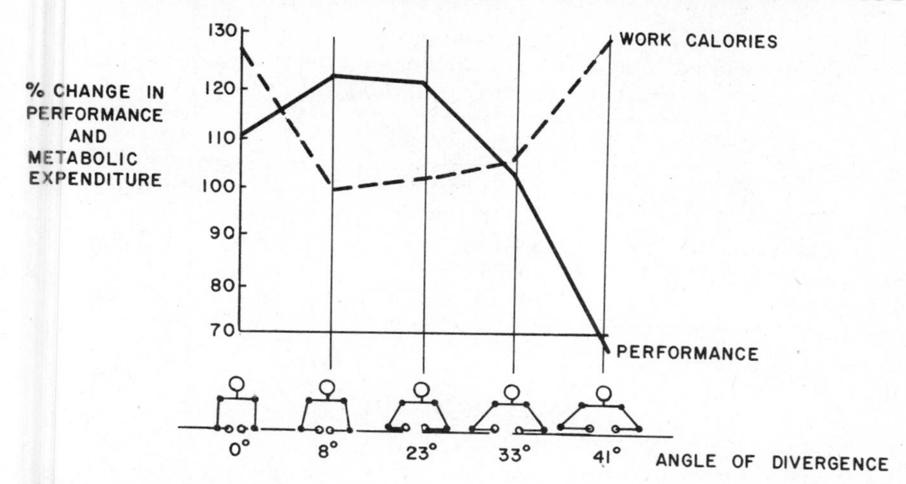

FIGURE 10.21 Performance and metabolic energy expenditure rates from study of 12 female grocery packers (Tichauer, 1978).

chapter and in Chapter 9. In a field study of 68 subjects working at each of five adjustable workstations, the preferred settings were recorded as well as the body postures (Grandjean, Hünting, and Nishiyama, 1982). These settings were: for keyboard height (home row above floor) 71–87 cm (mean 79 cm); for screen height (center above floor) 92–116 cm (mean 103 cm); for viewing angles (eye to screen center) +2 to −26° (mean −9°); and for visual distance (eye to screen) 61–93 cm (mean 79 cm). The body posture chosen was usually a backward leaning one with trunk-seat inclinations of 97–121° (mean 104°) and a head inclination of 34–65° (mean 51°).

These observations, and the biomechanical criteria given earlier, clearly show the importance of adjustability. The sitting posture of the worker determines how the workplace should be organized. When leaning backward, as in the preceding study to about a mean of 104°, a restful posture for the back can be obtained (Chapter 9). To maintain an elbow flexion angle of about 90°, and a shoulder flexion angle as low as possible, a slightly inclined, adjustable table is required. The field of vision determines the neck flexion angle. As discussed, this angle should preferably not exceed 20°, with as little rotation as possible. This requires a screen that is adjustable in height and distance, as well as in inclination (by itself or by the table). The keyboard should be such that wrist extension, flexion, and ulnar deviation is minimal.

It is also important to provide sufficient space for the legs of the worker. Thus, an adjustable table bottom height (see Table 9.6), as well as at least 60 cm of free space at the level of the knees and at least 80 cm at the level of the feet, is recommended.

Table 10.2 summarizes the recommendations of Grandjean et al. (1982). The controls for adjusting the dimensions of the workstation should, of course, be easy to handle.

TABLE 10.2
Recommended Ranges of Adjustability for VDT Workstations (after Grandjean, Hünting, and Nishiyama, 1982)

Item	Dimension	Measured at
Keyboard height	70–85 cm	Middle row to floor
Screen height	90–115 cm	Center to floor
Screen inclination	88–105°	To horizontal plane
Keyboard distance	10–26 cm	Middle row to table edge
Screen distance	50–75 cm	Screen to table edge

10.3 SUMMARY

It has been shown that localized musculoskeletal injury can be caused by sustained or frequent exertions. At first, there may be simple, acute muscle fatigue. If the causal conditions persist, however, inflammation of tendons, synovia, bursa, and adjacent joint structures result in more severe and chronic symptoms and functional limitations.

Several job-design considerations are necessary to prevent such cumulative trauma. First, there must be the recognition that muscular strength is quite limited in certain postures, and any exertion in these "weaker" postures will require the muscles to work close to or at their maximum capacity, producing high fatigue rates. Second, muscles cannot sustain static contractions over about 15 to 20% of their strength without fatigue. This is particularly relevant to the specification of work conditions requiring prolonged unusual postures—for example, stooped-over, reaching above the head, leaning to one side, standing on one foot, and so on. Also, when handling a heavy tool or object, because it is held at the end of the arm it can create high load moments at the shoulder, resulting in rapid muscle fatigue and associated shoulder tissue cumulative trauma.

The length of time provided for muscles to rest between contractions is important in prolonging the endurance time. Unfortunately, if the load on the muscle is high relative to its strength, and the contraction must be sustained for more than 5 to 7 sec, rest allowances of 20 to 50 times the contraction period may be necessary to allow recovery from the resulting fatigue process. Once again, any static exertion of even a few seconds should be of limited strength demand—less than 20% of maximum—or should have specific muscle rest breaks of long duration between each exertion.

Last, it has been shown that fatiguing muscle loads are often associated with job postural requirements, which in turn are dependent upon the degree of matching between the workplace/machine control and display layout dimensions and individual worker anthropometry. This leads to the recommendation that future workplace and machine designs must have a large range of adjustability to accommodate the normal anthropometric variations in worker populations. This aspect was discussed in more detail in the previous chapter on seat design (Chapter 9).

REFERENCES

Andersson, G. B. J., R. Ortengren, and A. Schultz, "Analysis and Measurement of the Loads on the Lumbar Spine During Work at the Table," *J. Biomech.*, **13,** 513–520 (1980).

Bjelle, A., M. Hagberg, and G. Michaelsson, "Occupational and Individual Factors in Acute Shoulder-Neck Disorders among Industrial Workers," *British J. of Indust. Med.*, **15**(4), 346–354 (1973).

Boussenna, M., E. N. Corlett, and S. T. Pheasant, "The Relationship between Discomfort and Postural Loading at the Joints," *Ergonomics,* **25**(4), 315–322 (1982).

Cailliet, R., *Shoulder Pain,* 2nd Ed., F. A. Davis, Philadelphia, 1981, pp. 38–53.

Caldwell, L. S. and R. S. Smith, "Pain and Endurance of Isometric Muscle Contractions," *J. Eng. Psych.*, **5**(1), 25–32 (1966).

Chaffin, D. B., "Localized Muscle Fatigue—Definition and Measurement," *J. Occup. Med.*, **15**(4), 346–354 (1973).

Engin, A. E., "On the Biomechanics of the Shoulder Complex," *J. Biomechanics,* **13,** 575–590 (1980).

Garg, A. and D. B. Chaffin, "A Biomechanical Computerized Simulation of Human Strengths," *AIIE,* **7**(March), 1–15 (1975).

Grandjean, E., "Fitting the Task to the Man," Taylor & Francis Ltd., London 1980, pp. 109–110.

Grandjean, E., W. Hünting, and K. Nishiyama, "Preferred VDT workstation settings, body posture and physical impairments," *J. Human Erg,* **11,** 45–53 (1982).

Habes, D., "Low-Back EMG and Pain in Stooped Posture," University of Michigan, *Center for Ergonomics* Technical Report (1980).

Hagberg, M., "Workload and Fatigue in Repetitive Arm Elevations," *Ergonomics,* **24**(7), 543–555 (1981a).

Hagberg, M., "Electromyographic Signs of Shoulder Muscular Fatigue in Two Elevated Arm Positions," *Amer. J. Phys. Med.*, **60**(3), 111–121 (1981b).

Herberts, P., R. Kadefors, G. Andersson, and I. Peterson, "Shoulder Pain in Industry: An Epidemiological Study on Welders," *Acta. Orthop. Scand.*, **52,** 229–306 (1981).

Herberts, P., R. Kadefors, and H. Broman, "Arm Positioning in Manual Tasks—An Electromyographic Study of Localized Muscle Fatigue," *Ergonomics,* **23**(7), 655–665 (1980).

Hunting, W., Th. Laubli, and E. Grandjean, "Constrained Postures of VDU Operators," in E. Grandjean and E. Vigliani, Eds., *Ergonomic Aspects of Visual Display Terminals,* Taylor and Francis, 1982, pp. 175–184.

Jorgensen, K., "Back Muscle Strength and Body Weight as Limiting Factors for Work in Standing Slightly-Stooped Position," *Scand. J. Rehab. Med.*, **2,** 149–153 (1970).

Konz, S., *Work Design,* Grid Publishing, Columbus, 1979, pp. 268 and 256.

Kumar, S. and W. G. S. Scaife, "A Precision Task, Posture, and Strain," *J. Safety Research,* **11**(1), 28–36 (1979).

Laubach, L. L., "Human Muscular Strength," in Webb Associates, Ed., *Anthropometric Source Book,* NASA No. 1024, U.S. National Aeronautics and Space Administration, Washington, D.C., 1978.

Lee, M. W., "A Stochastic Model of Muscle Fatigue in Frequent Strenuous Work Cycles," unpublished doctoral dissertation, The University of Michigan, University Microfilms, Inc., Ann Arbor, Michigan, and London, 1979.

Nolan, M. F., "Internal Rotator–Adductor Tendonitis," *Physical Therapy*, **55**(5), 544–545 (1979).

Owen, C. A., "The Painful Shoulder," *J. Occ. Med.*, **11**(2), 85–90 (1969).

Perry, T., "Normal Upper Extremity Kinesiology," *Physical Therapy*, **58,** 265–278 (1978).

Quiring, D, P. and J. H. Warfel, *The Extremities,* Lea and Febiger, Philadelphia, 1967, pp. 12–30.

Rohmert, W., *Maximalkräfte von Männern im Bewegungsraum der Arme und Beine,* Westdeutscher Verlag, Köln, Germany, 1966.

Schanne, F. A., "Three-Dimensional Hand Force Capability Model for the Seated Operator," unpublished doctoral dissertation, The University of Michigan, Ann Arbor, Michigan, 1972.

Schutz, R. K., "Cyclic Work-Rest Exercise's Effect on Continuous Hold Endurance Capability," unpublished doctoral dissertation, The University of Michigan, University Microfilms, Inc., Ann Arbor, Michigan, and London, 1972.

Snyder, R. G., D. B. Chaffin, and D. R. Foust, *Bioengineering Study of Basic Physical Measurements Related to Susceptibility to Cervical Hyperextension–Hyperflexion,* Tech. Rep. UM-HSRI-B1-75-6, Highway Safety Research Institute, The University of Michigan, 1975.

Thordsen, M. L., K. H. E. Kroemer, and L. L. Laubach, "Human Force Exertions in Aircraft Control Locations," unpublished report, Aerospace Medical Research Laboratory, Wright-Patterson Air Force Base, Ohio, 1972.

Tichauer, E. R., "Potential of Biomechanics for Solving Specific Hazard Problems," *Proceedings of ASSE 1968 Conference,* Amer. Soc. Safety Eng., Park Ridge, Ill., 1968, pp. 149–187.

Tichauer, E. R., *The Biomechanical Basis of Ergonomics,* Wiley-Interscience, New York, 1978, p. 38.

11

HAND TOOL DESIGN GUIDELINES

It is often said that the prosperity of human existence today developed because our ancient ancestors took the time to develop appropriate tools. These early tools, often crude by today's criteria, allowed our ancestors to effectively obtain food and protect themselves from predators. In fact, the economic and political stability of various early cultures often depended directly on the sophistication in hand tools available to the group.

Unfortunately today we sometimes behave as though the design and selection of hand tools is irrelevant in our complex industrial operations. Managers and engineers appear to have become preoccupied with the specification and design of large-scale systems, often leaving the hand tool problem until the moment a job or plant is about to begin production. This delay can and often does mean that the appropriate hand tool is not systematically chosen or designed. The result, realized perhaps after years of a plant's operation, is both substandard production compared with competitors and high injury and illness rates. The prevention of the latter problem is of particular concern in this book, but it is also believed that reduced biomechanical stresses on a worker's hands, arms, and shoulders by proper tool design and use can improve productivity and quality.

11.1 THE NEED FOR BIOMECHANICAL CONCEPTS IN HAND TOOL DESIGN

The human hand is a beautifully designed organ, capable of an infinite variety of configurations and functions. Unfortunately, we often require certain

TABLE 11.1
Some Reported Occupational Factors of Cumulative Trauma Disorders of the Upper Extremity (Armstrong et al., 1982)

Disorder	Reported Occupational Risk Factors
Carpal Tunnel Syndrome (CTS)	1. Accustomed and unaccustomed repetitive work with the hands 2. Work that involves repeated wrist flexion or extreme extension—particularly in combination with forceful pinching 3. Repeated forces on the base of the palm and wrist
Tenosynovitis and peritendonitis crepitans of the abductor and extensor pollicus tendons of the radial styloid (DeQuervain's disease)	1. More than 2,000 manipulations per hour 2. Performance of unaccustomed work 3. Single or repetitive local strain 4. Direct local blunt trauma 5. Simple repetitive movement that is forceful and fast 6. Repeated radial deviation of the wrist—particularly in combination with forceful exertions of the thumb 7. Repeated ulnar deviation of the wrist—particularly in combination with forceful exertions of the thumb
Tenosynovitis of finger flexor tendons	Exertions with a flexed wrist
Tenosynovitis of the finger extensor tendons	Ulnar deviation of the wrist outward rotation
Epicondylitis	Radial deviation of the wrist with inward wrist rotation
Ganglionic cysts	1. Sudden or hard unaccustomed use of tendon or joint 2. Repeated manipulations with extended wrist 3. Repeated twisting of the wrist
Neuritis in the fingers	Contact with hand tools over a nerve in the palm or sides or the fingers

types of manual exertions which, when frequently repeated daily for one's working life, can cause a progressive deterioration of functional capacity and pain. In other words, we create *cumulative trauma* to the musculo-skeletal system of the upper extremity and particularly to hand and wrist tissues. In reviewing the literature on this matter, Armstrong et al. (1982) listed a variety of occupational risk factors associated with common hand and wrist disorders (Table 11.1). The incidence of these types of cumulative trauma disorders is not generally known, but in industries where a large proportion of the work involves repetitive manual exertions it is not uncommon for plant-wide incident rates to exceed 10 injuries per 100 workers annually.

From inspection of the occupational risk factors listed in Table 11.1 it should be clear that many of them are directly related to the design of handtools and to the methods employed when using hand tools. In this regard, a review of hand injury statistics in the United States by Ayoub, Purswell, and Hoag (1975) showed that hand tools were involved in approximately 9% of all reported disabling injuries, and about three-fourths of these involved manual tools. It should be noted, however, that these estimates are probably low, because injury produced by cumulative trauma has only recently been recognized for compensation and thus for reporting purposes.

Some of the major biomechanical concepts in hand tool design were presented in Chapter 6 (Section 6.5.2). Basically, biomechanical studies show that forceful grip exertions of the hand rely on muscle contractions in the forearm, with muscle forces being transferred to the bones of the fingers via long tendons. The level of muscle exertion and tendon tension largely depend on the grip configuration, and to a lesser extent on hand and wrist anthropometry.

Further, these biomechanical studies disclose that the angle of the wrist during grip-type exertions directly affects the amount of intrawrist supporting forces acting normal to the direction of the tendons and their synovia. From this it is proposed that the wrist be kept relatively straight during forceful gripping to avoid large intrawrist forces. It is believed that large intrawrist forces disrupt the integrity of the synovia, resulting in tendon and synovium inflammation, swelling, and later entrapment of the median nerve within the carpal tunnel (i.e., Carpal Tunnel Syndrome). At present, many hand tool design guidelines are based on the biomechanical principles presented in Chapter 6. Some specific design guidelines for various hand tools are now presented, based largely on these biomechanical principles.

11.2 HAND TOOL SHAPE AND SIZE CONSIDERATIONS

The shape and size of a hand tool have a direct effect on both a person's performance capability (especially grip strength) and biomechanical stress on the upper extremity. Several guidelines are presented to reduce such problems.

11.2.1 Tool Shape to Avoid Wrist Deviation

Allowing the hand and forearm to remain in alignment during forceful grip exertions often necessitates specific tool handle configurations. Several examples follow:

Bent Pliers Design. In his work with Western Electric, Tichauer found that many wiring operations required a worker to grip a pliers with the wrist in a deviated posture. He found that the wrist posture was dictated by the layout of the work and the shape of the pliers (Tichauer, 1978). In some jobs it was possible to perform the job with a more aligned hand and forearm by bending the handle of the pliers, as depicted at the top of Figure 11.1. In a comparative study of two different types of pliers used by 80 new employees, over 60% of those using the common straight-handled pliers developed wrist related disorders at the end of 12 weeks, while only 10% of those using the new bent-handle design were affected (Tichauer, 1978).

Bent Knife Handle Design. In a similar situation, Armstrong performed biomechanical job evaluations and found that in certain poultry processing operations a common straight-handled knife that required extreme wrist flexion and ulnar deviation was used (Armstrong, 1982). On such jobs, the cumulative trauma incident rate for the wrist and hand was approximately 17 per 100 workers each year. This was about 50% higher than the plant average. A biomechanical job analysis resulted in the design of a knife with a pistol grip handle, as depicted in Figure 11.2. EMG studies of muscle forces supported the adoption of this type of knife. The studies also showed that such a knife would reduce the need to continually grip the sometimes slippery knife between cuts, since it was restrained in the hand by the wrap-around handle configuration. By relaxing the hand between cuts, muscle fatigue incidents are reduced.

Cylindrical Versus Pistol Shaped Tools. Many objects are fastened together today by screws driven into holes with the aid of pneumatic or electric motorized hand tools. The driving torque of the tool creates a tendency for the tool to rotate in a worker's hands unless firmly gripped. Grip forces of near maximal strength are not uncommon in such activities. If the wrist is forced into a deviated posture during such exertions, an increased risk of cumulative trauma is probable. To avoid this, biomechanical job evaluations are warranted to decide whether a cylindrical driver is more appropriate than a pistol-shaped driver. This decision very much depends on the orientation and location of the workpieces relative to the worker's arms. Figure 11.3 illustrates these considerations reported by Armstrong (1983).

11.2.2 Tool Shape to Avoid Shoulder Abduction

If the shape of a tool requires extreme wrist deviation, a person often will raise the arm to reduce the stress on the wrist. In essence, there is a physiological trade-off between the biomechanical stresses on the wrist and the stresses on the shoulder joint. A small amount of arm abduction at the shoulder, up to 20° from the vertical, will not normally create an excessive load moment on the shoulder. Abduction greater than this, however, rapidly increases the shoulder load moment. Of course if a heavy hand tool is involved, abduction of the arm simply compounds the moment requirement at the

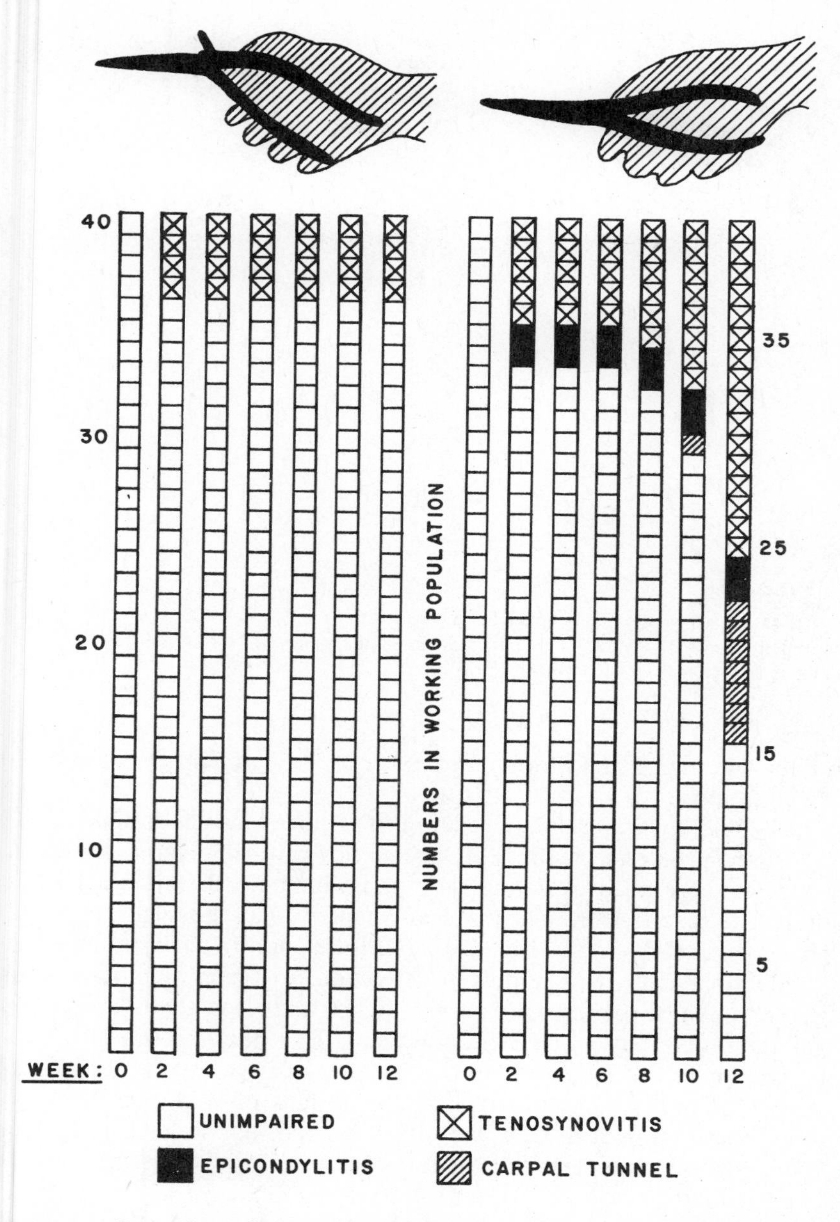

FIGURE 11.1 Comparison of two groups of trainees in electronics assembly shows that it is better to bend pliers then to bend the wrist. With bent pliers and wrists straight, workers become stabilized during second week of training. With wrist in ulnar deviation, a gradual increase in disease is observed. Note the sharp increase of losses in tenth and twelfth weeks of training: only 15 of 40 workers in the sample remained unimpaired (adapted from Tichauer, 1978).

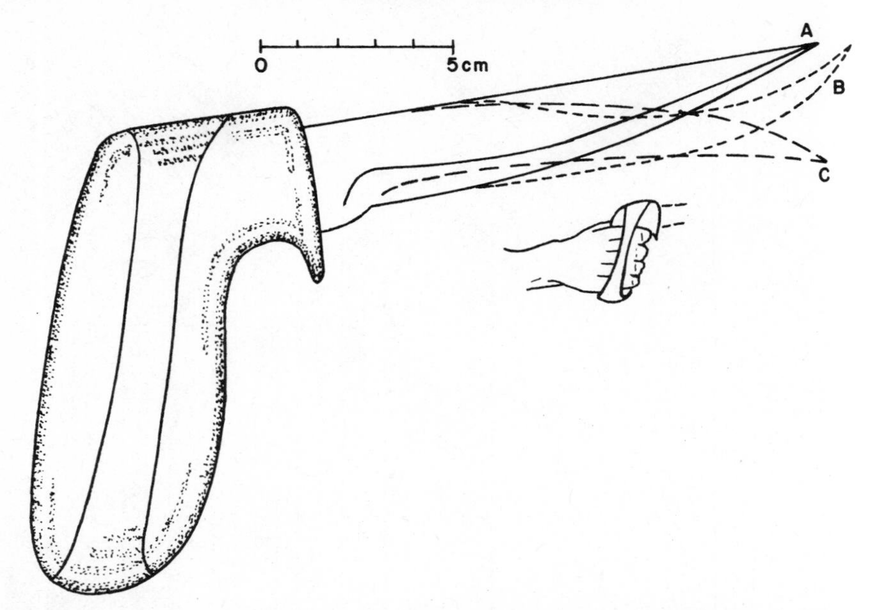

FIGURE 11.2 A recommended knife handle with three blades for reducing both wrist deviations and the tendency for the knife to fall out of the hand while removing the bones from turkey carcasses (Armstrong, 1982).

shoulder due to the hand tool alone, since its weight acts at the end of the upper extremity.

This problem was studied in relation to shoulder muscle fatigue (Chaffin, 1973). Both surface-electrode EMG frequency shifts and subjective discomfort ratings were used to define the fatigue limit for the shoulder in these studies. The results were discussed in Chapter 10 as they also relate to the height of a workbench (see Figure 10.11 for details). In general, if the shoulder abduction angle was about 30°, the time to reach significant muscle fatigue was over three times longer than when abducted at 60°, and six times longer than when 90° of abduction existed. This confirmed an earlier recommendation of Tichauer regarding the design of soldering iron configurations. Figure 11.4 shows two different designs of soldering irons—conventional with straight handle and off-set or bent-handle design developed by biomechanics engineers at Western Electric during the early 1960's to reduce the amount of shoulder and wrist deviation.

11.2.3 Tool Shape to Assist Grip

It also should be noted in Figure 11.4b that the bent soldering iron has a flared handle that provides a comfortable finger and thumb stop to reduce grip forces and prevent the hot tool from slipping in sometimes sweaty hands.

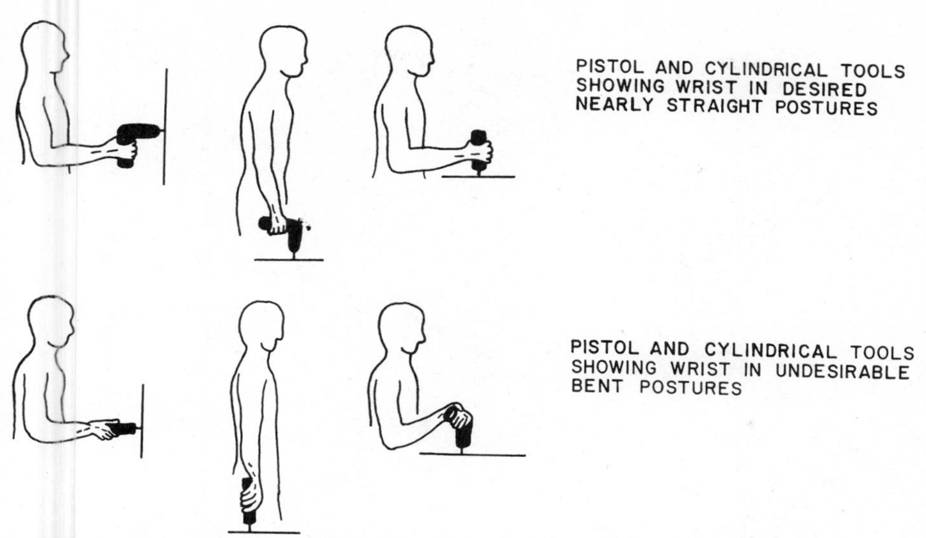

FIGURE 11.3 Based on biomechanical considerations, the handles of powered drivers should sometimes be pistol grip type, sometimes cylindrical. Choosing tools to avoid the bent wrist postures depicted in the lower three activities requires analysis of the work situation, with particular concern for the elevation and orientation of the work surface (adapted from Armstrong, 1983).

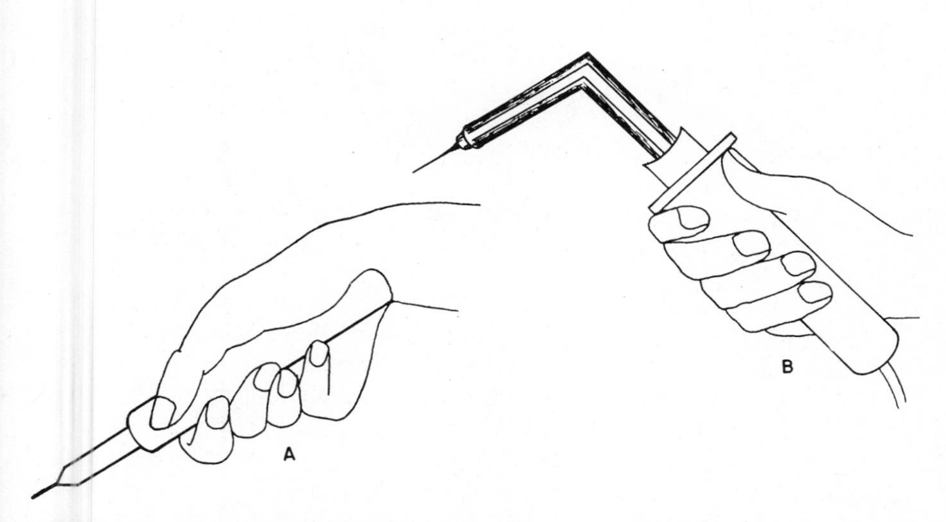

FIGURE 11.4 (a) Traditional straight-handled soldering iron oriented to solder object laying on work bench. (b) Bent-handle soldering iron in same orientation, but with forearm in more natural (less shoulder abduction and wrist deviation) posture suggested by Tichauer (Chaffin, 1973).

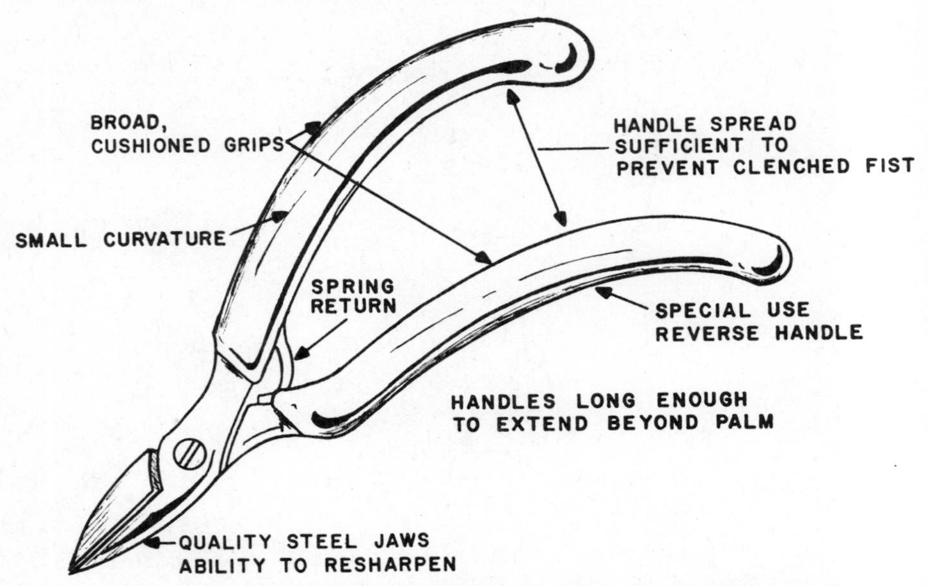

FIGURE 11.5 The sketch (adapted from Bob Brown, Hewlett Packard Business Computer Group) shows important ergonomic factors to consider when selecting tools requiring high grip forces.

Such a flared handle also reduces the chance of material splattering onto the hand, and it shields some of the hand against radiant heat.

The need to shape a tool to promote grip strength is also evident in the wire cutter design shown in Figure 11.5, advocated by Hewlett-Packard (1982). The use of long padded handles distributes the forces on the fingers and palmar tissues, avoiding sensitive areas. In general, any tool that must be forcefully squeezed should be designed with handles that avoid concentrating grip forces on a couple of fingers or in the center of the palm. In this latter regard, the center of the palm, though well suited to retain small objects placed in it, is poorly designed to withstand direct force application. This anatomical limitation is caused by the existence of (1) the median nerve, (2) arteries, and (3) synovium for the finger flexor tendons directly under the skin of the palm, as illustrated in Figure 11.6. Repeated grip forces concentrated in the center of the palm are known to cause injury to all of these fragile tissues, and must therefore be avoided.

The length of a tool handle should be sufficient to distribute the forces on either side of the palm and across digits two through five. Anthropometric studies of the breadth of male and female hands (Webb, 1978) suggest that the palmar force-bearing area be at least 9.0 cm long to assure that palmar forces are largely supported by the muscles on each side of the palm. In this context, of course, the handle should have only a small curvature, or even

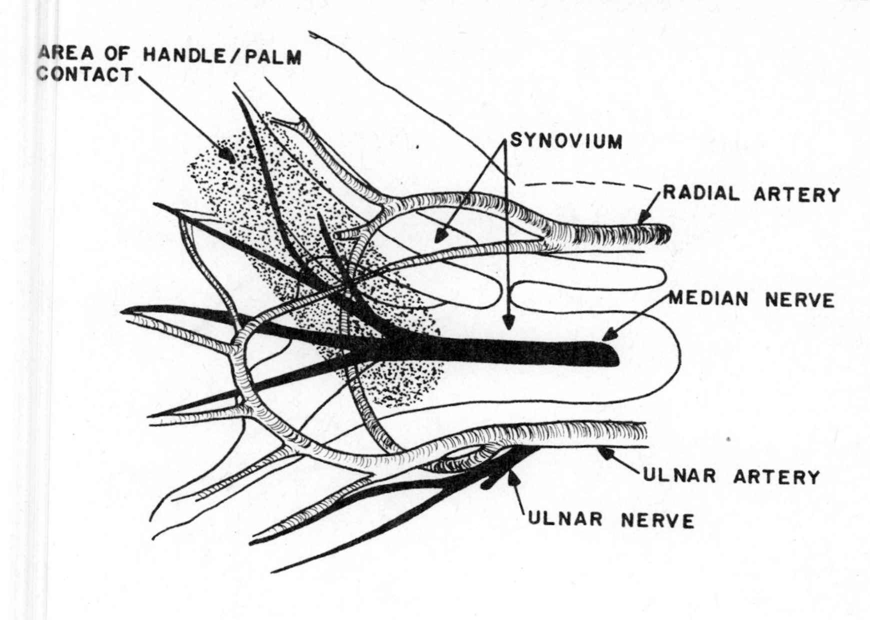

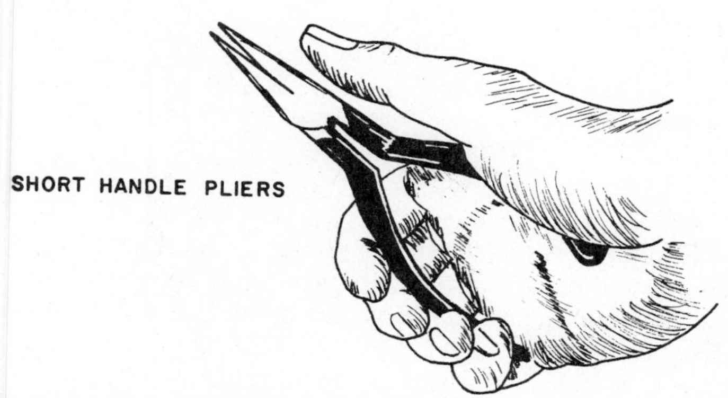

FIGURE 11.6 Small tool handle interaction with tissues in center of palm (Greenberg and Chaffin, 1976).

a reverse curvature in special applications as illustrated in Figure 11.5, to assure such a force distribution across the palm.

11.2.4 Size of Tool Handle to Facilitate Grip

Grip strength and the resulting stress on finger flexor tendons vary with the size of the object being grasped. If the handle force is applied to the distal

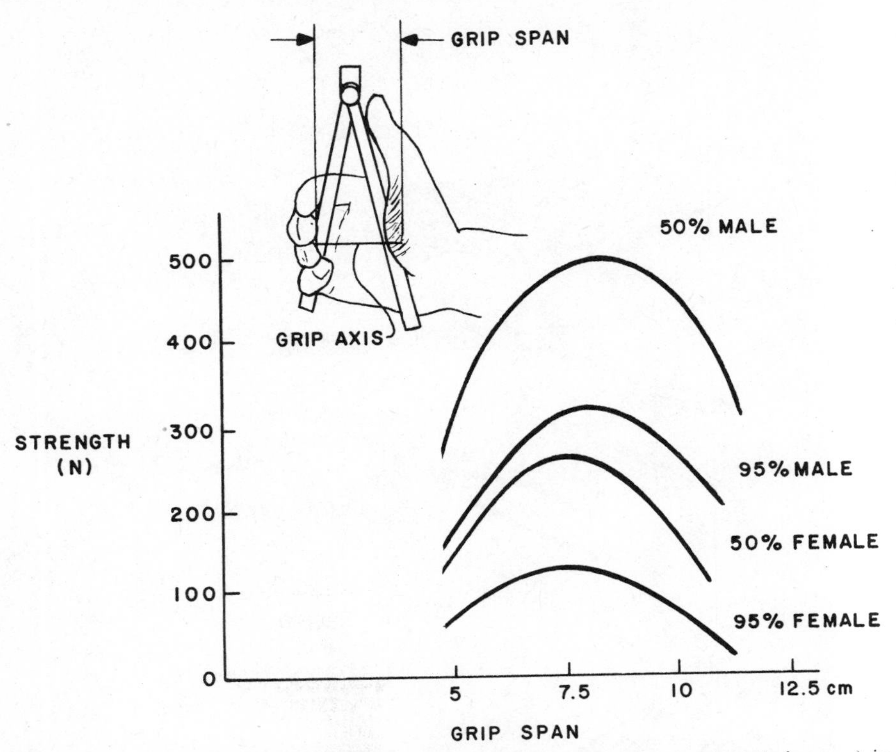

FIGURE 11.7 Maximum linear grip strengths for various handle openings (grip span) in a random population sample. The grip span was measured at the illustrated grip axis—the middle of the hand (Greenberg and Chaffin, 1976).

segment of the fingers, as is the case when grasping a large tool, the tendon forces can be two or three times larger than when a comparable force is applied to more proximal finger segments (see biomechanical discussion in Section 6.5). Conversely, if an object is very small, the fingers cannot effectively apply force to it, partially because the finger flexor muscles become extremely shortened, losing their contractile tension-producing capability. This latter muscle-performance limitation applies especially when attempting to forcefully grasp a small object or tool handle with a flexed wrist that further shortens the finger flexor muscles.

The effect of these manual force limitations is illustrated in Figure 11.7, in which grip strength varies significantly with grip span measured at the center of the hand. A maximum grip strength is achieved at about 7.5 to 8.0 cm. What is also apparent is the large variation in the population that can be expected, with women demonstrating approximately one-half the grip strength of men. These results suggest that (1) occasional manual grip must be restricted to values below about 90 N to accommodate the large variance in population strengths, and (2) grip span should be restricted to approximately 5 to 8 cm for maximum strength.

In this latter regard, if a tool handle is round, as in the pneumatic tools illustrated in Figure 11.3, the optimal grip span dimension is smaller than that depicted in Figure 11.7. Grip strengths on round objects were determined by Ayoub and Lo Presti (1971). The maximum strengths were recorded when the diameter of the object was about 4 cm. In other words, the shape of the cross section of an object affects grip performance as well as the size of the object.

11.2.5 Finger Clearance Considerations

Though the preceding discussion has emphasized tool size and shape parameters as they affect upper extremity stresses and performance, it is also appropriate to design tools or other manually manipulated objects with finger clearance in mind. For instance, what size opening is necessary on the ends of a tote box to assure that a large hand (possibly wearing gloves) can be inserted far enough into it to provide a full power grip (i.e., all four fingers wrap around the object and oppose the thumb)?

This is a concern of hand anthropometry. Several linear and circumferential dimensions for the hand are illustrated in Table 11.2, based on studies compiled in the *NASA Anthropometric Source Book* (Webb, 1978). According to Hertzberg (1972), if a person is wearing woolen or leather gloves, 0.75 cm should be added to these dimensions, and if wearing arctic-style mittens, the dimensions should be increased by 2.5 cm.

11.3 HAND TOOL WEIGHT AND USE CONSIDERATIONS

In attempting to reduce the requirement that a hand tool be manually operated by forceful squeezing, turning, or prying, many tools are now electrically or pneumatically powered. Unfortunately, some of the powered tools become quite heavy, especially when their weight includes the power cord or hose. Combined weights of 50 N are not uncommon for commercial drills, sanders, or buffers. It must be also recognized that the effect of the weight alone is further aggravated by additional muscle actions necessary to precisely position and stabilize a tool during operation.

From the biomechanical discussions presented in Chapter 6 and the arm load limits presented in Chapter 10, it should be clear that it is not possible to recommend a specific weight limit. If a hand tool needs to be supported for long periods of time directly in front of the body, one limit would be applicable. If the tool is to be continually used with the arm flexed or abducted at the shoulder, a lower value would be required. If the tool is used only occasionally for a few seconds, then higher values would be appropriate. A careful review of the arm loading considerations presented in Chap-

TABLE 11.2
Hand Anthropometry (adapted from Webb, 1978)

Hand Dimension	Male Size (cm) Mean	Male Size (cm) SD	Male Size (cm) 95th percentile	Female Size (cm) Mean	Female Size (cm) SD	Female Size (cm) 95th percentile
Palm breadth PB	8.7	0.4	9.5	7.6	0.4	8.4
Palm circumference PC	21.5	1.1	23.2	18.3	0.9	19.8
Palm thickness PT	3.0	0.2	3.3	2.5	0.2	2.9
Finger length FL	12.6	0.8	13.9	ND	ND	ND
Hand breadth HB	10.4	0.5	11.2	9.2	0.6	10.1
Hand circumference HC	25.4	1.1	27.4	ND	ND	ND

ND: Not Determined.

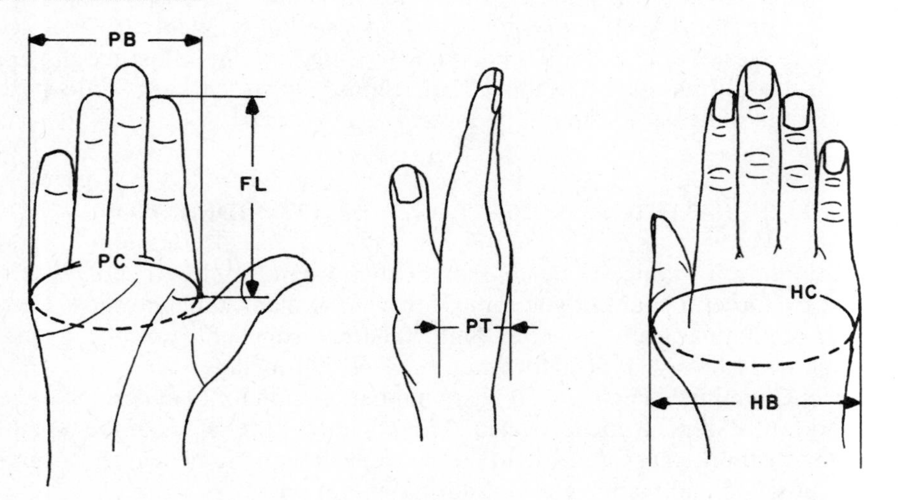

ter 10 should be of some assistance in resolving a design problem of this type.

If it is determined that a hand tool has significant weight or is used with the arm raised in flexion or abduction, then either a tool balancer device or padded arm (elbow) supports or both should be considered to reduce the load moment on the shoulder. Tool balancers work by counterbalancing the weight of a tool with a long spring (often coiled in a spool attached to a

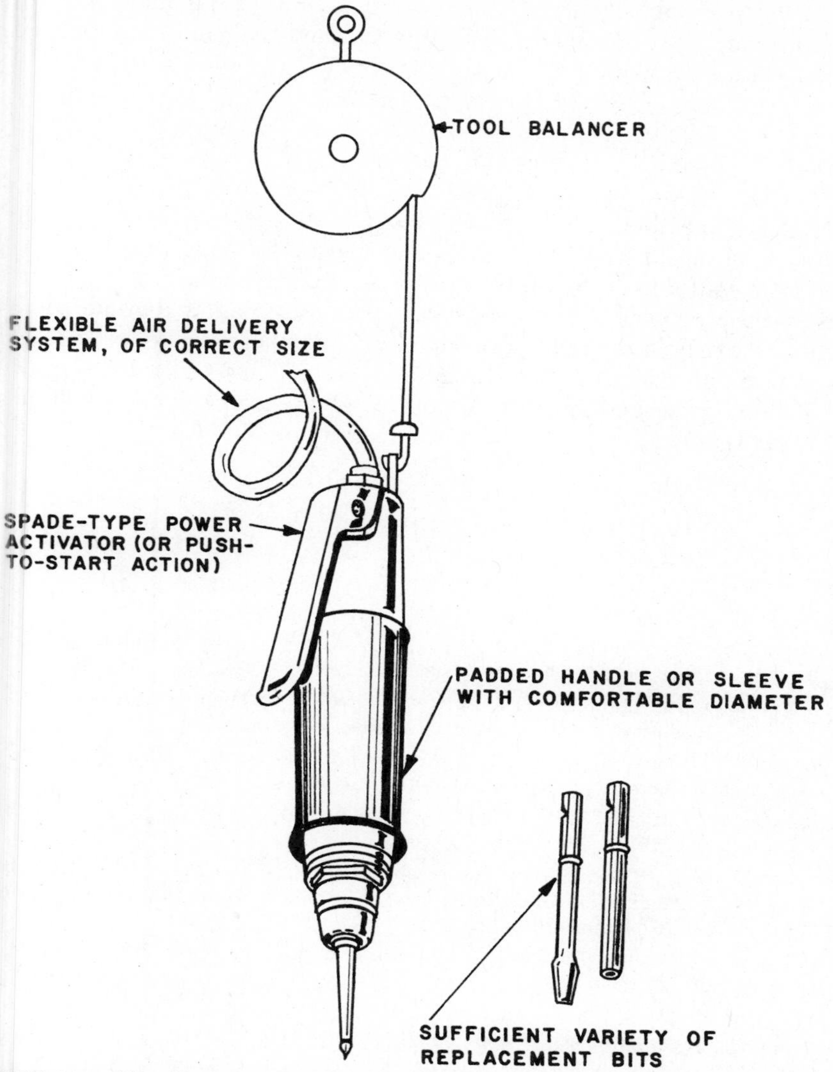

FIGURE 11.8 The sketch (adapted from Bob Brown, Hewlett Packard Business Computer Group) shows important ergonomic factors to consider when selecting power tools to be suspended from tool balancer.

cable) suspended over the work area and attached to the tool, as depicted in Figure 11.8. The successful use of most simple tool balancers requires (1) that the work area be limited in size to that below the tool balancer, though some "swinging" of the tool is possible, and (2) that the tool is used in one general orientation—vertical or horizontal, but not both. The use of a tool balancer permits heavy tools to be used with minimum effort, and often provides a convenient means to store hand tools in a sometimes congested work area while not in use.

11.4 SUMMARY

Improperly specified and designed hand tools are prevalent in industry today, resulting in frequent injuries and lost performance capability. This chapter has attempted to apply various biomechanical principles in the design and use of hand tools. It is assumed in this chapter that adequate information exists in the field of occupational biomechanics today to develop innovative approaches to the design and use of hand tools. Through such innovation, upper extremity musculoskeletal disease in industry will be significantly reduced.

REFERENCES

Armstrong, T. J., "An Ergonomics Guide to Carpal Tunnel Syndrome," *AIHA Ergonomic Guide Series,* Amer. Ind. Hygiene Assoc., Akron, Ohio (1983).

Armstrong, T. J., J. A. Foulke, B. S. Joseph, and S. A. Goldstein, "Investigation of Cumulative Trauma Disorders in a Poultry Processing Plant," *AIHAJ* **43**(2), 103–116 (1982).

Ayoub, M. M. and P. Lo Presti, "The Determination of an Optimum Size Cylindrical Handle by Use of Electromyography," *Ergonomics,* **4**(4), 503–518, 1971.

Ayoub, M. M., J. Purswell, and L. Hoag, *Research Requirements on Hand Tools,* The University of Oklahoma, NIOSH Technical Report (1975).

Chaffin, D. B., "Localized Muscle Fatigue—Definition and Measurement," *J. Occup. Med.,* **15**(4), 346–354 (1973).

Hertzberg, H. T. E., "Engineering Anthropometry," in Van Cott, H. P. and R. G. Kinkade, Ed., *Human Engineering Guide to Equipment Design,* McGraw-Hill, New York, republished by U.S. Government Printing Office, 1972, pp. 515.

Hewlett Packard Health and Safety Professionals, *Preventive Measures,* **3**(5), 1–3 (1982).

Tichauer, E. R., *The Biomechanical Basis of Ergonomics,* Wiley—Interscience, New York, 1978, pp. 41–43.

Webb Associates, *Anthropometric Source Book Vol. II,* NASA Reference 1024, Washington, D.C., 1978, pp. 43–47, 212, and 229–242.

12

GUIDELINES FOR WHOLE-BODY AND SEGMENTAL VIBRATION

Various technical developments in society have led to an increase in the number of people exposed to vibration in some form. As biological effects have become obvious, efforts have been made to reduce such exposure. Experimental data have been gathered and have resulted in a set of vibration criteria to limit vibration exposure. This chapter reviews the effects of vibration exposure on man. Measurement methods are summarized and some of the exposure criteria outlined. These are specified in ISO Standard 2631 (1978) for whole-body vibration exposure and in Draft Standard ISO/DIS 5349 (1980) for vibration transmitted to the hands and arms of people using vibrating tools.

12.1 DEFINITION AND MEASUREMENT

Vibration is the oscillatory motion of bodies. All bodies with mass and elasticity are capable of vibration, hence most engineering machines and structures as well as the human body experience vibration to some degree. Two general classes of vibration are distinguished in the literature. *Free vibration* takes place when a system oscillates under the action of internal forces only—that is, the system vibrates at one or more of its natural frequencies. *Forced vibration* is caused by external forces. If the frequency of excitation coincides with the natural frequency of the system, resonance that can cause large oscillations occurs. Because of energy dissipation due to friction and other resistances, damping occurs in all structures.

Oscillatory movements can repeat themselves regularly or irregularly. Regular repetition is called periodic motion and the repetition time (oscillation period) is called frequency. The simplest form of *periodic motion* is referred to as harmonic motion and results in sine wave. Since the instantaneous value of *harmonic vibration* can be described at any time by a mathematical expression, these types of oscillations are called *deterministic*. The opposite of deterministic is a *stochastic* or *random vibration*. Such phenomena occur frequently and can be described in spite of their irregular functions by means of averaging procedures.

When measuring oscillatory motions, certain properties are of interest. While many different descriptions have been used, the simplest are the *peak value* and the *average value* (Figure 12.1). The peak value indicates the maximum stress, but does not take the duration or time function of the motion into account. Thus, it is used mainly to study short-term motions—for example, in shock or impact loading.

The average value is easy to measure with various electronic instruments, but has only a limited relationship to the physical quantity of the vibration. It is found by the time integral

$$\overline{X} = \frac{1}{T}\int_0^T |X|\, dt \tag{12.1}$$

where T is the time of a cycle and X is the instantaneous amplitude of the signal.

A more relevant measure of vibration properties is the *root mean square* (*RMS*) *value*—that is, the square root of the mean squared values of the motions of a body (Figure 12.1). The RMS value is proportional to the energy content of the vibration:

$$X_{\mathrm{RMS}} = \sqrt{\frac{1}{T}\int_0^T X^2(t)\, dt} \tag{12.2}$$

The frequency content of an oscillatory motion is important in characterizing vibration. For periodic motions, a *frequency spectrum* is obtained by dividing the motion into sine wave components. Each such component is represented by discrete lines in a frequency diagram, the length of the line being proportional to the amplitude of the component vibration (Figure 12.2).

Frequency analysis provides a graphical representation of the vibration amplitude in different frequency bands. Usually a narrow band-pass filter is used with either a *constant absolute band width* (Figure 12.3) or a *constant relative band width*. The center frequency of the filter is then swept through the entire frequency range of interest either in steps or continuously (Figure 12.3). Filters with constant absolute band width may have a band width of

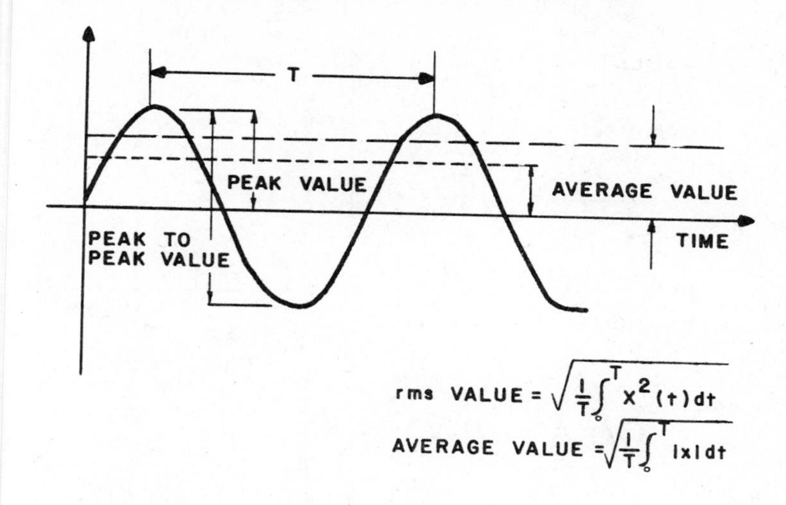

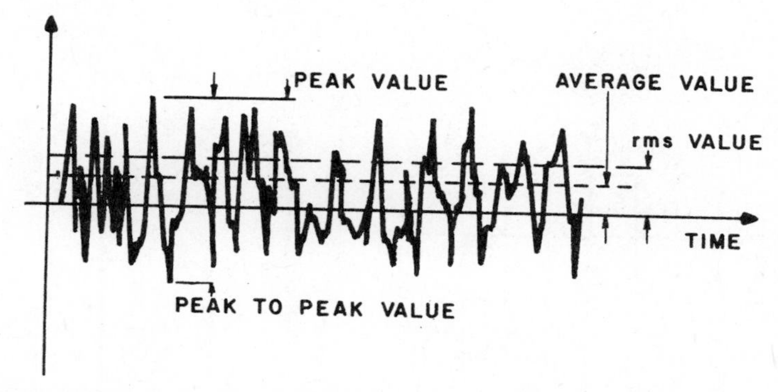

FIGURE 12.1 Descriptions of oscillatory motions. (Adapted from Bruel and Kjaer, 1982.)

3 Hz, 10 Hz, and so on, independent of its center frequency. Filters with constant relative band widths maintain a constant ratio between their center frequencies and the band widths (Figure 12.3). Both of these filters are used—the constant relative band width type generally to study vibrations in machines, since its resolution is superior at lower frequencies; and the constant absolute band width filter to study vibrations over a large spectrum, as its resolution is superior at higher frequencies. The band width of the filter is critical to the resolution. Narrow band-pass filters increase the level of information, but the necessary time to reliably analyze the vibration increases with narrow filters. To obtain high resolution in the lowest possible time, a wide band-pass filter analysis is usually used at first, followed by a

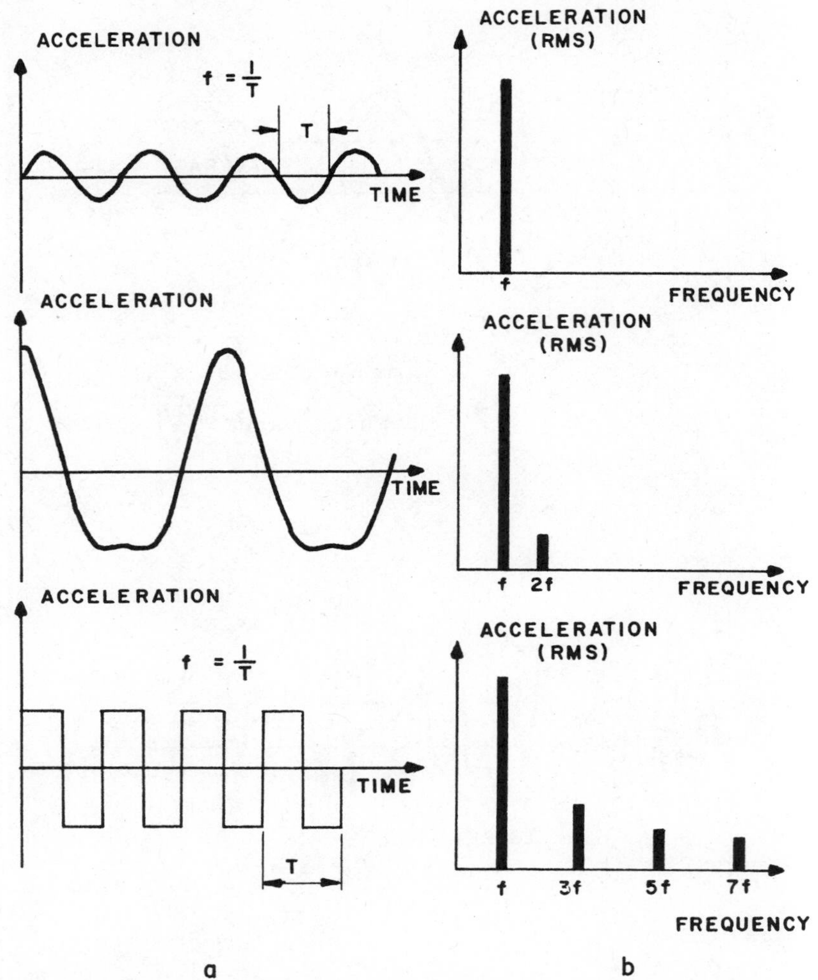

FIGURE 12.2 Periodic signal described as (a) time functions; (b) frequency functions. (Adapted from Bruel and Kjaer, 1973.)

narrow band-pass filter analysis in the frequencies of particular interest. (Figure 12.4)

The ideal filter allows passage of all frequency components within the band and completely excludes those outside it. In reality, some "cut-off" error is always present, allowing frequencies outside the ideal filter to be included (Figure 12.5). The filter is therefore usually characterized by the band width between two points in the same curve that are 3 dB below the nominal peak of the curve—the so called *3 dB band width*. Band width is also defined as the width of the ideal filter that allows the same effect when

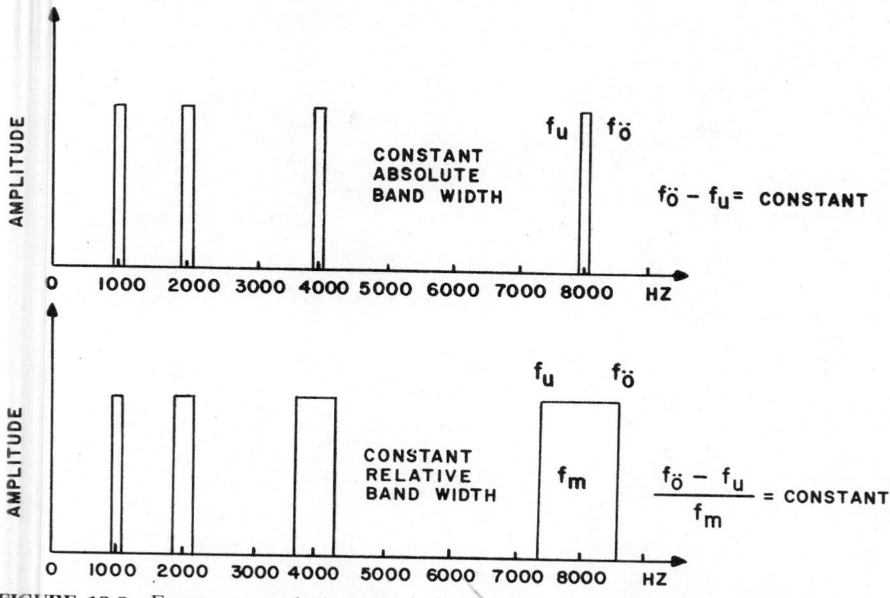

FIGURE 12.3 Frequency analysis with constant absolute band-width and constant relative band-width. (Adapted from Bruel and Kjaer, 1973.)

the input signal has an absolute weighted spectrum—that is, white noise. That measurement is called *effective band width* (Figure 12.5).

Whole-body vibration in man is usually measured using $\frac{1}{3}$ octave frequency bands. Sometimes for segmental vibration, 1 octave bands are used. The constant relative band width is usually given as a percentage—23% for third-octave bands and 70% for one-octave bands. For reporting purposes the center frequencies to be analyzed are standardized in octave-band analysis.

A common method of presenting the amplitude of vibration is in decibel (dB) units relative to a certain reference level. The mathematical relationship according to ISO/DIS 1683.2 is:

$$N(\mathrm{dB}) = 20 \log_{10} \frac{a}{a_0} \tag{12.3}$$

where

N is the number of decibels
a is the measured vibration amplitude
a_0 is the reference level amplitude

The recommended reference levels are given in Table 12.1. Without stating the reference value, the dB level of vibration is meaningless.

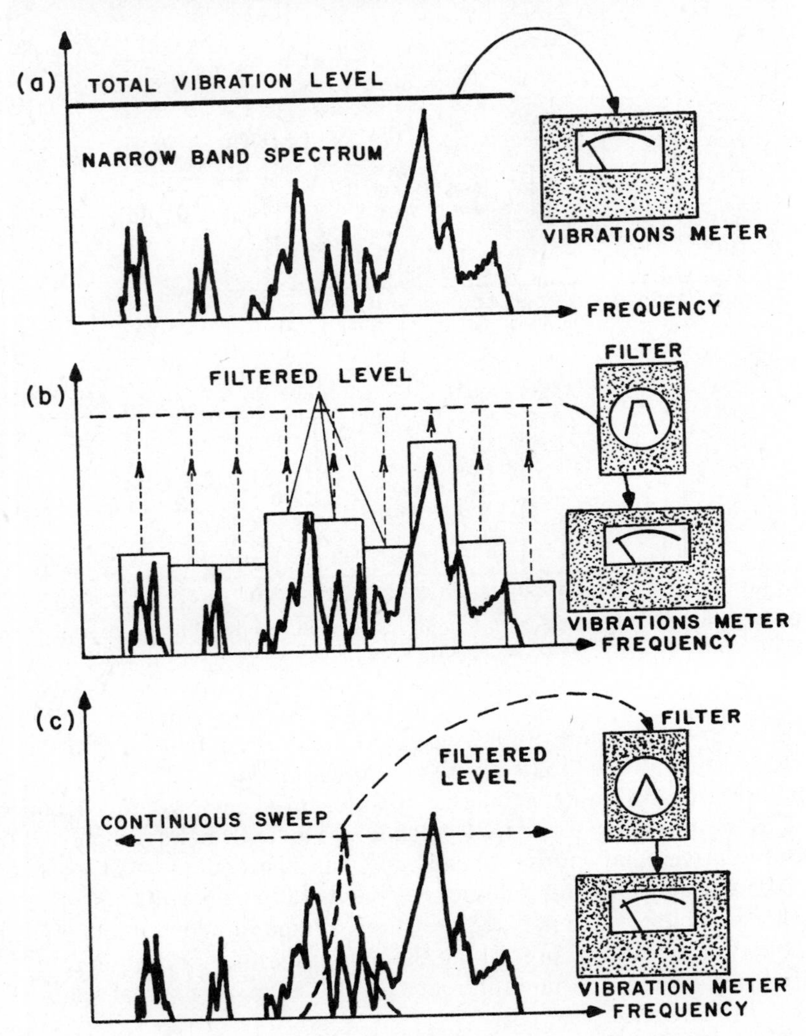

FIGURE 12.4 Frequency analysis using (a) wide frequency band filter; (b) narrow band filter in steps; (c) continuous narrow band filter sweep. (Adapted from Bruel and Kjaer, 1982.)

The shape and frequency of an oscillation is the same regardless of whether motion displacement, velocity, or acceleration parameters are used. There is a phase difference, however, between the three parameters (Figure 12.6). In harmonic vibrations the relationship is simple, depending only on the frequency. When the phase difference is ignored—for example, when measuring the level of vibration—velocity V can be obtained by dividing the acceleration by a factor proportional to the frequency (Figure 12.6), and

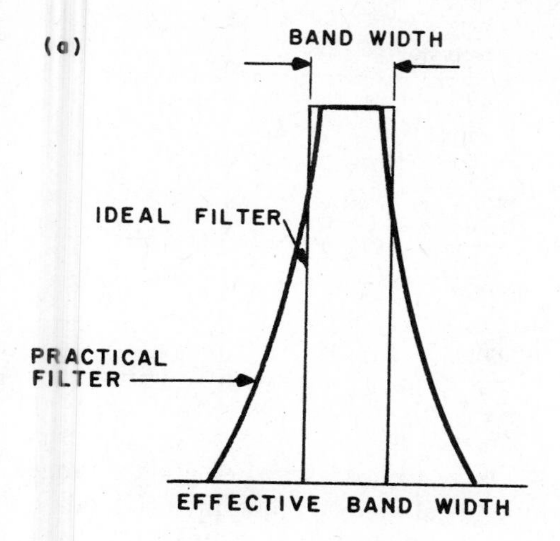

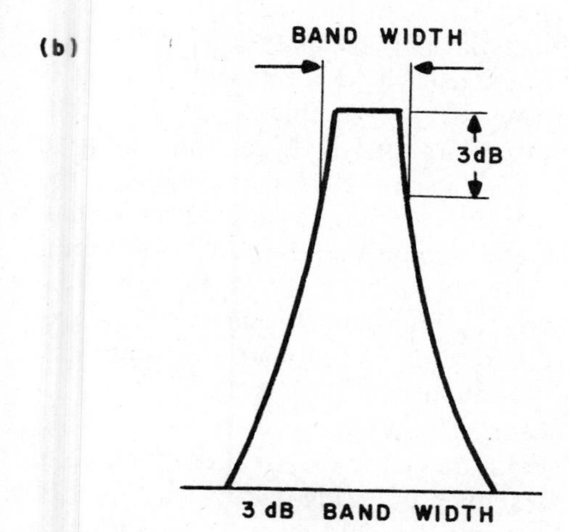

FIGURE 12.5 Ideal and practical filter with definitions of (a) effective band width, (b) 3 dB band-width. (Adapted from Bruel and Kjaer, 1982.)

motion displacement D can be determined by dividing acceleration by a factor proportional to the square of the frequency:

$$V = \frac{a}{2\pi f} = \int a \, dt \tag{12.4}$$

$$D = \frac{a}{4\pi^2 f^2} = \int v \, dt \tag{12.5}$$

To measure vibrations, *accelerometers* are most often used. The acceleration data can then be easily transferred to velocity and movement data using electronic integrators. The advantage of accelerometers is that they

TABLE 12.1
Decibel Reference Level

Quantity	Definition	SI Reference Level
Sound pressure level Vibratory acceleration level	$L_a = 20\log_{10}(a/a_0)$ dB	$a_0 = 10^{-6}$ m/sec^2
Vibratory velocity level Vibratory displacement level	$L_v = 20\log_{10}(v/v_0)$ dB	$v_0 = 10^{-9}$ m/sec
Vibratory force level	$L_F = 20\log_{10}(F/F_0)$ dB	$f_0 = 10^{-6}$ N

are small and sensitive to motion over large frequencies. For greater range of frequencies, *piezoelectric* accelerometers are most commonly used today. There are several different types of two principally different constructions—*compression-type* and *shear-type* (Figure 12.7). In the first, it is the mass that influences the piezoelectric element through compression; in the second, the influence is through shear. The sensitivity of the accelerometer is important. As large an output signal (m/sec^2) as possible is desirable, but this requires a large accelerometer with a large mass. A general rule is to choose accelerometers with a mass of $\frac{1}{10}$ or less of the dynamic mass of the object to be measured, and to use preamplifiers to amplify the output signal. The frequency range of the accelerometer should be adjusted to the range of interest. The low-frequency limit is usually given by the amplifier system characteristics (typically about 0.1 Hz), the high limit by the resonant frequency of the accelerometer. Up to one-third of the resonant frequency can usually be used, creating an error of less than 12% (Figure 12.8).

Vibration in man is often measured in acceleration units (m/sec^2) RMS with a third-octave frequency band width and over the frequency range of interest. The weighted vibration level is given in "weighted m/sec^2" or dB. Recall that the weighted vibration level in dB is 20 times the logarithm to the base ten of the ratio of a weighted acceleration to the reference acceleration. Or, mathematically, the acceleration is weighted in accordance with one of four frequency weightings as follows: 0.1–0.63 Hz, 1–80 Hz when measuring whole-body X and Y directions, 1–80 Hz for whole-body Z direction, and 8–1000 Hz for segmental in all directions. The acceleration is often time-weighted, using time constants of 8:1 or linear integration of 60 sec (peak). The reference acceleration is usually 1 m/sec^2.

12.2 THE EFFECTS OF VIBRATION ON THE HUMAN BODY

Vibrations influence the human body in several different ways. The reaction to a vibration exposure is greatly dependent on the frequency, amplitude,

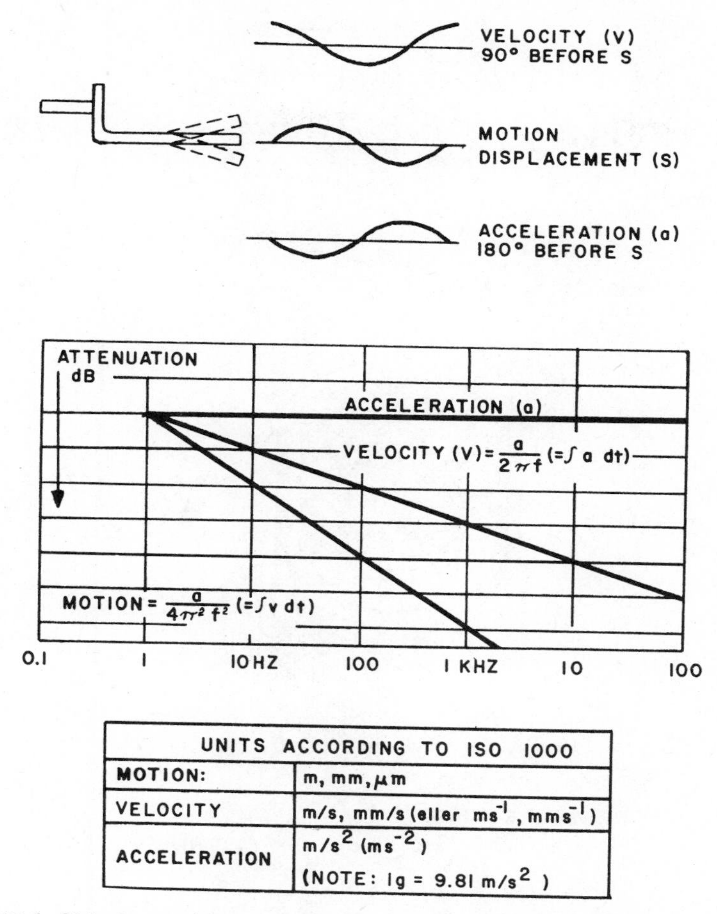

FIGURE 12.6 Velocity, motion, and acceleration are three vibration parameters (adapted from Bruel and Kjaer, 1982.)

and duration of exposure, but several other factors are also important, such as the direction of vibration input, the location of different body segments, the mass of the different body segments, the level of fatigue, and the presence of external body supports.

Human response to vibration can be both mechanical and psychological in nature. Harmful mechanical effects are due to the induced strain on different tissues, as a result of motions and deformations within the body. Mechanical damage to tissue can occur as well as physiological changes without any permanent organ damage. Unspecific psychological stress reactions also occur. While the mechaniccal effects are generally frequency-dependent, causing resonance within various organ systems, the psycho-

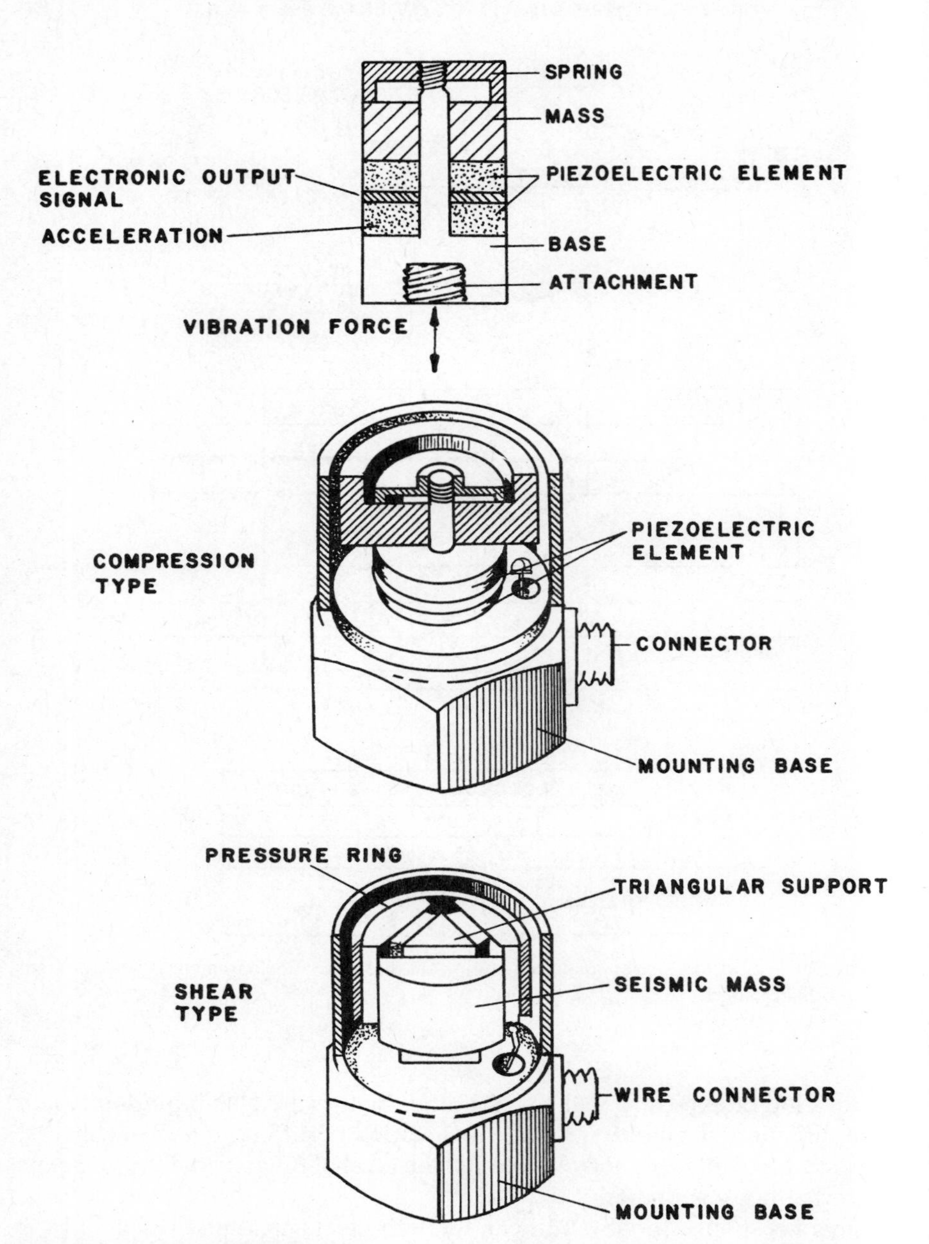

FIGURE 12.7 Illustration of compression-type and shear-type piezoelectric accelerometers. (Adapted from Bruel and Kjaer, 1982.)

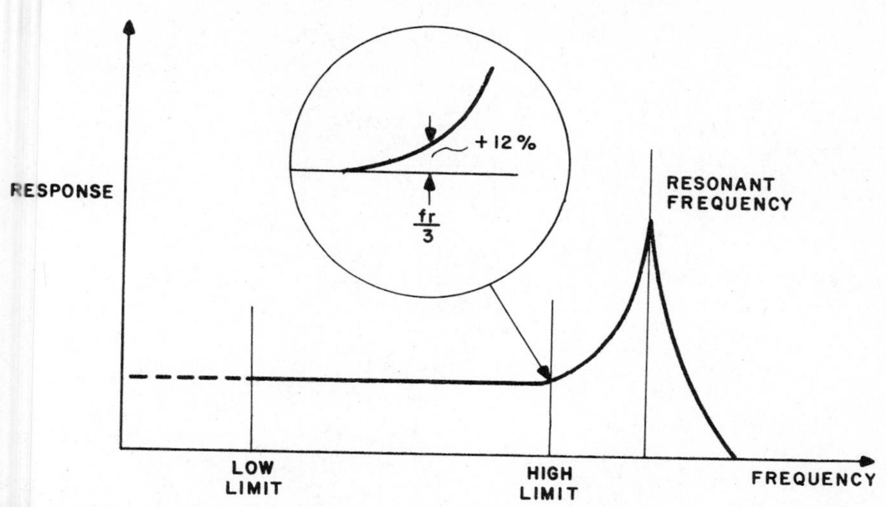

FIGURE 12.8 Frequency range of an accelerometer. Low-frequency limit determined by preamplifier; high-frequency limit one-third of the resonant frequency of the accelerometer. (Adapted from Bruel and Kjaer, 1982.)

logical reactions are not always so, nor are they well-correlated with other aspects of the physical vibrations, as discussed later in Section 12.3.

12.3 WHOLE-BODY VIBRATION

12.3.1 General Effects

The human body can be considered to be a dynamic mechanical system. It can be modeled as a linear system, however, only within a limited frequency range (up to 100 Hz) and when deformations are small. From an exposure point of view, the low-frequency range is the most interesting; in that range the body can be approximated as a linear lumped-parameter system of masses, springs, attenuators, and links (Figure 12.9). As indicated in this model, resonance occurs for different parts of the system at different frequencies. Exposure to vertical vibrations in the 5–10 Hz range generally causes resonance in the "thoracic-abdominal" system, at 20 to 30 Hz in the head–neck–shoulder system, and at 60 to 90 Hz in the eyeball. Above 100 Hz, the model is not useful. All values given refer to exposure in the longitudinal (vertical erect) direction. For other types of exposure, different resonance frequencies should be expected.

When vibrations are attenuated in the body, the vibration energy is absorbed by tissues and organs. The muscles are important in this respect. Vibration leads to both voluntary and involuntary contractions of muscles,

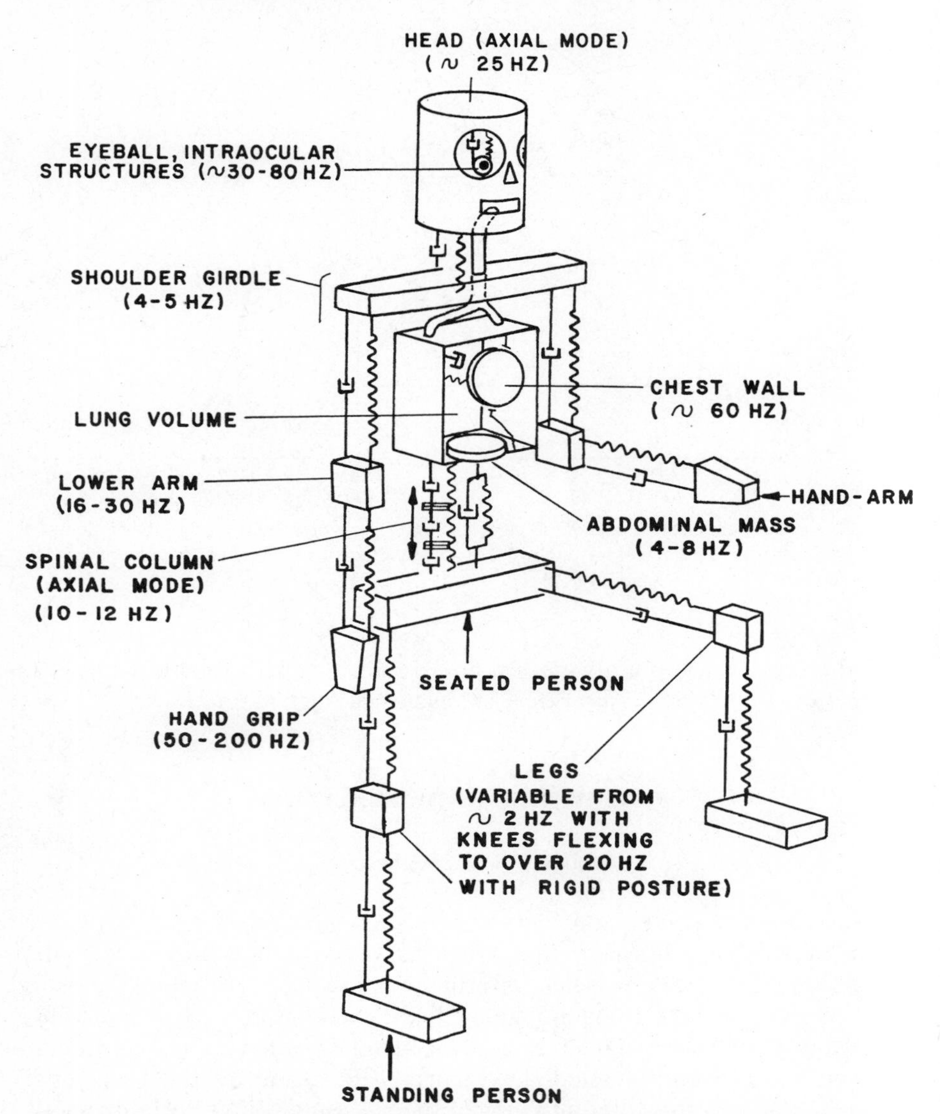

FIGURE 12.9 Mechanical system representing the human body subjected to vertical vibration (adapted from Rasmussen, 1982; von Gierke, Nixon, and Guignard, 1975.)

and can cause local muscle fatigue, particularly when the vibration is at the resonant-frequency level. Further, mechanical stimulation of muscles can cause reflex contractions, which will reduce motor performance capabilities. Electromyographic studies have documented stretch reflex responses from the skeletal muscles when vibrated; smooth muscle contractions have been recorded in the ureters, heart muscle, and peripheral blood vessels.

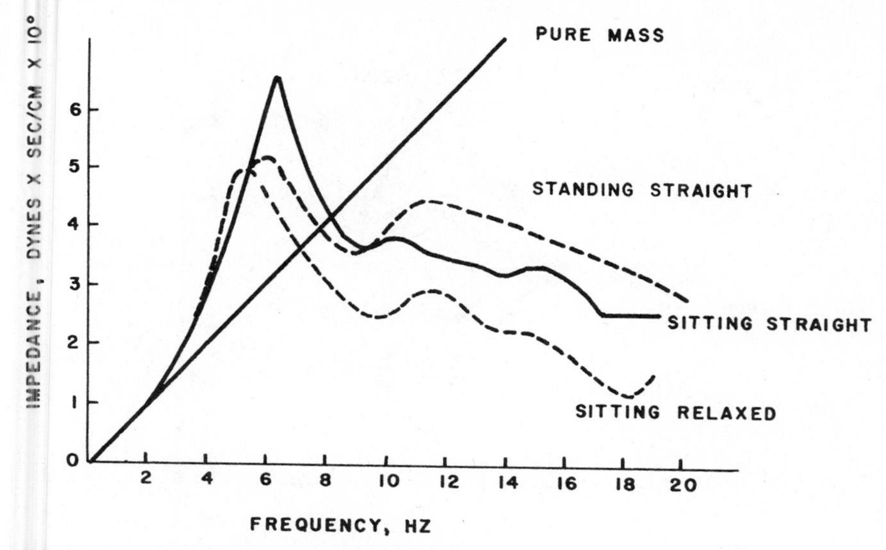

FIGURE 12.10 The mechanical impedance as a function of frequency and body posture. (Adapted from Rao and Ashley, 1976.)

The amount of mechanical energy transmission due to vibration is, as mentioned earlier, dependent on body position and muscle contractions. It can be described in terms of *mechanical impedance,* defined as the relationship of transmitted force to the velocity at which the force acts on the body. In other words, impedance is the ratio of a transformed system excitation (input) to a transformed response (output). Figure 12.10 illustrates the mechanical impedance of the body as a function of frequency and body posture (Rao & Ashley, 1976).

Vibration transmission can also be characterized by transfer functions, describing the relationship of input acceleration and measured output acceleration at a point in the body. In frequency ranges where attenuation is low, resonances cause increases in the transfer-function magnitude. Figure 12.11 shows data obtained when standing and sitting subjects are exposed to vertical vibrations. In a standing subject, the first resonance occurs at the hip, shoulder, and head at about 5 Hz. With the subjects sitting, resonance occurs at the shoulders and also to some degree at the head at 5 Hz. Further, a significant resonance from shoulder to head occurs at about 30 Hz.

In this latter frequency range, the amplitude of the head motion can be as much as three times that of the shoulders. Frequencies above 10 Hz generally induce less motion in the segments as the frequency increases. The attenuation of vibration throughout the human body is illustrated in Figure 12.12. The figure shows that the attenuation of a 50-Hz vibration from foot to head is about 30 dB, from hand to head, about 40 dB.

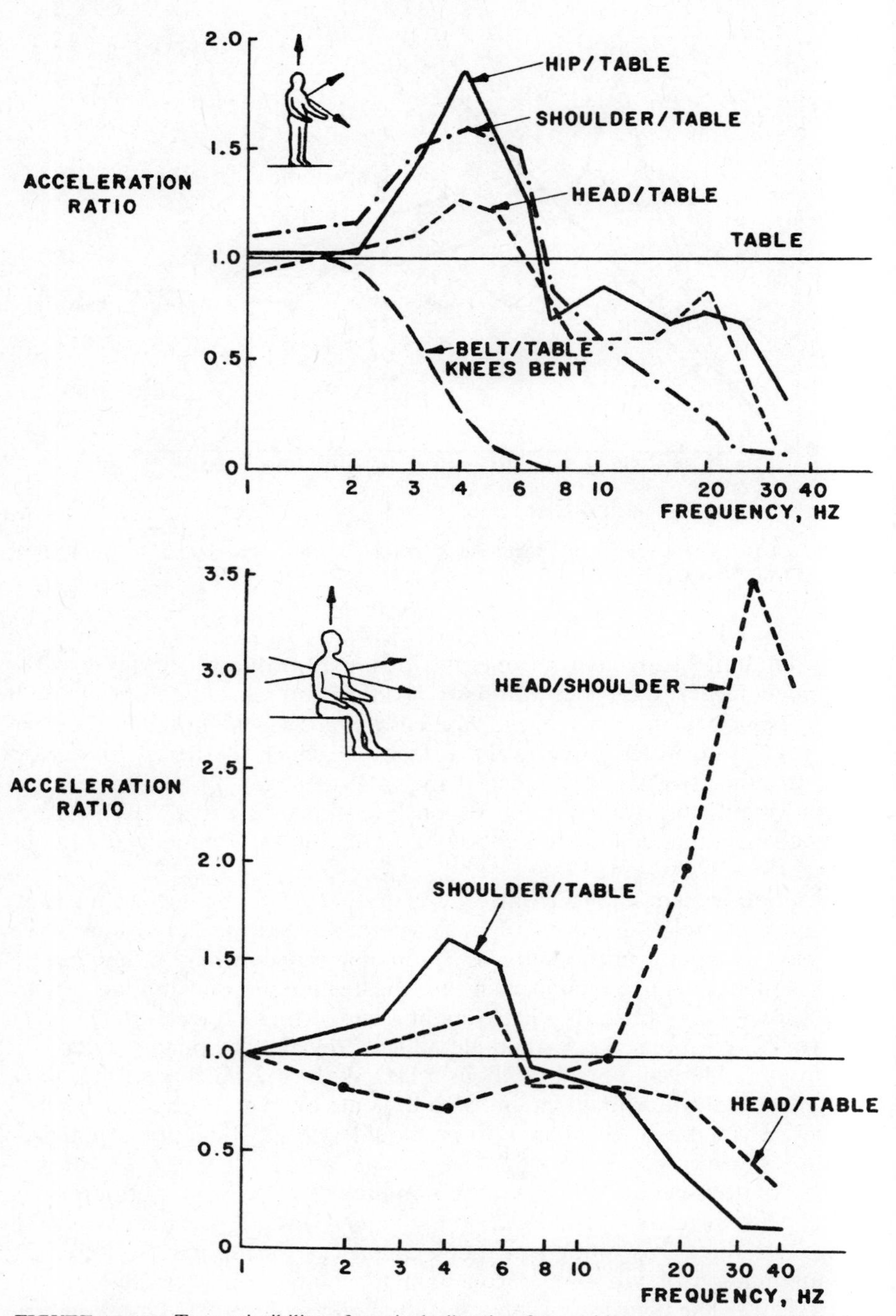

FIGURE 12.11 Transmissibility of vertical vibration from table to various parts of the body of (a) standing; (b) sitting human subject as a function of frequency. (Adapted from Rasmussen, 1982.)

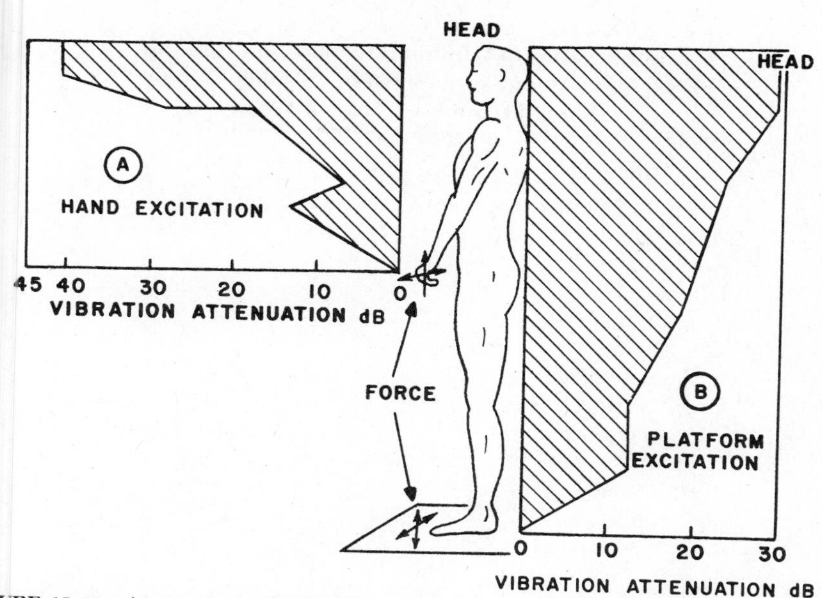

FIGURE 12.12 Attenuation of vibration of 50 Hz along the human body. The attenuation is expressed in decibels below values at the point of excitation. For excitation of (a) hand; (b) platform on which the subject stands. (Adapted from Rasmussen, 1982.)

The influence of whole-body vibration on the nervous system is largely unknown. Central processes can be influenced, however. Observations indicate that the general state of consciousness is influenced by vibrations. Low frequency vibrations (1–2 Hz) with moderate intensities induce sleep, while higher frequencies, particularly random ones, have the opposite effect (i.e., induce arousal). Unspecific psychological stress reactions have also been noted (Guignard, 1965; von Gierke, 1964), as well as degraded visual and motor effects on functional performance. Some symptoms of vibration exposure at low frequencies are given in Table 12.2, along with the frequency ranges at which the symptoms are most predominant.

12.3.2 Effects on the Spine

In recent years, considerable thought has been given to the effect of whole-body vibration on the spine.

Although conclusive evidence is not available, there are many studies suggesting an increased risk of low-back pain in drivers of tractors (Rosegger and Rosegger, 1960; Dupuis and Christ, 1972); of trucks (Kelsey and Hardy, 1975; Gruber, 1976; Frymoyer et al, 1979); of buses (Gruber and Ziperman, 1974; Kelsey and Hardy, 1975); and of airplanes (Fitzgerald and Crotty, 1972; Schulte-Wintrop and Knosche, 1970). These studies also suggest that low-

TABLE 12.2
Symptoms Due to Whole-Body Vibration and the Frequency Range at which They Usually Occur (Adapted from Rasmussen, 1981).

Symptoms	Frequency (Hz)
General feeling of discomfort	4–9
Head symptoms	13–20
Lower jaw symptoms	6–8
Influence on speech	13–20
"Lump in throat"	12–16
Chest pains	5–7
Abdominal pains	4–10
Urge to urinate	10–18
Increased muscle tone	13–20
Influence on breathing movements	4–8
Muscle contractions	4–9

back pain occurs at an earlier age in subjects exposed to vibration. Kelsey and Hardy (1975) found that truck driving increased the risk of disc herniation by a factor of four, while tractor driving and car commuting (20 miles or more per day) increased the risk by a factor of two. Studies of vibration-exposed populations also indicate that radiographic changes occur in the spines of these subjects (Rosegger and Rosegger, 1960; Dupuis and Christ, 1972; Gruber, 1976).

It is well known that the spine fractures (in compression) when subjected to strong vertical acceleration. Vibrations at lower acceleration levels have been suggested as causes of fatigue failures of different component structures of the spine; these vibrations also interfere with the nutrition of the disc, predisposing it to degenerative changes. A summary of data can be found in a report by Sandover (1981).

12.4 SEGMENTAL VIBRATION

As with the body as a whole, the hand–arm system can be considered a mechanical system of masses, springs, and viscous shock absorbers. A simplified model is illustrated in Figure 12.13. The skin and subcutaneous tissues are modeled as mass M1, parts of the hand as mass M2, and the rest of the hand and wrist and parts of the lower arm as mass M3. Although simplified, the model contributes to the understanding of how the hand–arm system reacts to vibration (Reynolds and Soedel, 1972; Mishoe and Suggs, 1974, 1977; Wood et al, 1978; Hempstock and O'Connor, 1978). As with whole-body vibration, the response to segmental vibration depends on frequency,

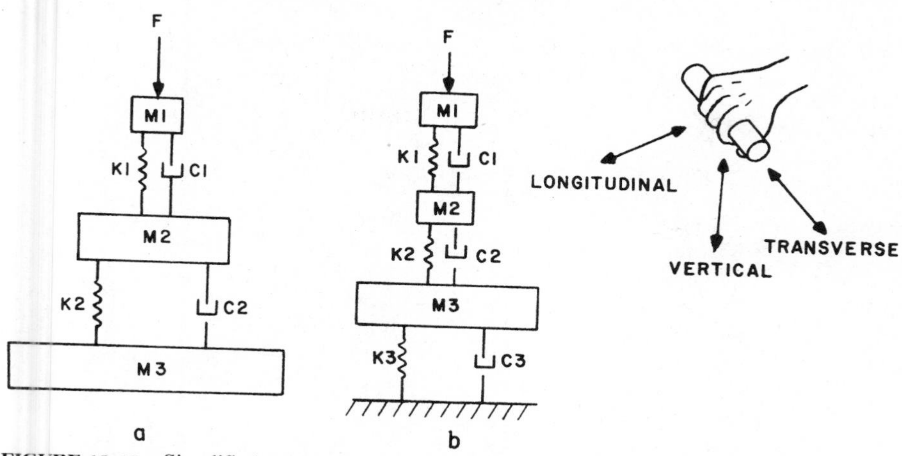

FIGURE 12.13 Simplified model of hand–arm system subjected to vibrations in (a) transverse direction; (b) longitudinal (distal–proximal) direction. (Adapted from Mishoe and Suggs, 1977.)

amplitude, acceleration, duration, direction, and point of application, and on body factors such as posture and muscle tensions.

12.4.1 Vibration Syndrome

Segmental vibration causes a symptom complex usually referred to as *vibration syndrome*. In some subjects only one symptom in the complex is manifested, while in others symptoms occur together. The symptoms originate from injuries to the blood vessels, nerves, bones, joints, and muscles. Injuries can occur after exposure times from months to decades, and are usually, at first, reversible. The most well-known of these symptoms is the so called *Reynaud's Syndrome* or *Traumatic Vasopastic Disease* (*TVD*) (Taylor, 1974). The syndrome is characterized by a sudden block in the blood circulation to the fingers, which become white, pale, cold, and sometimes painful. Tactile sensitivity is reduced, preventing precision work. The symptoms can be present for only a few minutes, sometimes for several hours. TVD is caused by smooth muscle constriction in the blood vessels of the fingers. It exists without vibrational exposure in up to 1% of the population (90% of the cases occur in women). Vibration exposure, however, increases the risk dramatically. The cause of TVD is believed to be in part a vascular disturbance due to changes in the blood vessel walls, and in part a nervous disturbance caused by a reflex contraction of the smooth muscles of the blood vessels. Other vibration-induced symptoms come from the peripheral nerves and consist of paresthesias, tingling sensations, and the like. A decreased nerve action-potential conduction velocity has been found, as well as a decrease in the ability to perform precise motor movements of the

TABLE 12.3
Occurrence of TVD in Different Occupational Groups (from Pelmear and Taylor, 1975)

		Vibration exposed			Controls		
Process	Population[a]	Total	No.	%	Total	No.	%
Sawing	Thetford	56	50	89	63	4	6
	Kielder	38	26	68	24	2	8
	Dumfries	29	14	48	23	1	4
Grinding	Cwmbran	51	30	59	70	3	4
	Bromsgrove	89	53	60	51	7	14
	Halesowen	21	3	14	18	3	17
	Leamington	74	8	11	77	3	4
	Old Cwmbran	36	8	22	—	—	—
	Darlaston	29	5	17	—	—	—
	Ayr	112	11	10	—	—	—
	Lincoln	59	17	29	—	—	—
	Smethwick	25	6	24	—	—	—
	Witton	15	2	13	—	—	—
	Strood	17	3	18	—	—	—
	Great Barr	7	0	0	—	—	—
Chipping	Birkenhead	98	56	57	26	5	19
	Rosyth	69	32	46	—	—	—
Swaging	Liverpool	2	1	50	13	1	6

[a] Refers to populations in different locations in U.K.

fingers. The frequency of TVD in some occupational groups is summarized in Table 12.3.

12.4.2 Transmission of Vibration in the Upper Extremity

Most hand-held tools generate random vibration over a wide frequency range (typically 2–2000Hz). Low-frequency vibrations of the upper extremity can be transmitted to the trunk and head and cause unspecific symptoms such as headache, vertigo, nausea, and psychological stress reactions. Attenuation of these types of vibration occurs mainly in the joints. The energy at higher frequencies is largely absorbed by superficial tissues, wherein cell damage can occur. When subjected to longitudinal vibration, the upper extremity can be considered to be a low-pass filter, the properties of which change at about 100 Hz, as illustrated in Figure 12.14 (Suggs, 1974; Pyykko et al., 1976). Attenuation occurs with 3 dB/octave in the 20–100 Hz frequency range. Between 100 and 630 Hz, the attenuation of the elbow and upper part of the arm increases by about 10 dB/octave, and in the wrist by about 6 dB. At 1000 Hz, vibration attenuation at the wrist is 40 dB, at the elbow 45–50

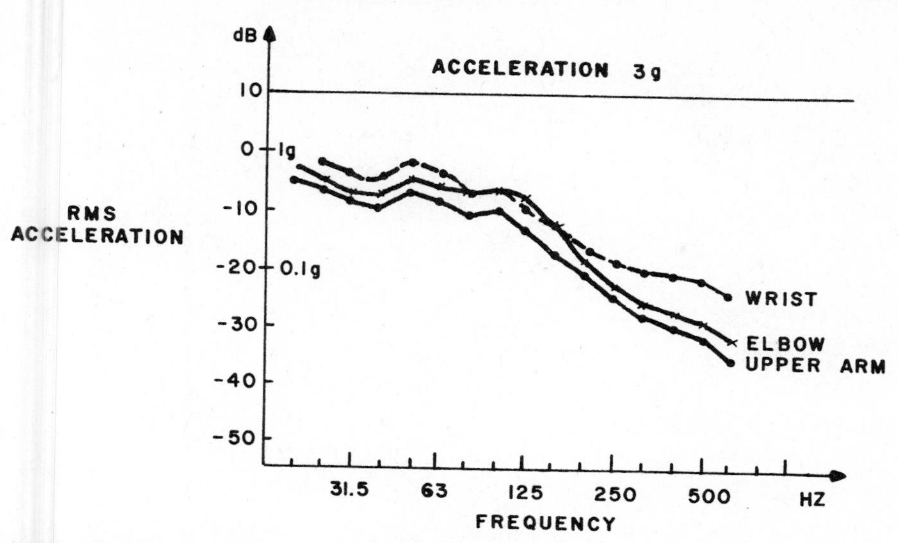

FIGURE 12.14 Attenuation of vibration in the upper extremity. (Adapted from Korhonen et al., 1977.)

dB. Iwata (1972) found that the vibration at the wrist was two to three times higher at 20 Hz than the input vibration—that is, resonance had occurred. At higher-frequency levels, above 1000 Hz, resonances of the hand–arm systems do not occur.

The transmission of vibrations in the upper extremity is linear. When the vibration of the hand-held tool increases by 10 dB, the vibration at the hand also increases by 10 dB. The hand grip force is important to the transfer function. When increased from 10 N to 40 N (12 dB), the hand vibration increases only by 3 to 5 dB. It appears that the transmitted vibration is proportional to the cube root of the hand grip force (Pykko et al., 1976).

12.5 VIBRATION EXPOSURE CRITERIA

Despite the variability and limitations of existing data, attempts have been made to formulate tentative standards. They should be regarded as recommended guidelines rather than firm design specifications.

12.5.1 Whole-Body Vibration

The guidelines that follow are based on ISO 2631-1978 and ISO 2631/DAM 1-1980, Limits are given in third-octave bands with frequencies from 1 to 80 Hz. Vibrations at levels lower than 1 Hz cause motion sickness and are poorly delineated in the guidelines. The level of vibration in the ISO guide-

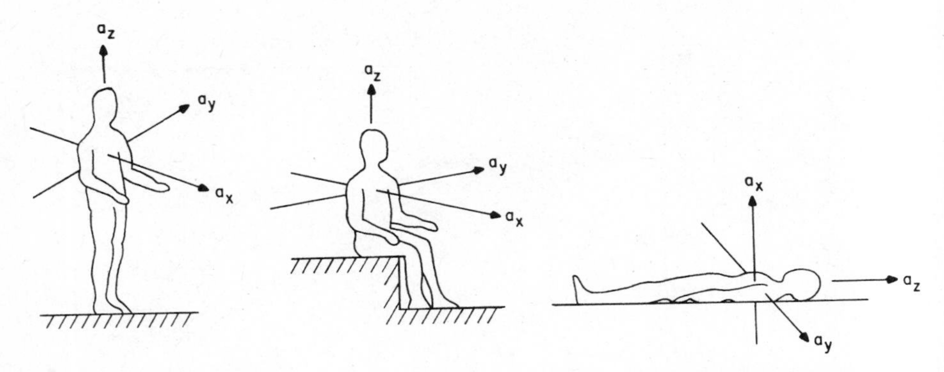

FIGURE 12.15 Coordinate system to evaluate whole-body vibration (From ISO 2631-1978.)

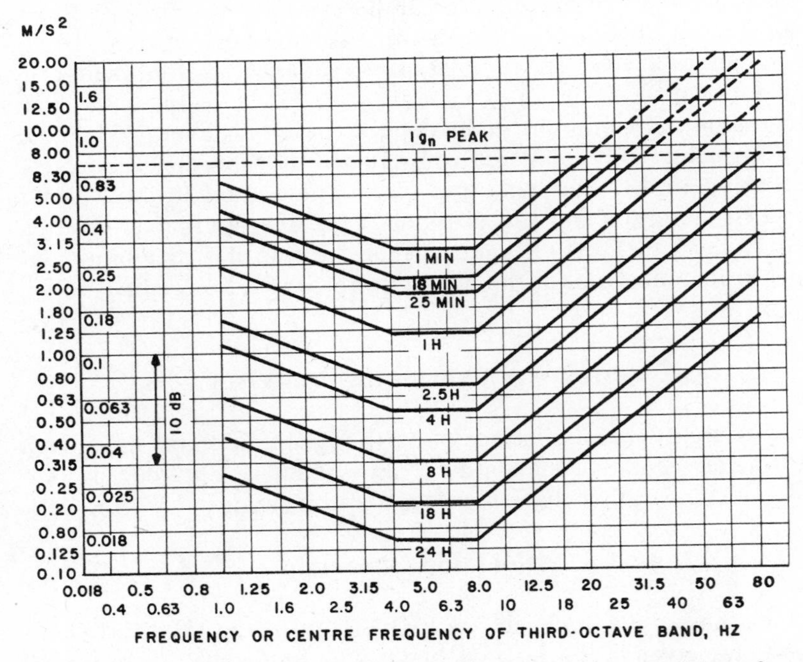

FIGURE 12.16 Limits for equal fatigue-decreased performance proficiency for vertical vibrations in third-octave band (Hz). (From ISO 2631-1978.)

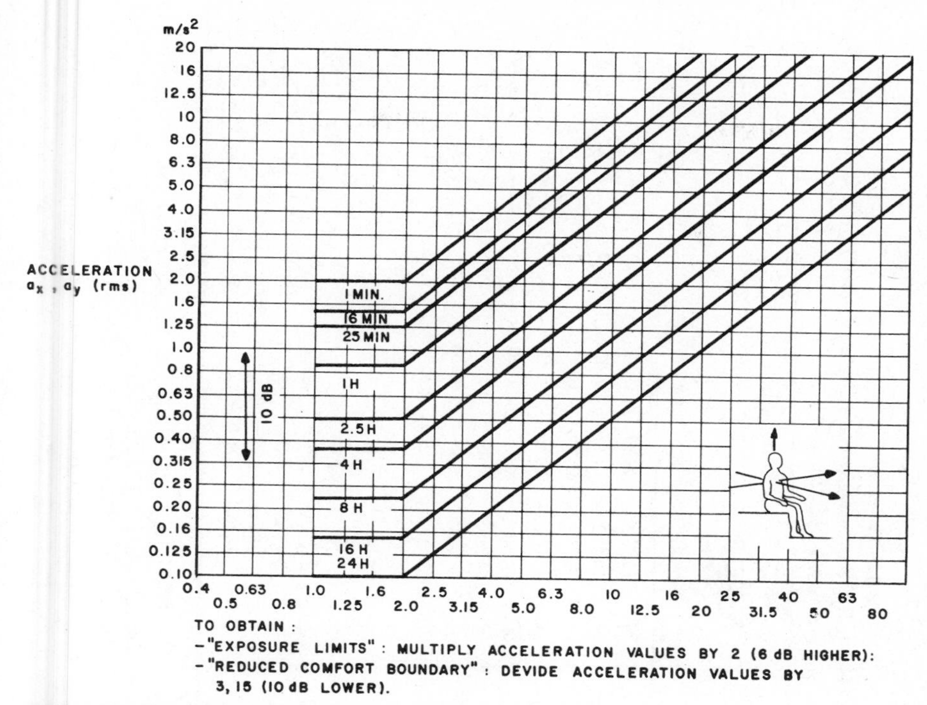

FIGURE 12.17 Limits for equal fatigue-decreased performance proficiency for horizontal vibrations in third-octave (Hz). (From ISO 2631-1978.)

lines is expressed as acceleration (m/sec^2) RMS. The measurement directions are presented in Figure 12.15. The limits are given in terms of (1) reduced comfort boundary, (2) fatigue-decreased proficiency boundary, and (3) exposure limit. Different values exist for vertical and horizontal vibrations.

The levels for fatigue decreased proficiency boundaries are given in Tables 12.4 and 12.5 for discrete exposure periods; and in Figures 12.16 and 12.17, as frequency acceleration diagrams. Reduced-comfort boundaries are obtained by dividing the fatigue-decreased proficiency limits by 3.15 (a decrease of 10 dB). Exposure limits are obtained by multiplying the proficiency limits by 2 (an increase of 6 dB).

The 1980 supplement to the standards suggests that to estimate reduced comfort and proficiency a filtered vibration signal over the entire frequency range should be used. A weighted value is then added to the frequency acceleration curves. Thus, in Figure 12.16 a wide-band filter should weigh the signals for a_x and a_y; in Figure 12.17, for a_z. The weighted values a_{xm} and a_{ym} are compared with the levels of the third-octave bands 1 to 2 Hz for a_x and a_y, while for a_z the comparison is made for the third-octave bands

TABLE 12.4
Numerical Acceleration Values (m/sec^2 RMS) for Horizontal Vibrations in Third-Octave Band (Hz) (Adapted from ISO 2631–1978)

Frequency (Center Frequency of Third-Octave Band), Hz	Exposure Times								
	24 hr	16 hr	8 hr	4 hr	2.5 hr	1 hr	25 min	18 min	1 min
1.0	0.100	0.150	0.224	0.355	0.50	0.85	1.25	1.50	2.0
1.25	0.100	0.150	0.224	0.355	0.50	0.85	1.25	1.50	2.0
1.6	0.100	0.150	0.224	0.355	0.50	0.85	1.25	1.50	2.0
2.0	0.100	0.150	0.224	0.355	0.50	0.85	1.25	1.50	2.0
2.5	0.125	0.190	0.280	0.450	0.63	1.06	1.6	1.9	2.5
3.15	0.160	0.238	0.356	0.560	0.8	1.32	2.0	2.36	3.15
4.0	0.200	0.300	0.450	0.710	1.0	1.70	2.5	3.0	4.0
5.0	0.250	0.375	0.560	0.900	1.25	2.12	3.15	3.75	5.0
6.3	0.315	0.475	0.710	1.12	1.6	2.65	4.0	4.75	6.3
8.0	0.40	0.60	0.900	1.40	2.0	3.35	5.0	6.0	8.0
10.0	0.50	0.75	1.12	1.80	2.5	4.25	6.3	7.5	10
12.5	0.63	0.95	1.40	2.24	3.15	5.30	8.0	9.5	12.5
16.0	0.80	1.18	1.80	2.80	4.0	6.70	10	11.8	16
20.0	1.00	1.50	2.24	3.55	5.0	8.5	12.5	15	20
25.0	1.25	1.90	2.80	4.50	6.3	10.6	16	19	25
31.5	1.60	2.36	3.55	5.60	8.0	13.2	20	23.6	31.5
40.0	2.00	3.00	4.50	7.10	10.0	17.0	25	30	40
50.0	2.50	3.75	5.60	9.00	12.5	21.2	31.5	37.5	50
63.0	3.15	4.75	7.10	11.2	16.0	26.5	40	45.7	63
80.0	4.00	6.00	9.00	14.0	20	33.5	50	60	80

TABLE 12.5

Numerical Acceleration Values (m/sec^2 RMS) for Vertical Vibrations in Third-Octave Band (Hz) (Adapted from ISO 2631–1978)

Frequency (Center Frequency of Third-Octave Band), Hz	Exposure Times								
	24 hr	16 hr	6 hr	4 hr	2.5 hr	1 hr	25 min	18 min	1 min
1.0	0.280	0.425	0.63	1.06	1.40	2.36	3.65	4.25	5.60
1.25	0.250	0.375	0.56	0.95	1.26	2.12	3.15	3.75	5.00
1.6	0.224	0.335	0.50	0.85	1.12	1.80	2.80	3.35	4.50
2.0	0.200	0.300	0.45	0.75	1.00	1.70	2.50	3.00	4.00
2.5	0.180	0.265	0.40	0.67	0.90	1.50	2.24	2.65	3.55
3.16	0.160	0.235	0.355	0.60	0.80	1.32	2.00	2.35	3.15
4.0	0.140	0.212	0.315	0.53	0.71	1.18	1.80	2.12	2.80
5.0	0.140	0.212	0.315	0.53	0.71	1.18	1.80	2.12	2.80
6.3	0.140	0.212	0.315	0.53	0.71	1.18	1.80	2.12	2.80
8.0	0.140	0.212	0.315	0.53	0.71	1.18	1.80	2.12	2.80
10.0	0.180	0.265	0.40	0.67	0.90	1.50	2.24	2.65	3.55
12.5	0.224	0.335	0.50	0.85	1.12	1.90	2.80	3.35	4.50
16.0	0.280	0.425	0.63	1.06	1.40	2.36	3.55	4.25	5.60
20.0	0.355	0.530	0.80	1.32	1.80	3.00	4.50	5.30	7.10
25.0	0.450	0.670	1.0	1.70	2.24	3.75	5.60	6.70	9.00
31.5	0.560	0.850	1.25	2.12	2.80	4.75	7.10	8.50	11.2
40.0	0.710	1.060	1.60	2.65	3.55	6.00	9.00	10.6	14.0
50.0	0.900	1.320	2.0	3.35	4.50	7.50	11.2	13.2	18.0
63.0	1.120	1.700	2.5	4.25	5.60	9.50	14.0	17.0	22.4
80.0	1.400	2.120	3.15	5.30	7.10	11.8	18.0	21.2	28.0

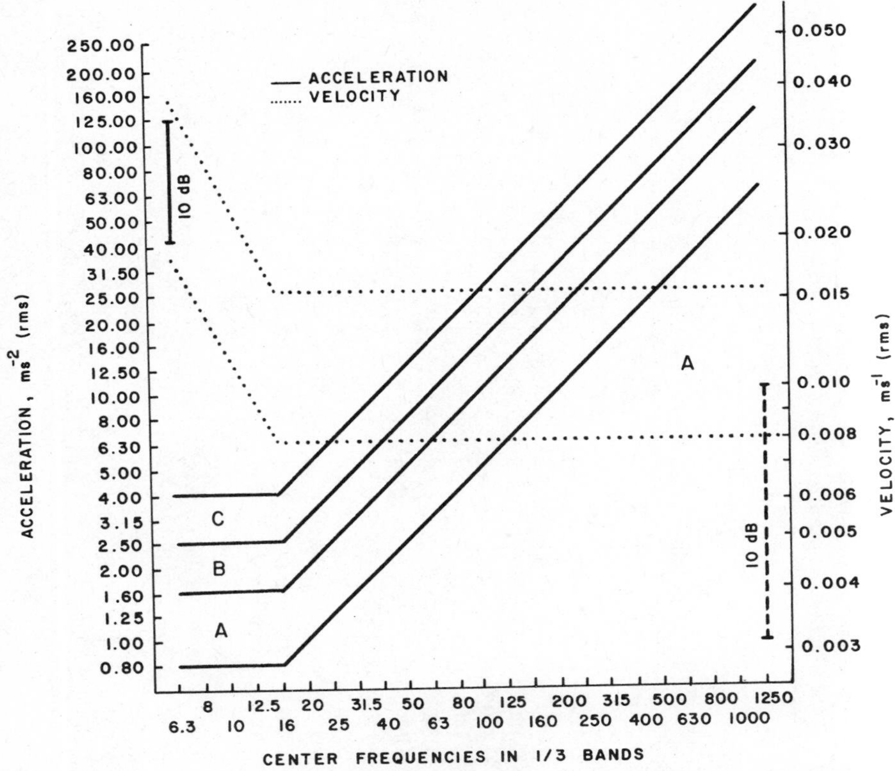

FIGURE 12.18 Limits for segmental vibration expressed as acceleration and velocity for third-octave band. A = 4–8 hr; B = 2–4 hr; C = 0.5–2 hr exposure. (From ISO/DIS 5345-1980.)

4 to 8 Hz. Should the weighted value exceed an acceptable level, a third-octave band analysis is made of the vibration signal for final evaluation. The reason for this amendment is that the research on which the recommendations are based was made using sine-wave vibration. The effect is that the exposure period is somewhat shorter than when third-octave band analysis alone is used. The result is the same, however, when the vibration curve is concentrated to one frequency only.

12.5.2 SEGMENTAL VIBRATION

Segmental vibration recommendations are based on ISO draft proposal 5349-1980. The suggested guideline levels are given in Figures 12.18 and 12.19 and Tables 12.6 and 12.7. They cover a frequency range of 8 Hz to 1 kHz for octave bands, and from 6.3 to 1250 Hz for third-octave bands. Potentially

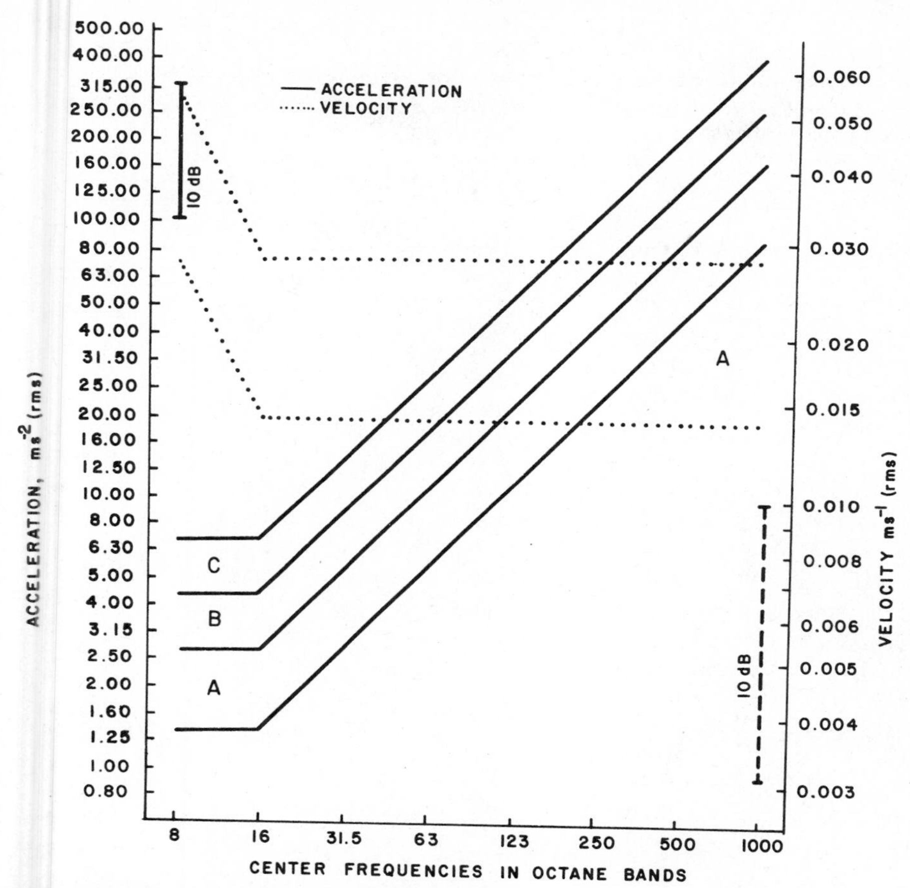

FIGURE 12.19 Limits for segmental vibration expressed as acceleration and velocity for octave band.

harmful and safe vibration levels are given as a range with an upper and lower limit. Exposures below the lower limit are considered safe, and above the upper limit as harmful. The range between the limits are divided into three zones—A, B, and C—each representing a recommended time interval for exposure. Zone A is 4 to 8 hr; Zone B, 2 to 4 hr; and Zone C, 0.5 to 2 hr. Figures 12.18 and 12.19 represent exposure curves for both third-octave and octave bands. Third-octave bands are recommended for analysis as they are likely to be more stringent when applied to the discrete frequency spectra often encountered in hand-held tools.

It is also common practice to use weighting filters for single-number readings. The overall weighting method has become the primary method to quote hand–arm vibration levels for regulatory purposes, industrial hygiene ap-

TABLE 12.6
Zone Limits for Segmental Vibration [The Acceleration Levels Are Expressed in (A) Third-Octave Band, (B) Octave Band (Adapted from ISO/DIS 5349–1980]

(A) Third-Octave Band

Center Frequency of Third-Octave Band	Acceleration, m/sec^2 RMS		
	Zone A	Zone B	Zone C
6.3	0.8–1.6	1.6–2.4	2.4–4.0
8	0.8–1.6	1.6–2.4	2.4–4.0
10	0.8–1.6	1.6–2.4	2.4–4.0
12.5	0.8–1.6	1.6–2.4	2.4–4.0
16	0.8–1.6	1.6–2.4	2.4–4.0
20	1–2	2–3	3–5
25	1.3–2.6	2.6–3.9	3.9–6.5
31.5	1.6–3.2	3.2–4.8	4.8–8
40	2–4	4–6	6–10
50	2.5–5	5–7.5	7.5–12.5
63	3.2–6.4	6.4–9.6	9.6–16
80	4–8	8–12	12–20
100	5–10	10–15	15–25
125	6.3–12.6	12.6–18.9	18.9–31.5
160	8–16	16–24	24–40
200	10–20	20–30	30–50
250	12.5–25	25–37.5	37.5–62.5
315	16–32	32–48	48–80
400	20–40	40–60	60–100
500	25–50	50–75	75–125
630	31.5–63	63–94.5	94.5–157.5
800	40–80	80–120	120–200
1000	50–100	100–150	150–250
1250	63–126	126–189	189–315

(B) Octave Band

Center Frequency of Third-Octave Band	Acceleration, m/sec^2 RMS		
	Zone A	Zone B	Zone C
8	1.4–2.8	2.8–4.2	4.2–7
16	1.4–3.8	3.8–4.2	4.2–7
31.5	2.7–5.4	5.4–8.1	8.1–13.5
63	5.4–10.8	10.8–16.2	16.2–27
125	10.7–21.4	31.4–32.1	32.1–53.5
250	21.3–42.6	42.6–63.9	63.9–106.5
500	42.5–85	85–127.5	127.5–212.5
1000	85–170	170–255	255–425

TABLE 12.7
Zone Limits for Segmental Vibration [Velocity Levels Are Expressed in (A) Third-Octave Band; (B) Octave Band (ISO/DIS 5349–1980)]

(A) Third-Octave Band

Center Frequency of Third-Octave Band	Velocity Level, m/sec RMS		
	Zone A	Zone B	Zone C
6.3	0.020–0.040	0.040–0.060	0.060–0.100
8	0.016–0.032	0.032–0.048	0.048–0.080
10	0.013–0.026	0.026–0.039	0.039–0.065
12.5	0.010–0.020	0.020–0.030	0.030–0.050
16			
20			
25			
31.5			
40			
50			
63			
80			
100			
125	0.008–0.016	0.016–0.024	0.024–0.040
160			
200			
250			
315			
400			
500			
630			
800			
1000			
1250			

(B) Octave Band

Center Frequency of Third-Octave Band	Velocity Level, m/sec RMS		
	Zone A	Zone B	Zone C
8	0.028–0.056	0.056–0.84	0.084–0.140
16	0.015–0.030	0.030–0.045	0.045–0.075
31.5			
63			
125	0.014–0.028	0.028–0.042	0.042–0.070
250			
500			
1000			

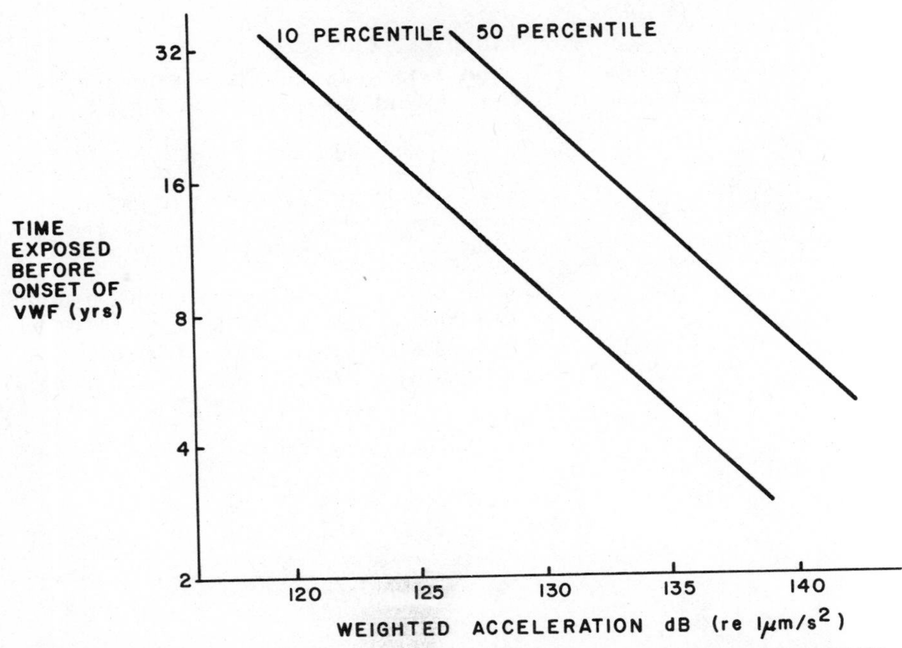

FIGURE 12.20 Time exposure before onset of Vibration White Finger Syndrome (VFW) as a function of overall weighted acceleration in dB. (Adapted from Rasmussen, 1982.)

plications, and for labeling of hand tools. The overall weighted acceleration is used in Figure 12.20 to illustrate the relationship of vibration level and time exposed until the Vibration White-Finger Syndrome becomes manifest. For research purposes, and for practical segmental vibration control work, where high resolution of the frequency spectrum is needed, third-octave band analysis is still to be preferred.

REFERENCES

Bruel and Kjaer, Sweden A. B., *Vibration Measurement*, Huddinge, Sweden, 1982, pp. 1–40 (in Swedish).

Bruel and Kjaer, Sweden A. B., *An Introduction to Vibration Measurement*, Huddinge, Sweden, 1973.

Dupuis, H. and W. Christ, "Untersuchung der Möglichkeit von Gesundeheits-Schädigungen im Bereich der Wirbelsäule bei Schlepperfahrern." *Max Plank Inst., Bad Kreuznach, Heft* A72/2, 1972.

Fitzgerald, J. G. and J. Crotty, "The Incidence of Backache Among Aircrew and Groundcrew in the RAF," *FPRC/1313*, 1972.

Frymoyer, J. W., M. H. Pope, J. Rosen, J. Goggin, D. Wilder, and M. Costanza, "Epidemiological Studies of Low Back Pain," Presented at 6th Annual Meeting, ISSLS, Göteborg, May–June, 1979.

von Gierke, H. E., "Biodynamic Responce of the Human Body." *Appl. Mech. Rev.*, **17**, 961–968 (1964).

von Gierke, H. W., C. W. Nixon, and J. C. Guignard, "Noise and Vibration," in M. Calvin and O. O. Gazenko, Eds., *Ecological and Physiological Foundation of Space Biology and Medicine*, Vol. 2, Book 1, NASA, Washington, D.C., 1975, pp. 355–405.

Gruber, G. J., "Relationships Between Wholebody Vibration and Morbidity Patterns Among Interstate Truck Drivers," Nat. Inst. Occ. Safety & Health, Cincinnati, *NIOSH Report* No. 77–167, 1976.

Gruber, G. J. and H. H. Ziperman, "Relationship Between Whole-Body Vibration and Morbidity Patterns Among Motor Coach Operators," *US DHEW (NIOSH)* Contract No. HSM-00-72-047, 1974.

Guignard, J. C., "Vibration," in J. A. Gilles, Ed., *A Textbook of Aviation Physiology*, Pergamon Press, Oxford, 1965, pp. 813–894.

Hempstock, T. I. and D. E. O'Connor, "Assessment of Hand Transmitted Vibration," *Ann. Occup. Hyg.*, **21**, 57–67 (1978).

International Standard ISO 2631–1978, *Guide for the Evaluation of Human Exposure to Whole-Body Vibration*, 1978.

ISO 2631/DAM 1, *Amendments to ISO 2631-1978*, International Organization of Standardization, 1980.

ISO Draft Proposal 5349, *Guide for the Measurement and Evaluation of Human Exposure to Vibration Transmitted to the Hand*, International Organization of Standardization, 1980.

Iwata, H., A. Matsuda, H. Takahashi, and S. Watabe, "Roentgenographic Findings in Elbows of Rock Drill Workers," *Acta Scholae Med. Univ. Gifu*, **19**, 393–404 (1971).

Kelsey, J. L. and E. J. Hardy, "Driving of Motor Vehicles as a Risk Factor for Acute Herniated Lumbar Intervertebral Disc," *Am. J. Epid.*, **102**, 63–73 (1975).

Korhonen, O., I. Pyykko, J. Starck, and J. Toivanen, *Local Vibration. Medical and Methodological problems, and Criteria*. Institute of Occupational Health, Helsinki, 1977, pp 1–83 (in Swedish).

Mishoe, J. W. and C. W. Suggs, "Hand-Arm Vibration Part I. Subjective Response to Single and Multi-Directional Sinusoidal and Non-Sinusoidal Excitation," *Journal of Sound and Vibration*, **35**(4), 479–488 (1974).

Mishoe, J. W. and C. W. Suggs, "Hand-Arm Vibration Part II. Vibrational Response of the Human Hand," *Journal of Sound and Vibration*, **53**(4), 545–558 (1977).

Pelmear, P. L. and W. Taylor, "The Results of Long-Term Vibration Exposure With a Review of Special Cases of Vibration White Finger," in W. Taylor and P. L. Pelmear, Eds., *Vibration White Finger in Industry*, London, Academic Press, 1975. pp. 83–110.

Pyykko, I., M. Farkkila, J. Toivanen, O. Korhonen, and J. Hyvarinen, "Transmission of Vibration in the Hand–Arm System With Special Reference to Changes in Compression Force and Acceleration," *Scand. J. Work Environm. Health*, **2**, 87–95 (1976).

Rao, B. K. N. and C. Ashley, "Subjective Effects of Vibration," in W. Tempest, Ed., *Infrasound and Low Frequency Vibration*, Academic Press, London, 1976.

Rasmussen, G., "Human Body Vibration Exposure and Its Measurement," *Technical Review, Bruel & Kjaer*, **1**, 1982. pp. 3–31.

Reynolds, D. D. and W. Soedel, "Dynamic Response of the Hand–Arm System to a Sinusoidal Input," *Journal of Sound and Vibration*, **21**(3), 339–353 (1972).

Rosegger, R. and S. Rosegger, "Arbeitsmedizinische Erkenntnisse beim Schlepperfahren," *Arch. Landtechn.*, **2**, 3–65 (1960).

Sandover, J., "Vibration Posture and Low-Back Pain Disorders of Professional Drivers," *Report No DHS 402,* University of Technology, Loughborough, England, 1981.

Schulte-Wintrop, H. C. and H. Knoche, "Backache in UH-ID Helicopter Crews," *AGARD-CP-255,* 1978.

Suggs, G. W., "Modelling of the Dynamic Characteristic of Hand–Arm System," in W. Taylor, Ed., *The Vibration Syndrome,* London, Academic Press, 1974, pp. 169–186.

Taylor, W., "The Vibration Syndrome: Introduction," in W. Taylor, Ed., *The Vibration Syndrome.* London, Academic Press, 1974, pp. 1–12.

Wood, L. A., C. W. Suggs, and C. F. Abrams, "Hand–Arm Vibration Part III: A Distributed Parameter Dynamic Model of the Hand–Arm System," *Journal of Sound and Vibration,* **57**(2), 157–169 (1978).

Wilder, D. G., B. B. Woodworth, J. W. Frymoyer, and M. H. Pope, "Vibration and the Human Spine," *Spine,* **7**, 243–254 (1982).

13

WORKER SELECTION AND TRAINING CRITERIA

13.1 WORKER SELECTION

13.1.1 Introduction

The purpose of worker selection and placement programs is to assign the right worker to a particular job. Ideally, all jobs should be such that worker selection or restricted placement is not required. In reality this is not the case and will not be so for some time due to the extreme variability in the performance capability of the population and in the lack of ergonomics in the design of work. Thus, selection becomes a means of reducing possible harmful physical effects of work created by mismatching of worker and job. It will become clear that the basis for worker selection is still uncertain. Further, job classification schemes to match subject capabilities is far from a simple technology, as discussed in Chapter 7.

Different options in worker selection, as well as their advantages and limitations, will be discussed in this chapter, as will general criteria on physical assessment. Some important definitions will be reviewed. They include (1) accuracy, (2) sensitivity, (3) specificity, and (4) predictive value.

Accuracy is a measure of a screening test's ability to provide a true measure of a quantity or quality.

Sensitivity is a measure of a test's accuracy in correctly identifying persons with a certain condition. It can be expressed as the fraction or percentage of all persons with a condition who will have a positive test:

$$\frac{\text{True Positives}}{\text{True Positives} + \text{False Negatives}} \times 100 \text{ percent}$$

TABLE 13.1
Predictive Value of a Positive Test as a Function of the Disease Prevalence.[a]

Prevalence of Disease (percent)	Predictive Value of Positive Test
1	16.1
2	21.2
5	50.0
10	67.9
15	77.0
20	82.6
25	86.4
50	95.0

[a] Laboratory test with 95 percent sensitivity and 95 percent specificity (from Galen, R.S., 1979, "Selection of Appropriate Laboratory Tests," *Clinician and Chemist*, Ed., B.S. Young, Washington, D.C.: The American Association for Clinical Chemistry, p. 76).

This means that if the sensitivity is 95 percent, 95 of a group of 100 with a certain condition will have a positive test (true positive), 5 a negative (false negative).

Specificity is a measure of a test's accuracy in correctly identifying persons who do not have the condition. It can be expressed as the fraction or percentage of negative tests in persons free of a condition (true negatives):

$$\frac{\text{True Negatives}}{\text{True Negatives} + \text{False Positives}} \times 100 \text{ percent}$$

A test with a 95 percent specificity applied to a group of 100 persons who do *not* have a disease will be negative in 95 cases (true negative) and positive in 5 (false positive).

Predictive value of a test is its ability to predict the presence of absence of a condition. It can be expressed as:

$$\frac{\text{True Positives}}{\text{True Positives} + \text{False Positives}} \times 100 \text{ percent}$$

The predictive value depends on the sensitivity and specificity of a test as well as on the actual frequency of a condition in the population. Table 13.1 illustrates the dependence of the predictive value on the disease prevalence. When the prevalence is 50 percent or higher, the predictive value is marginally influenced.

13.1.2 HISTORY AND PHYSICAL EXAMINATION

The purpose of taking a subject's medical and work history and performing a physical examination is to identify *present* health problems to assure that the person receives adequate treatment. In addition, the information is needed if one is to identify *high-risk people* in terms of future susceptibility to musculo-skeletal problems. This requires that the examiner know what factors are critical, as well as being able to identify them clinically. Radiographic examinations and quantitative physical pre-employment screening procedures will be discussed later in the chapter to assist in this latter endeavor.

The most important part of the medical history deals with previous musculo-skeletal problems. In fact, some physicians mantain that this is the only useful indicator in the medical history (Taylor, 1968; Glover, 1980). Low-back pain is probably the condition most frequently discussed in the current literature regarding pre-employment examinations. Recurrent episodes of low-back pain appear to be almost part of the natural history of the disease. Rowe (1963) found that 83% of those with low-back pain had recurrent attacks, while recurrencies in patients with sciatica was 75% (Rowe, 1965). Similar findings were made by Dillane et al (1966), Horal (1969), Hirsch et al (1969), Leavitt et al. (1971), Gyntelberg (1974), Pedersen (1981), Troup et al. (1981), and Bierring-Sørensen (1983). Horal and Rowe found age to be a contributing factor; recurrencies occured more frequently at higher age.

Chaffin and Park (1973) found a threefold increase in the risk of having a back pain episode in subjects with a previous history of back pain. Bergquist-Ullman and Larson (1977) report that 62% of a group of 217 workers with acute low-back pain had recurrencies within a year, and 18% more in two years.

Bierring-Sørensen (1983) performed a cross-sectional survey of 558 men and 583 women, 30–60 years of age. He found that subjects who in the year immediately before the study had either (1) many episodes of back pain, (2) many sickness absence days, (3) a short interval between episodes, or (4) an aggravated course of low-back pain had significantly increased risk of low back pain in the year following examination. Pedersen (1981), Troup et al. (1981), and Bierring-Sørensen (1983) found a history of sciatica to be a risk indicator, while Dillane et al. (1966) did not find this to be so. Magora and Taustein (1969) found that persons who had had sciatica had more and longer sickness absence periods, a finding confirmed by Andersson et al. (1983).

Pedersen (1981) found that a history of more than three previous episodes of low-back pain indicated a poor prognosis should a new episode occur, in that the course of any new episode was both longer and more severe. Troup et al. (1981) found that truly accident-related previous back pain was a negative prognostic factor in recurrencies. Acute onset of low-back pain similarly appears to cause a longer duration of the pain episode (Bergquist-Ullman and Larsson, 1977; Pedersen, 1981; Bierring-Sørensen, 1983).

Some optimism can be found in these studies, but there are also reports indicating that simply obtaining a medical history for prior back pain is not enough to reduce back injury rates (Snook et al., 1978). Although no statistics have been found for musculo-skeletal problems other than for the back, it is our belief that a positive history of such problems is a sign of increased future risk.

The physical examination can reveal signs of dysfunction indicative of a risk of future problems. General observation of posture, ranges of motion, and so on do not seem to be reliable indicators; however, age and sex have some indirect influence on susceptibility (see, for example, Andersson, 1981). As discussed in previous chapters, the risk of musculo-skeletal injury increases with age, and is greater in women than in men. The variability in individual risk, however, is so great within age and gender groups that to deny a person employment based solely on age or sex would be difficult to support.

13.1.3 Radiographic Pre-Employment Examination

Pre-employment X-rays of the lumbar spine are used widely. The usefulness of this approach is under debate, however. Further, there is serious concern about whether lumbo-sacral x-rays are safe. The correlation between most of the radiographic abnormalities identified on X-rays and the risk of low-back pain in weak. The exceptions are spondylolisthesis (an anterior displacement of one vertebra on the adjacent vertebra below usually due to a skeletal defect) and severe degenerative changes that can be more easily predicted, though still being debatable as the basis for denying a person a job. In general, there is disagreement on the importance of most radiographic defects and abnormalities (Splithoff, 1953; Runge, 1954, 1958; Hult, 1954; Fullenlowe and Williams, 1957; Fischer et al., 1958; Ross, 1962; Bond et al., 1964; Connell, 1968; Moreton, 1969; Horal, 1969; LaRocca and MacNab, 1969; Wiltse, 1969; 1971 Torgerson and Dotter, 1976; Magora and Schwartz, 1976, 1978). A number of studies have indicated a reduction in low-back sickness absence when using radiographic screening programs (Becker, 1961; Crookshank and Warshaw, 1961; Kelly, 1965; McGill, 1968; Kosiak, Aurelius, and Hartfield, 1968; Redfield, 1972; Leggo and Mathiasen, 1973). All of these control programs, however, used other preventive measures then simply X-rays at the same time. Montgomery (1976) in a review article questioned the evidence and advised against the use of pre-employment back X-rays for determining future risk. LaRocca and MacNab (1969); Hadler (1978); and Rockey et al. (1979) came to a similar conclusion. In 1973, the American College of Radiology, a group of Orthopaedic Surgeons, and the American Occupational Medical Association joined together in a conference to evaluate the effectiveness of such X-rays when compared with the potential radiation hazard. They concluded that the use of X-rays as the *sole* criterion for selection of workers was not justified, and that more concern was needed to protect workers from unnecessary radiation in such examinations (American College of Radiology, 1973).

The general conclusion is that the use of radiographs as a screening tool is not indicated. When a previous history of back pain is present, when clinical signs indicate possible disease, or where spondylolithesis is suspected from clinical examination, radiographs are valuable in determining further risk.

13.1.4 Quantitative Physical Pre-Employment Screening

13.1.4.1 Anthropometry

It should be obvious from the preceding chapters that the worker must be matched to the work task in terms of reach and space requirements. In essence, if a specific reach requirement is documented for a job it is necessary to evaluate the applicants capability to perform such a reach.

13.1.4.2 Range of Motion

Once again, job requirements should be adjusted to the subject's range of motion. Preferably, joints should not be bent or stretched to near their normal limits of motion, and at least not held there for long periods of time, especially if under load. Abnormal mobility, as found on examination, is more often a sign of present or past illness (injury) than a future risk factor. Theoretically, a lack of mobility of one joint can cause comparatively greater stresses on another joint; for example, limited ranges of motion of the hips and knees can require more motion of the spine in a particular job. A rigid joint is also less well adapted to absorb shock.

It is a well-know fact that joint motion is impaired following injury. Biering-Sørensen (1983) found that lesser range of spine motion in subjects with previous low-back pain was a sign of increased risk of future pain episodes, while in previously healthy subjects lesser spine mobility was associated with decreased risk.

13.1.4.3 Muscle Strength

The use of pre-employment strength testing to reduce the incidence and severity of musculo-skeletal problems has met with considerable enthusiasm over the past decade or so. The obvious goal is to assure that only people with sufficient strength to safely perform a job will be assigned to that job. Muscle strength evaluation methods and equipment have been described in Chapters 4 and 5.

There is some epidemiological support for the idea that strength testing is a useful means of reducing back injury rates. Chaffin and Park (1973) found a sharp increase in the low-back injury rates of subjects performing jobs requiring strength that was greater or equal to their isometric strength-test values. In fact, the risk was three times higher for the weaker subjects.

A *second* longitudinal study was performed by Chaffin et al. (1977) to further determine the value of strength testing, and the back-injury incidence rate was again found to be almost three times higher in the overstressed group. In a later paper, Chaffin et al. (1978) suggested that specific placement and selection programs should be undertaken by industry, based on strength performance criteria.

Another study involved the application of strength tests and a simultaneous biomechanical job analysis in the rubber industry (Keyserling et al., 1980). Subjects were strength tested and jobs biomechanically classified. The subjects were then assigned to jobs in such a way that some were overstressed and others were understressed. The medical records were followed for one year to determine musculo-skeletal problems over that period. Although the follow-up period was short, a significant effect of job matching based on strength criteria was obtained. Troup et al. (1981) in a prospective study found reduced dynamic strength of back flexor muscles to be a consistent predictor for recurrence or persistence of back pain.

It appears that strength testing can be a preventive approach, at least for low-back injuries. The testing procedure should be specific to the job being sought by the worker, however. It remains unclear whether a well-matched worker will develop a problem from a heavy physical job should the exposure period be extended beyond one year or more. Muscle strength and tissue resistance to future stress may not be directly related. In general, a worker should show that he or she has the strength required for a job before being employed.

13.1.4.4 General Fitness

There is some evidence that individuals in good health have a lower risk of chronic low-back pain than others, and that recovery after a pain episode is faster. Cady et al. (1979) evaluated five strength and fitness measurements and recorded the occurrence of back injuries in 1652 firefighters from 1971 to 1974. The prospective measurements included flexibility, isometric lifting strength, and cardio-pulmonary tests. The firefighters were divided into three groups—least fit, middle fit, and most fit. A graded and statistically significant protective effect was found for added levels of fitness and conditioning. Although other factors could have contributed to the result, it is certainly worthy of note. Svensson et al. (1983) found low-back pain to be more common in men who were physically less active in their leisure time. This relative inactivity can be due of course to back problems, and not a risk factor per se.

13.1.4.5 Criteria For Physical Assessment In Worker Selection

Chaffin (1983) has suggested that the following criteria be considered in choosing between methods of preemployment screening

1. Is it safe to administer?
2. Does it give reliable, quantitative values?
3. Is it related to specific job requirements?
4. Is it practical?
5. Does it predict risk of future injury or illness?

Each of these criteria is discussed briefly in the context of preemployment selection procedures (See also Chapter 4.)

As outlined, current opinion is that the risk of taking spine radiographs outweighs the possible advantage. Thus, radiographs should be considered only when further exploratory measures are indicated by the medical history and physical examination.

Various types of physical strength and endurance tests must be carefully evaluated to assure that they are safe. Previous history of musculo-skeletal problems or cardiovascular problems may contraindicate testing of a specific person.

Reliability can only be determined by test–retest procedures, which should preceed any new test program. Specificity is important and requires a careful evaluation of the job's physical requirements to insure that a test is related to the job.

Conditions can vary from one workplace to another. A history can always be obtained, but when quantitative testing is used, the following conditions should be met (Chaffin, 1983):

1. Hardware capable of simulation of different work situations.
2. Minimal administration time.
3. Minimal time for instruction and learning.

The most difficult criterion is the last. Does the test predict risk of future injury or illness? To answer this requires careful evaluation of epidemiological data, collected prospectively and for many years. Hopefully such studies will be performed in the future to validate the many varied procedures now being advocated.

13.2 PRE-EMPLOYMENT TRAINING

Job training is an important means of reducing injury rates. It should include training in the use of a particular tool—machine and other—and training in how to use it with minimum risk and stress to the body.

The importance of preemployment training is generally accepted. Concerning the prevention of low-back pain, there is at least some evidence that such training can be of value, although not all studies have reached the same conclusions. Further, training and other measures have often been used in combination, so the effect of training alone is difficult to assess. Blow and Jackson (1971) found that training in materials handling reduced back injuries in a short-term prospective study. Schooling people on how to lift in "back" schools has also been found to be effective, at least as a secondary preventive measure (Lidström and Zachrisson, 1973; Bergquist-Ullman and Larsson, 1977; White 1980; Nordin et al., 1981). Other investigators, however, have found back injuries to be equally common in employees of companies with or without training programs (Snook et al., 1978). Undoubtedly, our knowledge in this respect will increase in the future.

NIOSH has in its *Work Practices Guide for Manual Lifting* (1981) suggested the content of a training program in lifting, given here with some minor changes and additions.

1. *What Should Be Taught.* The aims of training for safety in lifting should be:

(a) To make trainees aware of the dangers of careless and unskilled lifting.

(b) To show them how to avoid unnecessary stress.

(c) To teach the worker individually to become aware of what he or she can handle comfortably without undue effort.

In order for the training to be effective, the instructors must be well versed in the sciences basic to manual materials handling and in safety engineering. The content of the course must also be suited to the educational background of the trainees. It should cover the following:

(a) *The risks to health of unskilled lifting:* Case histories from the organization concerned provide the best illustration.

(b) *The basic biomechanics of lifting:* The body as represented by a system of levers.

(c) *The effects of lifting on the body:* The basic anatomy of the spine and the muscles and joints of the trunk; the contribution of intratrunkal pressure while lifting.

(d) *Individual awareness of the body's strengths and weaknesses:* How to estimate one's comfortable lifting capacity.

(e) *How to avoid the unexpected:* Recognition of the physical factors that might contribute to an accident.

(f) *Handling skill:* Safe lifting postures; minimizing the load moment effects; timing for smooth and easy lifting.

(g) *Handling aids:* Platforms, stages or steps, trestles, handles, wheels, and shoulder pads should be demonstrated.

(h) *Warnings:* What to be aware of when lifting.

2. *How it should be Taught:*

It is not enough to teach safe lifting practices by slides or films. The trainee should be involved from the start in a practical way. Therefore, classes must be small. Training should not be restricted to a classroom. Lifting technique should be demonstrated and *practiced at the work* site. Supervisors, as well as trainees, should be involved in the training program.

The following are the minimum requirements for a training program:

(a) Begin with a poster campaign drawing attention to the need to be concerned about musculoskeletal injuries; change the posters every few days.

(b) Organize a training session for managers and supervisors.

(c) Present the training course to small groups of employees in a classroom and on the job.

(d) Have the plant doctor and the team of instructors tour the plant, discussing any points put to them by the workers.

(e) Initiate a biomechanical survey of jobs to identify their high stress demands.

(f) Set up procedures to assure that mechanical aids are maintained in good condition and workers know how to use them.

(g) Continue to make plant tours at regular intervals to discuss new job physical requirements.

In summary, there is a vital need for training both in the classroom and on the site, to involve all levels of workers including supervisors, and to be monitored routinely throughout the factory in order for the program to be effective.

13.3 SUMMARY

It is clear from this chapter that worker selection and training are possible methods of preventing musculoskeletal disease in industry. Considering the magnitude of musculoskeletal disease, all methods of prevention need to be used in spite of insufficient scientific evidence on the ultimate effectiveness of a given approach.

REFERENCES

Andersson, G. B. J., "Epidemiologic Aspects of Low-Back Pain in Industry," *Spine*, **6**, 53–60 (1981).

Andersson, G. B. J., H-O. Svensson, and A. Odén, "The intensity of work recovery in low back pain", *Spine* 1984. In press.

American College of Radiology, "Conference of Low-Back X-rays in Pre-Employment Physical Examinations," *Proceedings of Meeting*, January 11–14, sponsored by NIOSH, contract HSM-00-72-153 (1973).

Becker, W. F., "Prevention of Low Back Disability," *J. Occup. Med.*, **3,** 329 (1961).

Bergquist-Ullman, M. and U. Larsson, "Acute Low Back Pain in Industry," *Acta Orthop. Scand.*, (Suppl)170(1977).

Bierring-Sörensen, F. "The Prognostic Value of the Low Back History and Physical Measurements," unpublished doctoral dissertation, University of Copenhagen, 1983.

Blow, R. J and J. M. Jackson, "Rehabilitation of Registered Dock Workers," *Proc. Roy. Soc. Med.* **64,** 753–760 (1971).

Bond, M. B., et al., "Low-Back X-rays. Criteria for Their Use in Preplacement Examinations in Industry," *J.O.M.*, **6,** 373–380 (1964).

Cady, L. D., D. P. Bischoff, E. R. O'Connel, P. C. Thomas, and J. H. Allan, "Strength and Fitness and Subsequent Back Injuries in Fire-Fighters," *J.O.M.*, **21,** 269–272 (1979).

Chaffin, D. B., and K. S. Park, "A Longitudinal Study of Low-Back Pain as Associated with Occupational Weight Lifting Factors," *AIHA J.*, **34,** 513–525 (1973).

Chaffin, D. B., G. D. Herrin, W. M. Keyserling, and J. A. Foulke, *Pre-employment Strength Testing*, NIOSH Technical Report, NIOSH Physiology and Ergonomics Branch, Cincinnati, 1977.

Chaffin, D. B., "Functional Assessment for Heavy Physical Labor," in M. H. Alderman, Ed., *Clinical Medicine for the Occupational Physician*, Dekker, New York, 1983. in press.

Chaffin, D. B., G. D. Herrin, and W. M. Keyserling, "Pre-employment Strength Testing," *J.O.M.*, **20,** 403–408 (1978).

Connell, M. A., "Bony Abnormalities of the Low Back in Relation to Back Injury," *Southern Med. J.* **61,** 482 (1968).

Crookshank, J. W. and L. M. Warshaw, "Detecting Potential Low-Back Disabilities," *Southern Med. J.*, **54,** 636 (1961).

Dillane, J. B., J. Fry, and G. Kalton, "Acute Back Syndrome—A Study from General Practice," *Br. Med. J.*, **2,** 82–84 (1966).

Fischer, F. J., M. M. Friedman, R. E. van Denmark, "Roentgenographic Abnormalities in Soldiers with Low Back Pain: A Comparative Study," *Am. J. Roentgenol.*, **79,** 673–676 (1958).

Fullenlowe, T. M. and A. J. Williams, "Comparative Roentgen Findings in Symtomatic and Asymtomatic Backs," *Radiology*, **68,** 572–574 (1957).

Glover, J. R., "Prevention of Back Pain." In M. Jayson, Ed., *The Lumbar Spine and Back Pain*, Pitman Medical, M. Jayson, Ed., Turnbridge Wells (Eng.), 1980.

Gyntelberg, F., "One Year Incidence of Low Back Pain Among Male Residents of Copenhagen Aged 40–59," *Dan. Med. Bull.*, **21,** 30–36 (1974).

Hadler, N. M., "Legal Ramifications of the Medical Definition of Back Disease," *Ann. Intern. Med.*, **89,** 992–999 (1978).

Hirsch, C., B. Jonsson, and T. Lewin, "Low Back Symptoms in a Swedish Female Population," *Clin. Orthop.*, **63,** 171–176 (1969).

Horal, J., "The Clinical Appearance of Low Back Pain Disorders in the City of Gothenburg, Sweden," *Acta. Orthop. Scand.* (Suppl.), 118(1969).

Hult, L., "Cervical, Dorsal, and Lumbar Spinal Syndromes," *Acta. Orthop. Scand.* (Suppl.),17(1954).

Kelly, F. J., "Pre-Employment Medical Exam Including Back X-ray," *J.O.M.*, **4,** 132 (1965).

Keyserling, W. M., G. D. Herrin, and D. B. Chaffin, "Isometric Strength Testing as a Means of Controlling Medical Incidents on Strenous Jobs," *J.O.M.*, **22,** 332–336 (1980).

Kosiak, M., J. R. Aurelius, and W. F. Hartfiel, "Backache in Industry," *J.O.M.*, **2,** 51 (1966).

LaRocca, H. and I. Macnab, "Value of Pre-Employment Radiographic Assessment of the Lumbar Spine," *Can. Med. Assoc. J.*, **101,** 49–54 (1969).

Leavitt, S. S., T. L. Johnston, and R. D. Beyer, "The Process of Recovery: Patterns in Industrial Back Injury Part 1. Costs and Other Quantitative Measures of Effort," *Ind. Med. Surg.*, **40**(8), 7–14 (1971).

Leggo, C. and H. Mathiasen, "Preliminary Results of a Pre-Employment Back X-ray Program for State Traffic Officers," *J.O.M.*, **15,** 973–974 (1973).

Lidström, A. and M. Zachrisson, M., "Physical Therapy on Low Back Pain and Sciatica: An Attempt at Evaluation," *Scand. J. Rehabil. Med.*, **2,** 37–42 (1970).

Magora, A. and A. Schwartz, "Relation between the Low Back Pain Syndrome and X-ray Findings. 1. Degenerative Osteo-Arthritis," *Scand. J. Rehabil. Med.*, **8,** 115–125 (1976).

Magora, A. and A. Schwartz, "Relation between the Low Back Syndrome and X-ray Findings. 2. Transition Vertebrae (Mainly Sacralization)," *Scand. J. Rehab. Med.*, **10,** 135–145 (1978).

Magora, A., and I. Taustein "An investigation of the problem of sick-leave in the patient suffering from low back pain." *Industr. Med. Surg.* **38**(11), 398–408 (1969).

McGill, C. M., "Industrial Back Problems. A Control Program," *J.O.M.*, **10,** 174–178 (1968).

Montgomery, C. H., "Preemployment Back X-rays." *J.O.M.*, **18,** 495–497 (1976).

Moreton, R. D., "So-called Normal Backs," *Industr. Med. Surg.*, **38,** 46–49 (1969).

NIOSH *Work Practices Guide for Manual Lifting*, U.S. Department of Health and Human Services, 1981.

Nordin, M., D. Spengler, and V. H. Frankel, "A Back Prevention Program in Industry," Presented at the International Society for the Study of the Lumbar Spine, Paris (1981).

Pedersen, P. A., "Prognostic Indicators in Low Back Pain," *J. Royal. Coll. Gen. Pract.*, **31,** 209–216 (1981).

Redfield, J. T., "The Low Back X-ray as a Pre-Employment Screening Tool in the Forest Products Industry," *J.O.M.*, **13,** 219–226 (1972).

Rockey, P. H., J. Fontel, and G. S. Omenn, "Discriminatory Aspects of Pre-employment Screening: Low-Back X-ray Examinations in the Railroad Industry, "*Am. J. of Law and Med.*, **5,** 157–214 (1979).

Ross, E., "Ergebnisse einer Röntgen-Reihenuntersuchung der Wirbelsäule bei 5000 Männlischen Jugendlishen," *Fortschr. Röntgenstr.*, **97,** 734–751 (1962).

Rowe, M. L., "Preliminary Statistical Study of Low Back Pain," *J.O.M.*, **5,** 336–341 (1963).

Rowe, M. L., "Disc Surgery and Chronic Low Back Pain," *J.O.M.*, **7,** 196–202 (1965).

Runge, C. F., "Roentgenographic examination of the lumbosacral spine in routine preemployment examinations." *J Bone Joint Surg.*, **36A**, 75–84 (1954).

Runge, C. F., "Pre-existing Structural Defects and Severity of Compensable Back Injuries," *Industr. Med. Surg.*, **27,** 249 (1958).

Snook, S. H., "The Design of Manual Handling Tasks," *Ergonomics,* **21,** 963–985 (1978).

Snook, S. H., Campanelli, R. A., and J. W. Hart, "A Study of Three Preventive Approaches to Low Back Injury," *J. O. M.*, 478–481 (1978).

Splithoff, C. A., "Lumbosacral Junction. Roentgenographic Comparison of Patients With and Without Backaches," *JAMA*, **152,** 1610–1613 (1953).

Svensson, H-O., and G. B. J. Andersson, "Low Back Pain in Forty to Forty Seven Year Old Men: Work History and Work Environment Factors," *Spine,* **8,** 272–276 (1983).

Svensson, H-O., A. Vedin, C. Wilhelmsson, and G. B. J. Andersson, "Low back pain in relation to other diseases and cardiovascular risk factors," *Spine,* **8,** 277–285 (1983).

Taylor, P. J., "Personal Factors Associated with Sickness Absence," *Brit. J. Industr. Med.*, **25,** 106–110 (1968).

Torgerson, B. R., and W. E. Dotter, "Comparative Roentgenographic Study of the Asymptomatic and Symptomatic Lumbar Spine," *J. Bone Joint Surg.*, **58A,** 850–853 (1976).

Troup, J. D. G., J. W. Martin, and D. C. E. F. Lloyd, "Back Pain in Industry. A Prospective Study," *Spine*, **6,** 61–69 (1981).

White, A. H., "Low Back Patient Goes to School,"

Wiltse, L. L., "Lumbar Strain and Instability," *Proceedings*, American Academy of Orthopaedic Surgeons Symposium on the Spine, 54–83, 1969.

Wiltse, L. L., "The Effect of the Common Anomalies of the Lumbar Spine Upon Disc Degeneration and Low Back Pain," *Orthop. Clin. North Am.*, **2,** 569–582 (1971).

14

SUMMARY

This text describes and analyzes a variety of common situations at the work place that cause mechanical trauma to a worker's musculoskeletal system. A major thesis is that mechanical trauma can be avoided by improvement in two areas. First, biomechanical knowledge of the basic mechanisms that cause injury from a given physical act by a person must be understood; second, the desire and ability of people in industry to aggressively seek the means to prevent such harm must be further developed. On both accounts, considerable progress has been made in the last few decades.

We attempted to improve understanding of both of these areas. To do so we brought together a variety of methodological topics. Kinesiological data regarding human muscle, bone, and joint function as they affect motor performance are described. Anthropometric measurement methods and data regarding human size and mobility are presented. Methods and data describing worker strengths are included, along with descriptions of new measurement techniques for quantifying human motion parameters and muscle actions. Last, traditional motion analysis methods are presented briefly, along with newer computerized models of the musculoskeletal system available to evaluate the mechanical demands of various manual tasks in industry. Together these methods provide the means to understand the effects of a mechanical mismatch between a worker and his or her manual job requirements. In addition, data from scientists using these methods is used in contemporary biomechanical models to demonstrate the effectiveness and need for such models in solving complex occupational health and safety problems.

It is not enough to describe how mechanical trauma can occur in the workplace, however, if one is to improve future working conditions. The last six chapters of the book describe specific work situations that often result in increased incidence and severity of musculoskeletal disorders in industry. There is an attempt in these applied chapters to present guidelines that can be used to evaluate and, hopefully, improve specific working conditions and work methods. Topics presented in these chapters describe: (1) the maximum loads that workers can lift, push, or pull, (2) the location of tools and machine controls that have to be manipulated, (3) the design of a sitting workplace to avoid postural fatigue, (4) some basic biomechanical principles regarding the design and use of hand tools, (5) whole-body and segmental vibration criteria, and (6) a basis for improved functional testing of workers engaged in specific manual activities.

Though occupational biomechanics expertise will continue to be needed to prevent musculoskeletal disorders due to gross whole-body exertions typified by manual materials handling activities, a growing body of knowledge and concern is developing for those workers who must maintain awkward body postures for prolonged periods of time when sitting or standing due to inappropriate seats, equipments, or workplace layout. Also, repetitive exertions when performed at high frequencies and/or in extreme postures are of growing concern to occupational biomechanics. Clearly, the effects of vibration, both in the presence of exertions (e.g., using a powered hand saw) or when sedentary (e.g., riding in a vehicle) are just beginning to be understood from a biomechanical viewpoint, and will continue to be of interest.

The discussion of vibration stress illustrates a larger challenge to those in occupational biomechanics in the future. Though, for the present, occupational biomechanics is largely concerned with the adverse effects of a variety of manual activities in industry, future studies necessarily will concern the effects of combined stresses from manual exertions performed while the worker is also exposed to heat or cold, toxic substances that alter neuromuscular and skeletal functions, and extreme perceptual-motor and cognitive job demands.

In summary, occupational biomechanics is an emerging discipline and field of practice. It is sustained and nurtured by a growing awareness that musculoskeletal disease in industry *cannot* be prevented by simple means. Comprehensive programs of prevention must begin with a recognition that certain types of physical stress of either an occasional or repetitive nature are hazardous to many people. These types of stresses must be eliminated by consideration of biomechanical criteria in job and equipment design, and by biomechanically sound instructions to workers. Also, it must be recognized that musculoskeletal functions vary greatly between people. Hence, in certain situations, improved functional testing will be necessary to assure that future high-risk workers are not exposed to physical stressors that cannot be eliminated, but to which they are particularly susceptible. The cost of these programs will not be cheap, but the growing human suffering and cost of musculoskeletal disease today dictates a larger commitment to musculoskeletal disease research and prevention programs than at present.

APPENDIX **A**

PART 1: ANATOMICAL AND ANTHROPOMETRIC LANDMARKS AS PRESENTED BY WEBB ASSOCIATES*

* In *Anthropometric Source Book*, Vol. 1, NASA Reference Publication 1024, NASA, Washington D.C., 1978, pp. III-80 to III-82.

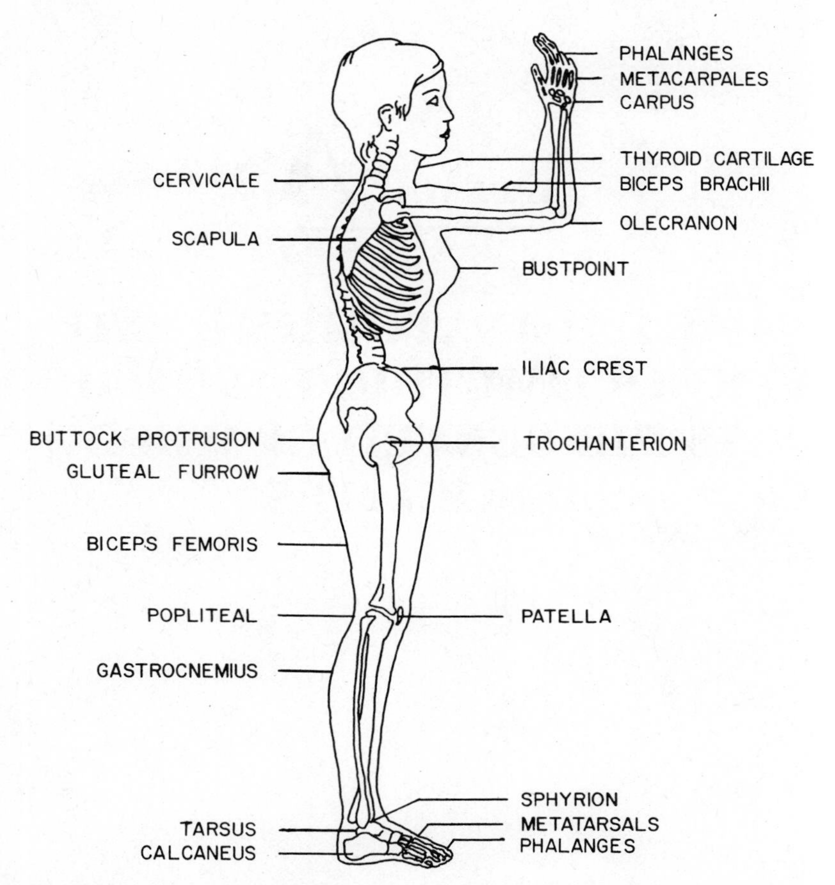

FIGURE A-1 Anatomical and anthropometric landmarks (Webb Associates, 1978).

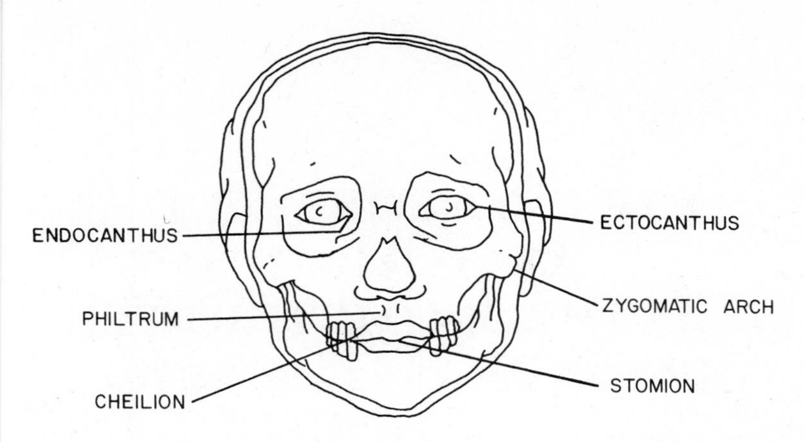

FIGURE A-2 Anthropometric landmarks of the head and face—front (Webb Associates, 1978).

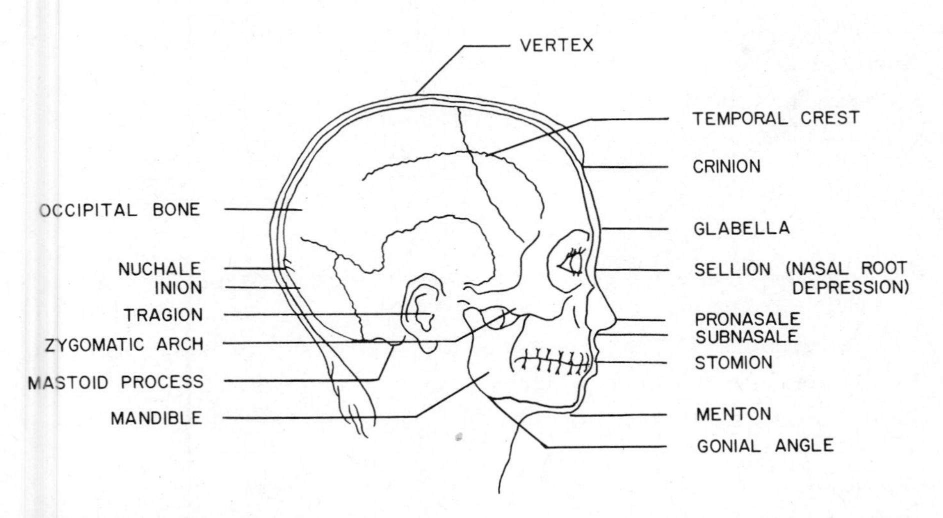

FIGURE A-3 Anthropometric landmarks of the head and face—side (Webb Associates, 1978).

PART 2: GLOSSARY OF ANATOMICAL AND ANTHROPOMETRIC TERMS (Webb Associates, 1978)

A

Abdominal extension level. The most anterior point on the curve of the abdomen in the midsagittal plane.

Abduct. To move away from the axis of the body or one of its parts.

Acromion. The most lateral point of the lateral edge of the spine of the scapula. Acromial height is usually equated with shoulder height.

Anterior. Pertaining to the front of the body; as opposed to posterior.

Auricular. Pertaining to the external ear.

Axilla. The armpit.

B

Bi. A prefix denoting connection with or relation to each of two symmetrically paired parts.

Biceps brachii. The large muscle on the anterior surface of the upper arm.

Biceps femoris. A large posterior muscle of the thigh.

Brow ridges. The bony ridges of the forehead that lie above the orbits of the eye.

Bustpoint. The most anterior protrusion of the right bra pocket.

Buttock protrusion. The maximum posterior protrusion of the right buttock.

C

Calcaneus. The heel bone.

Canthus. A corner or angle formed by the meeting of the eyelids.

Carpus. The wristbones, collectively.

Cervicale. The protrusion of the spinal column at the base of the neck caused by the tip of the spine (q.v.) of the 7th cervical vertebra.

Cheilion. The corners of the mouth formed by the juncture of the lips.

Coronal plane. Any vertical plane at right angles to the midsagittal plane.

Crinion. The point in the midplane where the hairline meets the forehead.

Cutaneous lip. The area between the upper lip and the nose.

D

Dactylion. The tip of the middle finger.

Deltoid muscle. The large muscle on the lateral border of the upper arm in the shoulder region.

Distal. The end of the body segment farthest from the head; as opposed to proximal.

E

Ectocanthus (also external canthus). The outside corner or angle formed by the meeting of the eyelids.

Endocanthus. The inside corner or angle formed by the meeting of the eyelids.

Epicondyle. The bony eminence at the distal end of the humerus, radius, and femur.

Extend. To move adjacent segments so that the angle between them is increased, as when the leg is straightened; as opposed to flex.

External. Away from the central long axis of the body; the outer portion of the body segment.

F

Femoral epicondyles. The bony projections on either side of the distal end of the femur.

Femur. The thigh bone.

Flex. To move a joint in such a direction as to bring together the two parts it connects, as when the elbow is bent; as opposed to extend.

Fossa. A depression, usually somewhat longitudinal in shape, in the surface of a part, as in a bone.

Frankfort plane. The standard horizontal plane or orientation of the head. The plane is established by a line passing through the right tragion (approximate earhole) and the lowest point of the right orbit (eye socket).

G

Gastrocnemius. The largest muscle in the calf of the leg.

Glabella. The most anterior point of the forehead between the brow ridges in the midsagittal plane.

Gluteal furrow. The furrow at the juncture of the buttock and the thigh.

Gonial angle. The angle at the back of the lower jaw formed by the intersection of the vertical and horizontal portions of the jaw.

H

Helix. The rolled outer part of the ear.

Humerus. The bone of the upper arm.

Humeral epicondyles. The bony projections on either side of the distal end of the humerus.

Hyperextend. To overextend a limb or other part of the body.

I

Iliac crest. The superior rim of the pelvic bone.

Inferior. Below, in relation to another structure; lower.

Inion. The summit of the external occipital protuberance; the most posterior bony protuberance on the back of the head.

Inseam. A term used in tailoring to indicate the inside length of a sleeve or trouser leg. It is measured on the medial side of the arm or leg.

Internal. Near the central long axis of the body; the inner portion of a body segment.

Interpupillary. Between the centers of the pupils of the eyes.

K

Knuckle. The joint formed by the meeting of a finger bone (phalanx) with a palm bone (metacarpal).

L

Lateral. Lying near or toward the sides of the body; as opposed to medial.

Lateral malleolus. The lateral bony protrusion of the ankle.

Larynx. The cartilaginous box of the throat that houses the voice mechanism. The "Adam's apple" is the most noticeable part of the larynx.

Lip prominence. The most anterior protrusion of either the upper or the lower lip.

M

Malleolus. A rounded bony projection in the ankle region. There is one on both the lateral and medial side of the leg.

Mandible. The lower jaw.

Mastoid process. A bony projection on the inferior lateral surface of the temporal bone behind the ear.

Medial. Lying near or toward the midline of the body; as opposed to lateral.

Menton. The point of the tip of the chin in the midsagittal plane.

Metacarpal. Pertaining to the long bones of the hand between the carpus and the phalanges.

Midaxillary line. A vertical line passing through the center of the axilla.

Midpatella. A point one-half the distance between the superior and inferior margins of the right patella.

Midsagittal plane. The vertical plane that divides the body into right and left halves.

Midshoulder. A point one-half the distance between the neck and the right acromion.

N

Nasal root depression. The area of greatest indentation where the bridge of the nose meets the forehead.

Nasal septum. The cartilaginous wall separating the right nostril from the left.

Navicular bone. The small bone of the hand just distal to the bend of the wrist on the thumb side.

Nuchale. The lowest point in the midsagittal plane of the occiput that can be palpated among the muscles in the posterior–superior part of the neck. This point is often visually obscured by hair.

O

Ocular. Pertaining to the eye.

Occipital bone. A curved bone forming the back and part of the base of the skull.

Olecranon. The proximal end of the ulna (the medial forearm bone).

Omphalion. The center point of the naval.

Orbit. The eye socket.

P

Patella. The kneecap.

Phalanges. The bones of the fingers and toes (singular, phalanx).

Philtrum. The vertical groove that runs from the upper lip to the base of the nasal septum.

Plantar. Pertaining to the sole of the foot.

Popliteal. Pertaining to the ligament behind the knee or to the part of the leg in back of the knee.

Posterior. Pertaining to the back of the body; as opposed to anterior.

Pronasale. The most anterior point on the nose.

Proximal. The end of a body segment nearest the head; as opposed to distal.

R

Radiale. The uppermost point on the lateral margin of the proximal end of the radius.

Radius. The bone of the forearm on the thumb side of the arm.

Ramus. The vertical portion of the lower jaw bone (mandible).

S

Sagittal. Pertaining to the anteroposterior median plane of the body, or to a plane parallel to the median.

Scapula. The shoulder blade.

Scye. A tailoring term to designate the armhole of a garment. Refers here to landmarks that approximate the lower level of the axilla.

Sellion. The point of greatest indentation of the nasal root depression.

Septum. A dividing wall between two cavities; the nasal septum is the fleshy partition between the two nasal cavities.

Sphyrion. The most distal extension of the tibia on the medial side of the foot.

Spine (or spinal process) of vertebrae. The posterior prominences of the vertebrae.

Sternum. The breastbone.

Stomion. The point of contact in the midsagittal plane between the upper and lower lip.

Stylion. The most distal point on the styloid process of the radius.

Styloid process. A long, spinelike projection of a bone.

Sub. A prefix designating below or under.

Submandibular. Below the mandible or lower jaw.

Subnasale. The point where the base of the nasal septum meets the philtrum.

Substernale. The point located at the middle of the lower edge of the breastbone.

Superior. Above, in relation to another structure; higher.

Supra. Prefix designating above or on.

Suprasternale. The lower point in the notch in the upper edge of the breastbone.

Surface distance. A measurement that follows the general contours of the surface of the body.

T

Tarsus. The collection of bones in the ankle joint, at the distal end of the tibia.

Temporal crest. A narrow bony ridge along the side of the head above the ear level that serves as a point of attachment for the temporal muscles.

Temporal muscles. The muscles of the temple region.

Thyroid cartilage. The bulge of the cartilage on the anterior surface of the throat; in men, the Adam's apple.

Tibia. The medial bone of the leg (shin bone).

Tibiale. The uppermost point of the medial margin of the tibia.

Tragion. The point located at the notch just above the tragus of the ear.

Trapezius muscle. The large muscle on each side of the back of the neck and shoulders, the action of which moves the shoulders.

Triceps. The muscle mass of the posterior upper arm.

Trochanterion. The tip of the bony lateral protrusion of the proximal end of the femur.

U

Ulna. One of the bones of the forearm on the little finger side of the arm.

V

Vertex. The top of the head.

Z

Zygomatic arch. The bony arch below the orbit of the skull extending horizontally along the side of the head from the cheekbone (the zygomatic bone) nearly to the external ear.

APPENDIX B

POPULATION WEIGHT AND MASS-CENTER DATA

TABLE B.1
Segment Weight Values Derived from Regression Equations Using Total Body Weight as the Independent Variable (from Webb Associates, 1978)

	Select Total Body Weight Increments[a]					
Segment	45.73 (100)	54.45 (120)	63.52 (140)	72.59 (160)	81.67 (180)	90.74 (200)
Head	3.87 (8.47)	4.12 (9.08)	4.40 (9.70)	4.68 (10.32)	4.95 (10.91)	5.23 (11.53)
Head and neck	4.75 (10.47)	5.24 (11.55)	5.72 (12.61)	6.20 (13.67)	6.69 (14.75)	7.17 (15.81)
Neck	1.27 (2.80)	1.40 (3.09)	1.53 (3.37)	1.66 (3.66)	1.80 (3.97)	1.93 (4.26)
Neck and torso	21.06 (46.44)	26.13 (57.62)	31.19 (68.77)	36.26 (79.95)	41.33 (91.13)	46.39 (102.29)
Thorax	9.30 (20.51)	11.62 (25.62)	13.94 (30.74)	16.26 (35.85)	18.58 (40.97)	20.90 (46.08)
Lumbar	5.48 (12.08)	6.85 (15.10)	8.22 (18.13)	9.58 (21.12)	10.95 (24.14)	12.31 (27.14)
Pelvis	5.01 (11.05)	6.26 (13.80)	7.50 (16.54)	8.75 (19.29)	10.00 (22.05)	11.25 (24.81)
Head, neck, and torso	24.75 (54.57)	30.14 (66.46)	35.53 (78.34)	40.92 (90.23)	46.31 (102.11)	51.70 (114.00)
Total arm	2.30 (5.07)	2.76 (6.09)	3.21 (7.08)	3.67 (8.09)	4.13 (9.11)	4.59 (10.12)
Upper arm	1.23 (2.71)	1.48 (3.26)	1.73 (3.81)	1.98 (4.37)	2.23 (4.92)	2.48 (5.47)
Forearm and hand	1.05 (2.31)	1.26 (2.78)	1.47 (3.24)	1.68 (3.70)	1.89 (4.17)	2.10 (4.63)
Forearm	0.70 (1.54)	0.87 (1.92)	1.04 (2.29)	1.22 (2.69)	1.39 (3.06)	1.56 (3.44)
Hand	0.32 (0.71)	0.37 (0.82)	0.42 (0.93)	0.47 (1.04)	0.52 (1.15)	0.57 (1.26)
Total leg	7.23 (15.94)	8.66 (19.09)	10.10 (22.27)	11.53 (25.42)	12.97 (28.60)	14.40 (31.75)
Thigh	4.24 (9.35)	5.29 (11.66)	6.34 (13.98)	7.39 (16.29)	8.45 (18.63)	9.50 (20.95)
Shank and foot	2.87 (6.33)	3.28 (7.23)	3.69 (8.14)	4.10 (9.04)	4.51 (9.94)	4.92 (10.85)
Shank	2.08 (4.59)	2.42 (5.34)	2.76 (6.09)	3.11 (6.86)	3.45 (7.61)	3.79 (8.36)
Foot	0.78 (1.72)	0.85 (1.87)	0.91 (2.01)	0.97 (2.14)	1.03 (2.27)	1.10 (2.43)

[a] Data given in kilograms, with pounds in parentheses.

TABLE B.2
Anatomical Location of Segment Centers of Gravity (Centers of Mass)

Body Part	Reference Dimensions	Location
Arm	5 mm proximal to distal end of deltoid M insertion and 24 mm distal to most proximal fibers of medical head of triceps.	In medial head of triceps adjacent to radial nerve and radial groove of humerus.
Forearm	11 mm proximal to most distal part of insertion of pronator teres M.	9 mm anterior to interosseous membrane, usually between flexor digitorum profundus and flexor pollicis longus MM or more toward flexor pollicis longus M or toward flexor digitorum sublimus M.
Hand	(In position of rest) 2 mm proximal to proximal transverse palmar crease in angle between the proximal transverse and the radial longitudinal creases.	On axis of III metacarpal, usually 2 mm deep to skin surface.
Thigh	29 mm below apex of femoral triangle and 18 mm proximal to the most distal fibers of adductor brevis M.	Deep to adductor canal and 13 mm medial to the linea aspera in the adductor brevis M (or in adductor magnus M or vastus medialis M).
Leg	35 mm below popliteus M and 16 mm above the proximal extremity of Achilles tendon.	At posterior part of tibialis posterior M (between flexor digitorum longus and flexor hallucis longus MM); 8 mm posterior to introsseous membrane.
Foot	66 mm from the center of the body of the talus; below the proximal halves of the second and third cuneiform bones.	In plantar ligaments or just superficial in the adjacent layer of deep foot muscles.
Shoulder mass	On a line perpendicular to the anterior face of the outer quarter of the blade of the scapula or near its axillary border 20.5 ± 9.0 mm from the bone and 78.0 ± 5.0 mm above the inferior angle; it falls within the axilla or in the adjacent thoracic wall.	

TABLE B.2 (*Continued*)

Body Part	Reference Dimensions
Head and neck	8 mm anterior to the basion on the inferior surface of the basioccipital bone or within the bone 24.0 ± 5.0 mm from the crest of the dorsum sellae; on the surface of the head, a point 10 mm anterior to the supratragic notch above the head of the mandible is directly lateral.
Head alone	A point in the sphenoid sinus averaging 4 mm beyond the antero-inferior margin of the sella. On the surface, its projections lie over the temporal fossa on or near the nasion–inion line at a point about 32 percent back from the nasion; equally distant above the zygomatic arch and behind the malar frontosphenoid process.
Thoxax	At the level of the disc between the ninth and tenth thoracic vertebrae or the level of either of the adjacent vertebral bodies at the anterior border of the column (anterior longitudinal ligament) or in the adjacent posterior mediastinum; on the surface, level below the nipples and above the transverse line between the pectoral and abdominal muscles, and the spine of the eighth thoracic vertebra.
Abdomino-pelvic mass	At the level of or below the disc between L4 and L5 in the posterior region of the vertebral body; between the umbilicus and the crest of the ilium.

REVIEW QUESTIONS

CHAPTER 1

1. List five different methodological developments that define the knowledge base for occupational biomechanics.
2. Give an example of a specific methodology listed in reply to question 1.
3. List five different contemporary applications of occupational biomechanical principles.
4. Describe how epidemiology has affected the growth of occupational biomechanics.
5. Describe how social and legal trends have affected occupational biomechanics.

CHAPTER 2

1. What are the main constituents of connective tissue?
2. What are the differences between ligaments, tendons, fascia, and skin?
3. Describe the process by which bone is formed.
4. Why is bone remodeling important to occupational biomechanics?
5. Describe the structure of skeletal muscle.
6. What is a motor unit?
7. Describe the process by which muscle contracts.
8. How is the necessary energy for the contractile process provided?

9. List different types of muscle contractions.
10. Describe the length–tension relationship of muscle.
11. Describe the force–velocity relationship of muscle.
12. What is a synovial joint?
13. List different suggested theories of joint lubrication.
14. What are the main parts of an intervertebral disc?
15. What is a spinal motion segment?

CHAPTER 3

1. List four different types of human measurements commonly included in engineering anthropometry textbooks.
2. What factors affect the prediction accuracy of a body segment link length?
3. What factors affect the use of link length data when predicting functional reach? (List in order of importance.)
4. What methods are used to estimate the mass of a human segment? (Evaluate each method.)
5. Describe and critique three different methods for estimating body segment mass center locations.
6. How are center location data often presented?
7. Define "moment of inertia," and present a diagram describing how it can be measured.
8. How is the moment of inertia of a body segment related to the segments' radius of gyration?

CHAPTER 4

1. Choose any specific type of arm motion, describe it using traditional joint motion terminology, and discuss the limitations of the traditional terminology you used.
2. List the presentation requirements for a well-described set of anthropometric data.
3. Choose an anthropometric source book from the references and critique the presentation of the data.
4. List the measurement requirements for a device used to estimate joint range-of-motion.
5. Evaluate two different methods of estimating joint range-of-motion.
6. Describe the effects of (1) age, (2) gender, (3) size, and (4) adjacent joint angle on the range-of-motion at a specific joint.

7. List and comment on five different test factors that could have a major effect on an individual's muscle strength performance during a test.
8. When testing for a specific static muscle strength moment, how does the angle of the joint affect the values? What explanation can you give for the effect?
9. What two biomechanical factors have a major effect on dynamic strength of a person?
10. Describe how a psychophysical strength estimate is obtained.
11. When predicting the strength moment capability of an individual performing a specific localized static exertion, what factors need to be considered in the prediction?
12. Briefly describe the expected effects of (1) gender, (2) age, (3) anthropometry, and (4) general strength conditioning on specific muscle strength performance.

CHAPTER 5

1. List several major criteria appropriate to the evaluation of a biomechanical measurement system.
2. Evaluate the use of accelerometers for measuring body motions.
3. Evaluate the use of goniometers for measuring body motions.
4. Evaluate the use of photogrammetric methods for measuring body motions.
5. Describe a contemporary EMG measuring system.
6. Describe how the EMG signal is used to estimate muscle forces and muscle fatigue.
7. Describe typical strength measurement system requirements for both static and dynamic measurements.
8. How are intra-abdominal pressure measurements obtained, and what assumptions exist in interpreting such measurements?
9. How are intradiscal measurements obtained, and what assumptions are important in interpreting such measurements?
10. Describe a typical force platform system and its use in occupational biomechanics.

CHAPTER 6

1. Why are biomechanical models necessary?
2. Solve for the external moments and reactive force acting at the elbow in Figure 6.2 for an average female holding 20 N in the hand. (Show force diagram and computations.)

3. Extend the analysis in 2 (above) to include the shoulder moment and reactive force with the arm extended horizontal, as in Figure 6.4.
4. Resolve problem 3 with $\theta_S = 30°$ and $\theta_E = 45°$ (see Figure 6.5).
5. Resolve problem 2 with the 20 N load acting with $\alpha = 45°$ (see Figure 6.6).
6. Solve for F_m and R internal force values in Figure 6.7 using $m = 5$ cm and the external load moment and reactive force values computed in problem 2. Discuss the limitations of these estimates.
7. Discuss how F_m would vary throughout the range-of-motion of the elbow when executing a high-speed motion, as referred to in Figure 6.15.
8. Discuss why the required coefficient-of-friction increases as the push/pull force on the hands increases. How does posture and walking speed affect this relationship?
9. Assume a 300 N load is being lifted by an average-size male in the posture depicted in Figure 6.23. Using equations 6.44, 6.45, 6.46, and 6.47, and an assumed hip moment of 160 N m, resolve the computations for the various spinal forces (see example of lifting 450 N in text).
10. Discuss the values obtained in problem 9 (above) in terms of injury risk. What factors contribute to the risk level the most?
11. Discuss the limitations of the simple (one muscle) low-back model used in problem 9.
12. Describe how dynamic load lifting affects the estimate of low-back stress. Suggest how such dynamic effects can be minimized.
13. Describe why pulling a heavy object appears to be more hazardous than pushing it.
14. Describe the theory of tendon residual strain as the basis for tendinitis in repeated manual exertions.
15. Explain how a deviated wrist posture could create injurious tendon–synovium stresses in the carpal tunnel.
16. Why do some authorities believe a person with a small wrist is more susceptible to carpal tunnel syndrome than a person with a large wrist?
17. If an average anthropometric woman is found to have a shoulder strength moment of 30 N m in posture d (Figure 6.42), what is the maximum hand load she could lift in that posture?

CHAPTER 7

1. Briefly describe the steps necessary to formally describe manual work in industry.

2. List the four basic principles that F.W. Taylor advocated to manage manual labor more effectively. Discuss each as it relates to occupational biomechanics.
3. What contributions did the Gilbreths make in the evolution of industrial motion study? Discuss each as it relates to occupational biomechanics.
4. Discuss the contributions and limitations of a traditional time and motion study when performing a contemporary occupational biomechanics analysis of a job.
5. Describe the job conditions (and give an example) wherein a physical stress checklist or survey is appropriate.
6. List the specific data necessary to perform a NIOSH load lifting analysis.
7. What are job/task conditions *not* included in the NIOSH analysis?
8. What data are necessary when performing a static strength analysis?
9. Describe how posture diagrams are used to evaluate awkward work postures. How are the data obtained?
10. What data are required to perform upper extremity analysis, and how are the data acquired?
11. How is a trunk flexion analysis performed on a job?

CHAPTER 8

1. Why are manual materials handling tasks of such interest today?
2. What four different types of literature and data were consulted by NIOSH experts in assembling the NIOSH *Guide to Manual Lifting?* Give an example of each.
3. What specific conditions must exist for the NIOSH lifting analysis to apply?
4. Define both the NIOSH *action limit* and *maximal permissible limit.* What changes are implied if a load lifted exceeds either limit?
5. When using the NIOSH *Guide to Manual Lifting*, what job conditions have the greatest effect on the resulting limits? Discuss the effect of each.
6. How does the height of a handle on a cart biomechanically affect the amount of force one can safely push or pull? Show a force diagram of several handle conditions and describe the effect of each.
7. What general factors must be considered when attempting to specify work conditions wherein a person would not be at high risk of slipping?
8. Why are asymmetric lifting conditions not recommended in industry?

CHAPTER 9

1. What are the main parts of the vertebral column and what are their general characteristics?
2. Name three different types of sitting, and explain how they differ biomechanically.
3. Compare the posture of the spine when standing and sitting. What are the main differences?
4. In what way does the design of the chair influence the load on the spine (as measured by disc pressure)?
5. Why should the table and chair be adjusted to each other, and to the worker? (Biomechanical argument.)
6. What are the advantages of semi-sitting?
7. List nine important seat design parameters.
8. Why and how should the backrest of an office chair be adjustable?
9. List important dimensions for the work surface, and explain why they are important.

CHAPTER 10

1. Explain how prolonged or highly repeated exertions in awkward postures (or with high external loads) are believed to cause cumulative trauma to the musculoskeletal system.
2. Why is the shoulder of concern in occupational biomechanics, and how is it affected by overhead work or work with the arm extended in front (flexed) or to the side (abducted)?
3. How were guidelines regarding arm/neck postures derived?
4. What is the concern when rotating the forearm about its long axis, and how can this concern be reduced?
5. How is arm push and pull strength affected by the position of a lever in front of the person?
6. What is the basis for a maximum neck forward flexion angle being specified for a prolonged posture requirement?
7. Why is the height of a workbench of concern in occupational biomechanics, and what general guidelines exist today?
8. What specific spatial conditions in a VDT workplace are believed to be important today? Explain why you chose each.

CHAPTER 11

1. Select five different common musculoskeletal disorders of the upper extremity, and define the pathology involved.

2. Present a biomechanical argument that could explain the need to maintain the wrist in a neutral posture while performing forceful grip exertions.
3. In evaluating a cylindrical powered hand tool (e.g., powered screw driver), what biomechanical design features would you consider important, and how could each be achieved?
4. Briefly explain how EMG data are used to evaluate alternative tool designs, and critically evaluate the effectiveness of such an approach.
5. Present biomechanical arguments that explain why certain shapes and sizes of the handle of a squeeze tool (e.g., pliers) are desirable.

CHAPTER 12

1. Define the two general classes of vibration.
2. How is vibration described (characterized)?
3. What are the advantages and disadvantages of using a narrow band-pass filter when measuring vibration?
4. Describe a common method used to present the amplitude of vibration.
5. How is vibration usually measured?
6. What are the general effects of whole-body vibration on man?
7. At what frequency levels do resonance occur in the trunk from vertical vibration when standing and sitting?
8. What is the main clinical syndrome occurring from segmental vibration?
9. What are the limits given in the ISO standards on whole-body vibration based on?

CHAPTER 13

1. Define sensitivity, specificity, and predictive value?
2. Why is worker selection important?
3. Describe the usefulness of radiography of the spine in pre-employment screening.
4. Why is pre-employment strength-testing useful?
5. Which criteria should be considered when choosing between methods of pre-employment screening (list and discuss why)?
6. What are the aims of training for safety in lifting?
7. List requirements for a training program in manual materials handling.

GLOSSARY

This glossary is based, in part, on glossaries from:

Frankel, V. H., and M. Nordin, *Basic Biomechanics of the Skeletal System*, Lea & Febiger, Philadelphia, 1981.

White, A. A, III, and M. M. Panjabi, *Clinical Biomechanics of the Spine*, Lippincott, Philadelphia, 1978.

Tichauer, E. R, *The Biomechanical Basis of Ergonomics*, Wiley-Interscience, New York, 1978.

A

Abduction. Motion away from the midline. Increases the angle between a limb and the sagittal plane.

Acceleration. The change in velocity of a body divided by the time over which change occurs.

Achilles tendon. Connects the principal plantar flexor muscles of the foot with the heel of the skeleton of the foot.

Action potential. For example, motor unit action potential is the nerve impulse propagating down a motoneuron and activating all muscle fibers of a motorunit (see Electromyography).

Adduction. Motion toward the midline. Decreases the angle between a limb and the sagittal plane.

Agonistic muscles. Muscles that initiate and carry out motion.

Anisotropy. The quality whereby a material exhibits unlike mechanical properties when loaded in different directions.

Antagonistic muscles. Muscles that oppose the actions of the agonistic muscles (oppose a movement).

Anthropometry. An empirical science defining the physical measures of a persons size and form.

Area moment of inertia. Quantity that takes into account the cross-sectional area and distribution of material around an axis during bending.

Arthritis. Degenerative joint disease (also inflammation of a joint).

Axial rotation. Rotation about an axis.

Axis of motion. Line about which all points move in a body in motion.

Axis of rotation. Line about which all points in a rotating body describe circles.

Axon. The long process of a nerve cell. Conducts impulse.

B

Bending. A loading mode in which a load is applied to a structure in a manner that causes it to bend about an axis, subjecting the structure to a combination of tension and compression.

Bending moment. A quantity at a point in a structure equal to the product of the applied force and the perpendicular distance from the point to the force line, usually measured in newton meters.

Biceps. Long twin-bellied muscle going from the shoulder blade to the proximal end of the radius, thus crossing and acting on both the shoulder and the elbow joints.

Bone remodeling. The ability of bone to adapt, by changing its size, shape, and structure, to the mechanical demands placed upon it.

Brachialis muscle. Short muscle originating in the lower third of anterior surface of humerus and inserting into the anterior part of the ulna close to elbow joint.

Bursa. Small bag filled with fluid, reducing friction between moving structures.

Bursitis. Inflammation of a bursa.

C

Capillary. Very small blood vessel that connects the smallest branches of the arteries with those of the veins.

Carpal tunnel. Channel on the palmar side of the wrist formed by the irregular small bones of the wrist and a tough ligament stretched across it. Through the carpal tunnel pass the flexor tendons of the fingers, the median nerve, and some blood vessels.

Carpus. The aggregate of eight small irregular bones forming the wrist.

Center of gravity. Equilibrium point of a supported body where all its weight is concentrated (see Center of mass).

Center of mass. That point at the exact center of an object's mass; often called the center of gravity (see Center of gravity).

Center of rotation. A point around which circular motion is described.

Chondrocyte. A cartilage cell.

Combined loading. Application of two or more loading modes to a structure.

Compression. A loading mode in which equal and opposite loads are applied normal to surface of the structure, resulting in shortening and widening of structure.

Concentric contraction. Increase of tension within a muscle, producing shortening. For example, the brachialis shortens when a weight is lifted by flexing the forearm.

Creep. Progressive deformation of soft tissues due to constant low loading over an extended period of time.

Cross-sectional area. Measure of the area of a piece of material cut at right angles to its longitudinal axis.

D

Deformation rate. The speed at which an applied load deforms a structure (see Speed of loading).

Degrees of freedom. The number of ways in which a body can move.

Deltoid. Large muscle of the shoulder that abducts and otherwise moves the upper arm about the shoulder joint against external loads.

Distal. In a limb: further away from the body. Elsewhere: further away from the central axis of the body.

Dorsiflexion. Bending upwards around an axis.

Dynamic work. "Work" according to the definition in mechanics. Defined as the product of a force multiplied by the distance through which its point of application moves.

Dynamics. The study of forces acting on a body in motion.

E

Elasticity. Property of a material which allows the material to return to its original shape and size after being deformed.

Electrogoniometer. Device to quantify in analog fashion or digitally an angle and changes of angle between body segments connected by a joint.

Electromyography. The recording of action potentials emitted from contracting muscles, EMG. Also used to measure action potentials from nerves.

Epicondylitis. Technical term for "tennis elbow."

Equilibrium. State of a body at rest in which the sums of all forces and moments are zero.

Ergonomics. A multidisciplinary activity dealing with the interactions between man and his total working environment, plus such traditional and environmental aspects as atmosphere, heat, light, and sun, as well as of tools and equipment of the workplace. (From American National Standard ANSI Z794.1-1972).

Excentric contraction. Increase of tension within a muscle while lengthening. For example, the brachialis exerts a force resisting the pull of gravity when extending a flexed forearm slowly.

Extension. The position of the joints of the extremities and back when one stands at rest, or the direction of motion that tends to restore this position; the opposite of flexion.

F

Fascia. Layer or sheet of connective tissue.

Fatigue fracture. A fracture typically produced by either low repetition of high loads or high repetition of relatively normal loads.

Flexion. Movement involving the bending of a joint whereby the angle between the bones is diminished; the opposite of extension (except at shoulder).

Force. An action that changes the state of rest or motion of a body to which it is applied.

Force couple. Two parallel forces of equal magnitude but opposite direction applied to a structure.

Free body. A structure considered in isolation for the purpose of studying the effect of forces acting on it.

Free body diagram. Diagram of an isolated portion of a structure used during free body analysis for the purpose of studying the effect of forces acting on the free body.

Friction force. A tangential force opposing motion which acts between two bodies in contact.

Frontal plane. The plane that passes through the longitudinal axis of the body.

G

Glycosaminoglycan (GAG). A long flexible chain of repeating disaccharide units that are the building blocks of proteoglycans.

Goniometer. Device measuring the angle and range of angular movement between two body segments connected by a joint.

Gravitational force. A force produced by gravitational attraction by the Earth on a body.

Ground reaction force. A gravitational force produced by the weight of an object against the surface on which it lies.

I

Instantaneous center of motion. The immovable point existing at an instant in time created by one segment (link) of a body rotating about an adjacent segment; all other points on the body rotate about this immovable point.

Interphalangeal joints. Joints connecting finger bones.

Ischial tuberosities. Two bony prominences forming the lowest point of the pelvis. On them rests the weight of the body when seated.

Isometric work. A muscle exerts a force (i.e. contracts) against resistance without producing any motion, for example, to hold a weight still with the extended arm. Isometric work, which results in increased demand for calories, is different from work in mechanics, defined as force multiplied by the distance an object moves.

J

Joint lubrication. A design feature of the joint which maintains the continuity of the thin film of synovial fluid between the joint surfaces, minimizing the contact and wear of the cartilaginous surfaces.

Joint reaction force. The internal reaction force acting at the contact surfaces when a joint in the body is subjected to external loads.

K

Kinematics. The branch of mechanics that deals with motion of a body without reference to force or mass.

Kinesiology. The study of human movements as a function of the construction of the musculoskeletal system.

Kinetics. The branch of mechanics that deals with the motion of a body under the action of given forces.

Kyphosis. Convexity of the spine. Normally observed in the thoracic region.

L

Lever arm. The perpendicular distance from the line of application of a force to the center of motion in a rigid structure, also known as the moment arm of the force.

Ligament. Connective tissue attaching bone to bone.

Load–deformation curve. A curve that plots the deformation of a structure when the structure is loaded in a known direction.

Locomotion. The act of moving the human body from place to place using the musculoskeletal system.

Longitudinal axis. A line through the longer part of a body about which a body rotates.

Lordosis. Concave curvature of the spine. Exists in the neck and in the lumbar region.

Lumbosacral joint. Joint between fifth lumbar vertebra and sacrum.

M

Matrix. The intercellular substance of a tissue.

Mathematical model. A set of mathematical equations that quantitatively describes the behavior of a given system.

Medial. Reference to that side of an anatomical structure that is closest to the midsagittal plane.

Median nerve. Large important nerve. Activates muscles that pronate the forearm and flex forearm, wrist, and fingers. The sensory part of the nerve provides feedback information from the thumb and the first two and one-half fingers.

Moment arm. See lever arm.

Moment. A quantity necessary to cause or resist rotation of a body, usually expressed in newton meters. (Special case is torque, which is a moment about a longitudinal axis.)

Motion segment. A unit of the spine representing inherent biomechanical characteristics of the ligamentous spine. Physically, it consists of two adjacent vertebrae and the interconnecting disc and ligament tissue, devoid of musculature.

Motor unit. The body of a nerve cell and its axon and all muscle fibers supplied by branches of one axon, that is, the functional "unit" of muscle.

Myoelectric signal. Electrical potential produced during contraction of muscle (see Electromyography).

N

Nerve root entrapment syndrome. Technical term for "pinched nerve."

Neutral axis. The central plane on which the tensile and compressive stresses and strains due to bending equal zero.

Newton. The unit of force in the Système International d'Unites. One newton is the amount of force required to give a 1-kg mass an acceleration of 1 meter per second per second.

P

Palpate. To locate by touch.

Pathology. The discipline dealing with the development and description of disease in terms of altered structure and function of the body.

Pectoralis major. Large triangular muscle. The base forms the origin, running parallel to the entire length of the breast bone. The apex inserts into the medial side of the humerus. Essentially an adductor of the upper arm.

Physiological cross-sectional area. The amount of muscle fiber in a given cross section of a muscle.

Physiological response. The normal response and adaptation of the living organ and its parts to stress.

Physiology. The science that deals with the normal function of the living organ and its parts.

Plantar flexion. Bending about the ankle joint in the direction of the sole of the foot.

Poplitea. The hollow at the back of the knee.

Pressure. The surface stress acting perpendicular to a unit area.

Pronation. The action of rotating the flexed forearm toward the midsagittal plane, so that the hands become prone, with palms down, back of hand up.

Proteoglycan. A macromolecule composed of glycosaminoglycans forming a hydrated gel; one of the primary structural components of cartilage.

Proximal. In a limb, closer to the body. Elsewhere, closer to the central axis of the body.

Pulmonary. Pertaining to the lung.

R

Range of motion. The range of translation and rotation of a joint for each of its degrees of freedom.

Repetitive loading. Repeated application of a load to a structure.

Rigid Body A collection of particles joined together rigidly. For practical purposes a body is said to be rigid if its deformation as compared to the other bodies in the system is small within a given range of the loads applied.

Rotation. Motion in which all points describe circular arcs about an immovable line or axis.

S

Sagittal plane. Anatomical reference plane vertically dividing the body into right and left portions.

Scoliosis. Lateral curvature of the spine.

Shear. A loading mode in which a load is applied parallel to the surface of the structure, causing internal angular deformation or slip.

Speed of loading. The rate at which load is applied to a structure (see Deformation rate).

Statics. The study of forces acting on a body in equilibrium.

Stiffness. A measure of resistance offered to external loads by a specimen or structure as it deforms.

Strain. Deformation (lengthening or shortening) of a body divided by its original length.

Strain gauge. A device that permits strain to be measured.

Strain rate. The speed at which a strain occurs.

Stress. Load per unit area which develops on a plane surface within a structure in response to externally applied loads.

Stress concentration. Any localized stress peak that cannot be predicted by simple strength of material theory.

Stress–strain curve. A curve generated by plotting the stress and the strain during compressive, tensile, or shear loading of a structure.

Supination. Process of rotating the flexed forearm outward so that hand becomes "supine," that is, "palms up."

Synovia. Membranes lining the inside of joint capsules and moving surfaces of joints. They secrete the synovial fluid, which lubricates joints.

T

Tangential. Relating to a straight line that is the limiting position of a secant of a curve through a fixed point.

Tendinitis. Also tendonitis. Inflammation of tendon (including tendon sheath).

Tendon. Connective tissue attaching muscle to bone.

Tendon sheaths. Tubular structures through which tendons run. They are lined with a synovial membrane and, therefore, not only guide but also lubricate the tendons.

Tenosynovitis. Inflammation of the tendon sheaths.

Tension. A loading mode in which equal and opposite loads are applied away from the surface of a structure, resulting in lengthening and narrowing.

Torsion. A loading mode in which a load is applied to a structure in a manner that causes it to twist about an axis, subjecting the structure to a combination of shear, tension, and compressive loads.

Translation. Parallel motion of one surface across another.

Transverse. Crosswise; in a horizontal direction.

Triceps. Three-headed large extensor muscle of the forearm. Originates from the back of the humerus and the shoulder blade and inserts into the proximal tip of the ulna.

U

Ultimate failure point. The point on the load–deformation curve past which complete failure of the structure occurs due to continued loading in the nonelastic region.

V

Vector. A quantity that has magnitude, direction, line of application, and point of application, commonly represented by a directed line segment.

Velocity. The displacement of a body divided by the time over which displacement occurs.

Viscoelasticity. The property of a material to show sensitivity to rate of loading or deformation.

W

Wolff's law. A law which states that bone is laid down where needed and resorbed where not needed.

Work. The amount of energy required to move a body from one position to another. Mechanical work is defined as the product of force applied to the distance moved in the direction of the force.

INDEX

Accelerometer, 113–114, 375–376
  advantages of, 114
  disadvantages of, 114
  piezoelectric, 376, 378
Acetylcholine (ACh), 33
Actin, 30, 31
ADP (adenosine diphosphate), 32, 42
Afferent nerve fibers, 27, 44
Anthropometric data:
  link length, 68–72
  mass center, 72–73
  moment of inertia, 73–75
  weight, 72
Anthropometric measurement methods:
  link length, 53–59
  mass center locations, 61–63
  moment of inertia, 63–67
  volume and weight, 59–61
Anthropometric Source Book (NASA), 365
Anthropometry, 53–77
  definition of, 5, 53
  of hand, 366
  joint motion and, 85–88
  muscle strength and, 104–107
  pre-employment screening, 403
  seated work and, 294–295, 296, 297, 298
ARBAN posture analysis system, 257
Arm, vibrations and, 386
Arm abduction and flexion, 332–338
  hand height and, 334–338
  major muscular actions for, 333
Arm motion, shoulder girdle and, 331
Asymmetric load handling, 279, 282–286
  definition of, 279
  foot position for, 282, 283
  mean strength values for, 283–285
  posture and, 283–284
ATP (adenosine triphosphate), 32–33, 42
Automation, limitations of, 263
Axial skeletal bone, 19
Axons, 27

"Back" schools, 406. *See also* Pre-employment training
Back supports, *see* Lumbar supports
Basic motion timestudy (BMT), 240
Bioinstrumentation, 6, 111–146
  criteria for, 111
  electromyography (EMG), 122–128
    criteria for, 125–127
    signal characteristics of, 123–125
  force platform system, 141–142
  intra-abdominal, 137–141
  intradiscal, 132–137
  motion analysis systems, 112–122
    accelerometers, 113–114, 375–376, 378
    goniometers, 82, 83, 112, 115–116
    photogrammetry, 82, 84, 117–120, 121
    video spot locators, 120–122
  muscle strength systems, 128–132
    dynamic, 132
    static, 128–132
Biomechanical Basis of Ergonomics, The (Tichauer), 10
Biomechanical models, 147–232
  dynamic, 176–190
    coplanar, of pushing, 187–190
    multiple-segment, of load lifting, 182–187
    single-segment, 178–182
  future developments in, 227–229
  low back, 191–213
  muscle strength prediction, 221–227
  necessity of, 147–148

Biomechanical models (*Continued*)
planar static, 148–171
of internal forces, 160–166
multiple-link, 166–171
of nonparallel forces, 158–160
single-body segment, 148–154
two-body segment, 155–158
wrist and hand, 213–221
Body diagram, for postural evaluation, 254, 255, 256
Body member movements, motion-time data system, 238
Bone(s):
age related changes in, 25
mechanical properties of, 23–25
anisotropic, 23, 24
energy-storage capacity, 23, 25
strength and stiffness, 23, 25
types of, 19
Bone cells, 19
Bone formation, 19–23
Bone fractures, 23–25
Bone hypertrophy, 25
Bone remodeling, 23
Bone structure, 19–23
cells, 19
epiphyseal plate, 20
epiphysis, 20
haversian systems, 22, 23
lamellar bone, 22
medullary cavity, 20
osteons, 23
periosteum, 20
woven bone, 20

Cancellous bone, 19, 22
Carpal tunnel syndrome (CTS), 214, 243, 356
Cartilage, 18–19
elastic and viscoelastic properties of, 47
intervertebral disc, 49–50
lubrication methods, 47–48, 49
osteoarthritis and, 48
synovial joint, 45–47
Cells:
bone, 19
vibration damage to, 386
Centripedal force, 179
Cervical lordosis, 289
Chair design, 313–317
arm rests, 304, 317
back muscle activity and, 305–307
backrest-seat inclination, 301–302, 304, 305, 306, 316–317
comfort rating of, 295–298
criterion for good, 293
dimension recommendations for, 314–317
disc pressure and, 304, 305, 306, 307
height, 315
knee flexion angle, 302, 307, 311
leg positions and, 309–313
length, 315–316
safety considerations, 317
sitting postures and, 293–294
Chondrocytes, 15
synovial joint, 45
Chronocycleograph, 236
Civil Service Commission, U.S., physical task checklist of, 246, 247
Coefficient of friction, 187. *See also* Friction
Collagen fibers, 15, 22
stress-strain curve for, 17
synovial joint, 46–47
Compact bone, 19, 23
Composite force system, 159
Compression-type piezoelectric accelerometer, 376, 378
Connective tissues, 14–26
cartilage, 18–19
fascia, 15, 16
ligaments, 15–18
skeletal muscle, 26–27
structure of, 14–15
tendons, 15–18
see also Bone(s); Joint(s); Muscles
Coriolis Force analysis, 182–187
Cortical bone, 19, 23
Creatine phosphate, 32, 42
Cyclegraph, 236
Cysts, ganglionic, 214, 356

Dendrites, 27
Depolarization, 33
Diaphysis, 19
Dimensional motion times (DMT), 240
Disc herniation, 201
Disc pressure:
measurement of, 136–137
during sitting, 302–305
chair design and, 304, 305, 306, 307
increased, 303
in standing and unsupported sitting postures, 303, 304
whole-body vibrations and, 383–384
see also Intervertebral discs
Dynamic biomechanical models, 176–190
co-planar, 187–190
of low back, 209–212
multiple segment, 182–187

need for, 186–187
single-segment, 178–182

Efferent nerve fibers, 27
Elastic cartilage, 19
Elastic fibers, 15
stress-strain curve for, 17
Elbow flexion:
planar static analysis of, 160–166
rotational strength performance and, 342–343
vibration attenuation and, 386–387
Elbow moment equation, 153–154
non-parallel force system, 159–160
planar static, 151–154
Elbow supports, in hand tool design, 366, 367
Electrogoniometer (Elgon), 82, 115
Electromyography (EMG), 122–128
criteria for, 125–127
muscle force ratio, 123–125
occupational activities studies, 125, 126
signal characteristics in, 123–125
Elemental motions, 236
standardized, 242
Elemental motion times, predetermined, 237
Elemental time standard system, 239
Elgon (electrogoniometer), 82, 115
EMG, *see* Electromyography
Engineering anthropometry, 5
Engineering Anthropometry Methods (Roebuck, Kroemer, and Thomson), 53
Epicondylitis, 356
Epimysium, 26, 28
Epiphysis, 19, 20
Equilibrium, conditions for, 150–152
Ergonomics, definition of, 9–10. *See also* Workplace layout
Exoskeleton-goniometer, 116
Extracellular matrix, 14

Fascia, 15–18
Fasciculi, 26, 28
Fatigue, Gilbreths' study of, 235. *See also* Muscle fatigue; Neck fatigue
Fatigue fractures, 23
Fibers, 14–15
stress-strain curves for, 17
twitch, 34, 35–36, 40
see also Muscle fibers
Fibroblasts, 15
Fibrocartilage, 19
Finger(s):
neuritis in, 356
vibration syndrome and, 385
Finger clearance, hand tool design for, 365
Flexometer, 82, 84
Foot-ground forces, 141
Foot slip potential, coplanar biomechanical models of, 187–190
Foot-slip prevention, 279, 280–281
Force(s):
centripedal, 179
characteristics of, 149
concurrent, 159
coplanar, 159
force platform system, 141–142
nonparallel, 158–160
Forearm rotational torque strength, flexed elbow and, 342–343
Forearm support, in hand tool design, 366, 367
Forward reaching, 338–345
Fractures, bone, 23–25
Free-body diagram, 151
Frequency analysis, vibration, 370–373
band width:
constant absolute, 370–371, 373
constant relative, 370, 371, 373
definition of, 372–373
effective, 373, 375
filter:
ideal, 372, 375
narrow band-pass, 370–371, 372
wide band-pass, 371–372
Friction:
coefficient of, 187
push and pull static strength and, 275, 279
Functional capacity tests, need for, 9

Ganglionic cysts, 214, 356
Glossary, 435–439
anatomical and anthropometric, 416–422
Golgi tendon organs, 41
Goniometer, 82, 83, 112, 115–116
advantages of, 116
disadvantages of, 116
exoskeleton-, 116
Goniometry, 82
Gravity, effect of, on musculoskeletal system, 59–60
Grip strength:
finger flexor muscles and, 363
palmar force-bearing area and, 362–363
tool shape and, 360–363
tool size and, 363
Ground substance, 14–15

Hamstring muscles, 292–292, 294
Hand anthropometry, 366
Hand disorders, 356

Hand height, arm abduction and flexion and, 334–338
Hand models, 213–221
Hand tool(s), vibrations and, 386–387
Hand tool design, 355–368
  biomechanical principles for, 357
  cumulative trauma disorders and, 356
  cylindrical *vs*. pistol shaped, 358, 361
  of knife handle, 358, 360
  need for, 355–357
  occupational risk factors and, 356–357
  of pliers, 358, 359
  shape and size, 357–365
    finger clearance and, 365
    grip strength and, 360, 362–365
    shoulder abduction and, 358, 360
    wrist posture and, 357–358, 359, 360, 361
  of soldering iron, 360, 361
  weight and use, 365–368
  of wire cutters, 362
Head, vibrations and, 386
High-risk workers, identification of, 401
Hyaline cartilage, 19
  synovial joint, 45–47

ILO, *see* International Labor Organization
Inertial property measurement, 63–67
  methods of:
    incremental submersion, 67
    pendulum, 65–66
    quick release, 67
  moment-of-inertia, 64–65, 67
  radius of gyration, 67–68
Information Sheet on Manual Lifting (ILO), 266
Internal forces, planar static analysis of, 160–166
International Labor Organization (ILO), *Information Sheet on Manual Lifting*, 266
Intervertebral discs:
  bending/rotation response of, 50
  compression response of, 50
  fluid content of, 49–50
  structure of, 48–51, 132–136
  see also Disc pressure; Spine
Intra-abdominal measurements, 137–141
  age and, 140
  applications and limitations of, 138–141
  early techniques for, 137–138
  with pressure transducers, 138
  with radio pills, 138, 139
Isometric muscle contraction, 34, 36, 38
Isotonic muscle contraction, 34, 36

Job analysis, *see* Work analysis systems
Job classification, *see* Worker selection
Job static strength analysis, 251–254
  coding form for, 252, 253
  data for, 251–252
Job training, reduction of low-back injury and, 405–407. *See also* Pre-employment training
Joint centers-of-rotation, 54–56, 57–58
  definitions, 57
  kinematic analysis of, 112
Joint lubrication, 47–48
Joint mobility, pre-employment screening for, 403
Joint motion:
  joint flexibility *vs*., 78–79
  measurement methods of, 81–84
    criteria for, 81
    flexometer, 82, 84
    goniometry, 82, 83, 112, 115–116
    photogrammetry, 82, 84, 117–120, 121
    spatial imaging, 82, 84
  ranges of, 85–87
    age and, 85
    anthropometric factors affecting, 85–88
    gender and, 85, 88
    two-joint muscle limitation of, 88–89
  terminology:
    standard anatomical, 79, 80
    triplanar system of, 79–81
Joints, 44–51
  articular, 44
  cartilaginous, 44
  definition of, 44
  fibrous, 44
  link length measurement methods, 53–59
  load moments at, 160
  synovial, 44
    structure of, 45–47
  vibrations and, 386

Kinematics, 3
Kinesiology, 3, 5
Kinetic chain, 155
Kinetics, 3
Krebs cycle, 32

Labor-management cooperative programs, 235
Landmarks, anatomical and anthropometric, 54–55, 413–415
Legs, sitting posture and, 309–313
Lifting:
  five-link model, 112–113
  ILO standards for, 266

NIOSH standards for, 274–275
psychophysical study of, 101, 103
strength test for, 96–97
see also Load lifting; Manual lifting analysis
Lifting hazards, 267–274
characteristics of, 267–268
control of:
administrative, 272, 274
engineering, 272
levels of, 268–272
Ligaments, 15–18
stress-strain curve for, 18
Linkage centers-of-rotation, 55–56, 57–58
Linkage system:
common, 59
definitions, 57–58
extremity to bone length ratios in, 56
joint centers-of-rotation, 54–56, 57–58
linkage centers-of-rotation, 55–56, 57
measurement of, 53–59
palpable bony landmarks in, 54–55
proportionality values in, 55, 56
torso, 193
torso joint to surface point relationships in, 56
used in biomechanical model, 177
Link length data:
for extremity lengths, 69–70
as population percentiles, 68, 70–71
as stature proportions, 68–69
for torso, 70–72
Load lifting:
from floor, 203–205
multiple-segment biodynamic model of, 182–187
see also Lifting
Load moments, 160
Long bone, 19
Low back fatigue, 348–349, 350
Low back injury:
job training and reduction of, 405–407
muscle strength testing for, 403–404
Low back models, 191–213
load lifting:
dynamic, 209–212
from floor, 203–205
push/pull, 212
three-dimensional, 206–209
Low back pain:
as employment risk factor, 401–402
incidence of, 7–8
seated work and, 298–299
whole-body vibrations and, 383–384
Lumbar lordosis, 289
seated work and, 299–301
Lumbar motion segment, 50
Lumbar support:
back muscle activity and, 307
definition of, 305
disc pressure and, 304
lumbar lordosis and, 299–302

Machine controls layout, 350–352
optimal, 343, 345
Manual lifting analysis, 246–251
variables in, 247–248
see also Work Practices Guide for Manual Lifting (NIOSH)
Manual materials handling, 263–288
asymmetric, 279, 282–286
lifting, 266–275
ILO standards, 266
NIOSH standards, 267–275
prevalence of, 263–264
pushing and pulling, 275–279
foot-slip prevention, 279, 280–281
system design characteristics for, 264
material/container, 265
task, 265
worker, 265
work practices, 266
training for, 406
Mass center locations, 425–426
data, 72–73, 74
kinematic analysis of, 112
Mass center measurement methods, body-segment, 61–63
moment subtraction, 62, 63
segmental zone, 62–63, 64
Materials handling, *see* Manual materials handling
Mechanical work-capacity evaluation, 5, 78–110
of joint motion, 78–89
methods for, 81–84
ranges of, 85–89
limitations of, 107–108
of muscle strength, 89–107
methods for, 91–97
ranges of, 97–103
Methods-Time Measurement (MTM) system:
coding conventions for, 242, 243
limitations of, 260
predicted move-time data, 237, 241–242
procedures in, 242
time-measurement unit (TMU) in, 237, 241
of tote box lifting, 242, 244
training requirements for, 237

Micromotion study, 236
Mobility, pre-employment screening for, 403
Moment-of-inertia data, 64–65, 67, 73–75
  computation of, 74–75
Motion-time analysis (MTA), 238
  predetermined, 237
Motion-time data for assembly work, 238
Motor unit, 27–28, 29
Motor unit potential, 122
MTM, *see* Methods-Time Measurement system
Muscle action, 43–44
  agonists in, 43
  antagonists in, 43
  coordinated, 43–44
  shunt, 44
  spurt, 44
  synergists in, 43
Muscle action potential, 33, 34, 122
Muscle contraction, 30–39
  basic mechanism of, 30–34
  concentric, 39
  contraction period, 34
  eccentric, 39
  electrical events in, 33–34
    acetylcholine (ACh) and, 33–34
    depolarization, 33
    muscle action potential, 33, 34
    nerve action potential, 33
  EMG prediction of, 124
  energy sources for:
    ADP (adenosine diphosphate), 32
    ATP (adenosine triphosphate), 32–33
    glycogen, 34, 42
    phosphocreatine, 32, 42
  fiber composition and, 35–36
  isometric, 34–36, 38
  isotonic, 34–36
  latent period, 34
  length-tension relationship and, 36–38, 39, 40
  proteins involved in, 30–32
    actin, 30, 31
    myosin, 30, 31
    tropomyosin B, 30
    troponin, 30
  relaxation period, 34
  stimulus response factors, 34–35, 37
  summation, 35
  tetanus, 35
  twitch, 34, 35–36, 40
  velocity tension relationship and, 38–39, 41
  whole-body vibrations and, 379–380
Muscle endurance, 325, 326, 327
Muscle endurance-force relationship, 325, 326
Muscle endurance time, 40, 42, 338, 339, 340
Muscle fatigue, 39–40, 42–43, 325–326
  EMG evaluation of, 125, 127
  endurance time and, 40, 42
  energy depletion and, 40
  fiber type and, 40, 43
  lactate accumulation and, 33, 40
  of low back, 348–349, 350
  of neck, 346–347, 348
  oxygen debt and, 33
  of shoulder, 325–326, 334
    elbow supports and, 338
    endurance times and, 338, 339, 340
    forearm supports and, 338
    hand tool design and, 360
Muscle fibers, 26–27, 44
  banding of, 28, 30, 31
Muscle force redundancy, 163
Muscle hypertrophy, 43
Muscle reflex, 44
Muscle spindles, 44
Muscle strength:
  age and, 104
  anthropometric factors affecting, 104–107
  definition of, 89–91, 128
  dynamic, 90
    body motion and, 94–95
    force velocity and, 95
    postural effect on, 95
  exertion instructions effect on, 90
  force-time relationship, 90
  gender and, 103–104
  maximum voluntary exertion levels, 89–90
  pre-employment testing of, 403–404
  static, 90
    as function of joint angle, 92–94
    postural effect on, 92–94
    values, 92
Muscle strength measurement systems:
  of dynamic strength, 132, 134
  of static strength, 128–132, 133, 251–254
    Backlund and Nordgren fixture for, 130
    of children, 130
    coding form for, 252, 253
    criteria for, 128–129
    data for, 251–252
    T. Stobbe fixture for, 129–130
    whole-body, 130–132
Muscle strength moments, 98–99
Muscle strength prediction modeling, 221–227
Muscle strength receptors, 44

Muscle strength values, 97–103
  dynamic, 101, 103
  static, 92, 97–101
    localized, 98–99, 100
    whole-body, 99–101, 102
Muscle structure, 26–30
  banding patterns, 28, 30, 31, 32
  epimysium, 26
  fascia, 26
  fasciculi, 26
  fibers, 28, 30
    nerve, 27, 29
    Sharpey's, 28
  motor unit, 27–28, 29
  myofibrils, 28
  myotendial junction, 28
  sarcomere, 30
Muscle tone, 44
Musculoskeletal injury:
  incidence of, 7–9
  dysfunction, 328–329
  fatigue, 325–326
  inflammation, 326–328
Musculoskeletal system:
  bone, 19–26
    mechanical properties of, 23–25
    remodeling of, 25–26
    structure of, 19–23
  cartilage, 18–19
  connective tissues of, 14–26
  discs, intervertebral, 48–51
  fascia, 15, 16
  joints, 44–51
    lubrication of, 47–48
    osteoarthritis of, 48
    synovial, 45–47
  ligaments, 15–18
  muscle, 26–44
    contraction of, 30–39
    fatigue of, 39–40, 42–43
    structure of, 26–30
  tendons, 15–18
Myoelectric activity, 123
Myosin, 30, 31

Neck fatigue, 346–347, 348
Neck/head posture, work limitations, 345–347
Nerve action potential, 33
Nerve cells, 27
Nerve-compression syndrome, 8
Nerve fibers, 27, 29, 44
Nervous system, vibrations and, 384, 385
Neuritis, in fingers, 356
Neuron, 27
NIOSH, *see Work Practices Guide for Manual Lifting* (NIOSH)
Nonparallel forces, static planar model of, 158–160

OA (osteoarthritis), 48
Occupational biomechanics:
  applications of, 10–11
  definition of, 1–2
  historical development of, 2–6
  methodological areas of:
    anthropometric, 5
    bioinstrumentation, 6
    biomechanical models, 5
    kinesiological, 3, 5
    mechanical work-capacity evaluation, 5–6
    motion-time analysis, 6
  support for, 7–10
    epidemiological, 7–9
    ergonomic, 9–10
    social/legal, 9
Oscillatory movements, *see* Vibrations
Ossification, 19–23
Osteoarthritis (OA), 48
Osteoblasts, 19
Osteoclast, 19
Osteocytes, 15, 19
Osteoid bone, 19
Ovaco Working Posture Analysis System (OWAS), 256–257
Overexertion, NIOSH report (1981) on, 263–264
Overhead reaching, 331–338
OWAS, *see* Ovaco Working Posture Analysis System (OWAS)
Oxygen debt, 33

Pain perception process, 325–326, 328
Parallel force system, 158
Peritendonitis crepitans, 356
Phosphocreatine, 32, 42
Photogrammetry, 82, 84, 117–120, 121
  advantages of, 121
  disadvantages of, 121
Physical examinations, pre-employment, 400–403
Physical stress checklists and surveys, 246, 248, 249
Piezoelectric accelerometers, 376, 378
Postural evaluation methods:
  ARBAN system, 257
  body diagram, 254, 255, 256
  OWAS system, 256–257
  posturegram, 255–256

Postural evaluation methods: (*Continued*)
TRAM system, 257
upper extremity, 257–258, 259
Postural Stability Diagram, 257
Posture, muscle strength and, 92–94, 95, 101
Posturegram, 255–256
Posture targeting, 254–255, 256
Predetermined work times, motion-time analysis system, 234, 240
Pre-employment examinations:
physical, 402
radiographic, 402–403
Pre-employment training, 405–407
aim of, 406
content of, 406
methods for, 407
Principles of Scientific Management (Taylor), 234–235
Psychophysical strength method, 97, 101
Pulling, *see* Pushing and pulling
Pushing and pulling, 158–160, 275–279
coplanar models of, 187–190
design limits for, 276, 278
foot position and, 276, 277, 282
foot-slip prevention and, 279
friction and, 275
handle height and, 276, 277
overexertion injuries and, 275
psychophysical study of, 101
in seated posture, 343, 344
static planar model of, 158–160

Radius of gyration, 67–68
computation of, 74–75
Reflex(es), muscle, 41
Residual strain, 215
Resting membrane potential, 122
Reynaud's Syndrome, segmental vibration and, 385
RMS (root mean square) value, of vibrations, 370
Robots, industrial, 263
Rohmert curve, 40
Root mean square (RMS) value of vibrations, 370

Sarcomere, 30, 32
Seated work, 289–323
anthropometric dimensions for, 294–295
sagittal, 296, 297, 298
transverse, 296
vertical, 297
chair design for, 293–294, 313–317
comfort aspects of, 295–298
disc pressure and, 302–305
legs and, 309–313
low back pain and, 298–299
lumbar support for, 299–302, 304
muscle activity and, 305–307
shoulder and, 309, 310, 311
sitting postures for, 289–294
spine and, 289–290, 291, 292, 298–309
work tables for, 317–319
see also Sitting; Sitting postures
Semisitting, 293–294, 311
Sensory nerve fibers, 27, 44
Shear-type piezoelectric accelerometer, 376, 378
Shoulder:
hand tool design considerations for, 358, 360
sitting posture and, 309, 310, 311
tendonitis of, 328, 329
Shoulder girdle, 331
Shoulder joint, 331–338
arm abduction and flexion and, 332–338
forward reaching and, 338–345
mobility of, 331
overhead reaching and, 331–338
rotator cuff muscles and, 332
stability of, 331–332
Shoulder pain, sources of, 334
Shunt muscle, 44
Sitting:
biomechanical aspects of, 289
definition of, 289
Sitting postures:
anterior (forward leaning), 290–291, 292
low-back pain and, 298–299
middle, 290
posterior (backward) leaning, 291, 292
types of, 290–291
Spatial imaging, 82, 84
Spine:
gross anatomy of, 289–290, 291, 292
low back biomechanical models of, 191–212
pre-employment radiographic examination of, 402–403
sitting posture and, 298–309
disc pressure and, 302–305
low back pain and, 298–299
lumbar lordosis and, 299–301
lumbar supports and, 299–302
muscle activity and, 305–307
whole-body vibrations and, 383–384
see also Intervertebral discs
Spondylolisthesis, 402
Spurt muscle, 44
Static biomechanical models:
co-planar:
multiple-link, 166–171
three-dimensional, 171–176

planar, 148–171
of internal forces, 160–166
of nonparallel forces, 158–160
single-body segment, 148–154
two-body segment, 155–158
Steady state, 33
Stooped posture, 347–350
Strain residual, 215
Synovial fluid, 16
Synovium, 16

Tables, *see* Work surface
Tendon(s):
finger flexor, 216–217
muscle attachment and, 28
muscle contraction and, 37–38, 40
strain of, 215–216
wrist, 213–221
Tendonitis, 8, 214, 327–329
of shoulder:
elevated arm work and, 334
high-velocity arm motions and, 335, 338
Tendon jerk, 44
Tennis elbow, 8
Tenosynovitis, 214, 356
Thoracic kyphosis, 289
Three-dimensional modeling:
low back, 206–209
static, 171–176
Time study, 233
Tool balancer devices, 366–367
shoulder muscle fatigue and, 341
Torso linkage system, 193
Trabeculae, bone, 19, 22
Training programs, worker, 272, 274. *See also* Pre-employment training
TRAM posture analysis system, 257
Traumatic Vasopastic Disease (TVD), segmental vibration and, 385–386
Tropomyosin B, 30
Troponin, 30
Trunk, vibrations and, 386
Trunk flexion analysis, 258–259. *See also* Intra-abdominal measurements
TVD (Traumatic Vasopastic Disease), segmental vibration and, 385–386
Twitch, 34
fast-, fibers, 35–36, 40
slow-, fibers, 35, 40

VDT (Video Display Terminal), workplace specifications, 350–352
Vibrations:
definitions of:
deterministic, 370
forced, 369
free, 369
harmonic, 370
periodic, 370
random, 370
stochastic, 370
effects of, on human body, 376–379
factors affecting, 376–377
mechanical, 377
psychological, 377, 379
exposure criteria for:
segmental, 392–396
whole-body, 387–392
measurement of:
in acceleration units, 376
with accelerometers, 114, 375–376, 378, 379
in decibel (dB) units, 373, 376
frequency analysis, 370–373, 374
parameters of, 374–375, 377
peak *vs.* average value, 370, 371
root mean square (RMS) value, 370, 371
segmental, 384–387
exposure criteria, 392–396
hand-arm system and, 384–385
upper extremity and, 386–387
vibration syndrome, 385–386
whole-body, 379–384
exposure criteria, 387–392
mechanical impedance and, 381
muscle contractions and, 379–380
nervous system and, 383
spine and, 383–384
symptoms of, 384
transfer function and, 381, 382, 383
Vibration syndrome, 385–386
Video Display Terminal (VDT), workplace specifications, 350–352
Video spot locator systems, 120–122
advantages of, 121
disadvantages of, 121–122
Volume measurement, body-segment, 59–61

Weight data, 72
population, 424
Weight measurement, body-segment, 59–61
Wolf's law, 25
Work analysis systems:
benefits in, 243–244
contemporary, 246–259
manual lifting analysis, 246–251
physical stress checklists and surveys, 246, 247, 248, 249
postural analysis, 254–258, 259

Work analysis systems: (*Continued*)
static strength analysis, 251–254
trunk flexion analysis, 258
future impact of, 260–261
general procedures in, 233–234, 245–246
historical perspective of, 233–236
Frank and Lillian Gilbreth, 235–236
Frederick W. Taylor, 234–235
limitations in, 244–245
traditional, 236–246
basic motion timestudy (BMT), 240
body member movements, 238
dimensional motion times (DMT), 240
elemental time standard, 239
Methods-Time Measurement (MTM), 237, 239, 241–243
motion-time analysis (MTA), 238
motion-time data for assembly, 238
predetermined human work times, 240
work-factor, 239
Work classification systems, 6
Worker selection, 399–405
age factors, 402
gender factors, 402
high-risk workers, identification of, 401
medical history and:
low back pain incidence in, 401
purpose of, 400–401
physical examination for, 402
purpose of, 399
radiographic examination for, 402–403
screening procedures:
anthropometric, 403
criteria for, 404–405
for general fitness, 404
for mobility, 403
for muscle strength, 403–404
test measures for:
accuracy, 399
predictive value, 400
sensitivity, 399–400
specificity, 400
see also Pre-employment training
Work-factor motion-time system, 239
Workplace layout, 324–354
forward reach, shoulder- and arm-dependent, 338–345
adjustable fixtures and, 341
elbow angles and, 338–343
elbow supports and, 338
forearm rotational torque strengths and, 342
forearm supports and, 338
seated push or pull exertions, 343
tool balancer devices and, 341
musculoskeletal injuries and, 324–329
overhead reach, shoulder-dependent, 331–338
stooped posture, neck/head limitations, 345–347
workbench height, 347–350, 351
see also Machine controls layout
Work Practices Guide for Manual Lifting (NIOSH), 266–275
criteria for developing, 266–267, 268
evaluation of, 274–275
hazardous lifting designations in, 267–272
action limits (AL), 268–269, 270–272
level classifications, 269
maximal permissible limits (MPL), 269, 270–272
recommendations for controlling, 272, 274
variables, 267–268
worker training recommendations of, 406–407
Work surface, 317–319
adjustability of, 318, 319
dimensions:
bottom height, 318
slope, 319
top height, 318–319
field of vision of, 318–319
height, 309
height limitations, 347–350
Wrist:
anatomy of, 214, 215
hand tool design considerations for, 357–358, 359, 360, 361
vibration attenuation and, 386–387
Wrist disorders, 214, 356
Wrist models, 213–221

X-ray examinations, pre-employment, 402–403